The Organic Chemistry of Biological Pathways

The Organic Chemistry of Biological Pathways©

John E. McMurry
Tadhg P. Begley
Both of CORNELL UNIVERSITY

ROBERTS AND COMPANY PUBLISHERS
Englewood, Colorado

ROBERTS AND COMPANY PUBLISHERS
4950 South Yosemite Street, F2 #197
Greenwood Village, Colorado 80111 USA
Internet: www.roberts-publishers.com
Telephone: (303) 221-3325
Facsimile: (303) 221-3326

ORDER INFORMATION
Phone: (800) 351-1161
Facsimile: (516) 422-4097

Distributed in Europe by Scion Publishing Ltd.
Internet: www.scionpublishing.com
Telephone: +44 (0) 1295 722873
Facsimile: +44 (0) 1295 722875

Publisher: Ben Roberts
Production Manager: Susan Riley at Side by Side Studios
Illustrator: Emiko-Rose Paul at Echo Medical Media
Cover Art: Emiko-Rose Paul and Quade Paul at Echo Medical Media
Cover and Text Designer: Mark Ong at Side by Side Studios
Copyeditor: Luana Richards
Compositor: Side by Side Studios
Printer and Binder: C & C Offset Printing Co., Ltd.

Chapter opener art created with The PyMOL Molecular Graphics System

Printed in China

The text of of this book is set in ITC Galliard, a typeface based on the work of Robert Granjon, a sixteenth-century letter cutter, whose typefaces are renowned for their beauty and legibility. The typeface was adapted to modern technology by Matthew Carter in 1978. The display type and chemical structures are set in Frutiger, which was originally designed by Adrian Frutiger in 1968 for signage in the Charles de Gaulle Airport in Paris but which has since become a model for a modern and legible sans serif typeface.

ISBN 0-9747077-1-6

Library of Congress Cataloging-in-Publication Data
McMurry, John.
The organic chemistry of biological pathways / John E. McMurry, Tadhg P. Begley.
p. ; cm.
Includes bibliographical references and index.
ISBN 0-9747077-1-6 (alk. paper)
1. Bioorganic chemistry. 2. Metabolism.
[DNLM: 1. Metabolism--physiology. 2. Enzymes--metabolism. QU 120 M478o 2004] I. Begley, Tadhg P. II. Title.
QP550.M38 2004
572'.4--dc22

2004020374

10 9 8 7 6 5 4 3 2 1

Author royalties from this book
are being donated to the Cystic Fibrosis Foundation.

Brief Table of Contents

Expanded Table of Contents

Foreword

In an era when many Chemistry Departments in front-rank universities have broadened their names to embrace the biological aspects of chemistry, there is a clear need for new modes of teaching at the chemistry–biology interface. To understand the phenomena of chemical biology, one needs to understand both the rules for how organic molecules behave and the logic of metabolism.

This text by John McMurry and Tadhg Begley on *The Organic Chemistry of Biological Pathways* provides a contemporary and authoritative treatment of the molecular logic of the chemistry of life. The book begins by defining the central concepts of organic molecule reactivity and the mechanisms of reactions, with examples drawn from biology. Then the major classes of small molecules found in primary metabolic pathways are introduced, and topics such as ionization states, stereochemistry, prochirality, and other essential features of structure and reactivity are clearly explained.

The core of the book examines the metabolism of lipids, carbohydrates, amino acids, and nucleotides, focusing on the small organic molecules in each class and explaining the organic chemistry that underlies metabolic transformations. Each chapter begins with the chemistry of degradative metabolism; for example, the conversion of lipids or specific amino acids back to the central pool of two-, three-, and four-carbon building blocks. Then the biosynthetic reactions for creating each small molecule are discussed; for example, the 20 amino acids used for protein synthesis or the nucleotides that are polymerized into RNA or DNA. In each chapter, the sequential chemical transformations of the metabolic pathways are clearly laid out in sufficient structural detail that this book can serve as an authoritative source for any reader/scholar interested in the chemistry of life. Yet, the presentation is also open and accessible to first-time readers in the subject.

Most metabolic transformations require cofactors, both organic coenzymes and metal ions, and this book sets forth the roles of these cofactors in each transformation with admirable clarity. For instance, the explanation of the electron availability and reactivity of oxo–iron intermediates that carry out most oxidative transformations of biomolecules is the clearest description I have ever encountered.

An additional chapter deals with the biosynthesis of some major classes of natural products. These include the β-lactam antibiotics of the penicillin and cephalosporin class, the antibiotic erythromycin as a member of the polyketide superfamily, prostanoid metabolites, morphine, and tetrapyrroles including hemes and coenzyme B_{12}. This chapter is also a model of organizational and mechanistic clarity that provides both an effective teaching module and a definitive summary for scholars seeking a comprehensive analysis of the logic Nature uses to build complex organic scaffolds.

The authors have chosen to organize the text not by organic reaction type but by metabolic pathway. Doing so allows them to explain not only the chemistry of any given step but also the molecular logic of the entire pathway. Readers thus gain a deeper understanding of how and why the organic molecules of life undergo their sequential transformations and interconversions. To integrate the information from earlier chapters, as well as tie in traditional modes of organic reaction classification, the authors conclude with a summary chapter of the most common reaction types encountered in metabolism.

As the knowledge base in chemical biology explodes, with inputs from genomics, proteomics, and metabolomics on the one hand and libraries obtained from combinatorial chemistry synthetic efforts on the other, there is a need to organize new concepts in novel ways. The chemical logic of reactions that go on in cells needs to be available to scientists from many backgrounds. This text provides a firm substratum for the many facets of chemical biology as it delineates the logic and mechanisms by which Nature interconverts the molecules of life.

Christopher T. Walsh
Department of Biological Chemistry
and Molecular Pharmacology
Harvard Medical School

Preface

It's probably easiest to introduce this book by first saying what it is *not*. This is not a comprehensive biochemistry book, it's not a mainstream organic chemistry book, and it's not a book on enzymology. Furthermore, this is not a monograph aimed at researchers. Yes, it does include several hundred references to the original literature, but most are very recent and have been selected to show the current status of a given topic rather than historical development.

Now, to say what this book *is*. This is a textbook, written for an audience of advanced undergraduates and graduate students in all areas of biorganic and biological chemistry. We assume that readers have a background in organic chemistry at the level of the typical two-semester college course, and we attempt to provide an accurate treatment of major biochemical pathways from the perspective of mechanistic organic chemistry. Although enzymes are of course crucial to the reactions in those pathways, our focus is always on the reactivity patterns of the substrate molecules and on the organic, arrow-pushing details of the individual reactions.

With the emphasis in biochemical research now focused primarily at the molecular level, with enzyme crystal structures now obtained routinely, and with enzyme–substrate interactions now visualized easily using simple computer programs, today's students have a unique opportunity to understand the chemical logic of living systems. We try to provide that understanding by showing how the reactions that take place in living organisms follow the same rules of chemical reactivity and occur by the same chemical mechanisms as reactions that take place in the laboratory. Biochemical transformations are not mysterious; there are logical and understandable reasons why they take place as they do. Biochemical reactions usually occur with more specificity and control than analogous laboratory reactions, but the principles behind biochemical and laboratory reactions are the same.

We begin this book with a brief review chapter on the fundamental organic reaction mechanisms commonly found in biochemical pathways. Following this brushup on reaction mechanisms is a general introduction to the main classes of biomolecules. Then comes the heart of the book: full chapters devoted to the major metabolic pathways of each major class of biomolecules—lipids, carbohydrates, amino acids, and nucleotides. In the course of these chapters, we cover the chemistry of the common coenzymes, learn how to access and visualize enzyme active sites, see examples of the common sorts of biological transformations, and ultimately begin to develop an understanding of the patterns found throughout bioorganic chemistry. After a brief discussion of natural product biosynthesis in Chapter 7, we conclude with a summary of the common reaction patterns and mechanisms used in nature to effect chemical transformations.

Good luck in your studies. We have enjoyed writing this book, and we hope that after reading it you come to a deeper appreciation and understanding of the beautiful and intriguing chemistry that takes place in living organisms.

John McMurry
Tadhg Begley
Department of Chemistry and Chemical Biology
Cornell University

Reviewers

Charles H. Clapp, Bucknell University
Alan H. Fairlamb, University of Dundee
Malcolm D. E. Forbes, University of North Carolina at Chapel Hill
Bernard T. Golding, University of Newcastle
Charles B. Grissom, University of Utah
Ahamindra Jain, University of California, Berkeley
Jennifer Kohler, Stanford University
Brian R. Linton, Bowdoin College
Hung-wen Liu, University of Texas at Austin
Lara K. Mahal, University of Texas at Austin
Ryan Mehl, Franklin and Marshall College
Walter G. Niehaus, Virginia Polytechnic Institute and State University
Ronald J. Parry, Rice University
Nigel Richards, University of Florida
Steven Rokita, University of Maryland
David H. Sherman, University of Michigan
James Stivers, Johns Hopkins School of Medicine
Sean Taylor, The Ohio State University
Peter Tipton, University of Missouri—Columbia
Michael S. VanNieuwenhze, University of California, San Diego
Robert H. White, Virginia Polytechnic Institute and State University
Susan White, Bryn Mawr College
Christian P. Whitman, University of Texas at Austin

Student reviewers

Amy Augustine
Keri Colabroy
Pieter Dorrestein
Brian Lawhorn
Colleen McGrath
Guangxing Sun

The Authors

John E. McMurry received his B.A. from Harvard University and his Ph.D. at Columbia University. Dr. McMurry is a Fellow of the American Association for the Advancement of Science, and an Alfred P. Sloan Research Foundation Fellow. He has received several awards, which include the National Institutes of Health Career Development Award, the Alexander von Humboldt Senior Scientist Award, and the Max Planck Research Award. In addition to *The Organic Chemistry of Biological Pathways*, he is also the author of *Organic Chemistry*, *Fundamentals of Organic Chemistry*, and *Chemistry* (with Robert Fay).

Tadhg P. Begley received his B.Sc. from National University of Ireland and his Ph.D. at the California Institute of Technology. Dr. Begley is the recipient of many awards, including the Merck Faculty Development Award and the Camille and Henry Dreyfus Teacher-Scholar Award. His research group uses the principles and techniques of organic chemistry to study complex organic transformations found in vitamin biosynthetic pathways. In addition to *The Organic Chemistry of Biological Pathways*, Dr. Begley has edited *Cofactor Biosynthesis: A Mechanistic Perspective.*

1 Common Mechanisms in Biological Chemistry

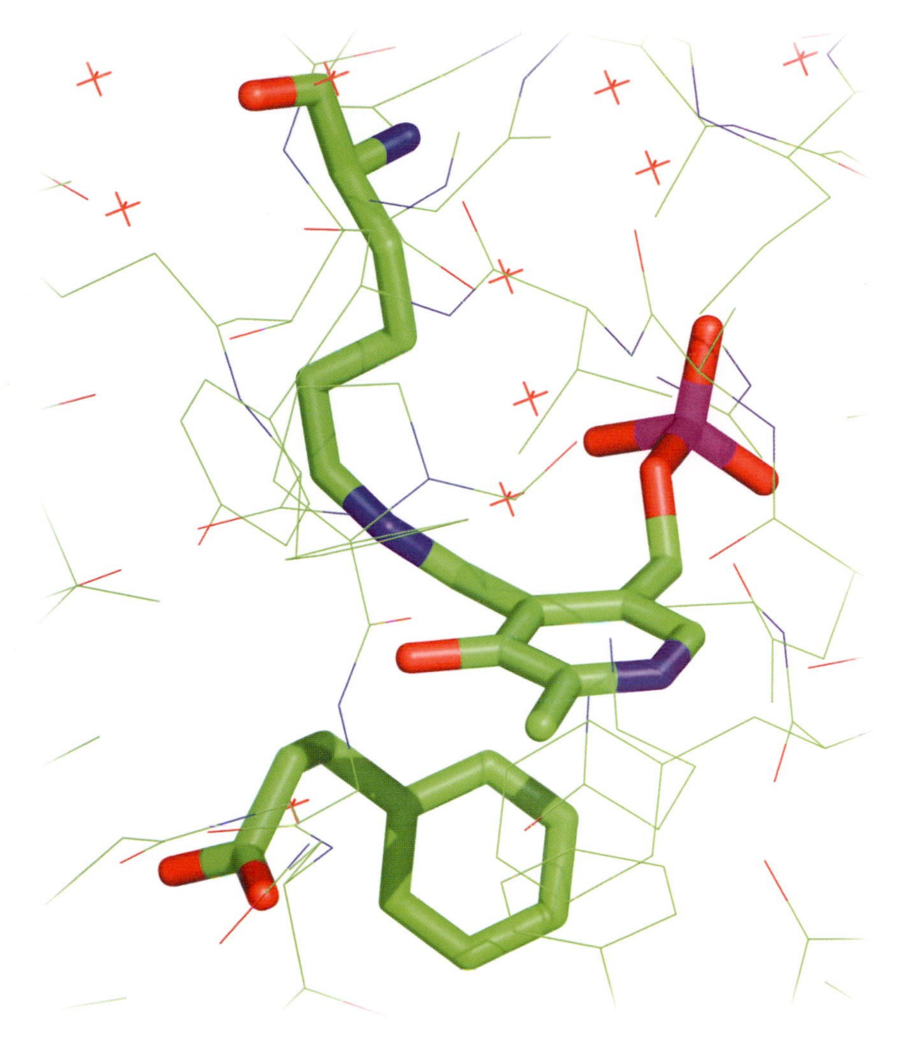

This scheme shows the active site of the enzyme that catalyzes the transamination reaction in the biosynthesis of phenylalanine from chorismate. A mechanistic understanding of biosynthetic pathways is a powerful tool for elucidating the chemical logic of living systems.

The final decades of the 20th century saw the beginning of a scientific revolution. Based on our newly acquired ability to manipulate, sequence, and synthesize deoxyribonucleic acid (DNA), the way is now open to isolate, study, and eventually modify each of the approximately 30,000 genes in our bodies. Medicines will become safer, more effective, and more specific; terrible genetic diseases such as sickle-cell anemia and cystic fibrosis will be cured; life spans will increase and the quality of life will improve as heart disease and cancer are brought under control.

None of these changes could occur without a detailed knowledge of chemistry, for it is our understanding of life processes at the molecular level that has made this revolution possible and that will sustain it. Biochemical processes are not mysterious. It's true that the many proteins, enzymes, nucleic acids, polysaccharides, and other substances in living organisms are enormously complex, but despite their complexity, they are still molecules. They are subject to the same chemical laws as all other molecules, and their reactions follow the same rules of reactivity and take place by the same mechanisms as those of far simpler molecules.

The focus of this book is on examining biochemical processes from a chemical perspective. We'll begin with a brief review of organic chemistry, looking first at the common functional groups found in biological molecules and then at some fundamental mechanisms by which organic molecules react. Following this general review of organic reactivity, we'll look at the structures and chemical characteristics of the main classes of biomolecules: carbohydrates, lipids, proteins, enzymes, and nucleic acids. Finally, we'll come to the heart of the matter: the organic chemistry of biological transformations. We'll dissect the details of important biochemical pathways to see both *how* and *why* these pathways occur. The result will be both a deeper understanding of biochemistry and a deeper appreciation for the remarkable subtleties by which living organisms function.

1.1 Functional Groups in Biological Chemistry

Chemists have learned through experience that organic compounds can be classified into families according to their structural features and that members of a given family have similar chemical reactivity. The structural features that make such classifications possible are called *functional groups.* A **functional group** is a group of atoms within a molecule that has a characteristic chemical behavior. Chemically, a functional group behaves in pretty much the same way in every molecule where it occurs. An ester (RCO_2R), for instance, usually undergoes a hydrolysis reaction with water to yield a carboxylic acid (RCO_2H) and an alcohol (ROH); a thiol (RSH) usually undergoes an oxidation reaction to yield a disulfide (RSSR); and so on. Table 1.1 lists some common functional groups found in biological molecules.

Table 1.1 Common Functional Groups in Biological Molecules

Structure*	Name	Structure*	Name
C=C	Alkene (double bond)	:N=C	Imine (Schiff base)
	Arene (aromatic ring)	:O:=C	Carbonyl group
C–OH	Alcohol	O=C–H	Aldehyde
C–O–C	Ether	O=C	Ketone
C–N:	Amine	O=C–OH	Carboxylic acid
C–SH	Thiol	O=C–O–C	Ester
C–S–C	Sulfide	O=C–S–C	Thioester
C–S–S–C	Disulfide	O=C–N̈	Amide
C–O–P(=O)(O⁻)–O⁻	Monophosphate	O=C–O–P(=O)(O⁻)–O⁻	Acyl phosphate
C–O–P(=O)(O⁻)–O–P(=O)(O⁻)–O⁻	Diphosphate		

* The bonds whose connections aren't specified are assumed to be attached to carbon or hydrogen atoms in the rest of the molecule.

Alcohols, ethers, amines, thiols, sulfides, disulfides, and phosphates all have a carbon atom singly bonded to an electronegative atom. Alcohols, ethers, and phosphates have a carbon atom bonded to oxygen; amines have a carbon atom bonded to nitrogen; and thiols, sulfides, and disulfides have a carbon atom bonded to sulfur. In all cases, the bonds are polar, with the carbon atom being electron-poor and thus bearing a partial positive charge ($\delta+$), while the electronegative atom is electron-rich and thus bears a partial negative charge ($\delta-$). These polarity patterns are shown in Figure 1.1.

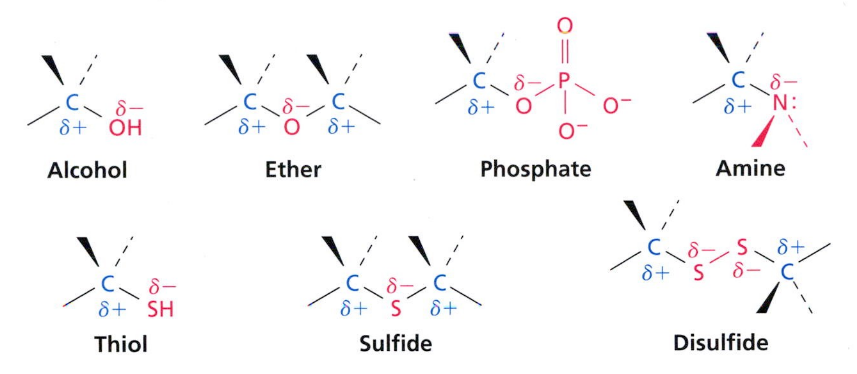

FIGURE 1.1 Polarity patterns of some common functional groups. The electronegative atom bears a partial negative charge ($\delta-$), and the carbon atom bears a partial positive charge ($\delta+$).

Note particularly in Table 1.1 the different families of compounds that contain the **carbonyl group, C=O**. Carbonyl groups are present in the vast majority of biological molecules. These compounds behave similarly in many respects but differ depending on the identity of the atoms bonded to the carbonyl-group carbon. Aldehydes have at least one hydrogen bonded to the C=O; ketones have two carbons bonded to the C=O; carboxylic acids have an —OH group bonded to the C=O; esters have an ether-like oxygen (−OR) bonded to the C=O; thioesters have a sulfide-like sulfur (−SR) bonded to the C=O; amides have an amine-like nitrogen ($-NH_2$, −NHR, or $-NR_2$) bonded to the C=O; and acyl phosphates have a phosphate group ($-OPO_3^{2-}$) bonded to the C=O. You might note that an acyl phosphate is structurally (and chemically) similar to a carboxylic acid anhydride.

As shown in Figure 1.2, carbonyl compounds are polar, with the electron-poor C=O carbon atom bearing a partial positive charge and the electron-rich oxygen atom bearing a partial negative charge.

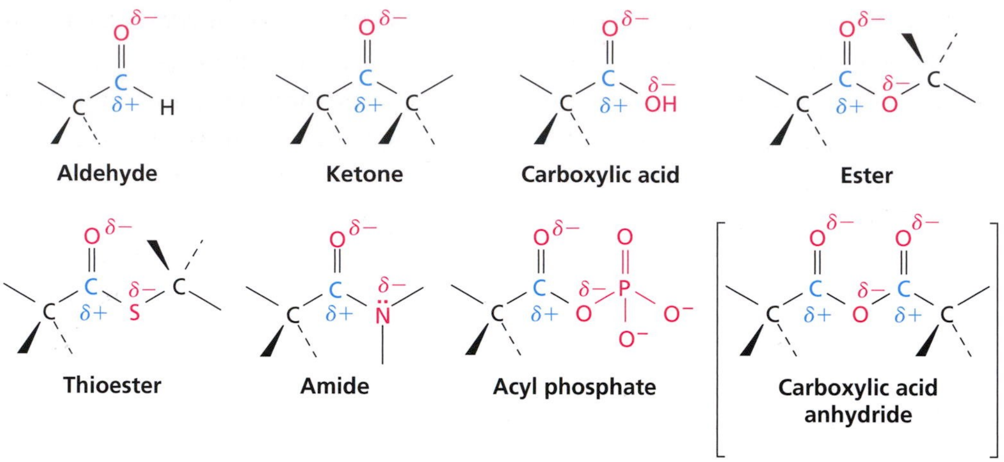

FIGURE 1.2 Polarity patterns in some carbonyl-containing functional groups. The carbonyl carbon atom is electron-poor ($\delta+$) and the oxygen atom is electron-rich ($\delta-$).

1.2 Acids and Bases; Electrophiles and Nucleophiles

Brønsted–Lowry Acids and Bases

Acids and bases are enormously important in biochemistry. The vast majority of biological transformations are catalyzed by acids or bases, and a thorough knowledge of acid–base chemistry is crucial for understanding how reactions occur.

According to the Brønsted–Lowry definition, an **acid** is a substance that donates a proton (hydrogen ion, H^+), and a **base** is a substance that accepts a proton. A carboxylic acid such as acetic acid, for example, can donate its —OH proton to a base such as methylamine in a reversible, **proton-transfer reaction**. The product that results by loss of H^+ from an acid is the **conjugate base** of the acid, and the product that results from addition of H^+ to a base is the **conjugate acid** of the base.

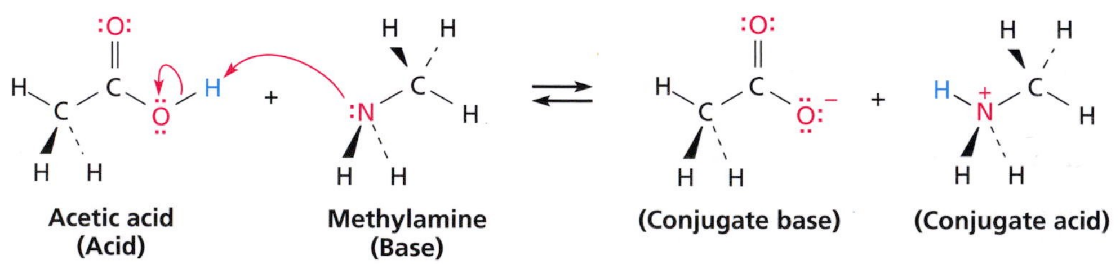

Note the standard convention used to show how this proton-transfer reaction occurs: A curved arrow (red) indicates that a pair of electrons moves *from* the atom at the tail of the arrow (the nitrogen in methylamine) *to* the atom at the head of the arrow (the acidic hydrogen in acetic acid). That is, the electrons used to form the new N—H bond flow from the base to the acid. As the N—H bond forms, the O—H bond breaks and its electrons remain with oxygen, as shown by a second curved arrow. *A curved arrow always represents the movement of electrons, not atoms.*

Acids differ in their ability to donate protons. Recall from general chemistry that the strength of an acid HA in water solution is expressed by its **pK_a**, the negative common logarithm of its **acidity constant, K_a**. A stronger acid has a smaller pK_a (or larger K_a); a weaker acid has a larger pK_a (or smaller K_a).

For the reaction: $HA + H_2O \rightleftharpoons A^- + H_3O^+$

$$K_a = \frac{[H_3O^+][A^-]}{[HA]} \text{ and } pK_a = -\log K_a$$

Stronger acid—smaller pK_a
Weaker acid—larger pK_a

Table 1.2 lists the pK_a's of some typical acids encountered in biochemistry. Note that the pK_a of water is 15.74, the value that results when K_w, the ion-product constant for water, is divided by the molar concentration of pure water, 55.5 M:

$$HA + H_2O \rightleftharpoons A^- + H_3O^+ \qquad \text{where } HA = H_2O,\ A^- = OH^-$$

$$K_a = \frac{[H_3O^+][A^-]}{[HA]} = \frac{[H_3O^+][OH^-]}{[H_2O]} = \frac{K_w}{55.5}$$

$$= \frac{1.00 \times 10^{-14}}{55.5} = 1.80 \times 10^{-16}$$

$$pK_a = -\log 1.80 \times 10^{-16} = 15.74$$

Note also in Table 1.2 that carbonyl compounds are weakly acidic, a point we'll discuss in more detail in Section 1.7.

Table 1.2 Relative Strengths of Some Acids

Functional group	Example	pK_a	
Carboxylic acid	$CH_3C(=O)OH$	4.76	Stronger acid
Imidazolium ion	H–$^{+}$N imidazole ring N–H	6.95	
Ammonia	NH_4^+	9.26	
Thiol	CH_3SH	10.3	
Alkylammonium ion	$CH_3NH_3^+$	10.66	
β-Keto ester	$CH_3C(=O)CH_2C(=O)OCH_3$	10.6	
Water	H_2O	15.74	
Alcohol	CH_3CH_2OH	16.00	
Ketone	$CH_3C(=O)CH_3$	19.3	
Thioester	$CH_3C(=O)SCH_3$	21	
Ester	$CH_3C(=O)OCH_3$	25	Weaker acid

Just as acids differ in their ability to donate a proton, bases differ in their ability to accept a proton. The strength of a base B in water solution is normally expressed using the *acidity* of its conjugate acid, BH^+.

For the reaction: $BH^+ + H_2O \rightleftharpoons B + H_3O^+$

$$K_a = \frac{[B][H_3O^+]}{[BH^+]}$$

so
$$K_a \times K_b = \left(\frac{[B][H_3O^+]}{[BH^+]}\right)\left(\frac{[BH^+][OH^-]}{[B]}\right)$$
$$= [H_3O^+][OH^-] = K_w = 1.00 \times 10^{-14}$$

Thus $K_a = \frac{K_w}{K_b}$ and $K_b = \frac{K_w}{K_a}$

so $pK_a + pK_b = 14$ and $pK_b = 14 - pK_a$

Stronger base—larger pK_a for BH^+
Weaker base—smaller pK_a for BH^+

These equations say that we can determine the basicity of a base B by knowing the K_a of its conjugate acid BH^+. A stronger base holds H^+ more tightly, so it has a weaker conjugate acid (larger pK_a); a weaker base holds H^+ less tightly, so it has a stronger conjugate acid (smaller pK_a). Table 1.3 lists some typical bases found in biochemistry. Note that water can act as either a weak acid or a weak base, depending on whether it donates a proton to give OH^- or accepts a proton to give H_3O^+. Similarly with imidazole, alcohols, and carbonyl compounds, which can either donate or accept protons depending on the circumstances.

Table 1.3 Relative Strengths of Some Bases

Functional group	Example (base)	Example (BH^+)	pK_a of BH^+	
Hydroxide ion	$:\ddot{O}H^-$	H_2O	15.74	Stronger base
Guanidino	$H_2NC(=\ddot{N}H)NHCH_2CH_3$	$H_2NC(={}^+NH_2)NHCH_2CH_3$	12.5	↑
Amine	$CH_3\ddot{N}H_2$	$CH_3\overset{+}{N}H_3$	10.66	
Ammonia	$:NH_3$	$^+NH_4$	9.26	
Imidazole	imidazole (:N= ring N—H)	H—N⁺ imidazolium (ring N—H)	6.95	
Water	$H_2\ddot{O}:$	H_3O^+	−1.74	
Alcohol	$CH_3\ddot{O}H$	$CH_3\overset{+}{O}H_2$	−2.05	
Ketone	$CH_3C(=\ddot{O}:)CH_3$	$CH_3C(={}^+OH)CH_3$	−7.5	Weaker base

Lewis Acids and Bases

The Brønsted–Lowry definition of acids and bases covers only compounds that donate or accept H^+. Of more general use, however, is the Lewis definition. A **Lewis acid** is a substance that accepts an electron pair from a base, and a **Lewis base** is a substance that donates an electron pair to an acid. For all practical purposes, Lewis and Brønsted–Lowry bases are the same: Both have lone pairs of electrons that they donate to acids. Lewis and Brønsted–Lowry acids, however, are *not* necessarily the same.

The fact that a Lewis acid must be able to accept an electron pair means that it must have a vacant, low-energy orbital. Thus, the Lewis definition of acidity is much broader than the Brønsted–Lowry definition and includes many species in addition to H^+. For example, various metal cations and transition-metal compounds, such as Mg^{2+}, Zn^{2+}, and Fe^{3+} are Lewis acids.

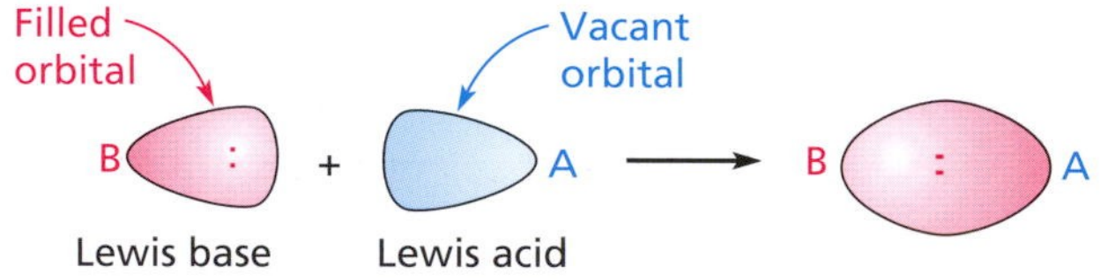

Filled orbital
Vacant orbital
B : + A → B : A
Lewis base
Lewis acid

Lewis acids are involved in a great many biological reactions, often as cofactors in enzyme-catalyzed processes. Metal cations such as Mg^{2+} and Zn^{2+} are particularly common, but complex compounds such as iron–sulfur clusters are also found. We'll see an example in Section 4.4 where citrate undergoes acid-catalyzed dehydration to yield *cis*-aconitate, a reaction in the citric acid cycle.

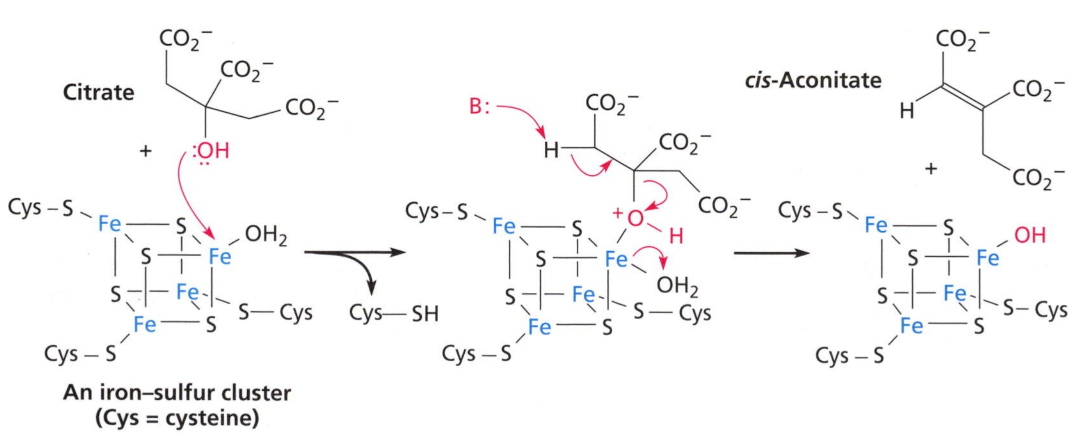

Citrate
CO_2^-
CO_2^-
CO_2^-
:OH
+
Cys–S
Fe
S
OH_2
S
Fe
S
Fe
S–Cys
Fe
S
Cys–S
Cys–SH
B:
H
CO_2^-
CO_2^-
CO_2^-
^+O
H
OH_2
S–Cys
cis-Aconitate
CO_2^-
CO_2^-
H
CO_2^-
OH
S–Cys

An iron–sulfur cluster (Cys = cysteine)

Electrophiles and Nucleophiles

Closely related to acids and bases are *electrophiles* and *nucleophiles*. An **electrophile** is a substance that is "electron-loving." It has a positively polarized, electron-poor atom and can form a bond by accepting a pair of electrons from an electron-rich atom. A **nucleophile**, by contrast, is "nucleus-loving." It has a negatively polarized, electron-rich atom and can form a bond by donating a pair of electrons to an electron-poor atom. Thus, electrophiles are essentially the same as Lewis acids and nucleophiles are the same as Lewis bases. In practice, however, the words "acid" and "base" are generally used when electrons are donated to H^+ or a metal ion, and the words "electrophile" and "nucleophile" are used when electrons are donated to a carbon atom.

Electrophiles are either positively charged or neutral and have a positively polarized, electron-poor atom that can accept an electron pair from a nucleophile/base. Acids (H^+ donors), trialkylsulfonium compounds (R_3S^+), and carbonyl compounds are examples (Figure 1.3).

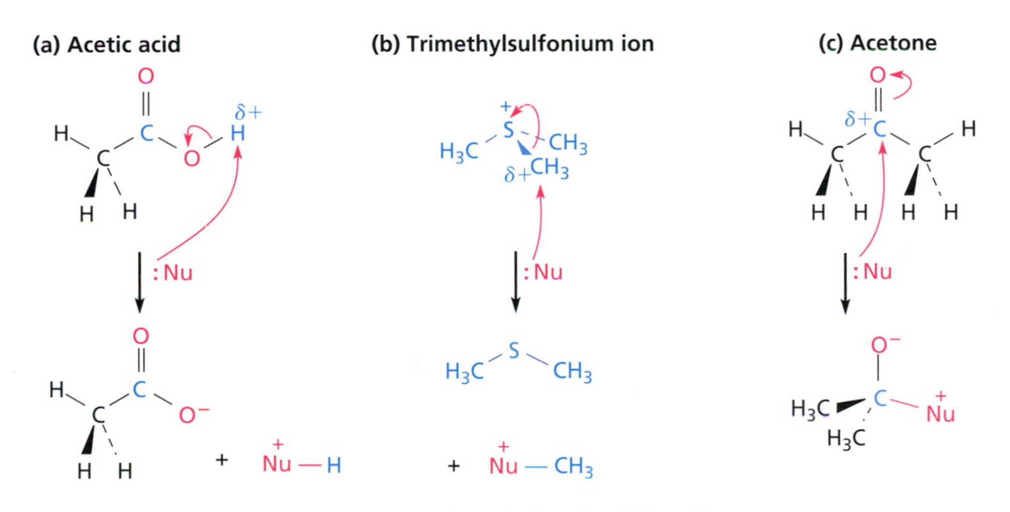

FIGURE 1.3 Some electrophiles and their reactions with nucleophiles (:Nu).

Nucleophiles are either negatively charged or neutral and have a lone pair of electrons they can donate to an electrophile/acid. Amines, water, hydroxide ion, alkoxide ions (RO^-), and thiolate ions (RS^-) are examples.

1.3 Mechanisms: Electrophilic Addition Reactions

Chemical reactions carried out in living organisms follow the same rules of reactivity as reactions carried out in the laboratory. The "solvent" is often different, the temperature is often different, and the catalyst is certainly different, but the reactions occur by the same fundamental mechanisms. That's not to say that *all* bioorganic reactions have obvious laboratory counterparts—some of the most chemically interesting biotransformations cannot be duplicated in the laboratory without an enzyme because too many side reactions would occur. Nevertheless, the chemical mechanisms of biotransformations can be understood and accounted for by organic chemistry. In this and the remaining sections of Chapter 1, we'll look at some fundamental organic reaction mechanisms, beginning with the electrophilic addition reactions of C═C bonds.

An **electrophilic addition reaction** is initiated by addition of an electrophile to an unsaturated (electron-rich) partner, usually an alkene, and leads to formation of a saturated product. In the laboratory, for example, water undergoes an acid-catalyzed electrophilic addition to 2-methylpropene to yield 2-methyl-2-propanol. The reaction takes place in three steps and proceeds through a positively charged, carbocation intermediate (Figure 1.4). In the first step, electrons from the nucleophilic C═C bond attack an electrophilic hydrogen atom of H_3O^+, forming a C—H bond. The intermediate carbocation then reacts with water as nucleophile, giving first a protonated alcohol and then the neutral alcohol after a proton-transfer step that regenerates H_3O^+. Note that the initial protonation takes place on the less highly substituted carbon of the double bond, leading to the more highly substituted, more stable, carbocation.

Biological examples of electrophilic addition reactions occur frequently in the biosynthetic routes leading to steroids and other terpenoids, although they are less common elsewhere. The electrophile in such reactions is a positively charged or positively polarized carbon atom, which often adds to a C═C bond within the same molecule. As an example, α-terpineol, a substance found in pine oil and used in perfumery, is derived biosynthetically from linalyl diphosphate by an internal electrophilic addition reaction. Following formation of an allylic carbocation by dissociation of the diphosphate (here abbreviated PPO),

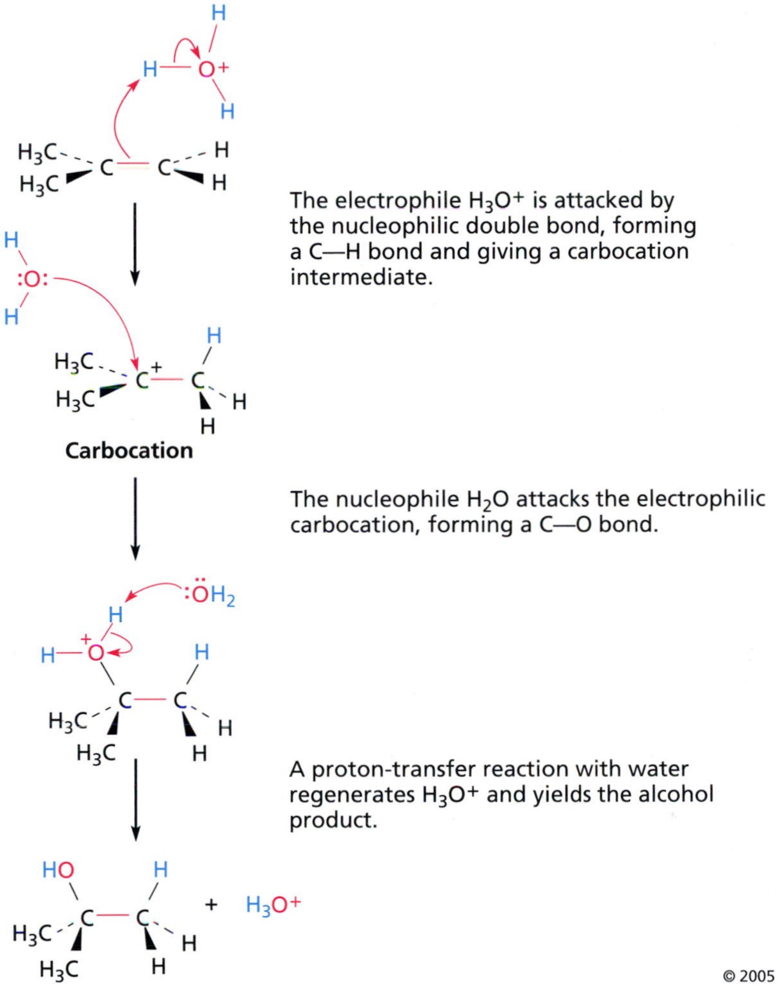

Carbocation

The electrophile H_3O^+ is attacked by the nucleophilic double bond, forming a C—H bond and giving a carbocation intermediate.

The nucleophile H_2O attacks the electrophilic carbocation, forming a C—O bond.

A proton-transfer reaction with water regenerates H_3O^+ and yields the alcohol product.

FIGURE 1.4 The mechanism of the acid-catalyzed electrophilic addition of water to 2-methylpropene. The reaction involves a carbocation intermediate.

electrophilic addition to the nucleophilic C═C bond at the other end of the molecule occurs, giving a second carbocation that then reacts with nucleophilic water. A proton transfer from the protonated alcohol to water yields α-terpineol (Figure 1.5). We'll see more such examples when we look at steroid biosynthesis in Section 3.5.

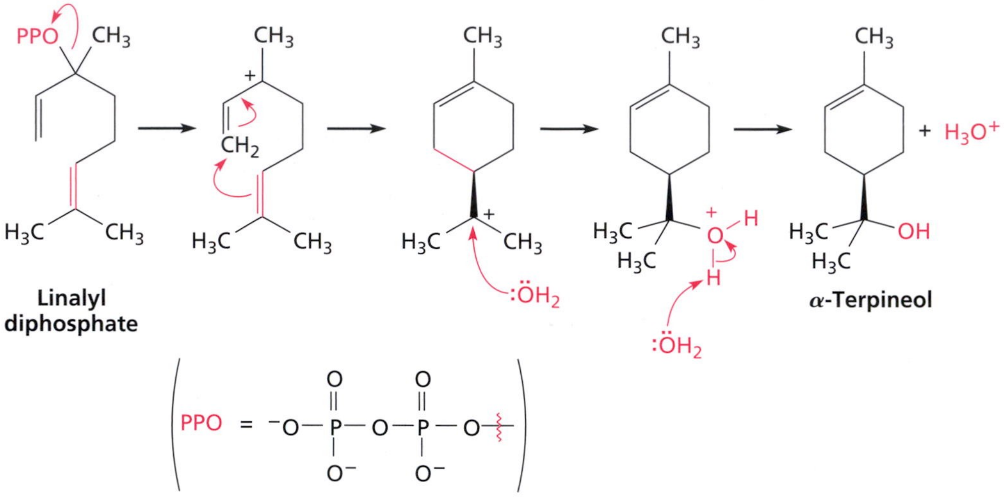

PPO CH3 | CH3 | CH3 | CH3 | CH3 + H3O+
H3C CH3 Linalyl diphosphate | CH2 H3C CH3 | H3C CH3 :ÖH2 | H3C H3C O H H :ÖH2 | H3C H3C OH α-Terpineol

(PPO = ⁻O—P(=O)(O⁻)—O—P(=O)(O⁻)—O—)

FIGURE 1.5 The biosynthesis of α-terpineol from linalyl diphosphate occurs by an electrophilic addition reaction.

1.4 Mechanisms: Nucleophilic Substitution Reactions

A **nucleophilic substitution reaction** is the substitution of one nucleophile (the *leaving group*) by another on a saturated, sp^3-hybridized carbon atom: Br^- by OH^-, for example. Nucleophilic substitution reactions in the laboratory generally proceed by either an $\mathbf{S_N1}$ **mechanism** or an $\mathbf{S_N2}$ **mechanism** depending on the reactants, the solvent, the pH, and other variables. S_N1 reactions usually take place with tertiary or allylic substrates and occur in two steps through a carbocation intermediate. S_N2 reactions usually take place with primary substrates and take place in a single step without an intermediate.

The mechanism of a typical S_N1 reaction is shown in Figure 1.6. As indicated, the substrate undergoes a spontaneous dissociation to generate a carbocation intermediate, which reacts with the substituting nucleophile to give product.

The mechanism of a typical S_N2 process is shown in Figure 1.7 for the reaction of hydroxide ion with (*S*)-2-bromobutane. The reaction takes place in a single step when the incoming nucleophile uses a lone pair of electrons to attack the

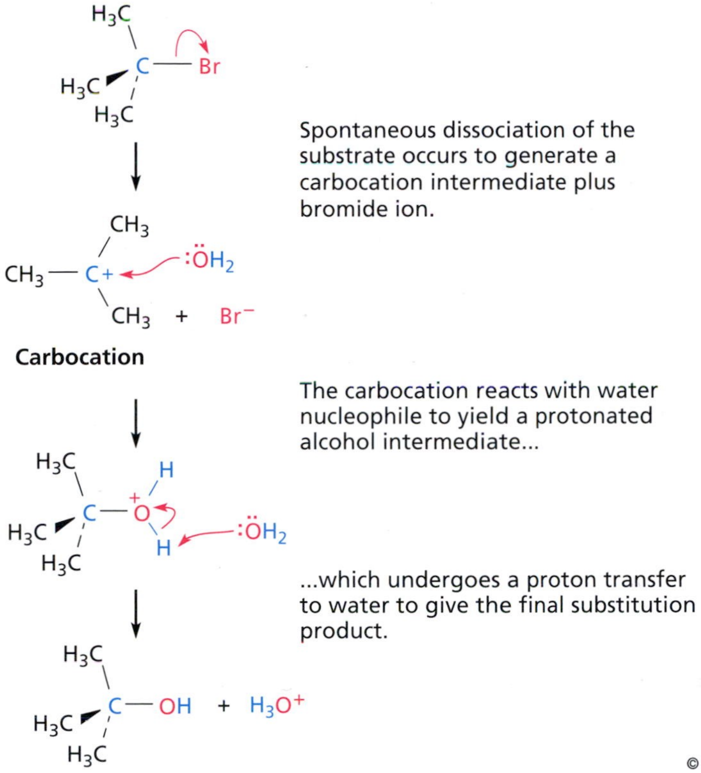

Spontaneous dissociation of the substrate occurs to generate a carbocation intermediate plus bromide ion.

Carbocation

The carbocation reacts with water nucleophile to yield a protonated alcohol intermediate...

...which undergoes a proton transfer to water to give the final substitution product.

FIGURE 1.6 Mechanism of the S_N1 reaction of 2-bromo-2-methylpropane with water to yield 2-methyl-2-propanol. The reaction occurs by a spontaneous dissociation to give a carbocation intermediate, which reacts with water.

electrophilic carbon atom of the alkyl halide from a direction 180° opposite the C—Br bond. As the OH^- comes in and a new O—C bond begins to form, the old C—Br bond begins to break and the Br^- leaves. Because the incoming and outgoing nucleophiles are on opposite sides of the molecule, the stereochemistry at the reacting center inverts during an S_N2 reaction. (*S*)-2-Bromobutane yields (*R*)-2-butanol, for example. (There is no guarantee that inversion will change the assignment of a stereocenter from *R* to *S* or vice versa, because the relative priorities of the four groups attached to the stereocenter may also change.)

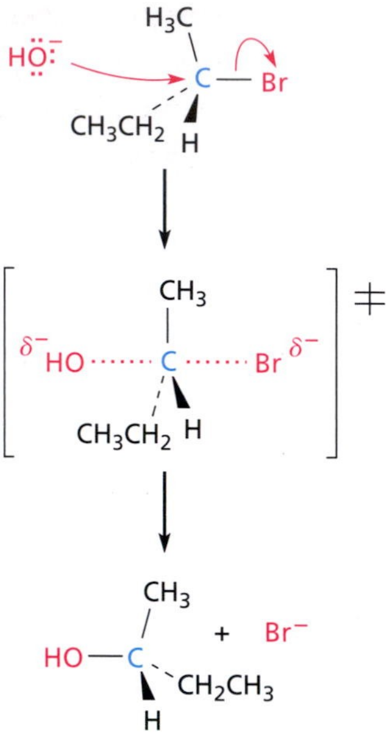

OH^- nucleophile uses a lone pair of electrons to attack the electrophilic alkyl halide carbon from a direction 180° oppoite the C—Br bond. The transition state has partially formed C—O and C—Br bonds.

The stereochemistry at carbon inverts as the C—H bond forms fully and the Br^- ion leaves.

FIGURE 1.7 Mechanism of the S_N2 reaction of (*S*)-2-bromobutane with hydroxide ion to yield (*R*)-2-butanol. The reaction occurs in a single step with inversion of stereochemistry at the reacting carbon atom.

Biochemical examples of both S_N1 and S_N2 reactions occur in numerous pathways. An S_N1 reaction, for instance, takes place during the biological conversion of geranyl diphosphate to geraniol, a fragrant alcohol found in roses and used in perfumery. Initial dissociation of the diphosphate gives a stable allylic carbocation, which reacts with water nucleophile and transfers a proton to yield geraniol.

An S_N1 reaction

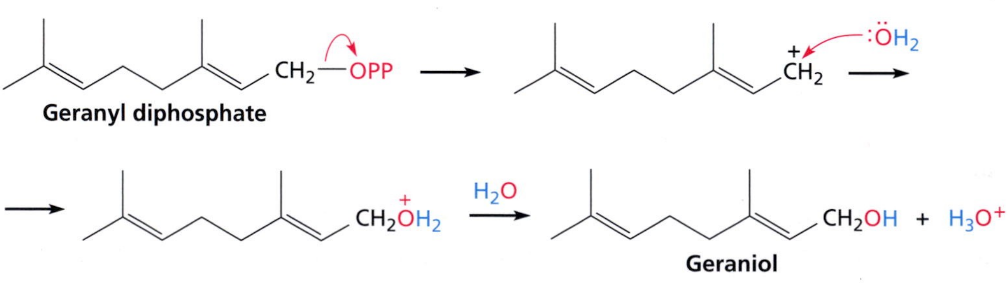

Geranyl diphosphate

Geraniol

An S_N2 reaction is involved in biological methylation reactions whereby a —CH_3 group is transferred from *S*-adenosylmethionine to various nucleophiles.

In the biosynthetic transformation of norepinephrine to epinephrine (adrenaline), for instance, the nucleophilic amine nitrogen atom of norepinephrine attacks the electrophilic methyl carbon atom of *S*-adenosylmethionine in an S_N2 reaction, displacing *S*-adenosylhomocysteine as the leaving group (Figure 1.8).

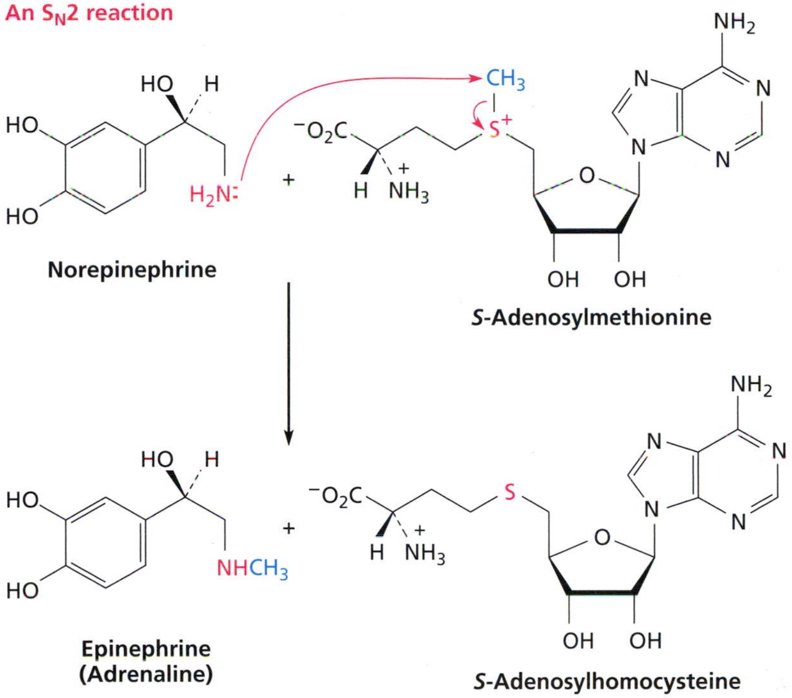

An S_N2 reaction

Norepinephrine

S-Adenosylmethionine

Epinephrine (Adrenaline)

S-Adenosylhomocysteine

FIGURE 1.8 The biosynthesis of epinephrine from norepinephrine occurs by an S_N2 reaction with *S*-adenosylmethionine.

1.5 Mechanisms: Nucleophilic Carbonyl Addition Reactions

Carbonyl groups are present in the vast majority of biological molecules, and carbonyl reactions are thus encountered in almost all biochemical pathways. In discussing carbonyl-group chemistry, it's useful to make a distinction between two general classes of compounds. In one class are aldehydes and ketones, which have their carbonyl carbon bonded to atoms (C and H) that can't stabilize a negative charge and therefore don't typically act as leaving groups in substitution reactions. In the second class are carboxylic acids and their derivatives, which have their carbonyl carbon bonded to an electronegative atom (O, N, or S) that *can* stabilize a negative charge and thus *can* act as a leaving group in a substitution reaction.

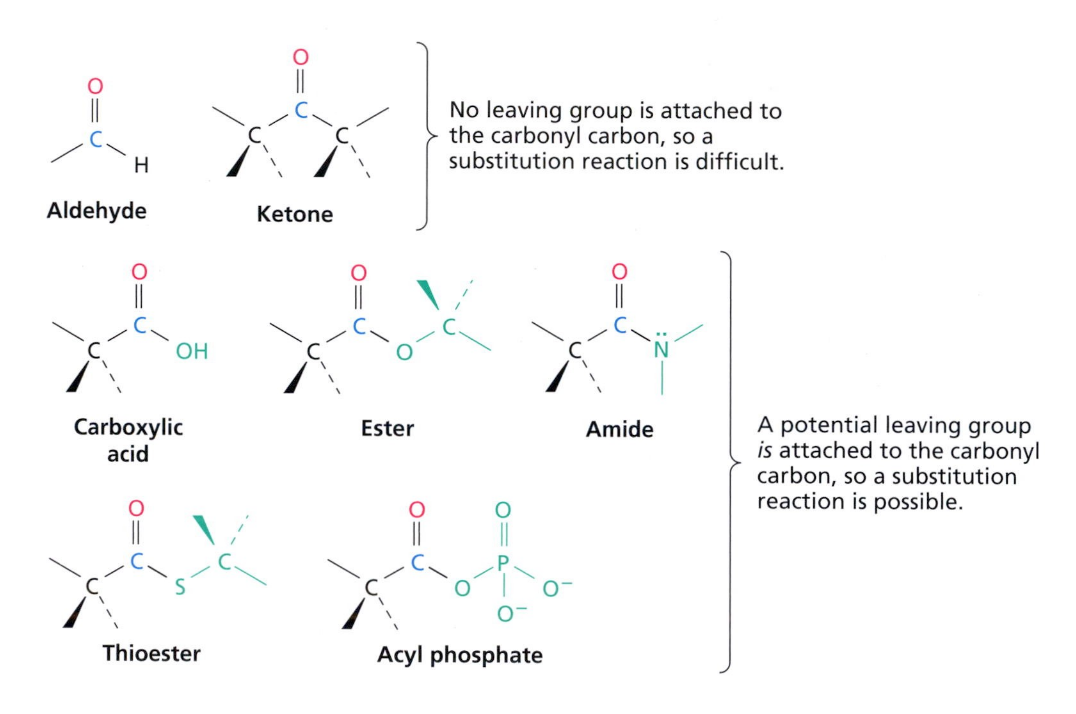

The C=O bond in every carbonyl compound, regardless of structure, is polarized with the oxygen negative and the carbon positive. As a result, the carbonyl oxygen is *nucleophilic* and reacts with acids/electrophiles, while the carbon is *electrophilic* and reacts with bases/nucleophiles. These simple reactivity patterns show up in almost all biological pathways.

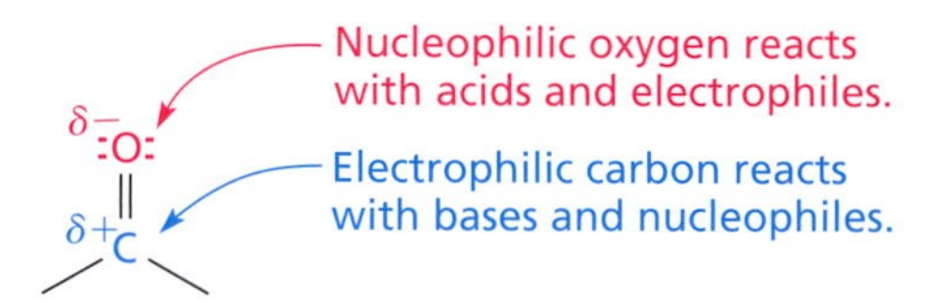

Nucleophilic Addition Reactions

A **nucleophilic addition reaction** is the addition of a nucleophile (:Nu) to the electrophilic carbon of an aldehyde or ketone. As an electron pair from the nucleophile forms a bond to the carbon, an electron pair from the C=O bond moves toward oxygen, giving an alkoxide ion, RO^-. The carbonyl carbon rehybridizes from sp^2 to sp^3 during the process, so the alkoxide product has tetrahedral geometry.

A nucleophilic addition reaction

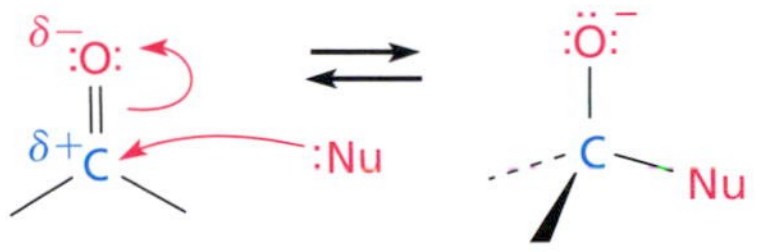

Once formed, the tetrahedral alkoxide ion can do any of several things, as shown in Figure 1.9. When a nucleophile such as hydride ion ($H:^-$) or a carbanion ($R_3C:^-$) adds, the alkoxide ion undergoes protonation to yield a stable alcohol.

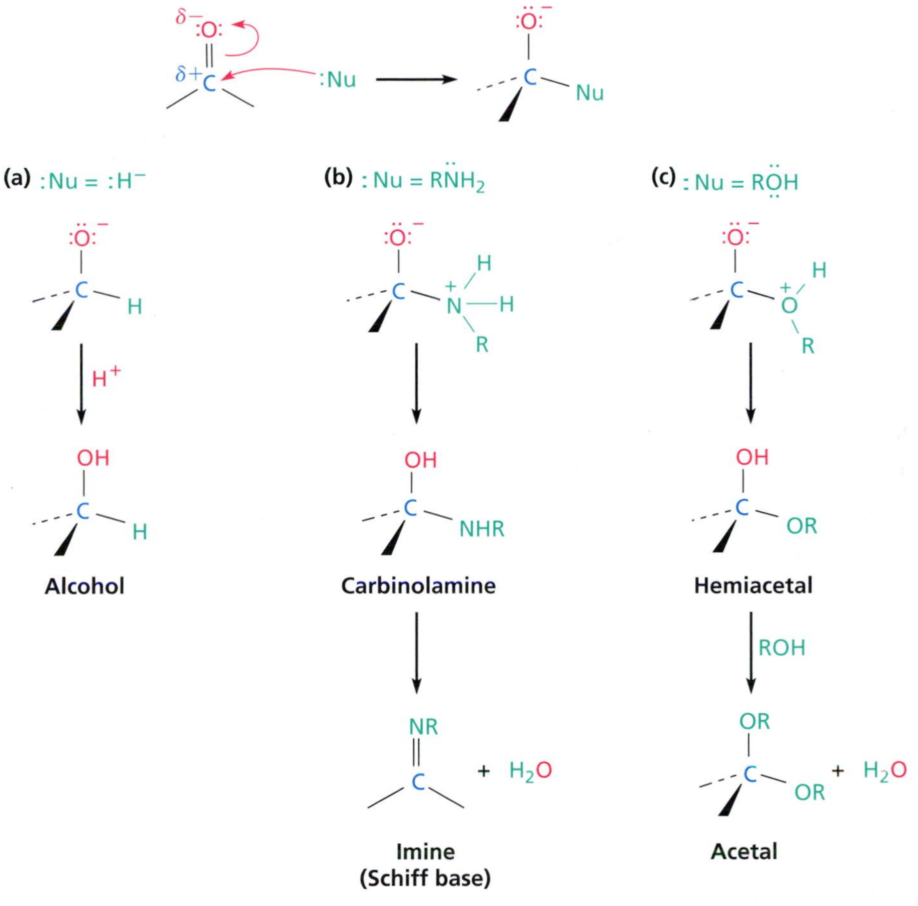

(a) $:Nu = :H^-$ — H^+ — **Alcohol**

(b) $:Nu = R\ddot{N}H_2$ — **Carbinolamine** — **Imine (Schiff base)** + H_2O

(c) $:Nu = R\ddot{O}H$ — **Hemiacetal** — ROH — **Acetal** + H_2O

FIGURE 1.9 Some typical nucleophilic addition reactions of aldehydes and ketones. (a) With a hydride ion as nucleophile, protonation of the alkoxide intermediate leads to an alcohol. (b) With an amine as nucleophile, proton transfer and loss of water leads to an imine. (c) With an alcohol as nucleophile, proton transfer leads to a hemiacetal, and further reaction with a second equivalent of alcohol leads to an acetal.

When a primary amine nucleophile (RNH_2) adds, the alkoxide ion undergoes a proton transfer to yield a **carbinolamine**, which loses water to form an **imine** ($R_2C{=}NR'$), often called a **Schiff base** in biochemistry. When an alcohol nucleophile (ROH) adds, the alkoxide undergoes proton transfer to yield a **hemiacetal**, which can react with a second equivalent of alcohol and lose water to give an **acetal** [$R_2C(OR')_2$]. In all the reactions that follow, note the role of acid and base catalysts.

Alcohol Formation

In the laboratory, the conversion of an aldehyde or ketone to an alcohol is generally carried out using $NaBH_4$ as the nucleophilic hydride-ion donor. In biological pathways, however, NADH (reduced nicotinamide adenine dinucleotide) or the closely related NADPH (reduced nicotinamide adenine dinucleotide phosphate) is the most frequently used hydride-ion donor. An example that occurs in the pathway by which organisms synthesize fatty acids is the conversion of acetoacetyl ACP (acyl carrier protein) to 3-hydroxybutyryl ACP. We'll look at the details of the process in Section 3.4.

Imine (Schiff Base) Formation

Imines are formed in a reversible, acid-catalyzed process that begins with nucleophilic addition of a primary amine to the carbonyl group of an aldehyde or ketone. The initial dipolar addition product undergoes a rapid proton transfer that removes an H^+ from N and places another H^+ on O to give a carbinolamine, which is protonated on the oxygen atom by an acid catalyst. The effect of this protonation is to convert —OH into a much better leaving group (—OH_2^+) so that it can be expelled by the electrons on nitrogen. Deprotonation of the resultant iminium ion then gives the imine product. The mechanism is shown in Figure 1.10.

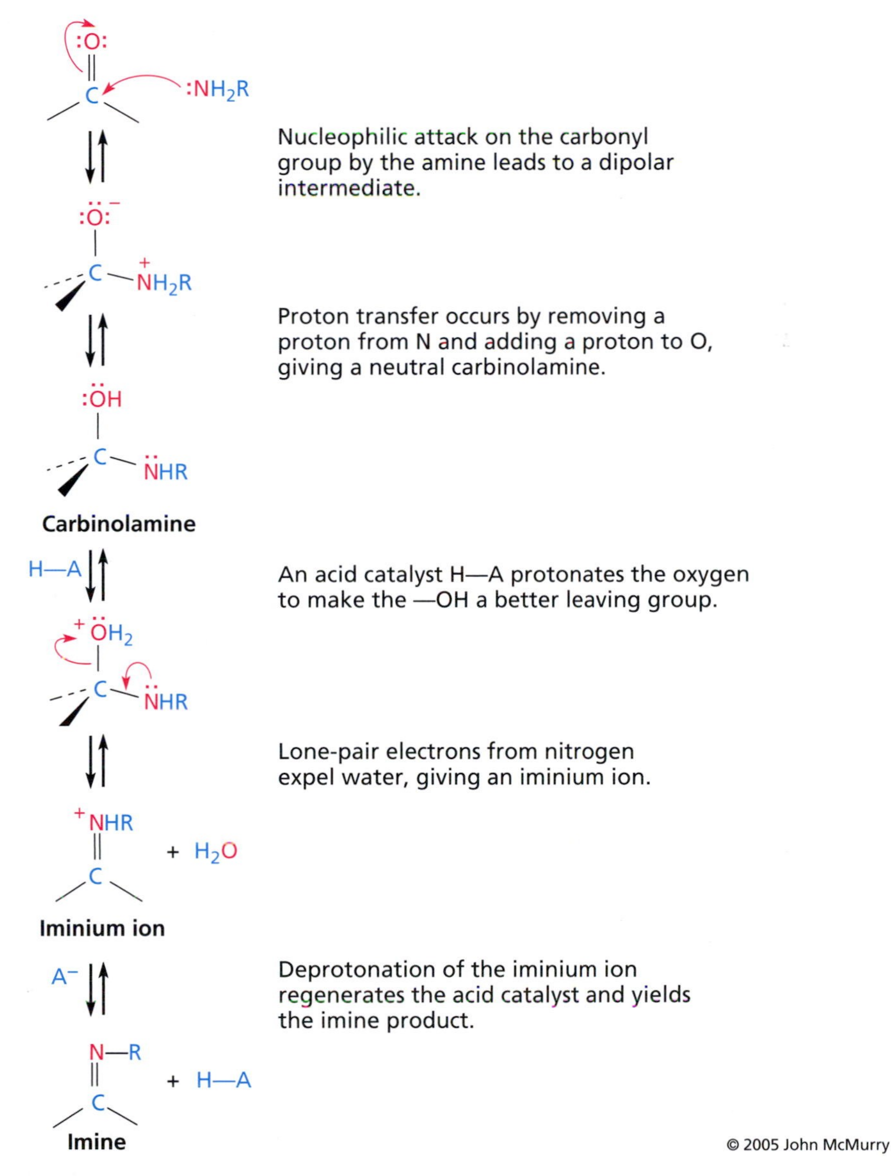

:NH2R
Nucleophilic attack on the carbonyl group by the amine leads to a dipolar intermediate.
NH2R
Proton transfer occurs by removing a proton from N and adding a proton to O, giving a neutral carbinolamine.
:OH
NHR
Carbinolamine
H—A
An acid catalyst H—A protonates the oxygen to make the —OH a better leaving group.
OH2
NHR
Lone-pair electrons from nitrogen expel water, giving an iminium ion.
NHR
+ H2O
Iminium ion
A−
Deprotonation of the iminium ion regenerates the acid catalyst and yields the imine product.
N—R
+ H—A
Imine

FIGURE 1.10 Mechanism of acid-catalyzed imine (Schiff base) formation by reaction of an aldehyde or ketone with a primary amine, RNH_2.

The conversion of a ketone to an imine is a step in numerous biological pathways, including the route by which many amino acids are synthesized in the body. For instance, the ketone pyruvate and the amine pyridoxamine phosphate, a derivative of vitamin B_6, form an imine that is converted to the amino acid alanine. We'll look at the details in Section 5.1.

Pyruvate + Pyridoxamine phosphate ⇌ An imine + H_2O

Acetal Formation

Acetals, like imines, are formed in a reversible, acid-catalyzed process. The reaction begins with protonation of the carbonyl oxygen to increase its reactivity, followed by nucleophilic addition of an alcohol. Deprotonation then gives a neutral hemiacetal, and reprotonation on the hydroxyl oxygen converts the —OH into a better leaving group so that it can be expelled by electrons on the neighboring —OR to produce an oxonium ion. This oxonium ion behaves much like the protonated carbonyl group in the first step, undergoing a second nucleophilic addition with alcohol. A final deprotonation then gives the acetal product. The mechanism is shown in Figure 1.11.

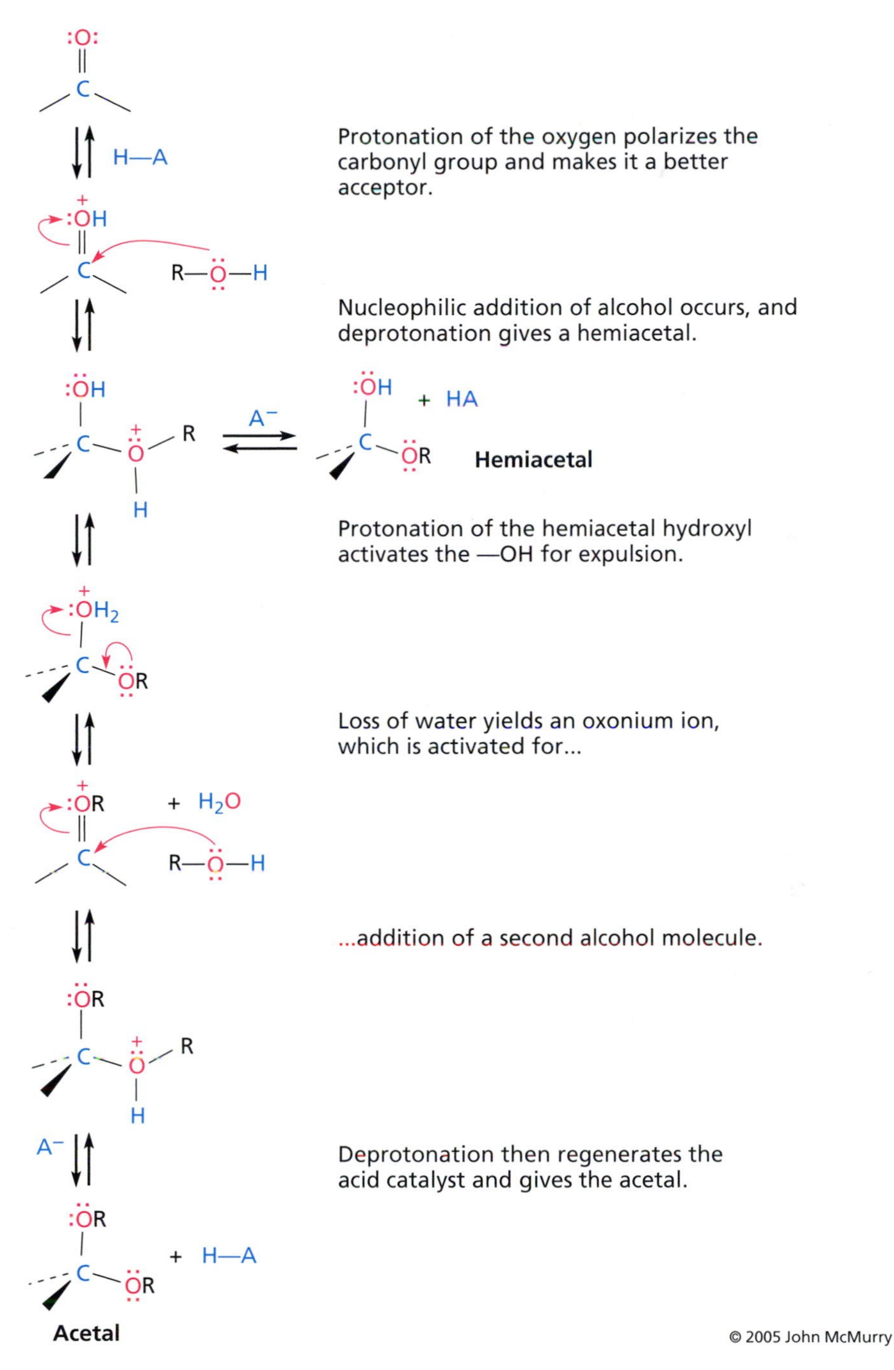

FIGURE 1.11 Mechanism of acid-catalyzed hemiacetal and acetal formation by reaction of an aldehyde or ketone with an alcohol.

The formation of hemiacetals and acetals is a central part of carbohydrate chemistry. Glucose, for instance, is in a readily reversible equilibrium between open (aldehyde + alcohol) and closed (cyclic hemiacetal) forms. Many glucose molecules can then join together by acetal links to form starch and cellulose (Section 4.1).

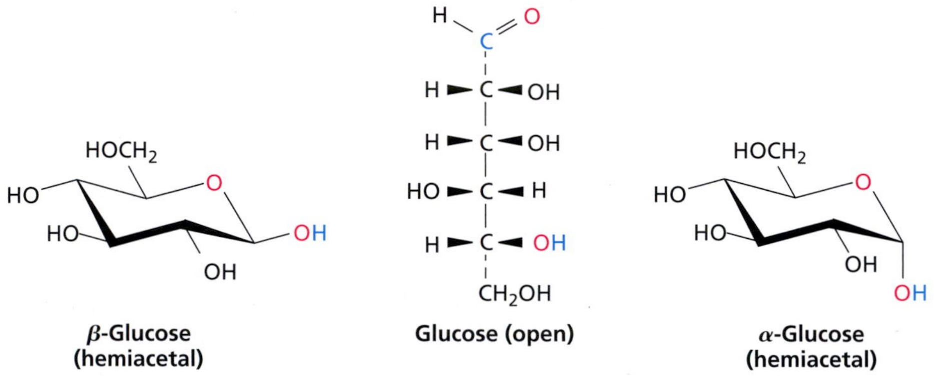

β-Glucose (hemiacetal) **Glucose (open)** **α-Glucose (hemiacetal)**

Conjugate (1,4) Nucleophilic Additions

Closely related to the direct (1,2) addition of a nucleophile to the C═O bond of an aldehyde or ketone is the **conjugate (1,4) addition**, or **Michael reaction**, of a nucleophile to the C═C bond of an α,β-unsaturated aldehyde or ketone (or thioester). The initial product, a resonance-stabilized **enolate ion**, typically undergoes protonation on the α carbon (the carbon *next to* the C═O) to give a saturated aldehyde or ketone (or thioester) product.

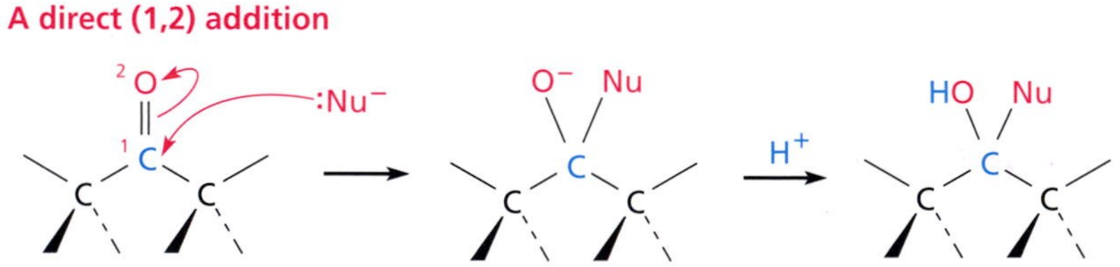

A direct (1,2) addition

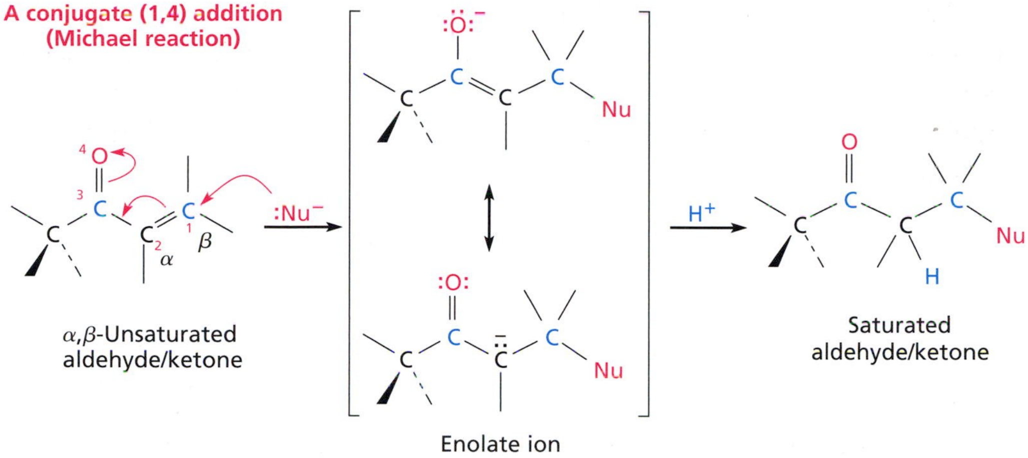

A conjugate (1,4) addition (Michael reaction)

α,β-Unsaturated aldehyde/ketone

:Nu⁻

Enolate ion

H⁺

Saturated aldehyde/ketone

Note that this conjugate addition reaction gives an overall result similar to that of the electrophilic alkene addition discussed in Section 1.3, but the mechanisms of the two processes are entirely different. Isolated alkenes react with *electrophiles* and form carbocation intermediates; α,β-unsaturated carbonyl compounds react with *nucleophiles* and form enolate-ion intermediates. Conjugate addition occurs because the electronegative oxygen atom of the α,β-unsaturated carbonyl compound withdraws electrons from the β carbon, thereby making it more electron-poor and more electrophilic than a typical alkene C=C bond.

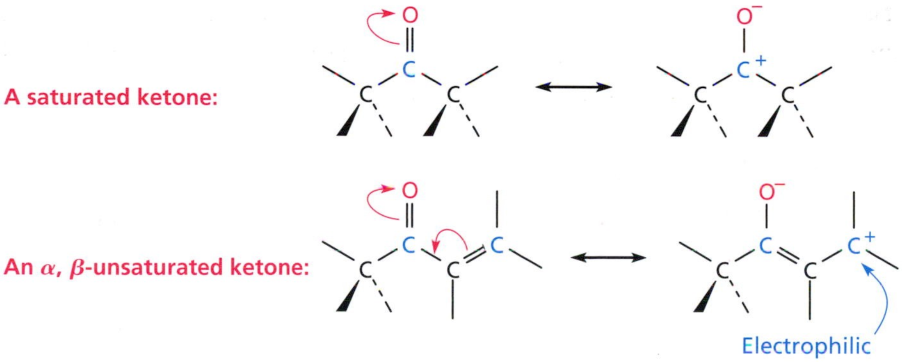

A saturated ketone:

An α, β-unsaturated ketone:

Electrophilic

Among the many biological examples of conjugate additions is the conversion of fumarate to malate by reaction with water, a step in the citric acid cycle by which acetate is metabolized to CO_2.

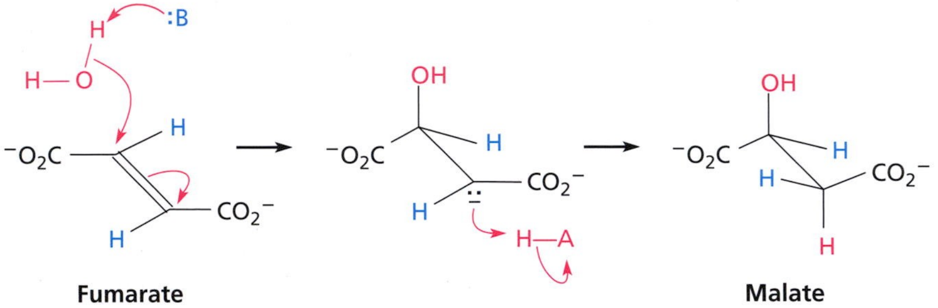

Fumarate Malate

1.6 Mechanisms: Nucleophilic Acyl Substitution Reactions

Carboxylic acids and their derivatives are characterized by the presence of an electronegative atom (O, N, S) bonded to the carbonyl carbon. As a result, these compounds can undergo **nucleophilic acyl substitution reactions**—the substitution of the leaving group bonded to the carbonyl carbon (:Y) by an attacking nucleophile (:Nu^-).

A nucleophilic acyl substitution reaction

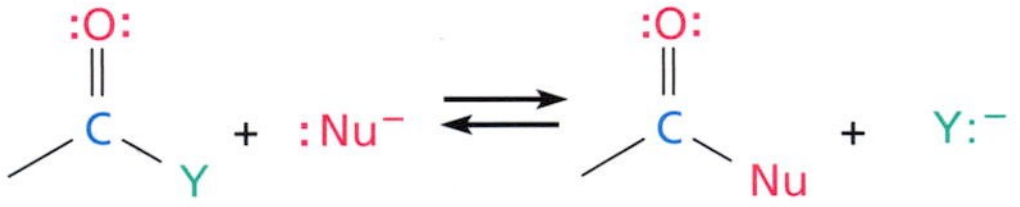

As shown in Figure 1.12 for the reaction of OH^- with methyl acetate, a nucleophilic acyl substitution reaction is initiated by addition of the nucleophile to the carbonyl carbon in the usual way. But the tetrahedrally hybridized alkoxide intermediate is not isolated; instead, it reacts further by expelling the leaving group and forming a new carbonyl compound. The overall result of nucleophilic acyl substitution is the replacement of the leaving group by the attacking nucleophile, just as occurs in the nucleophilic *alkyl* substitutions discussed in Section 1.4. Only the *mechanisms* of the two substitutions are different.

Both the addition step and the elimination step can affect the overall rate of a nucleophilic acyl substitution reaction, but the first step is generally rate-limiting. Thus, the greater the stability of the carbonyl compound, the less reactive it is. We therefore find that, of the carbonyl-containing functional groups commonly

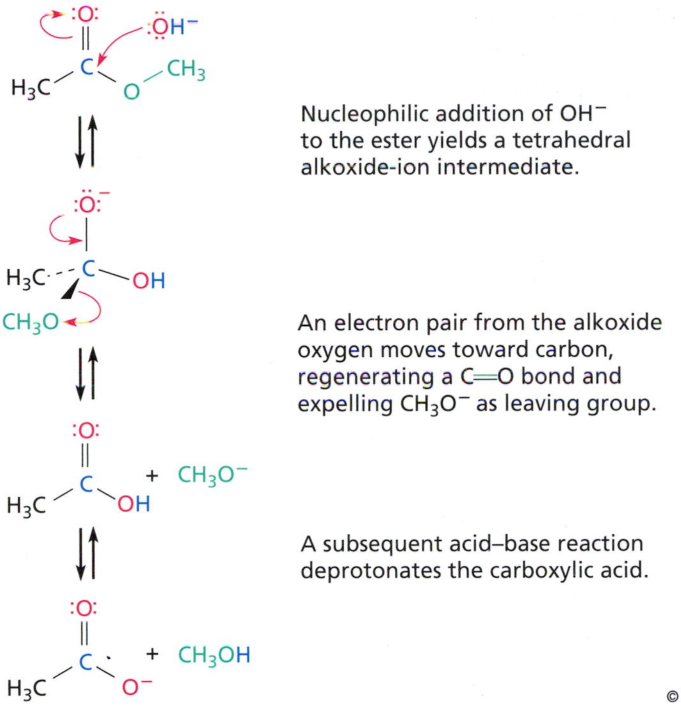

Nucleophilic addition of OH^- to the ester yields a tetrahedral alkoxide-ion intermediate.

An electron pair from the alkoxide oxygen moves toward carbon, regenerating a C=O bond and expelling CH_3O^- as leaving group.

A subsequent acid–base reaction deprotonates the carboxylic acid.

FIGURE 1.12 Mechanism of the nucleophilic acyl substitution reaction of OH^- with methyl acetate to give acetate. The reaction occurs by a nucleophilic addition to the carbonyl group, followed by expulsion of the leaving group in a second step.

found in living organisms, amides are the least reactive because of resonance stabilization; esters are somewhat more reactive; and thioesters and acyl phosphates are the most reactive toward substitution. (Acyl phosphates are generally further activated for substitution by complexation with a Lewis acidic metal cation such as Mg^{2+}.)

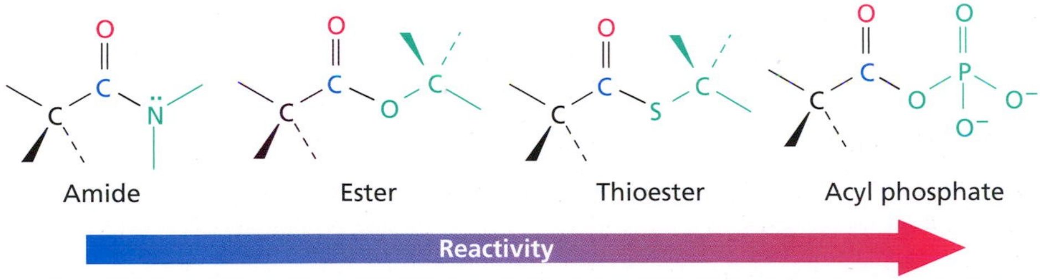

Amide Ester Thioester Acyl phosphate

Reactivity

Nucleophilic acyl substitution reactions occur frequently in biochemistry. For example, the carboxypeptidase-catalyzed hydrolysis of the C-terminal amide bond in proteins is one such process.

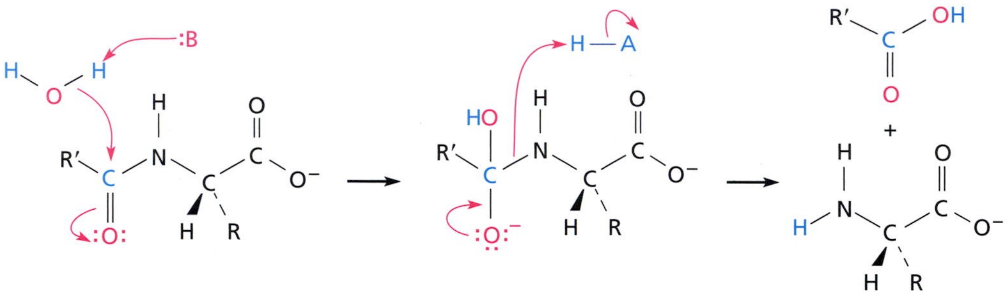

1.7 Mechanisms: Carbonyl Condensation Reactions

The third major reaction of carbonyl compounds, **carbonyl condensation**, occurs when two carbonyl compounds join to give a single product. When an aldehyde is treated with base, for example, two molecules combine to yield a β-hydroxy aldehyde product. Similarly, when an ester is treated with base, two molecules combine to yield a β-keto ester product.

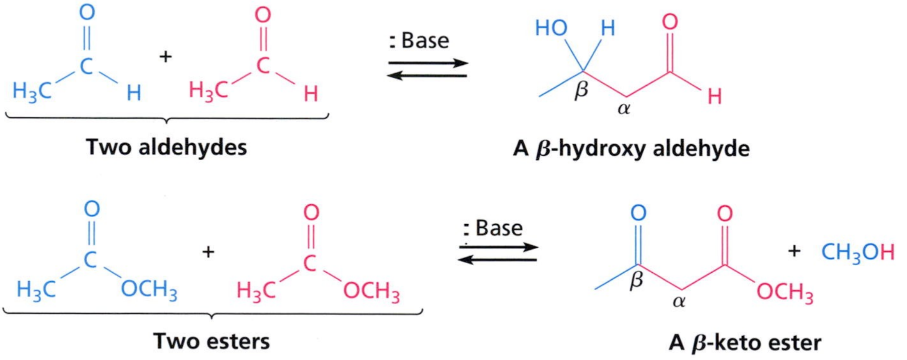

Two aldehydes

A β-hydroxy aldehyde

Two esters

A β-keto ester

Carbonyl condensation results in bond formation between the carbonyl carbon of one partner and the α carbon of the other partner. The reactions occur because the α hydrogen of a carbonyl compound is weakly acidic and can therefore be removed by reaction with a base to yield an enolate ion. Like other anions, enolate ions are nucleophiles.

The acidity of carbonyl compounds is due to resonance stabilization of the enolate ion, which allows the negative charge to be shared by the α carbon and the electronegative carbonyl oxygen. As shown in Table 1.4, aldehydes and ketones are the most acidic monocarbonyl compounds, with thioesters, esters,

Table 1.4 Acidity Constants of Some Carbonyl Compounds

Carbonyl compound	Example	pK_a	
Carboxylic acid	CH_3COH (C=O)	4.7	Stronger acid
1,3-Diketone	$CH_3CCH_2CCH_3$ (both C=O)	9.0	
β-Keto ester	$CH_3CCH_2COCH_3$ (both C=O)	10.6	
1,3-Diester	$CH_3OCCH_2COCH_3$ (both C=O)	12.9	
Aldehyde	CH_3CH (C=O)	17	
Ketone	CH_3CCH_3 (C=O)	19.3	
Thioester	CH_3CSCH_3 (C=O)	21	
Ester	CH_3COCH_3 (C=O)	25	
Amide	$CH_3CN(CH_3)_2$ (C=O)	30	Weaker acid

and amides less so. Most acidic of all, however, are 1,3-*dicarbonyl* compounds (β-dicarbonyl compounds), which have an α position flanked by *two* adjacent carbonyl groups.

The condensation of an aldehyde or ketone is called the **aldol reaction** and takes place by the mechanism shown in Figure 1.13. One molecule reacts with base to give a nucleophilic enolate ion, which then adds to the second molecule in a nucleophilic addition reaction. Protonation of the initially formed alkoxide ion yields the neutral β-hydroxy aldehyde or ketone product. Note that the condensation is reversible: A β-hydroxy aldehyde or ketone can fragment on treatment with base to yield two molecules of aldehyde or ketone.

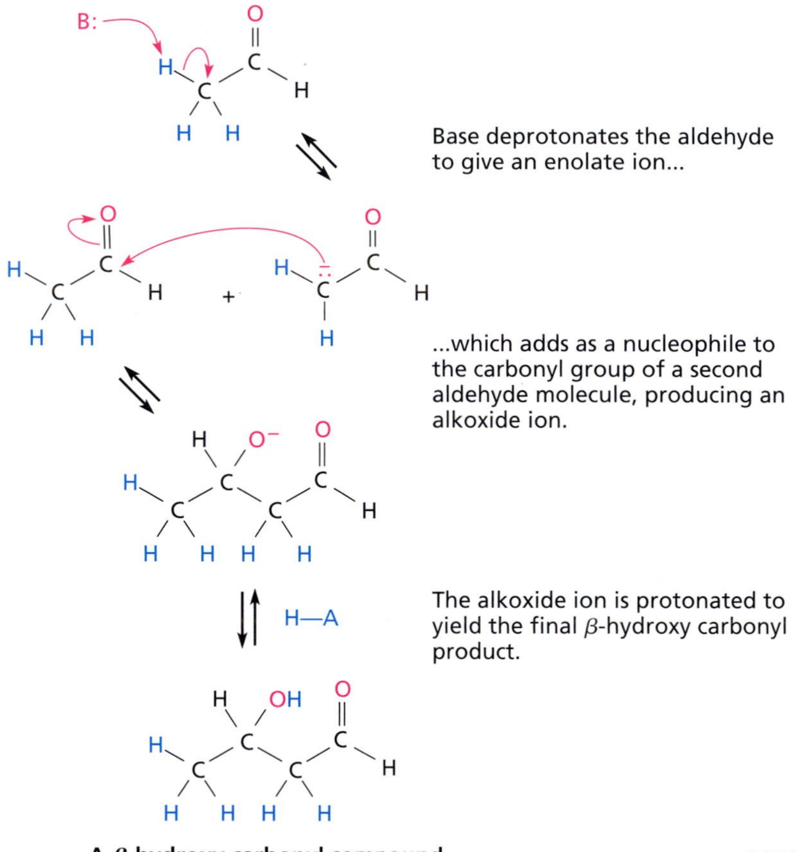

Base deprotonates the aldehyde to give an enolate ion...

...which adds as a nucleophile to the carbonyl group of a second aldehyde molecule, producing an alkoxide ion.

The alkoxide ion is protonated to yield the final β-hydroxy carbonyl product.

H—A

A β-hydroxy carbonyl compound

FIGURE 1.13 Mechanism of the aldol reaction, a reversible, base-catalyzed condensation reaction between two molecules of aldehyde or ketone to yield a β-hydroxy carbonyl compound. The key step is nucleophilic addition of an enolate ion to a C=O bond.

The condensation of an ester is called the **Claisen condensation reaction** and takes place by the mechanism shown in Figure 1.14. One molecule reacts with base to give a nucleophilic enolate ion, which adds to the second molecule in a nucleophilic acyl substitution reaction. The initially formed alkoxide ion

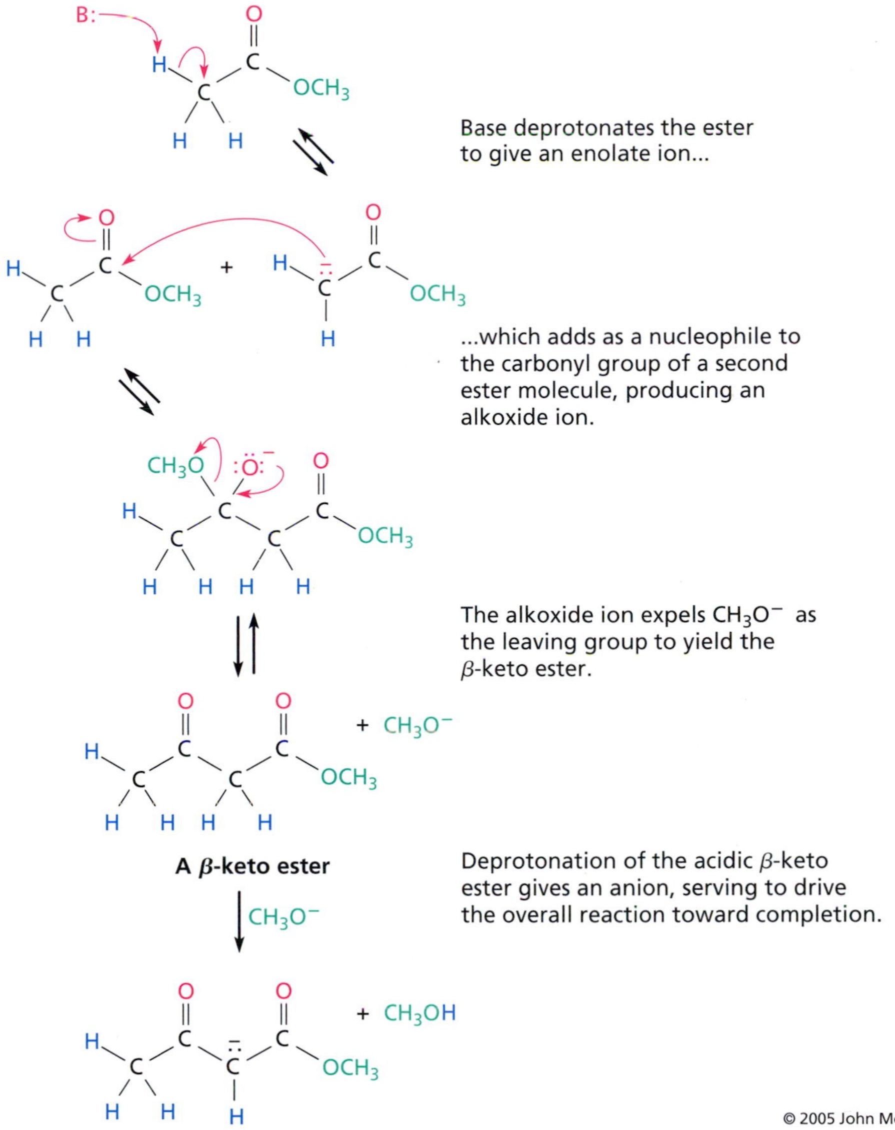

Base deprotonates the ester to give an enolate ion...

...which adds as a nucleophile to the carbonyl group of a second ester molecule, producing an alkoxide ion.

The alkoxide ion expels CH_3O^- as the leaving group to yield the β-keto ester.

A β-keto ester

Deprotonation of the acidic β-keto ester gives an anion, serving to drive the overall reaction toward completion.

FIGURE 1.14 Mechanism of the Claisen condensation reaction, a reversible, base-catalyzed condensation reaction between two molecules of ester to yield a β-keto ester. The key step is nucleophilic acyl substitution by an enolate ion, with expulsion of an alkoxide leaving group.

expels CH_3O^- as the leaving group to regenerate a C═O bond and form the β-keto ester product. As with the aldol reaction, the Claisen condensation is reversible: A β-keto ester can fragment on treatment with base to yield two molecules of ester.

Carbonyl condensation reactions are involved in nearly all biochemical pathways and serve as the primary biological method for forming and breaking carbon–carbon bonds. For instance, one step in the biosynthesis of glucose from pyruvate is the aldol reaction of glyceraldehyde 3-phosphate with dihydroxyacetone phosphate. As another example, the biological pathway for terpenoid and steroid biosynthesis begins with a Claisen condensation of the thioester acetyl CoA (coenzyme A) to give acetoacetyl CoA. We'll look at the details of glucose biosynthesis in Section 4.5 and the details of steroid biosynthesis in Section 3.6.

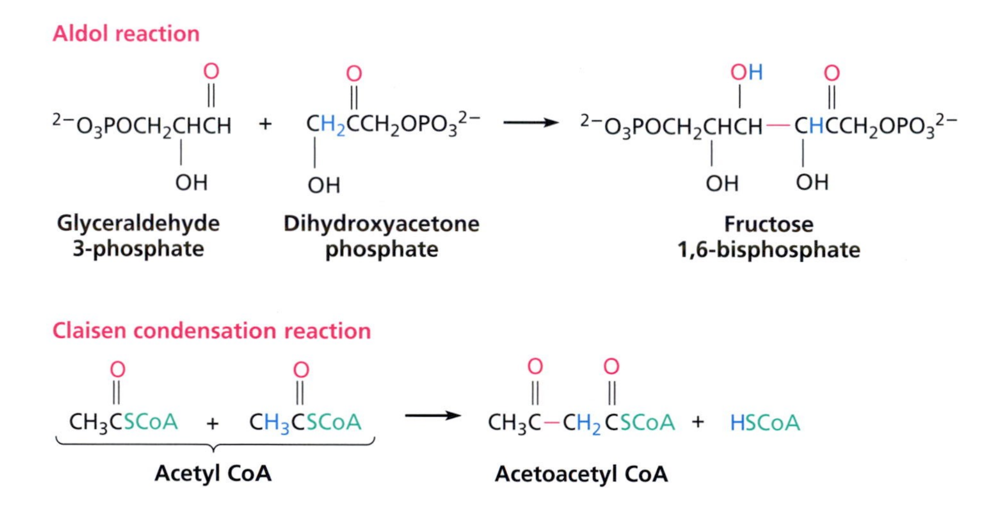

1.8 Mechanisms: Elimination Reactions

The elimination of HX to yield an alkene appears to be the simple reverse of an electrophilic addition of HX. In fact, though, **elimination reactions** are a good deal more complex than additions and can occur by any of several mechanisms. In the laboratory, the three most common processes are the E1, E2, and E1cB reactions, which differ in the timing of C—H and C—X bond-breaking. In the E1 reaction, the C—X bond breaks first to give a carbocation intermediate, which then undergoes base abstraction of H^+ to yield the alkene (the exact reverse of the electrophilic addition reaction described in Section 1.3). In the E2 reaction, base-induced C—H bond cleavage is simultaneous with C—X bond cleavage, giving the alkene in a single step. In the E1cB reaction (cB for "conjugate base"), base abstraction of the proton occurs first, giving a carbanion intermediate that undergoes loss of X^- in a subsequent step to give the alkene.

E1 Reaction: C–X bond breaks first to give a carbocation intermediate, followed by base removal of a proton to yield the alkene.

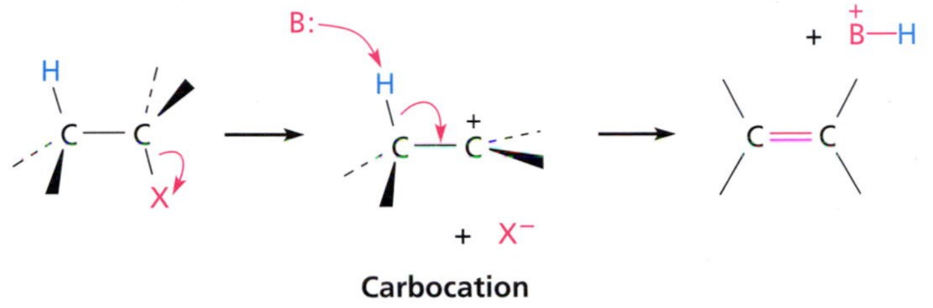

Carbocation

E2 Reaction: C–H and C–X bonds break simultaneously, giving the alkene in a single step without intermediates.

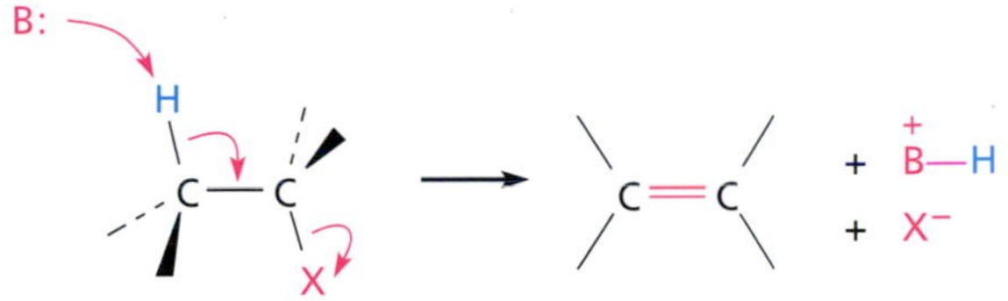

E1cB Reaction: C–H bond breaks first, giving a carbanion intermediate that loses X^- to form the alkene.

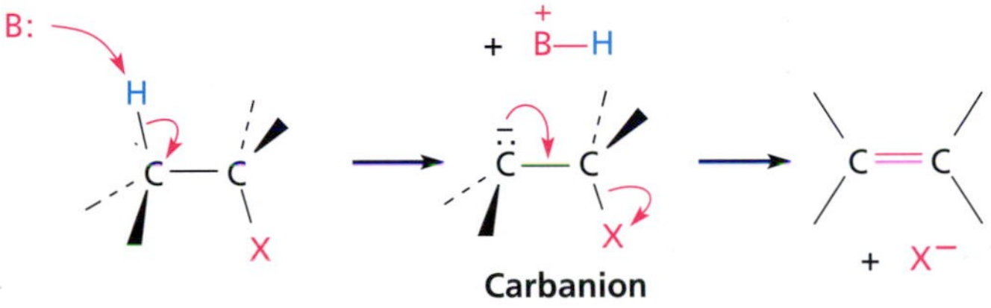

Carbanion

Examples of all three mechanisms occur in different biological pathways, but the E1cB mechanism is particularly common. The substrate is usually an alcohol (X = OH) or protonated alcohol ($X = OH_2^+$), and the H atom that is removed is usually made acidic, particularly in E1cB reactions, by being adjacent to a carbonyl group. Thus, β-hydroxy carbonyl compounds (aldol reaction products) are frequently converted to α,β-unsaturated carbonyl compounds by elimination reactions. An example is the dehydration of a β-hydroxy thioester to the corresponding unsaturated thioester, a reaction that occurs in fatty-acid biosynthesis (Section 3.4). The base in this reaction is a histidine residue in the enzyme, and the elimination is assisted by complexation of the —OH group to the protonated histidine as a Lewis acid. Note that the reaction occurs with *syn* stereochemistry, meaning that the —H and —OH groups in this example are eliminated from the same side of the molecule.

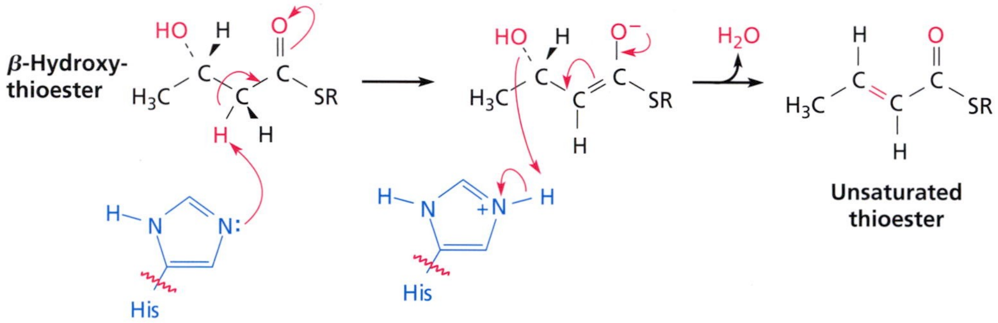

β-Hydroxy-thioester

Unsaturated thioester

1.9 Oxidations and Reductions

Oxidation–reduction, or **redox**, chemistry is a large and complicated, but extremely important, topic. Rather than attempt a complete catalog of the subject at this point, however, let's focus on the mechanisms of two of the more commonly occurring biological redox processes: the oxidation of an alcohol and the reduction of a carbonyl compound. We'll look at the mechanisms of other kinds of biological oxidations in later chapters as the need arises when we discuss specific pathways.

In the laboratory, alcohol oxidations generally occur through a mechanism that involves attachment to oxygen of a leaving group X, usually a metal in a high oxidation state such as Cr(VI) or Mn(VII). An E2-like elimination reaction then forms the C═O bond and expels the metal in a lower oxidation state. Note that the C—H hydrogen is removed by base as H^+ during the elimination step.

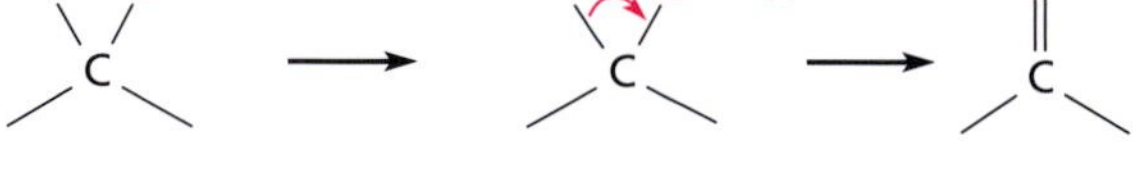

Although a similar mechanism does occasionally occur in biological pathways, many biological oxidations of an alcohol occur by a reversible hydride-transfer mechanism involving one of the coenzymes NAD^+ (oxidized nicotinamide adenine dinucleotide) or $NADP^+$ (oxidized nicotinamide adenine dinucleotide phosphate). As shown in Figure 1.15, the reaction occurs in a single step without intermediates when a base B: abstracts the acidic O—H proton, the electrons from the O—H bond move to form a C═O bond, and the hydrogen attached to carbon is transferred to NAD^+. Note that the C—H hydrogen is transferred as H^-, in contrast to the typical laboratory oxidation where it is removed as H^+. Note also that the hydride ion adds to the C═C—C═N^+ part of NAD^+ in a conjugate nucleophilic addition reaction, much as water might add to the C═C—C═O part of an α,β-unsaturated ketone (Section 1.5).

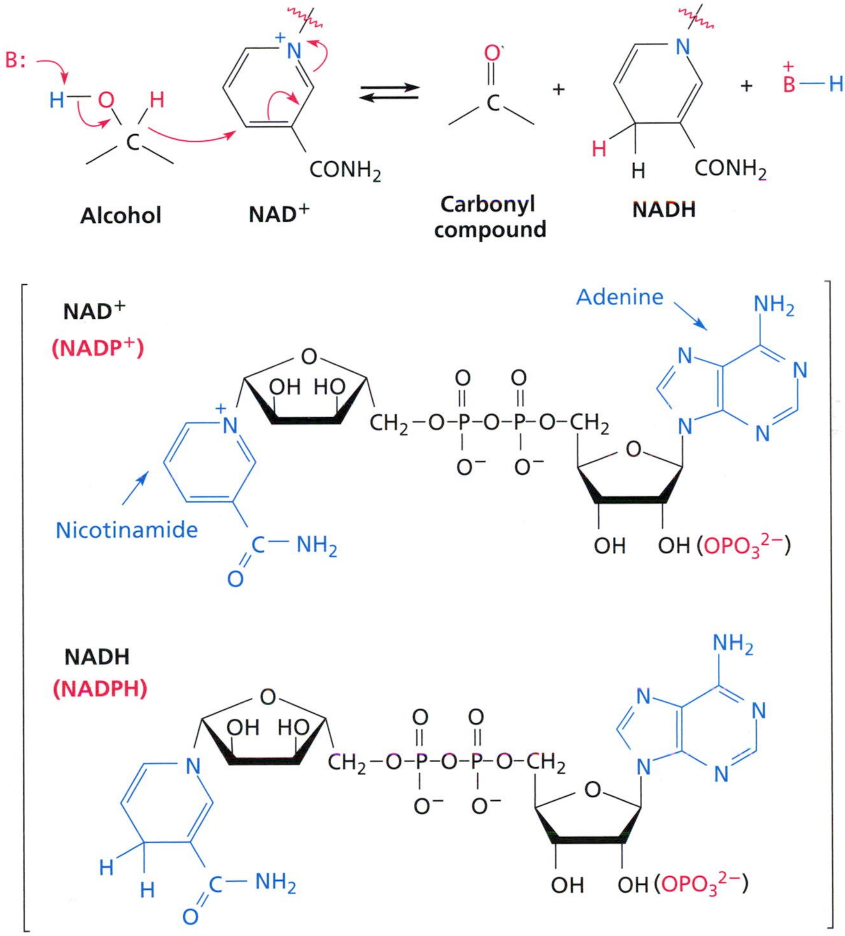

Alcohol | NAD^+ | Carbonyl compound | NADH

NAD^+ ($NADP^+$)

Adenine

Nicotinamide

NADH (NADPH)

FIGURE 1.15 Mechanism of alcohol oxidation by NAD^+.

Expulsion of hydride ion is not often seen in laboratory chemistry because, as noted at the beginning of Section 1.5, H^- is a poor leaving group. One analogous process that does occur in laboratory chemistry, however, is the Cannizzaro reaction, which involves the disproportionation of an aromatic aldehyde on treatment with base. Benzaldehyde, for example, is converted to a mixture of benzyl alcohol and benzoic acid by reaction with NaOH. Hydroxide ion first adds to the aldehyde carbonyl group to give an alkoxide intermediate, which transfers a hydride ion to a second molecule of aldehyde. The first aldehyde is thereby oxidized, and the second aldehyde is reduced.

Cannizzaro reaction

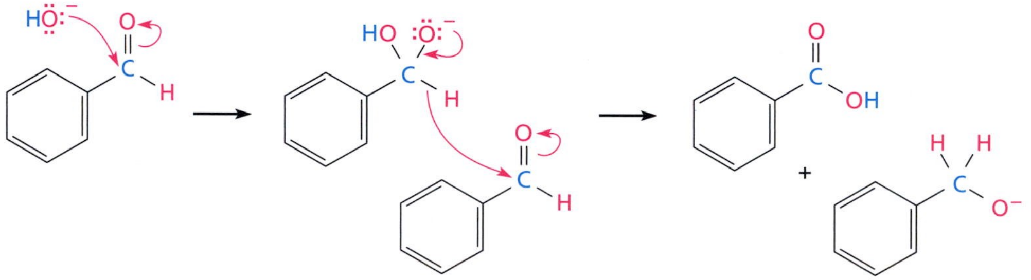

Biological reductions are the reverse of oxidations. As noted in Section 1.5, NADH transfers a hydride ion to the carbonyl group in a nucleophilic addition reaction, and the alkoxide intermediate is protonated.

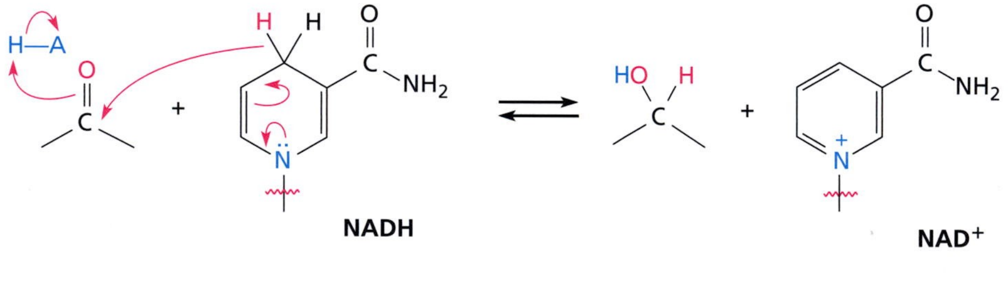

NADH **NAD$^+$**

Problems

1.1 Which of the following substances can behave as either an acid or a base, depending on the circumstances?

(a) CH_3SH (b) Ca^{2+} (c) NH_3 (d) CH_3SCH_3
(e) NH_4^+ (f) $H_2C{=}CH_2$ (g) $CH_3CO_2^-$ (h) $^+H_3NCH_2CO_2^-$

1.2 Which C=C bond do you think is more nucleophilic, that in 1-butene or that in 3-buten-2-one? Explain.

$CH_3CH_2CH{=}CH_2$ — **1-Butene**

$CH_3C(=O)CH{=}CH_2$ — **3-Buten-2-one**

1.3 Rank the following compounds in order of increasing acidity:

(a) Acetone, $pK_a = 19.3$ (b) Phenol, $pK_a = 9.9$
(c) Methanethiol, $pK_a = 10.3$ (d) Formic acid, $K_a = 1.99 \times 10^{-4}$
(e) Ethyl acetoacetate, $K_a = 2.51 \times 10^{-11}$

1.4 Rank the following compounds in order of increasing basicity:

(a) Aniline, $pK_{a(BH^+)} = 4.63$ (b) Pyrrole, $pK_{a(BH^+)} = 0.4$
(c) Dimethylamine, $pK_{a(BH^+)} = 10.5$
(d) Ammonia, $K_{a(BH^+)} = 5.5 \times 10^{-10}$

1.5 Protonation of acetic acid by H_2SO_4 might occur on either of two oxygen atoms. Draw resonance structures of both possible products, and explain why protonation occurs preferentially on the double-bond oxygen.

Acetic acid $H_3C{-}C(={:}\ddot{O}{:}){-}\ddot{O}{-}H \xrightleftharpoons{H_2SO_4}$

1.6 Protonation of a guanidino compound occurs on the double-bond nitrogen rather than on either of the single-bond nitrogens. Draw resonance structures of the three possible protonation products, and explain the observed result.

$H_2\ddot{N}{-}C(={:}NH){-}\ddot{N}H{-}R$

A guanidino compound

1.7 Predict the product(s) of the following biological reactions by interpreting the flow of electrons as indicated by the curved arrows:

(a)

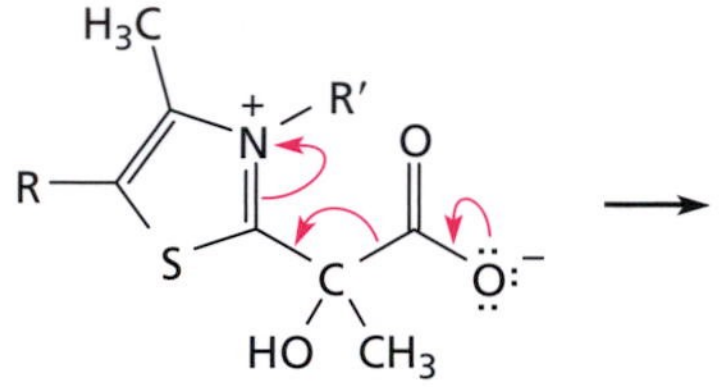

(b)

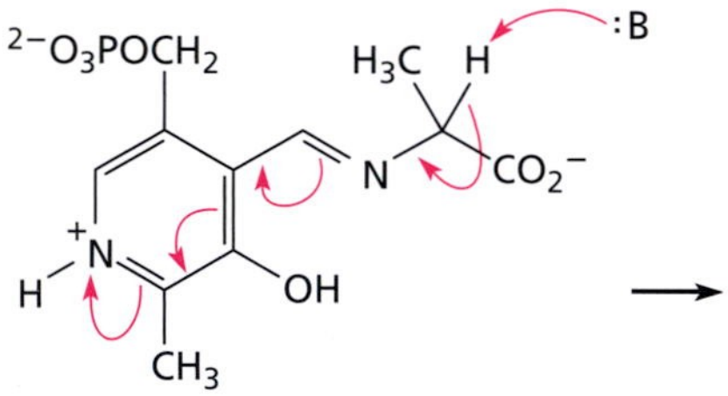

(c)

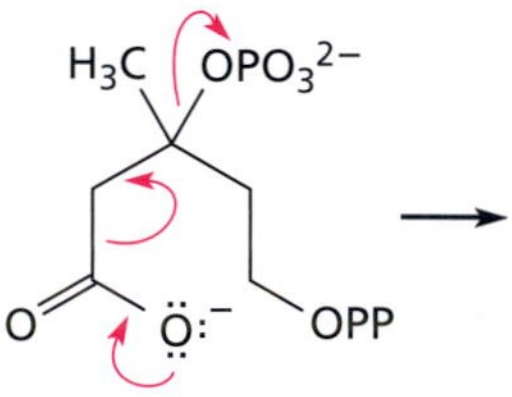

1.8 Complete the following biological mechanisms by adding curved arrows to indicate electron flow:

(a)

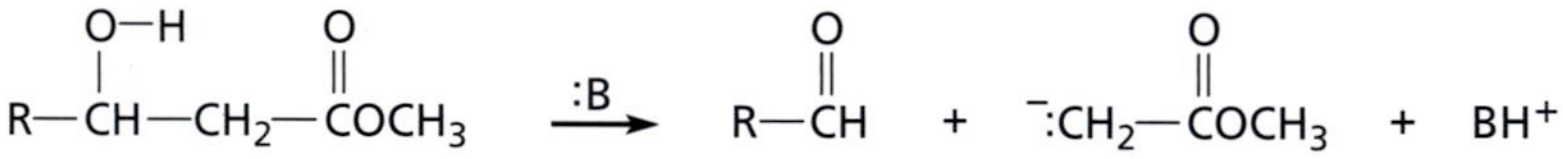

$$R{-}CH(OH){-}CH_2{-}C(=O)OCH_3 \xrightarrow{:B} R{-}CH{=}O + {}^{-}{:}CH_2{-}C(=O)OCH_3 + BH^+$$

(b)

$$R{-}C(=O){-}SCoA + CH_2(R''){-}C(=O){-}R' \xrightarrow{:B} R{-}C(O^-)(SCoA){-}CHR''{-}C(=O){-}R' + BH^+ \longrightarrow R{-}C(=O){-}CHR''{-}C(=O){-}R' + CoAS^-$$

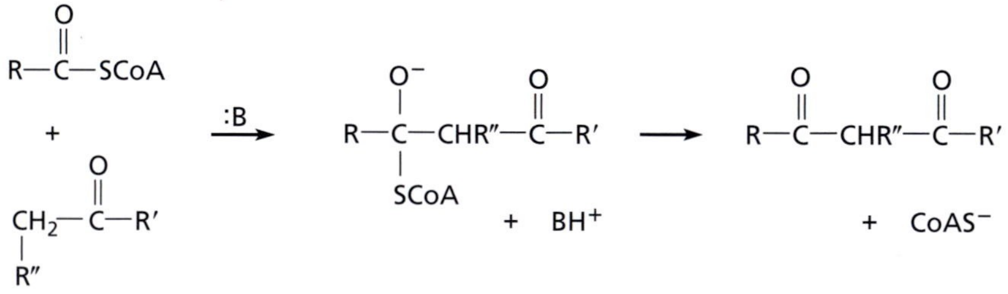

(c)

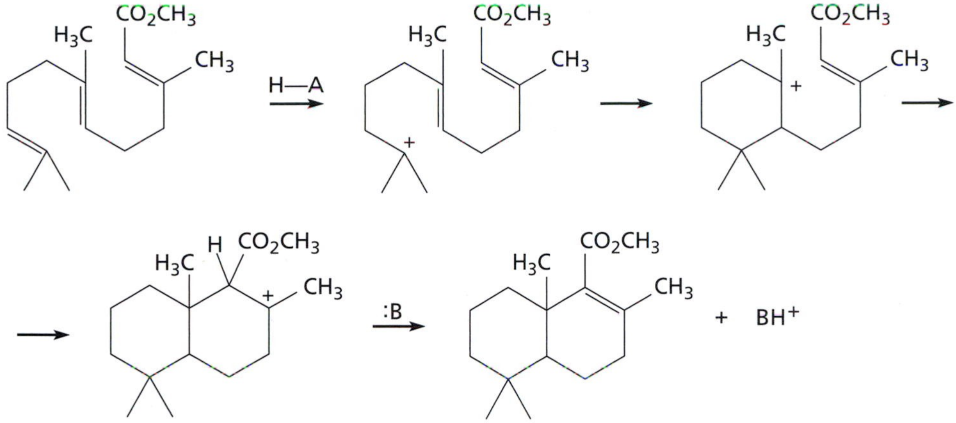

H_3C CO_2CH_3 CH_3 $\xrightarrow{H-A}$ $\longrightarrow$ $\longrightarrow$

$\longrightarrow$ H_3C H CO_2CH_3 CH_3 $\xrightarrow{:B}$ H_3C CO_2CH_3 CH_3 + BH^+

1.9 Propose a mechanism for the following step in the β-oxidation pathway for degradation of fatty acids. HSCoA is the abbreviation for coenzyme A, a thiol.

$$RCH_2CH_2\overset{O}{\overset{\|}{C}}CH_2\overset{O}{\overset{\|}{C}}SCoA \xrightarrow{HSCoA} RCH_2CH_2\overset{O}{\overset{\|}{C}}SCoA + CH_3\overset{O}{\overset{\|}{C}}SCoA$$

1.10 The conversion of fructose 1,6-bisphosphate to glyceraldehyde 3-phosphate plus dihydroxyacetone phosphate is a step in the glycolysis pathway for degrading carbohydrates. Propose a mechanism.

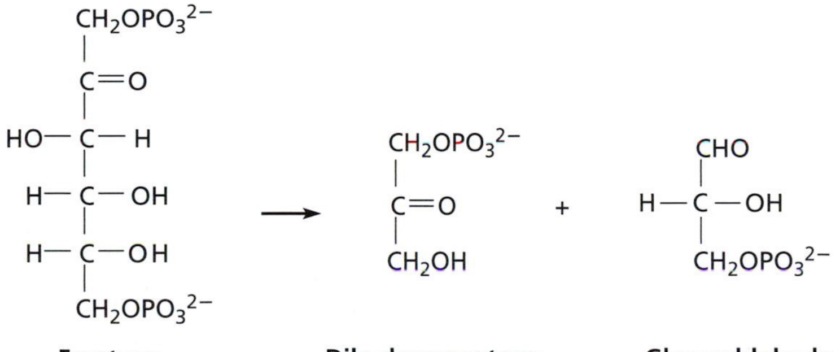

$CH_2OPO_3^{2-}$, C=O, HO—C—H, H—C—OH, H—C—OH, $CH_2OPO_3^{2-}$ $\longrightarrow$ $CH_2OPO_3^{2-}$, C=O, CH_2OH + CHO, H—C—OH, $CH_2OPO_3^{2-}$

Fructose 1,6-bisphosphate **Dihydroxyacetone phosphate** **Glyceraldehyde 3-phosphate**

1.11 Propose a mechanism for the conversion of 3-phosphoglycerate to phosphoenolpyruvate (PEP), a step in the glycolysis pathway.

$$HOCH_2\overset{OPO_3^{2-}}{\overset{|}{C}}HCO_2^- \longrightarrow H_2C{=}\overset{OPO_3^{2-}}{\overset{|}{C}}CO_2^-$$

2-Phosphoglycerate **Phosphoenolpyruvate**

1.12 The biological conversion of an aldehyde to a thioester occurs in two steps: (1) nucleophilic addition of a thiol to give a hemithioacetal, and (2) oxidation of the hemithioacetal by NAD^+. Show the structure of the intermediate hemithioacetal, and propose mechanisms for both steps.

$$RCH_2CHO \xrightarrow{R'SH} \text{Hemithioacetal} \xrightarrow{NAD^+} RCH_2C(=O)SR'$$

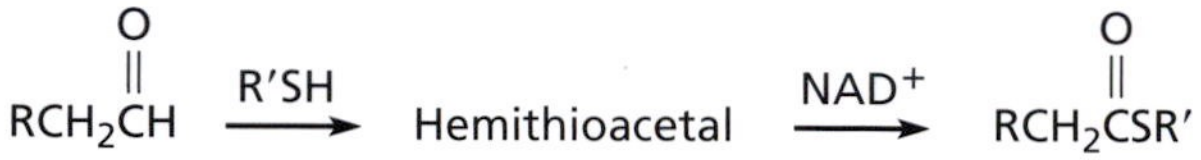

1.13 The loss of CO_2 (decarboxylation) from a β-keto acid happens frequently in biological chemistry and takes place by a mechanism that is closely related to a retro-aldol reaction. Propose a mechanism for the following reaction that occurs in the citric acid cycle.

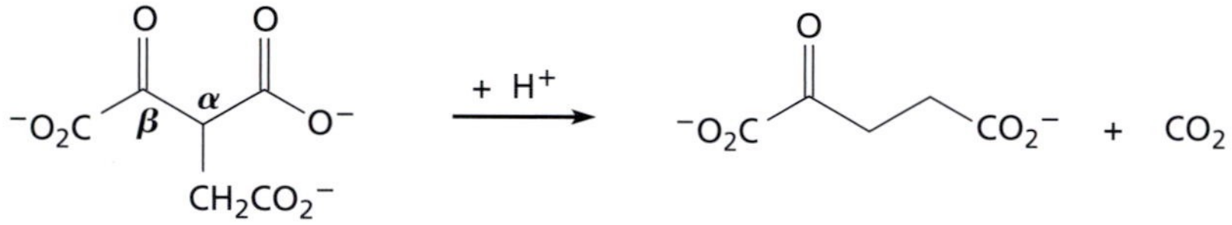

^-O_2C, β, α, O^-, $CH_2CO_2^-$ $\xrightarrow{+\ H^+}$ ^-O_2C, CO_2^- + CO_2

A β-keto acid **α-Ketoglutarate**

1.14 One of the biological pathways by which an amine is converted to a ketone involves two steps: (1) oxidation of the amine by NAD^+ to give an imine, and (2) hydrolysis of the imine to give a ketone plus ammonia. Glutamate, for instance, is converted by this process into α-ketoglutarate. Show the structure of the imine intermediate, and propose mechanisms for both steps.

$^+NH_3$, ^-O_2C, CO_2^- $\xrightarrow{NAD^+}$ Imine $\xrightarrow{H_2O}$ ^-O_2C, CO_2^- + NH_3

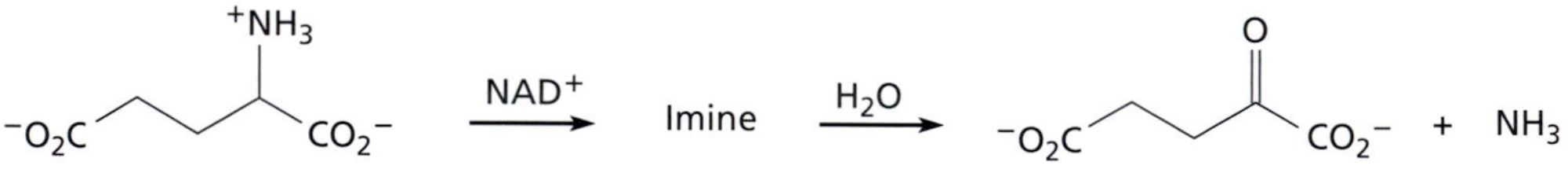

Glutamate **α-Ketoglutarate**

1.15 The following reaction is part of the sequence by which pyruvate is converted to acetyl CoA. Propose a mechanism, and tell what kind of reaction is occurring.

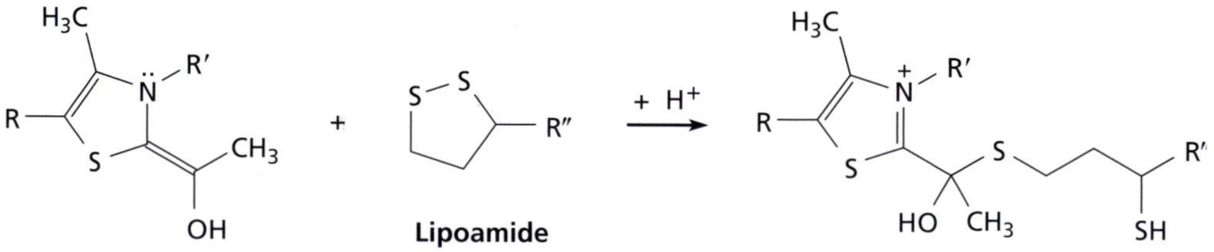

H_3C, R', N, R, S, CH_3, OH + S, S, R'' $\xrightarrow{+\ H^+}$ H_3C, R', N, R, S, S, R', HO, CH_3, SH

Lipoamide

1.16 The amino acid methionine is formed by a methylation reaction of homocysteine with *N*-methyltetrahydrofolate. The stereochemistry of the reaction has been probed by carrying out the transformation using a donor with a "chiral methyl group" in which deuterium (D) and tritium (T) isotopes of hydrogen are present. Does the methylation reaction occur with inversion or retention of configuration? What mechanistic inferences can you draw?

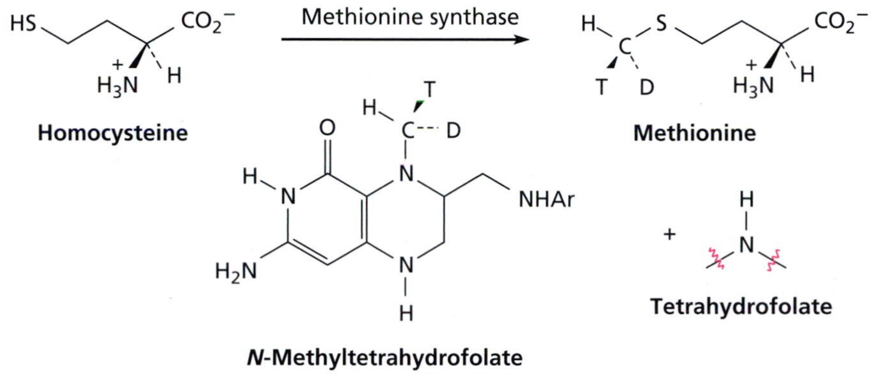

HS, CO_2^-, $H_3\overset{+}{N}$, H

Homocysteine

Methionine synthase

H, T, C, D, O, N, H, N, NHAr, H_2N, N, H

***N*-Methyltetrahydrofolate**

H, S, C, T, D, CO_2^-, $H_3\overset{+}{N}$, H

Methionine

\+ H, N

Tetrahydrofolate

2 Biomolecules

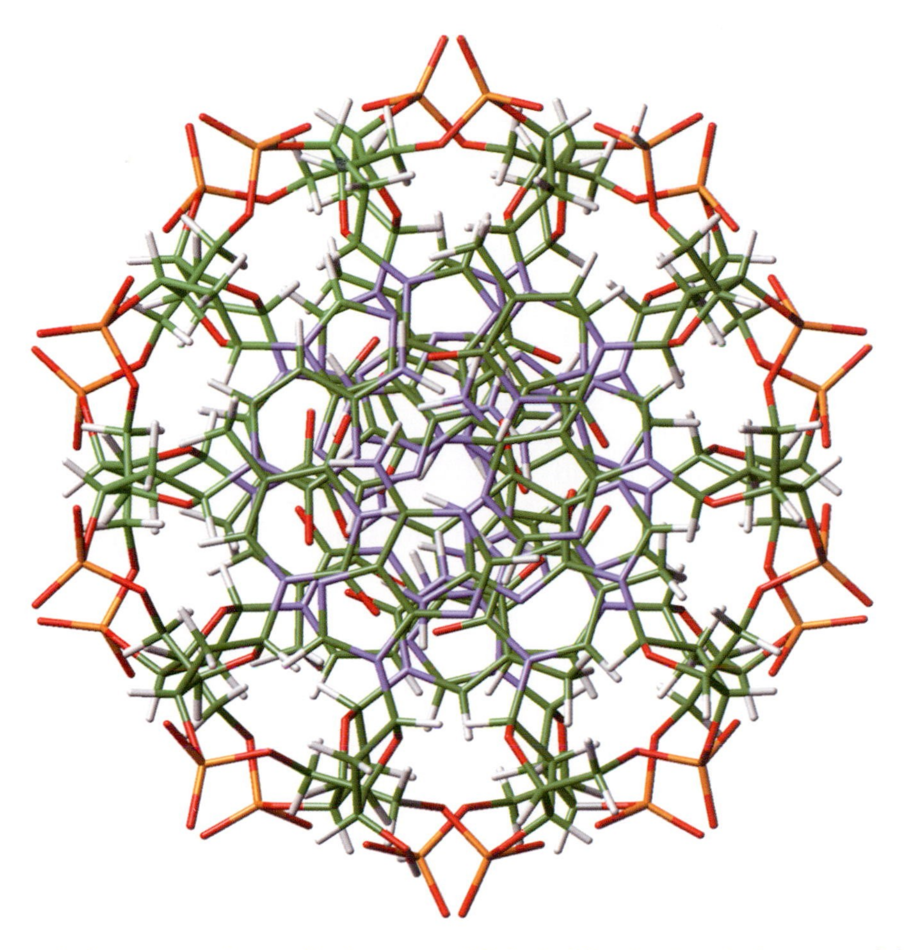

Structure is the great translator of the language of biological function into the language of chemistry. The structure of the DNA double helix shown above is the most significant single structure in biology. While its discovery in 1953 initiated a revolution in our understanding of the chemistry of all living systems, not even the most visionary of biologists could have predicted that 50 years later more than 500 genomes would be sequenced, including all 3 billion bases of the human genome.

Having seen the common organic reaction mechanisms in Chapter 1, let's take a quick look in this chapter at the common classes of biological molecules. Much of this material may be familiar to you, but some of it may also need reviewing because introductory organic chemistry courses often cover this material lightly. Section 2.1 on stereochemistry is likely to need special attention, particularly the part on prochirality.

2.1 Chirality and Biological Chemistry

The vast majority of biological molecules have a handedness, or **chirality**. Because of this molecular handedness, highly specific interactions between molecules are possible, leading to the extraordinary reaction specificity often observed in enzyme-catalyzed processes.

The most common cause of chirality in a molecule is the presence of a carbon atom bonded to four different groups—a so-called *stereocenter*, or **chirality center**. The amino acid alanine, for instance, is chiral because it has an —H atom, a

—CH_3 group, an —NH_2 group, and a —CO_2H group bonded to the central carbon atom. (In all the molecular models in this book, we'll follow the standard color convention: H = ivory, C = black, O = red, N = blue, and S = yellow.)

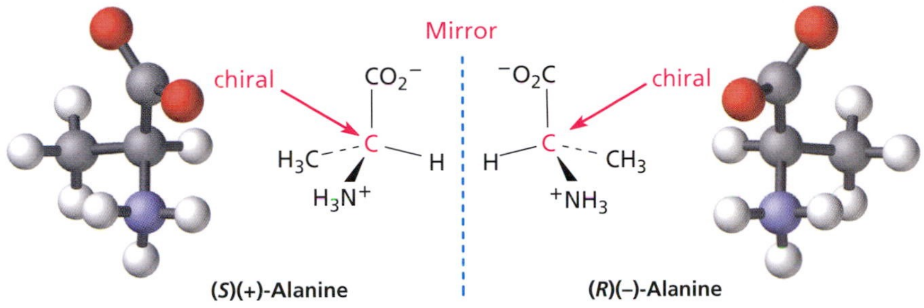

(*S*)(+)-Alanine (*R*)(−)-Alanine

Enantiomers

Chiral molecules have two nonsuperimposable mirror-image forms called **enantiomers**, which are identical in their physical properties except for their interaction with *plane-polarized light*. When a beam of plane-polarized light passes through a solution of chiral molecules, the **levorotatory** enantiomer rotates the plane of polarization in a left-hand, counterclockwise direction symbolized (−), and the **dextrorotatory** enantiomer rotates the plane of polarization in a right-hand, clockwise direction symbolized (+). A 1:1 mixture of the two enantiomers is symbolized (±) and is called a **racemic mixture**. Racemic mixtures show no evident rotation of plane-polarized light because the exactly opposite rotations of the two enantiomers cancel.

The three-dimensional arrangement, or **configuration**, of a chirality center is defined by first using the **Cahn–Ingold–Prelog sequence rules** to assign relative priorities to the four attached groups. The molecule is then oriented so that the group of lowest priority (4) points directly back away from the viewer, with the remaining three groups now radiating toward the viewer like spokes on a steering wheel. If a curved arrow from the highest to second-highest to third-highest priority substituents ($1 \rightarrow 2 \rightarrow 3$) is clockwise, the chirality center has an ***R* configuration** (Latin *rectus*, "right"). If the arrow from $1 \rightarrow 2 \rightarrow 3$ is counterclockwise, the chirality center has an ***S* configuration** (Latin *sinister*, "left").

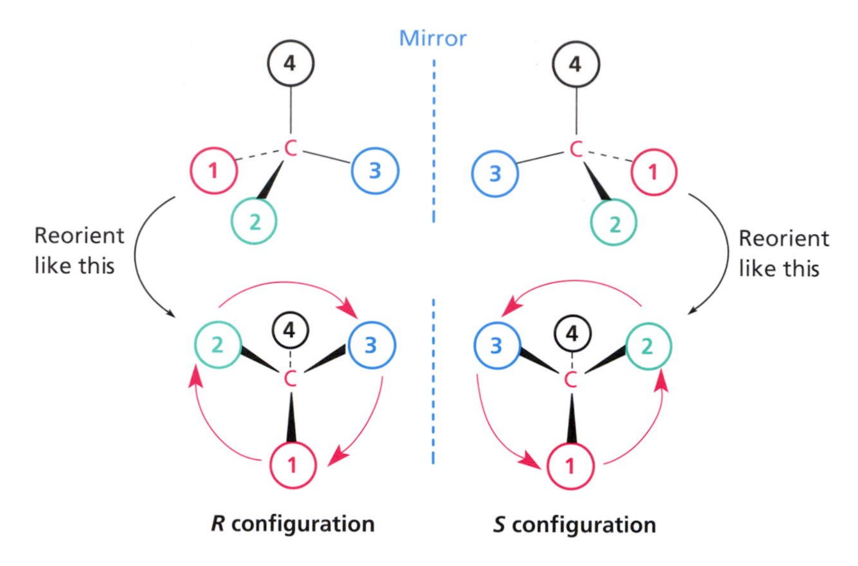

The Cahn–Ingold–Prelog sequence rules for assigning priorities are as follows:

Rule 1 Assign priorities to the four atoms directly attached to the chirality center in order of decreasing atomic number. The atom with highest atomic number is ranked 1; the atom with lowest atomic number (usually H) is ranked 4. If different isotopes of an element are present, assign priorities according to mass number. Thus, the order of atoms typically found in biological molecules is $S > P > O > N > C > {}^2H > {}^1H$.

Rule 2 If a tie exists after rule 1, compare atomic numbers of the second atoms in each substituent, continuing outward as necessary through the third or fourth atoms until the first point of difference is found. For example, $—CH_3$ and $—CO_2H$ are tied by rule 1, but $—CO_2H$ is higher in priority by rule 2 because O (the second atom outward in $—CO_2H$) is higher than H (the second atom outward in $—CH_3$).

Rule 3 Multiple-bonded atoms are equivalent to the same number of single-bonded atoms. For instance:

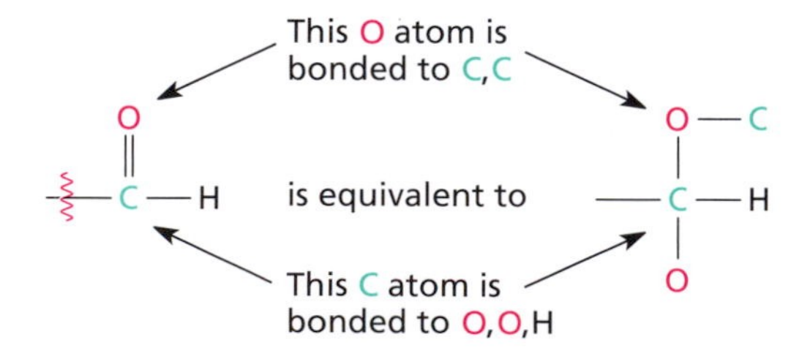

Check for yourself that the structure shown for (+)-alanine at the beginning of this section does in fact have an *S* configuration and that the structure shown for (−)-alanine has an *R* configuration.

Diastereomers, Epimers, and Meso Compounds

Alanine has only one chirality center and only two mirror-image stereoisomers. Many biological molecules, however, have more than one chirality center and thus have numerous stereoisomers. As a general rule, a molecule with *n* chirality centers can have up to 2^n stereoisomers (though it may have fewer, as we'll see below). Take, for instance, the simple sugar 2,3,4-trihydroxybutanal, which has two chirality centers and $2^2 = 4$ stereoisomers (Figure 2.1). The four stereoisomers comprise two pairs of enantiomers: the 2*S*,3*S* / 2*R*,3*R* pair (named erythrose) and the 2*R*,3*S* / 2*S*,3*R* pair (named threose). But what is the relationship between two isomers that aren't enantiomers—2*S*,3*S* and 2*R*,3*S*, for instance?

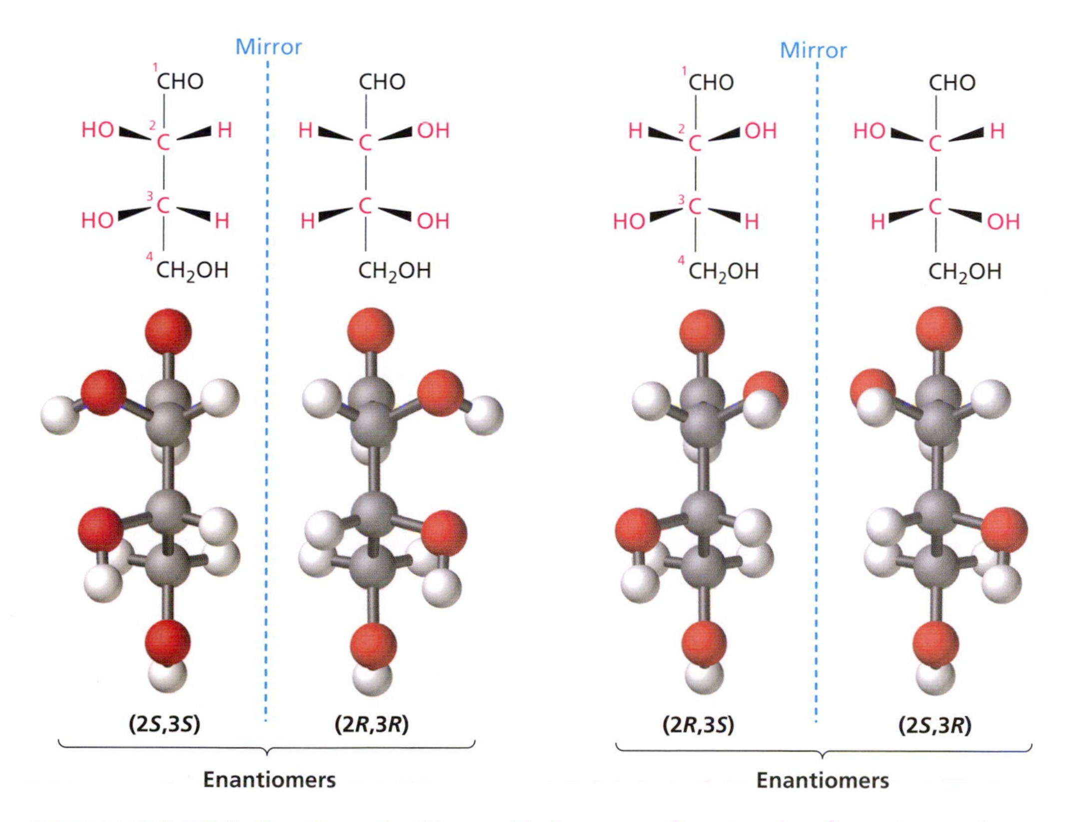

FIGURE 2.1 2,3,4-Trihydroxybutanal, with two chirality centers, has two pairs of enantiomers, for a total of four stereoisomers.

Nonmirror-image stereoisomers, such as the 2*S*,3*S* and 2*R*,3*S* trihydroxybutanals, are called **diastereomers**. Whereas enantiomers have opposite configurations at *all* chirality centers, diastereomers have opposite configurations at *some* centers but the same configuration at others.

In the special case where two diastereomers differ at only one chirality center but are the same at all others, we say that the compounds are **epimers**. Cholestanol and coprostanol, for instance, are both found in human feces and both have nine chirality centers. Eight of the nine are identical, but the one at C5 is different. Thus, cholestanol and coprostanol are *epimeric* at C5. Along related lines, the function of enzymes called *epimerases* is to change the stereochemistry at one chirality center, thereby producing an epimer of the reactant.

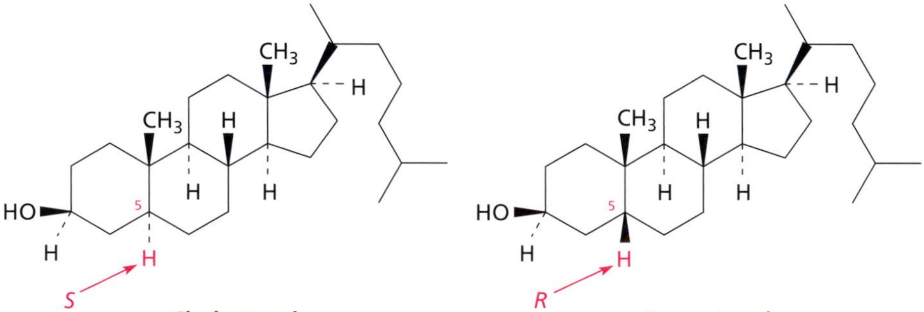

Cholestanol Coprostanol

Epimers

As a final example of a compound with more than one chirality center, look at tartaric acid (2,3-dihydroxybutanedioic acid). Even though tartaric acid has two chirality centers and a potential for four stereoisomers, it in fact has only *three*. The 2*R*,3*S* and 2*S*,3*R* structures are enantiomers, but the 2*S*,3*S* and 2*R*,3*R* structures are *identical*. That is, a 180° rotation of either one produces the other. This identity arises because the molecule has a symmetry plane that cuts through the C2–C3 bond and makes the top half of the molecule a mirror image of the bottom half (Figure 2.2). Such compounds that contain chirality centers yet have a symmetry plane in one conformation and are achiral overall are called **meso compounds**.

A summary of the types of stereoisomers is given in Table 2.1.

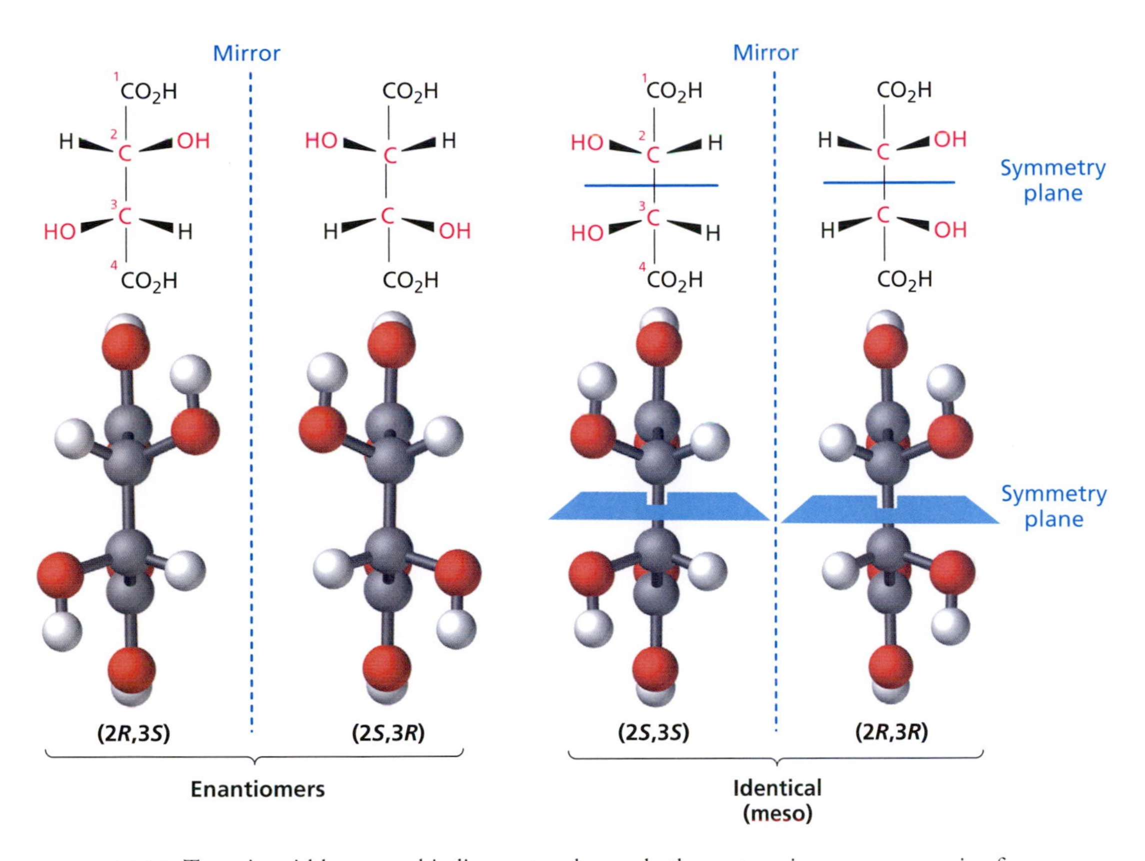

FIGURE 2.2 Tartaric acid has two chirality centers but only three stereoisomers—one pair of enantiomers and one meso form. In the conformation shown, the meso form has a plane of symmetry that makes one half of the molecule a mirror image of the other half.

Table 2.1 Summary of the Types of Stereoisomers

Name	Description
Enantiomers	Mirror-image stereoisomers; differ at all chirality centers
Diastereomers	Nonmirror-image stereoisomers; differ at one or more, but not all chirality centers
Epimers	Nonmirror-image stereoisomers; differ at only one chirality center
Meso compound	Contains two or more chirality centers but is achiral overall because of a symmetry plane

Prochirality

Closely related to the concept of chirality, and equally important in biological chemistry, is the notion of *prochirality*. A molecule is said to be **prochiral** if it can be converted from achiral to chiral in a single chemical step. For instance, an achiral ketone like 2-butanone is prochiral because it gives the chiral alcohol 2-butanol on reduction. That is, the sp^2 carbon of 2-butanone has three different groups attached, but the same sp^3 carbon on 2-butanol has four different groups attached.

$$CH_3C(=O)CH_2CH_3 \longrightarrow CH_3CH(OH)CH_2CH_3$$

2-Butanone (Prochiral) **2-Butanol** (Chiral)

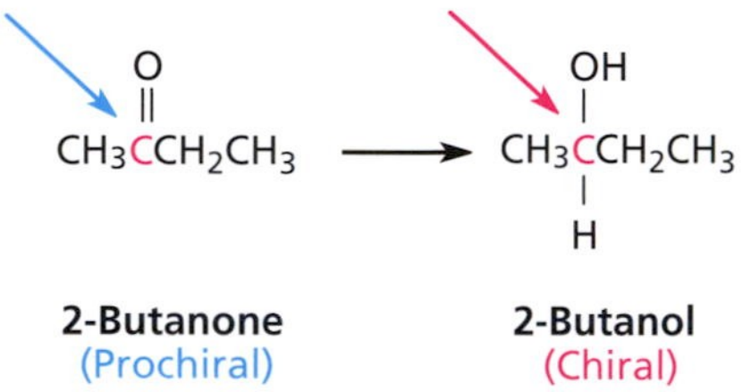

Which enantiomer of 2-butanol is produced in the reduction depends on which face of the planar carbonyl group is attacked. To distinguish between the possibilities, we use the stereochemical descriptors *re* and *si*. Assign priorities to the three different groups attached to the trigonal, sp^2-hybridized carbon, and imagine curved arrows from the highest to second-highest to third-highest priority substituents. The face on which the arrows curve clockwise is designated *re*, and the face on which the arrows curve counterclockwise is designated *si*. In this particular example, reduction from the *re* face gives (S)-2-butanol and reduction from the *si* face gives (R)-2-butanol.

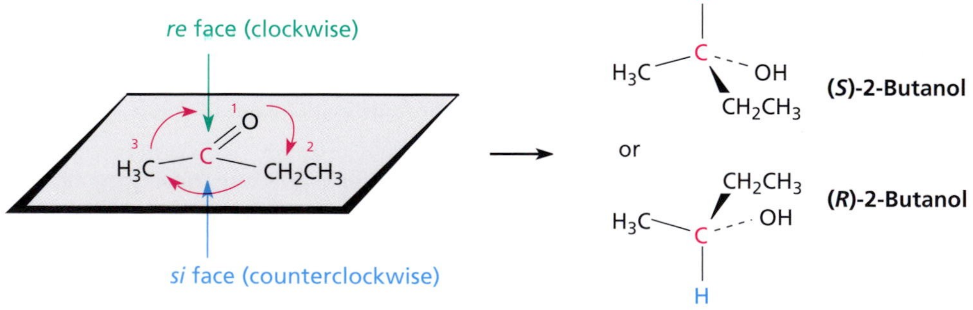

re face (clockwise)

si face (counterclockwise)

(S)-2-Butanol

or

(R)-2-Butanol

In addition to compounds with planar, sp^2-hybridized atoms, compounds with tetrahedral, sp^3-hybridized atoms can also be prochiral. An sp^3-hybridized atom is said to be a **prochirality center** if, by changing one of its attached groups, it becomes a chirality center. The —CH_2OH carbon atom of ethanol, for instance, is a prochirality center because changing one of its attached —H atoms converts it into a chirality center.

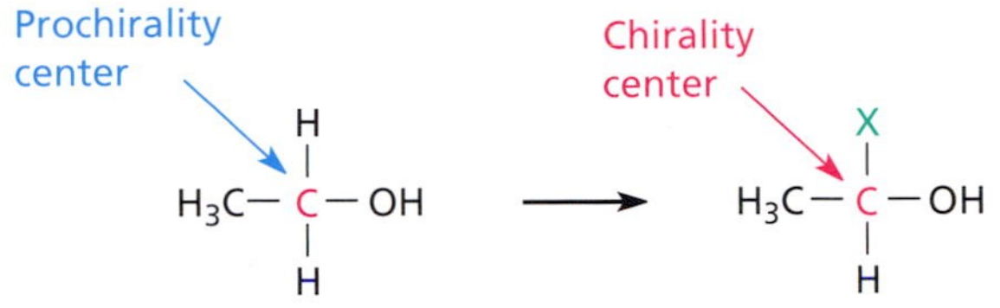

Prochirality center → Chirality center

To distinguish between the two identical atoms (or groups) on a prochirality center, we imagine a change that will raise the priority of one atom (or group) over the other without affecting its priority with respect to other attached groups. On the —CH_2OH carbon of ethanol, for instance, we might imagine replacing one of the ^1H atoms (protium) by ^2H (deuterium). The newly introduced ^2H atom is higher in priority than the remaining ^1H atom, but it remains lower in priority than other groups attached to the carbon. Of the two identical atoms in the original compound, that atom whose replacement leads to an *R* chirality center is said to be ***pro-R***, and that atom whose replacement leads to an *S* chirality center is ***pro-S***.

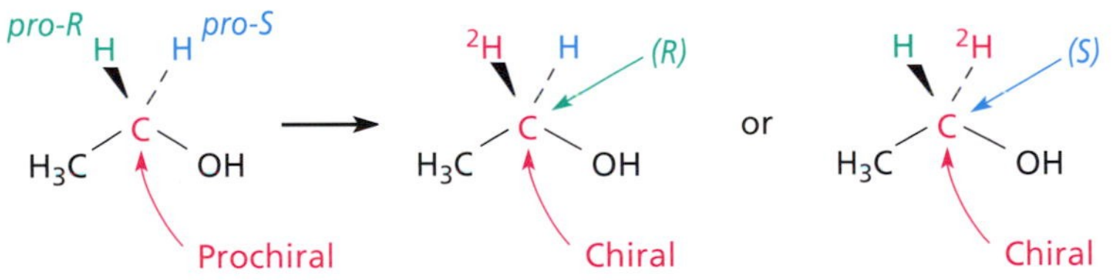

Prochiral → Chiral (R) or Chiral (S)

Numerous biological reactions involve prochiral compounds. One of the steps in the citric acid cycle, for instance, is the addition of H_2O to fumarate to give malate. Addition of —OH occurs on the *si* face of a fumarate carbon and gives (*S*)-malate as product.

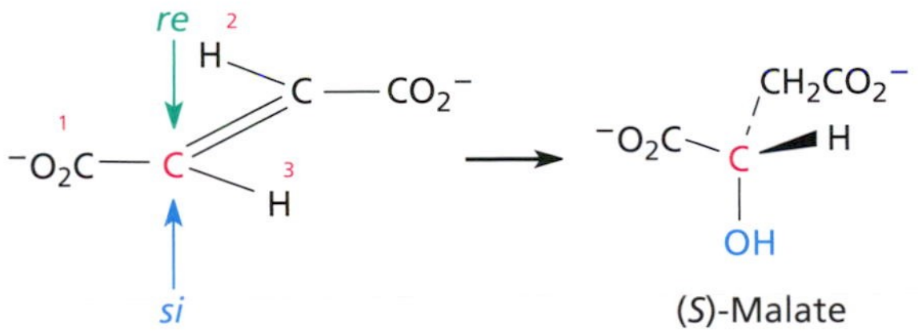

(*S*)-Malate

As another example, it has been shown by studies with deuterium-labeled substrates that the oxidation of ethanol with NAD^+ catalyzed by yeast alcohol dehydrogenase occurs with exclusive removal of the *pro-R* hydrogen from ethanol and with addition only to the *re* face of NAD^+.

H_3C H$_R$ H$_S$ OH + *si* *re* $CONH_2$ H NAD^+ ⇌ O H_3C C H + $CONH_2$ H$_R$ H$_S$ NADH

2.2 Biomolecules: Lipids

A **lipid** is a small, naturally occurring molecule that has limited solubility in water and can be isolated from an organism by extraction with a nonpolar organic solvent. Fats, oils, waxes, some vitamins and hormones, and most nonprotein cell-membrane components are examples. Of the many kinds of lipids, we'll be concerned primarily with a few representative ones: *triacylglycerols*, *terpenoids*, *steroids*, and *prostaglandins*. Other kinds of lipids, such as the phospholipids that make up the bilayer around cell membranes, are equally important, but their biochemistry is related to that of triacylglycerols and we won't cover them specifically.

Triacylglycerols

Animal fats and vegetable oils are the most widely occurring lipids. Both are *triglycerides*, or **triacylglycerols**—triesters of glycerol with three long-chain carboxylic acids called **fatty acids**. The fatty acids are generally unbranched and have an even number of carbons from 12 to 20. If one or more double bonds are present, they usually have (*Z*), or cis, geometry. More than 100 different fatty acids are known, and about 40 occur widely. Table 2.2 lists some common fatty acids.

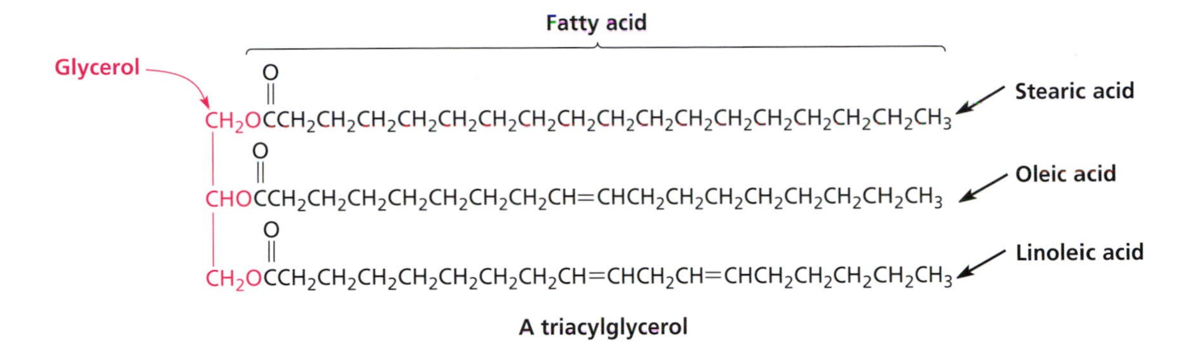

Table 2.2 Structures of Some Common Fatty Acids

Name	Number of carbons	Melting point (°C)	Structure
Saturated			
Lauric	12	43.2	$CH_3(CH_2)_{10}CO_2H$
Myristic	14	53.9	$CH_3(CH_2)_{12}CO_2H$
Palmitic	16	63.1	$CH_3(CH_2)_{14}CO_2H$
Stearic	18	68.8	$CH_3(CH_2)_{16}CO_2H$
Arachidic	20	76.5	$CH_3(CH_2)_{18}CO_2H$
Unsaturated			
Palmitoleic	16	−0.1	(Z)-$CH_3(CH_2)_5CH{=}CH(CH_2)_7CO_2H$
Oleic	18	13.4	(Z)-$CH_3(CH_2)_7CH{=}CH(CH_2)_7CO_2H$
Linoleic	18	−12	(Z,Z)-$CH_3(CH_2)_4(CH{=}CHCH_2)_2(CH_2)_6CO_2H$
Linolenic	18	−11	(all Z)-$CH_3CH_2(CH{=}CHCH_2)_3(CH_2)_6CO_2H$
Arachidonic	20	−49.5	(all Z)-$CH_3(CH_2)_4(CH{=}CHCH_2)_4CH_2CH_2CO_2H$

As indicated in Table 2.2, unsaturated fatty acids generally have lower melting points than their saturated analogs because their irregular shapes make it more difficult for them to nestle together uniformly in crystals. A similar trend is found with triacylglycerols: Vegetable oils, which have a higher proportion of unsaturated fatty acids, have generally lower melting points; animal fats, which have a lower proportion of unsaturated fatty acids, have generally higher melting points.

In animals, fat functions primarily as a long-term reservoir for energy storage, providing approximately six times as much metabolic energy as an equal mass of hydrated glycogen. We'll see the chemical details of fat metabolism in Section 3.3 and the details of fat biosynthesis in Section 3.4.

Other Lipids: Terpenoids, Steroids, and Prostaglandins

Derived primarily from plants, bacteria, and fungi, **terpenoids** are a group of lipids with a vast diversity of structure: Some are larger and some are smaller, some are hydrocarbons and some have oxygen or nitrogen, some are open-chain and some are cyclic (Figure 2.3). More than 22,000 different terpenoids are known, and all are related despite their apparent structural diversity. They are derived biosynthetically from the simple 5-carbon molecule isopentenyl diphosphate by pathways whose details we'll examine in Section 3.5.

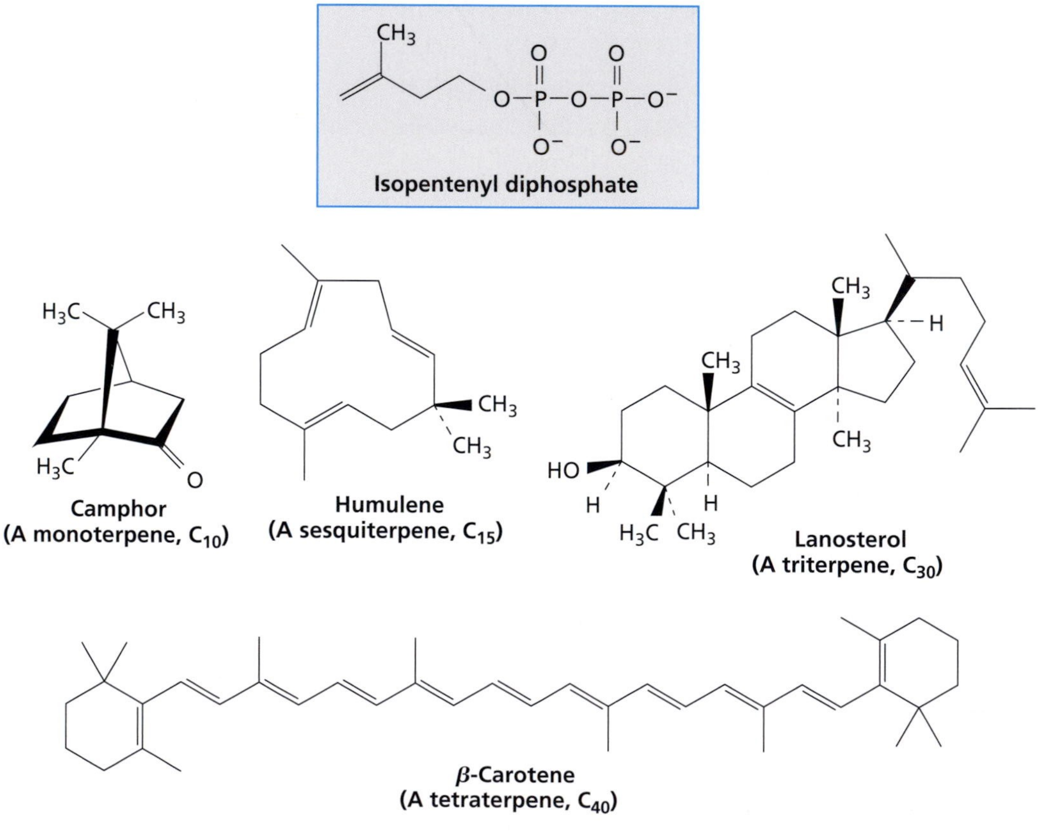

Isopentenyl diphosphate

Camphor (A monoterpene, C_{10})

Humulene (A sesquiterpene, C_{15})

Lanosterol (A triterpene, C_{30})

β-Carotene (A tetraterpene, C_{40})

FIGURE 2.3 Structures of some representative terpenoids.

Steroids are lipids whose structures are based on a tetracyclic skeleton of three 6-membered rings joined to one 5-membered ring. The rings are labeled A–D beginning at the lower left, and the carbons are numbered beginning in ring A. The substituent attached to the 5-membered ring at C17 is called the *side chain*, and there is always a hydroxyl or carbonyl group at C3.

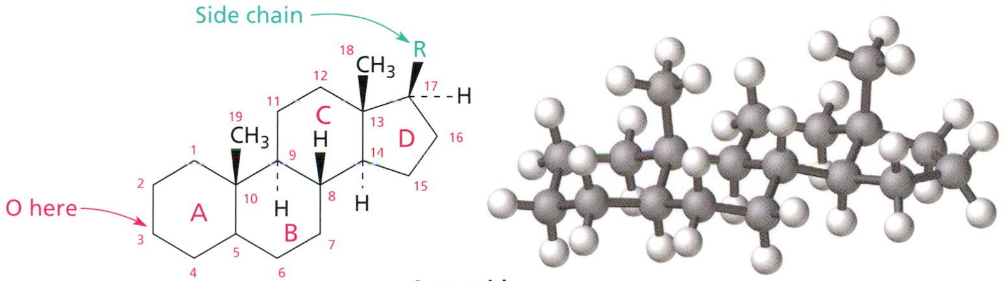

A steroid

Steroids have many structural variations and many biological functions (Figure 2.4). Cholesterol, for instance, is the biological precursor to all steroid hormones and is itself a component of cell membranes. Bile acids, such as cholic acid,

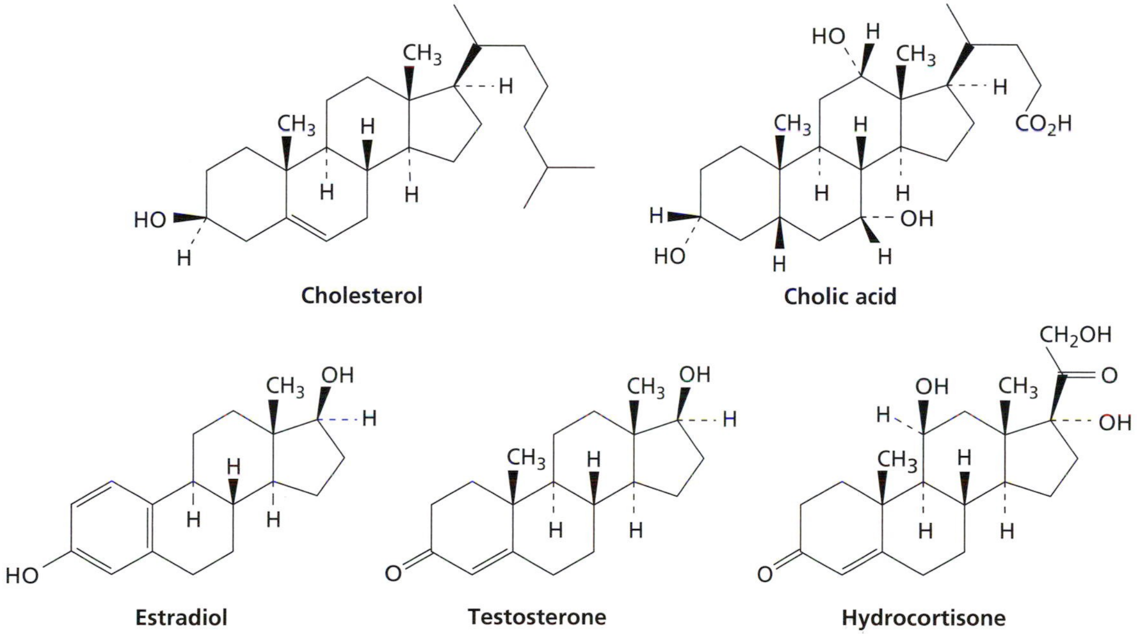

Cholesterol **Cholic acid**

Estradiol **Testosterone** **Hydrocortisone**

FIGURE 2.4 Structures of some common steroids.

help to emulsify water-insoluble foods during digestion. Sex hormones, such as estradiol and testosterone, control tissue growth, maturation, and reproduction. Adrenocortical hormones, such as hydrocortisone, control a variety of physiological functions, including regulation of glucose metabolism and control of inflammation. We'll see in Section 3.6 how steroids arise biosynthetically.

Prostaglandins and related compounds are a group of C_{20} carboxylic acids derived biologically from arachidonic acid (5,8,11,14-eicosatetraenoic acid) through a pathway we'll explore in Section 7.3. The several dozen known prostaglandins have an extraordinary range of physiological effects, including an ability to lower blood pressure, affect blood clotting and kidney function, control inflammation, and stimulate uterine contractions during childbirth. Closely related compounds are involved in producing the asthmatic response.

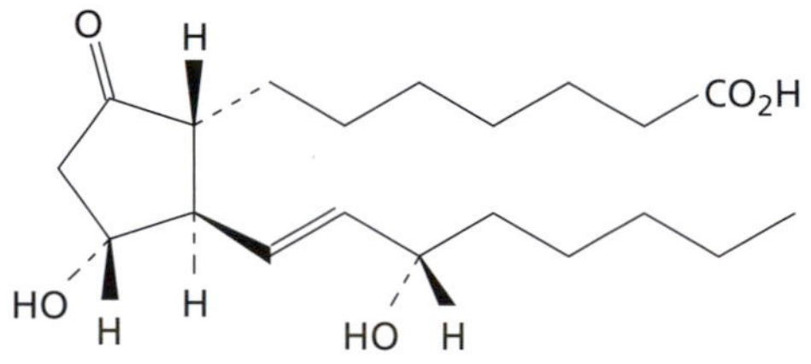

Prostaglandin E_1

2.3 Biomolecules: Carbohydrates

Carbohydrates, commonly called *sugars*, are classed as either simple or complex. A **simple carbohydrate**, or **monosaccharide**, is a straight-chain, polyhydroxy aldehyde or ketone, such as glucose, fructose, ribose, sedoheptulose, and many others. All have numerous chirality centers, and many stereoisomers of each are therefore possible. A **complex carbohydrate** consists of two or more simple sugars joined together by an acetal link. Sucrose, a disaccharide, and cellulose, a polysaccharide, are common examples (Figure 2.5).

Monosaccharides are further described by the nature of their carbonyl group and how many carbons they contain. Aldehydo sugars are **aldoses**, and keto sugars are **ketoses**, where the *-ose* suffix designates a carbohydrate. Glucose is an *aldohexose*, fructose is a *ketohexose*, ribose is an *aldopentose*, and so on.

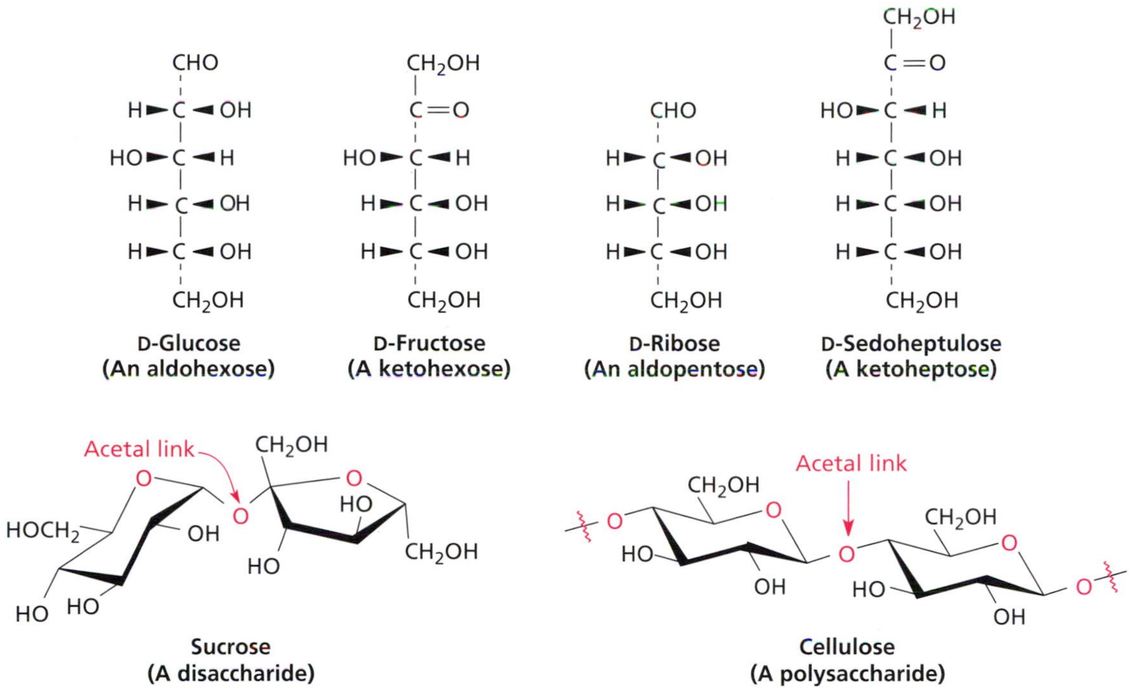

D-Glucose (An aldohexose)

D-Fructose (A ketohexose)

D-Ribose (An aldopentose)

D-Sedoheptulose (A ketoheptose)

Sucrose (A disaccharide)

Cellulose (A polysaccharide)

FIGURE 2.5 Structures of some typical carbohydrates.

Carbohydrate Stereochemistry

The standard method for depicting open-chain carbohydrate chemistry is to use **Fischer projections**, which represent a tetrahedral carbon atom by crossed lines. Horizontal lines represent bonds coming out of the page, and vertical lines represent bonds receding into the page. Carbohydrates are drawn in their open-chain form, with individual carbons stacked on top of one another and with the carbonyl carbon placed at or near the top (Figure 2.6).

An aldotetrose has 2 chirality centers and $2^2 = 4$ stereoisomers (2 pairs of enantiomers); an aldopentose has 3 chirality centers and 8 stereoisomers (4 pairs of enantiomers); and an aldohexose has 4 chirality centers and 16 stereoisomers (8 pairs of enantiomers). For historical reasons, those enantiomers whose chirality center farthest from the carbonyl group has an —OH group pointing toward the *right* in Fischer projection is referred to as a D **sugar**, while those

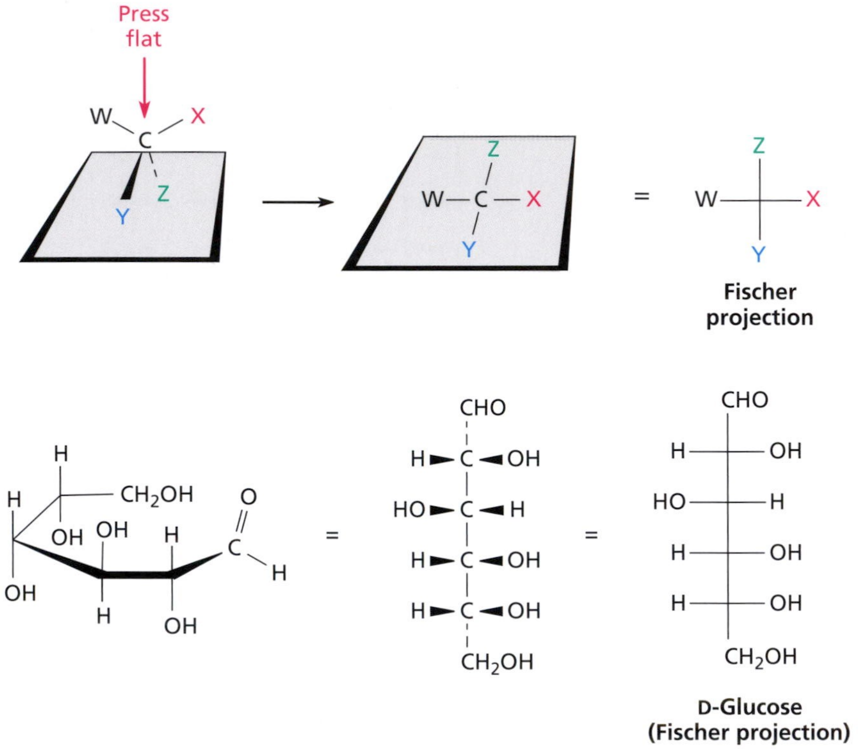

FIGURE 2.6 Fischer projections. Carbohydrates are drawn in their open-chain form, with individual carbons stacked on top of one another and the carbonyl carbon placed at or near the top.

enantiomers whose lowest chirality center has an —OH group pointing toward the *left* in Fischer projection is an L **sugar**. The names and structures of the two D aldotetroses, the four D aldopentoses, and the eight D aldohexoses are shown in Figure 2.7. Each of the D structures has a mirror-image L enantiomer, with opposite stereochemistry at all chirality centers.

Monosaccharide Anomers

The open-chain structures shown in Figure 2.7 are useful for indicating relative stereochemistry but don't show the molecules as they really exist. In fact, monosaccharides exist as cyclic hemiacetals, formed by reversible, acid-catalyzed nucleophilic addition of a hydroxyl group somewhere in the chain to the aldehyde or ketone carbonyl group. When cyclization occurs, a new chirality center is generated

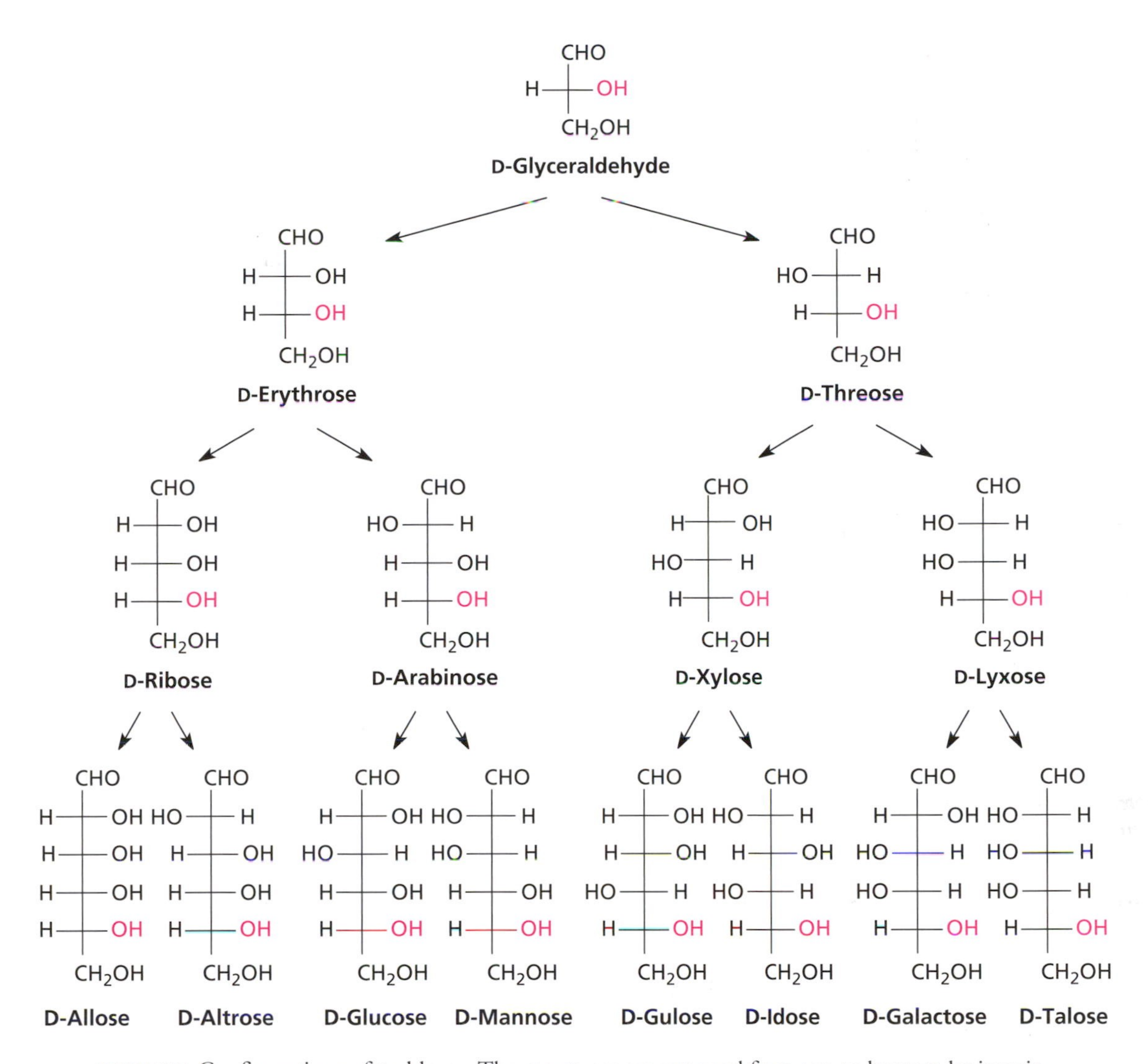

FIGURE 2.7 Configurations of D aldoses. The structures are arranged from top to bottom by imagining that insertion of a new chirality center just below the aldehyde carbon generates the pair of structures indicated. Each of the D sugars shown also has a mirror-image L enantiomer.

at the former carbonyl carbon, producing two diastereomeric products called **anomers**. The compound with its newly generated —OH *cis* to the oxygen atom at the lowest chiral center at the other end of the chain in a Fischer projection is the ***α* anomer**; the compound with its new —OH *trans* to the oxygen atom at the lowest chiral center in a Fischer projection is the ***β* anomer** (Figure 2.8). D-Glucose,

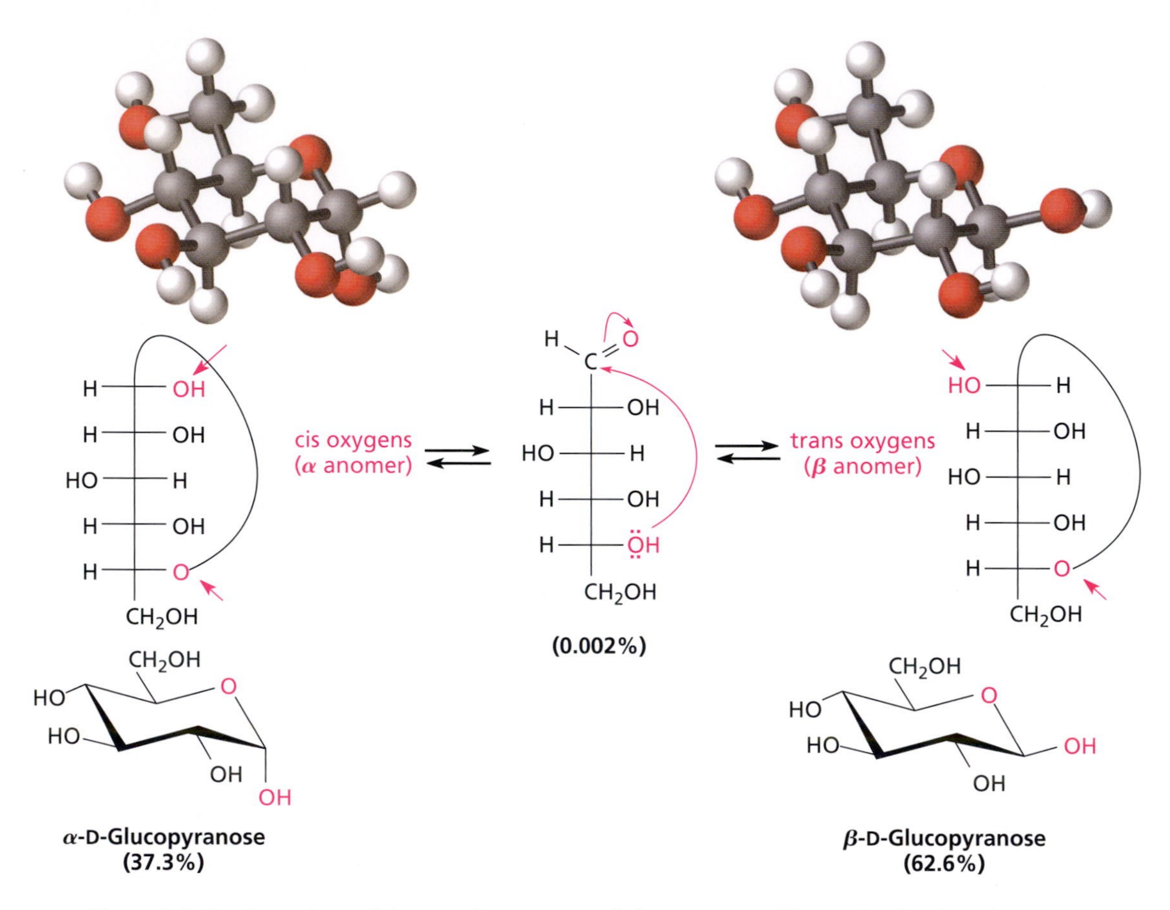

Figure 2.8 Configurations of the α and β anomers of glucopyranose. The molecule whose anomeric −OH group is *cis* to the oxygen atom on the lowest chiral center at the other end of the chain (red) in a Fischer projection is the α anomer. The molecule whose anomeric −OH group is *trans* to the oxygen atom on the lowest chiral center at the other end of the chain in a Fischer projection is the β anomer.

for instance, exists at equilibrium in water solution as a 37:63 mixture of α and β anomers. The 6-membered cyclic hemiacetal structure for a sugar is called the **pyranose** form, so the full names of the two anomers are α-D-glucopyranose and β-D-glucopyranose.

Some monosaccharides also exist in a 5-membered cyclic hemiacetal form called a **furanose** form. D-Fructose, for example, exists in water solution as 70% β-pyranose, 2% α-pyranose, 0.7% open-chain, 23% β-furanose, and 5% α-furanose form results from addition of the —OH at C6 to the carbonyl group, while the furanose form results from addition of the —OH at C5.

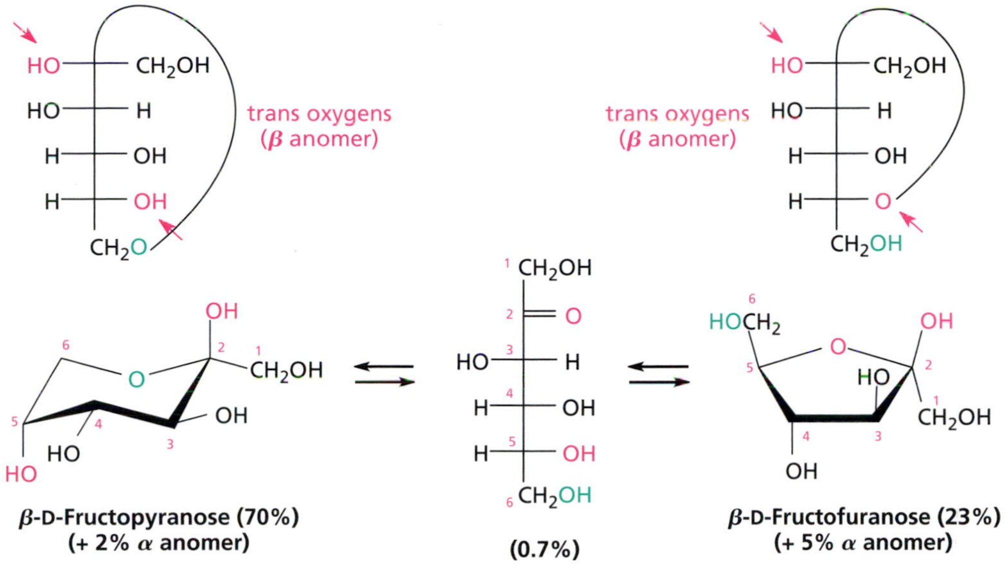

trans oxygens (β anomer)

trans oxygens (β anomer)

β-D-Fructopyranose (70%)
(+ 2% α anomer)

(0.7%)

β-D-Fructofuranose (23%)
(+ 5% α anomer)

Disaccharides and Polysaccharides

We saw in Section 1.5 that hemiacetals undergo acid-catalyzed reaction with alcohols to yield acetals. If the hemiacetal is a cyclic monosaccharide, the acetal product is called a **glycoside**. If the alcohol is itself a sugar, the product glycoside is a **disaccharide**, such as lactose or sucrose. Continuation of the process a large number of times yields a **polysaccharide**, such as the cellulose found in all plants or the amylose found in starch (Figure 2.9).

The glycoside link between monosaccharide units can have either α or β stereochemistry at the anomeric center of the hemiacetal sugar, and the link can be formed to any —OH group of the second sugar. Reaction at the C4 —OH group is particularly common but not required. Note the structural variety shown by the examples in Figure 2.9. In lactose, two different sugars (galactose and glucose) are joined by a β glycoside link between the anomeric carbon at C1 of galactose and the —OH at C4′ of glucose. (The "prime" in 4′ indicates that 1 and 4′ refer to different sugars.) In sucrose, glucose and fructose are linked between anomeric centers on both. The glucose unit is α, while the fructose unit is β. Cellulose and amylose both have many hundreds of glucose units with (1→4′) links, but they differ in their stereochemistry at the anomeric centers.

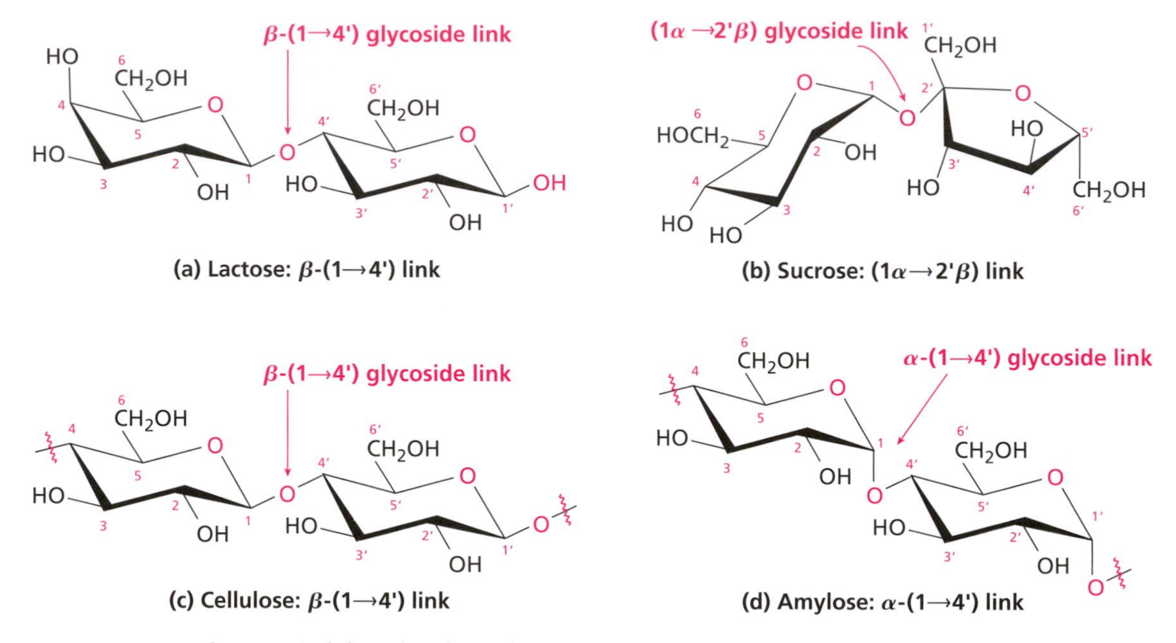

FIGURE 2.9 Some typical disaccharides and polysaccharides. (a) In lactose, galactose and glucose are joined by a β-($1\rightarrow4'$) glycoside link. (b) In sucrose, glucose and fructose are joined by a ($1\alpha\rightarrow2'\beta$) link between anomeric centers. (c) Cellulose [β-($1\rightarrow4'$)] and (d) amylose [α-($1\rightarrow4'$)] are polymers of glucose.

Deoxy Sugars and Amino Sugars

In addition to the pure carbohydrates just discussed, a variety of carbohydrate-derived substances are also important in biological chemistry. A **deoxy sugar**, for instance, has an oxygen atom "missing." 2-Deoxyribose, a constituent of DNA, is of course the most common example. Note that 2-deoxyribose exists in water solution as a complex equilibrium mixture of both furanose and pyranose forms.

Oxygen missing

α-D-2-Deoxyribopyranose (40%)
(+ 35% β anomer)

(0.7%)

α-D-2-Deoxyribofuranose (13%)
(+ 12% β anomer)

An **amino sugar** has an —OH group replaced by an —NH_2. The acetamide derivative of D-glucosamine (D-2-amino-2-deoxyglucose), for instance, is the monomer unit from which insect *chitin* is derived. Amino sugars are also constituents of antibiotics such as streptomycin and of the glycosaminoglycans and glycoproteins that make up cartilage and other connective tissue. Often, the connection between polysaccharide and protein parts of a glycoprotein is an α glycoside link between an *N*-acetylgalactosamine and the —OH group of a serine or threonine amino acid in the protein chain. In addition, there are many *N*-linked glycoproteins in which the carbohydrate is bonded to asparagine.

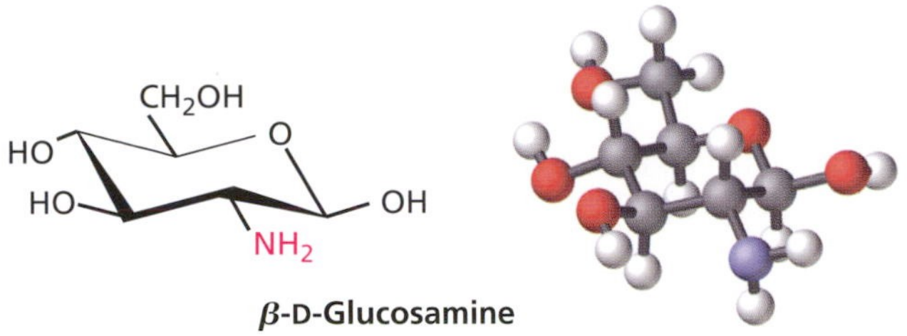

CH_2OH HO HO O OH NH_2

β-D-Glucosamine

O CH_2OH O HO NH O=C CH_3 O—CH_2CH—NH C=O

α-D-*N*-Acetyl-galactosamine

Protein chain

Serine

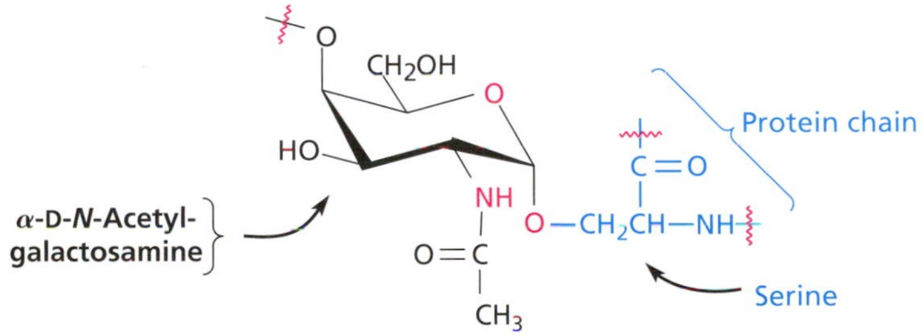

2.4 Biomolecules: Amino Acids, Peptides, and Proteins

The word *protein* is derived from the Greek *proteios*, meaning "primary"—an apt description for a class of compounds of primary importance to all living organisms. Proteins are of many different types and have many different functions; it has been estimated, in fact, that the human body contains more than 100,000 different proteins. All are made up of *amino acids* linked together by amide bonds to form long chains.

Amino Acids

An **amino acid**, as the name implies, is a difunctional molecule that contains both a basic amino group and an acidic carboxyl group. Because it's both basic and acidic, an amino acid undergoes an intramolecular acid–base reaction and exists in water solution primarily in a dipolar form called a **zwitterion**.

$$H_3C-CH(NH_2)-COOH \rightleftharpoons H_3C-CH(\overset{+}{N}H_3)-COO^-$$

(uncharged) **Alanine** (zwitterion)

In acid solution, an amino acid zwitterion acts as a base and the carboxylate is protonated to yield an ammonium cation. In base solution, a zwitterion acts as an acid and the ammonium group is deprotonated to yield a carboxylate anion. Note that the carboxylate group rather than the amino group is the base that is protonated, while the ammonium group rather than the carboxyl group is the acid that is deprotonated.

In acid solution

$$R-CH(\overset{+}{N}H_3)-COO^- + H_3O^+ \rightleftharpoons R-CH(\overset{+}{N}H_3)-COOH + H_2O$$

In base solution

$$R-CH(\overset{+}{N}H_3)-COO^- + OH^- \rightleftharpoons R-CH(NH_2)-COO^- + H_2O$$

The structures, abbreviations, and pK_a values of the 20 amino acids commonly found in proteins are shown in Table 2.3 in the form that predominates within cells at a physiological pH of 7.3. All 20 are **α-amino acids**, meaning that the amino group is a substituent on the α carbon—the one next to the carbonyl group. Nineteen of the twenty are primary amines (—NH_2) and differ only in the identity of the **side chain** attached to the α carbon. Proline is the only secondary amine and the only amino acid whose nitrogen atom is part of a ring.

Side chain

A primary α-amino acid

Proline, a secondary α-amino acid

In addition to the 20 amino acids found in proteins, more than 700 other nonprotein amino acids are also found in nature. γ-Aminobutyric acid (GABA), for instance, is found in the brain and acts as a neurotransmitter; homocysteine is found in blood and is linked to coronary heart disease; and thyroxine is found in the thyroid gland where it acts as a hormone.

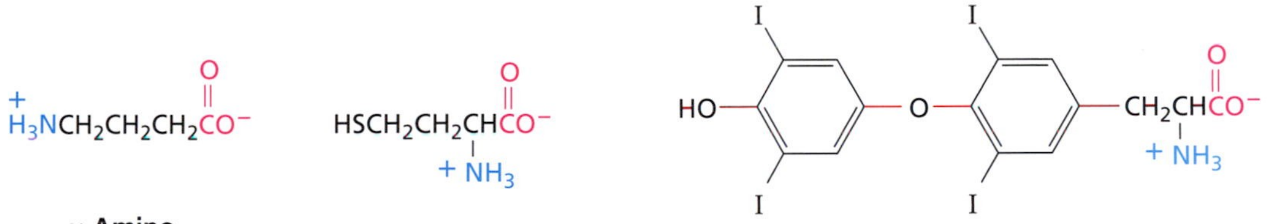

$H_3\overset{+}{N}CH_2CH_2CH_2CO^-$ (with C=O)

γ-Aminobutyric acid

$HSCH_2CH_2CH(\overset{+}{N}H_3)CO^-$ (with C=O)

Homocysteine

Thyroxine

Table 2.3 The 20 Common Amino Acids in Proteins

Name	Abbreviations		MW	Structure	pK_a α-CO_2H	pK_a α-NH_3^+	pK_a side chain
Neutral Amino Acids							
Alanine	Ala	A	89	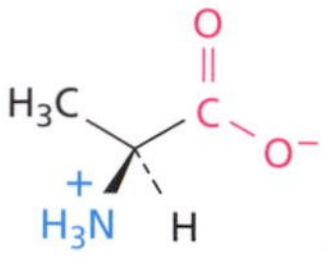	2.34	9.69	—
Asparagine	Asn	N	132	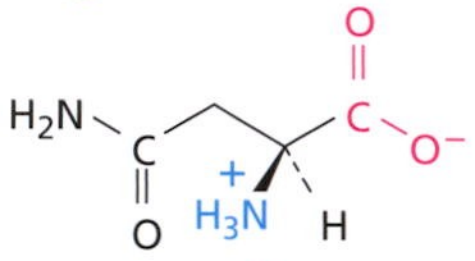	2.02	8.80	—
Cysteine	Cys	C	121	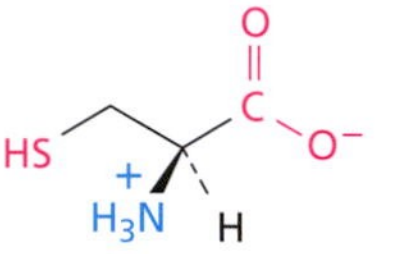	1.96	10.28	8.18
Glutamine	Gln	Q	146	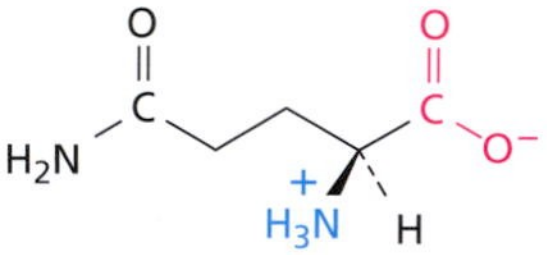	2.17	9.13	—
Glycine	Gly	G	75	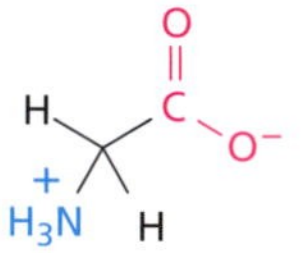	2.34	9.60	—
Isoleucine	Ile	I	131	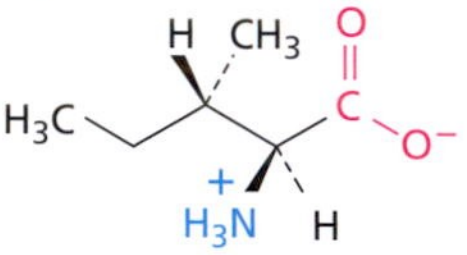	2.36	9.60	—
Leucine	Leu	L	131	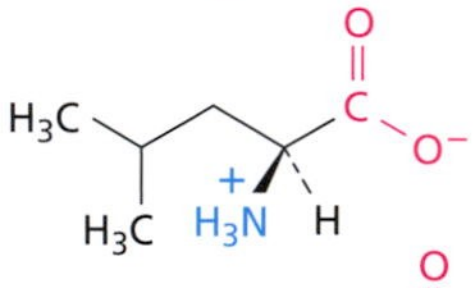	2.36	9.60	—
Methionine	Met	M	149	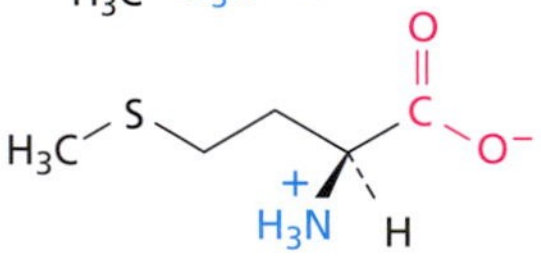	2.28	9.21	—
Phenylalanine	Phe	F	165	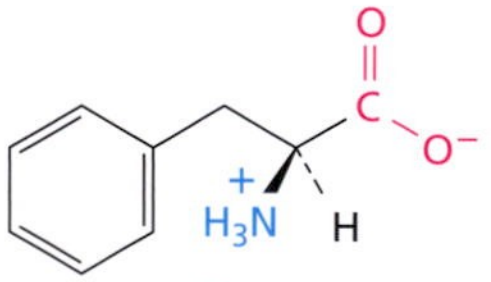	1.83	9.13	—
Proline	Pro	P	115	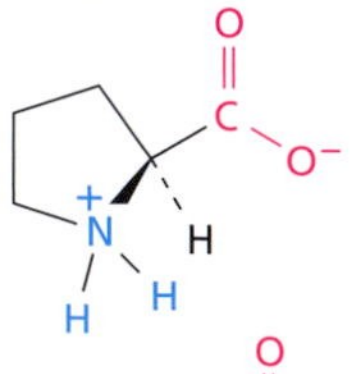	1.99	10.60	—
Serine	Ser	S	105	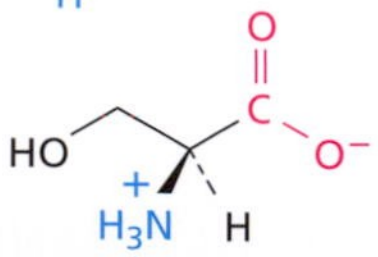	2.21	9.15	—

Table 2.3 The 20 Common Amino Acids in Proteins *(continued)*

Name	Abbreviations		MW	Structure	pK_a α-CO_2H	pK_a α-NH_3^+	pK_a side chain
Neutral Amino Acids							
Threonine	Thr	T	119	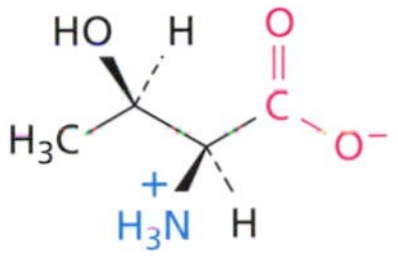	2.09	9.10	—
Tryptophan	Trp	W	204	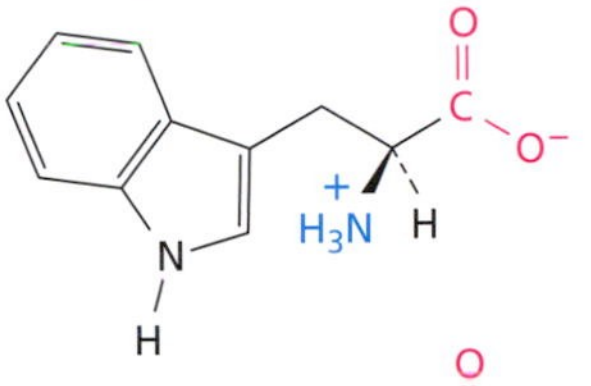	2.83	9.39	—
Tyrosine	Tyr	Y	181	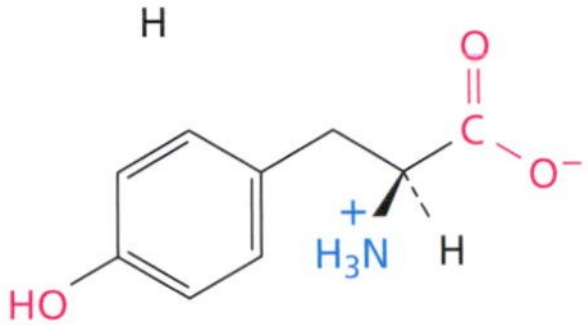	2.20	9.11	10.07
Valine	Val	V	117	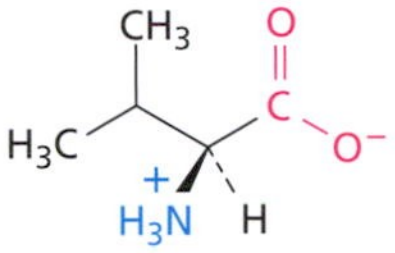	2.32	9.62	—
Acidic amino acids							
Aspartic acid	Asp	D	133	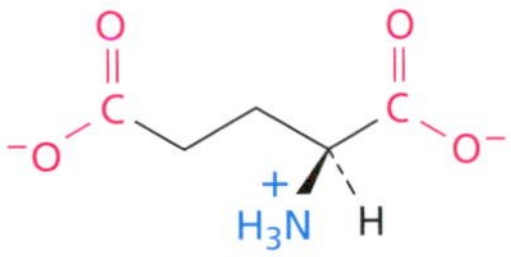	1.88	9.60	3.65
Glutamic acid	Glu	E	147	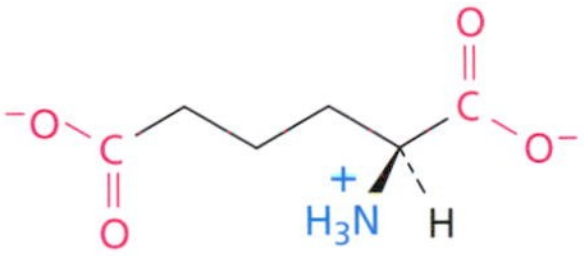	2.19	9.67	4.25
Basic amino acids							
Arginine	Arg	R	174	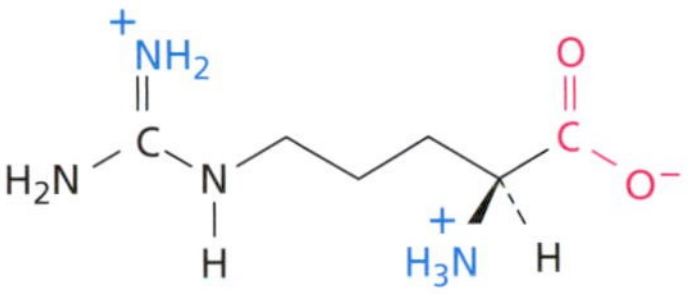	2.17	9.04	12.48
Histidine	His	H	155	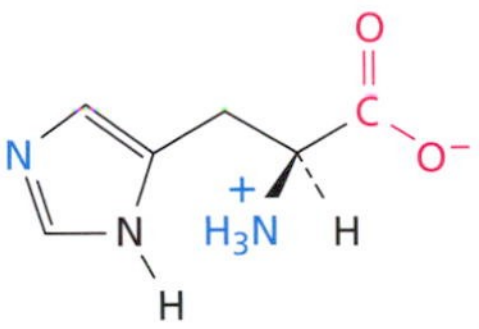	1.82	9.17	6.00
Lysine	Lys	K	146	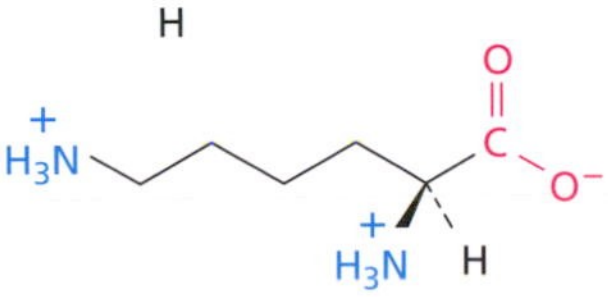	2.18	8.95	10.53

With the exception of glycine, the α carbon atoms of amino acids are chirality centers. In Fischer projections, the naturally occurring enantiomers are represented by placing the $—CO_2^-$ at the top and the side chain down as if drawing a carbohydrate, and then putting the amino group on the left. Because of their stereochemical similarity to L sugars, the naturally occurring amino acids are often referred to as L amino acids. You might note that 18 of the 19 chiral L amino acids have *S* stereochemistry, but cysteine is *R*. This apparent anomaly arises, not because cysteine is stereochemically different from the others but because the $—CH_2SH$ group of cysteine is the only side chain that is higher in priority than $—CO_2^-$ according to Cahn–Ingold–Prelog sequence rules. This reversal in priority also reverses the *R/S* assignment.

CO_2^- / $H_3\overset{+}{N}$—C—H / CH_3
L-Alanine
(*S*)-Alanine

CO_2^- / $H_3\overset{+}{N}$—C—H / CH_2OH
L-Serine
(*S*)-Serine

CO_2^- / $H_3\overset{+}{N}$—C—H / CH_2SH
L-Cysteine
(*R*)-Cysteine

[CHO / HO—C—H / CH_2OH]
L-Glyceraldehyde

The 20 common amino acids are classified as neutral, acidic, or basic, depending on the nature of their side chains. Fifteen have neutral side chains, two have an acidic $—CO_2H$ in their side chain, and three have a basic amine group in their side chain. Note that cysteine and tyrosine, though usually classified as neutral, nevertheless have weakly acidic side chains that can be deprotonated at sufficiently high pH. At pH 7.3 within cells, the side-chain $—CO_2H$ groups of aspartic acid and glutamic acid are deprotonated, and the basic side-chain nitrogens of lysine and arginine are protonated. Histidine, however, which contains a heterocyclic imidazole ring in its side chain, is not quite basic enough to be protonated at pH 7.3.

Peptides and Proteins

Amino acids join together in long chains by forming amide bonds, or **peptide bonds**, between the α $—NH_2$ group of one and the α $—CO_2H$ of another.

Chains with fewer than 50 or so amino acids are often called **peptides**, while the term **protein** is generally used for larger chains. The long, repeating sequence of —N—CH—CO— atoms that make up the chain is called the **backbone**, and individual amino acids are usually called **residues**. By convention, a peptide or protein is written with the **N-terminal amino acid** (the one with the free amino group) on the left and the **C-terminal amino acid** (the one with the free carboxyl group) on the right. Note how, when the chain is in an extended, stretched-out conformation, the side-chain groups alternate between top and bottom, and between up and down.

Amide bonds

Many

An amide nitrogen is nonbasic because the electron lone pair on N is shared by resonance with the neighboring carbonyl group. This resonance interaction gives the C—N bond a certain amount of double-bond character and restricts rotation around it. As a result, the carbon, nitrogen, and four attached atoms lie in a plane, with the N—H bond oriented 180° to the C═O bond.

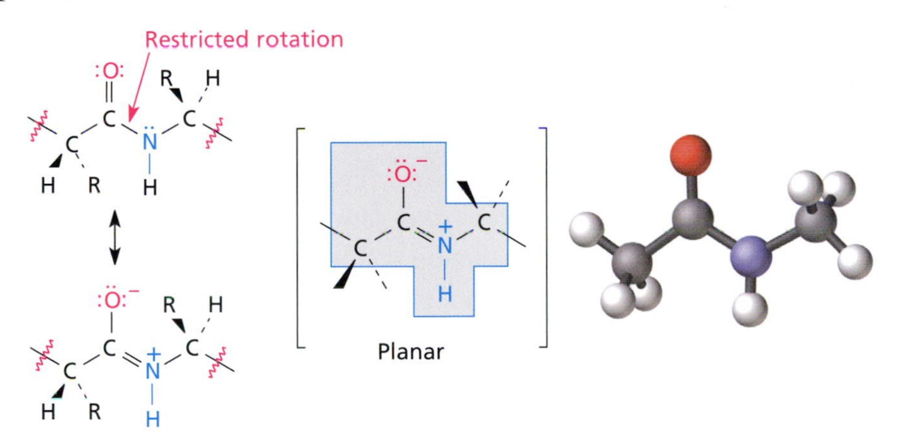

Although amide bonding is the primary means of joining amino acids, a second kind of link occasionally occurs when the side-chain —SH groups of two cysteine residues are joined by a disulfide bond, RS—SR. A disulfide bond between cysteines in two different chains links the otherwise separate chains together, while a disulfide bond between two cysteines in the same chain forms a loop.

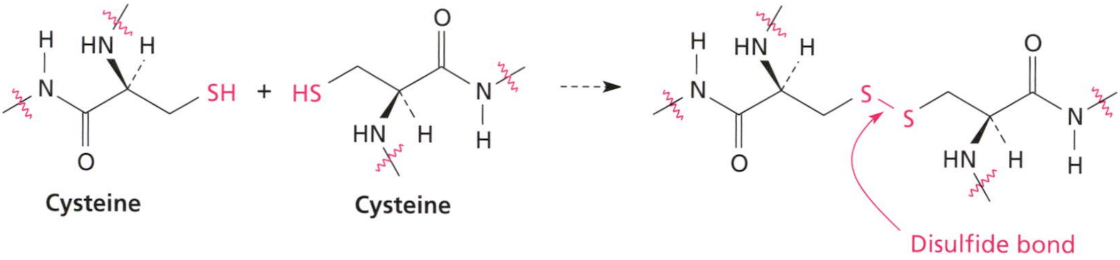

Cysteine Cysteine

Disulfide bond

We'll not discuss the topic of protein structure in any detail, but as a reminder, there are four levels of structure used to describe a protein: The **primary structure** lists a protein's amino acid sequence; the **secondary structure** describes how local regions of the protein orient into regular patterns; the **tertiary structure** describes how the entire protein molecule folds into an overall three-dimensional shape; and the **quaternary structure** describes how individual protein molecules associate to form large aggregates.

The primary structures of small peptides are generally determined either by an automated protein sequenator, using Edman-degradation chemistry to sequentially cleave amino acids from the N-terminal end of the chain, or by mass spectrometry. Larger proteins, however, are often sequenced indirectly from a knowledge of the *nucleotide* sequence of the corresponding gene.

Secondary and tertiary structures are determined by X-ray crystallography because no method yet exists for predicting the precise way in which a given protein sequence will fold. The most common secondary structures, though not the only ones, are the *α helix* and the *β pleated-sheet*. An **α helix** is a right-handed coil of the protein backbone, much like the coil of a telephone cord (Figure 2.10). Each full turn of the helix contains 3.6 amino acid residues, with a distance

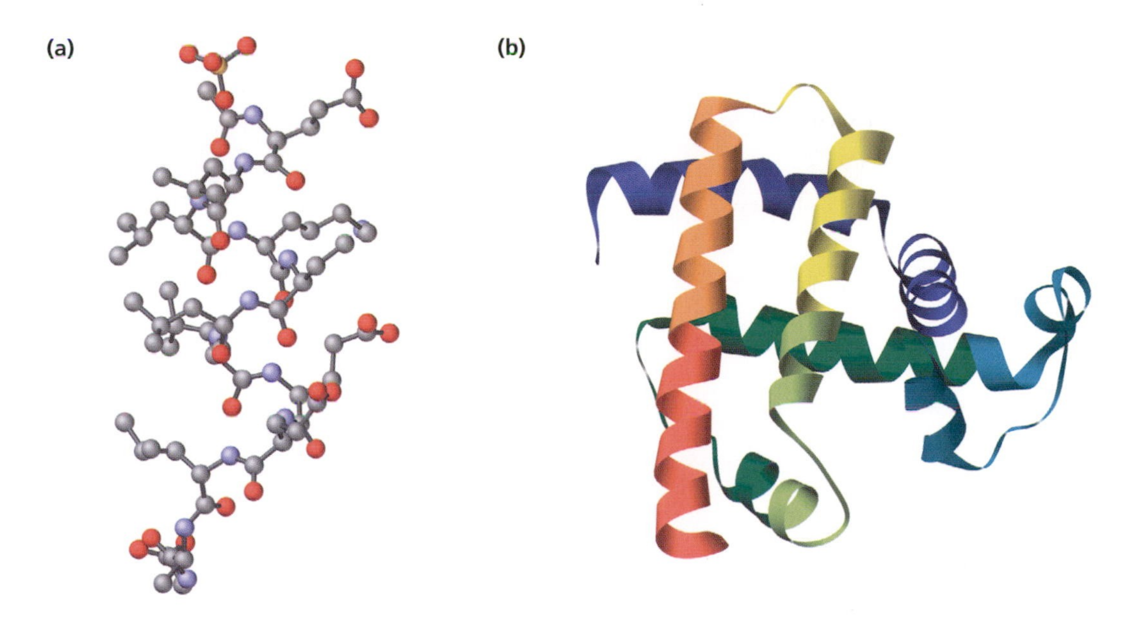

FIGURE 2.10 (**a**) The α-helix secondary structure of proteins is stabilized by hydrogen bonds between the N—H group of one residue and a C═O group four residues away. (**b**) The structure of myoglobin, a protein with extensive helical segments, shown as ribbons.

between coils of 5.4 Å (540 pm). The structure is stabilized by hydrogen bonds between amide N—H groups and C═O groups four residues away, and each N—H····O distance is 2.8 Å. The α helix is an extremely common secondary structure, and almost all enzymes contain numerous helical regions.

A **β pleated-sheet** differs from an α helix in that the peptide chain is extended rather than coiled and the hydrogen bonds occur between residues in nearby chains (Figure 2.11). The neighboring chains can run either in the same direction (parallel) or in opposite directions (antiparallel), although the antiparallel arrangement is more common and energetically somewhat more favorable. Small β-sheet regions occur frequently in globular proteins.

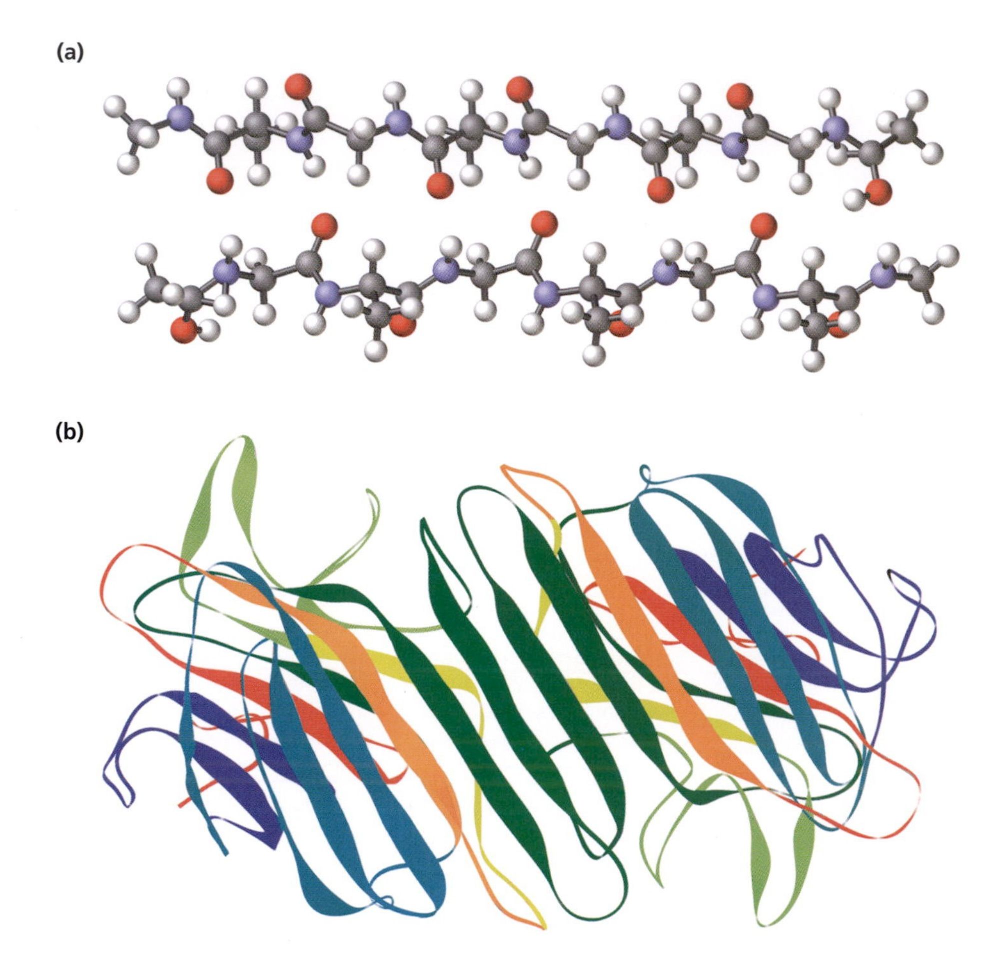

FIGURE 2.11 (**a**) The β pleated-sheet secondary structure of proteins is stabilized by hydrogen bonds between parallel or antiparallel chains. (**b**) The structure of concanavalin A, a protein with extensive regions of antiparallel β-sheets, shown as ribbons.

2.5 Biomolecules: Nucleic Acids

Deoxyribonucleic acid (DNA) and **ribonucleic acid (RNA)** are the carriers and processors of genetic information. Both are large biological polymers made of individual **nucleotide** units joined together to form a long chain. Each nucleotide is composed of a **nucleoside** bonded to a phosphate group, and each nucleoside is composed of an aldopentose sugar linked through its anomeric carbon to the nitrogen atom of a cyclic amine base.

The sugar component in DNA is 2-deoxyribose, and the sugar in RNA is ribose. DNA contains four different amine bases, two substituted purines (adenine and guanine) and two substituted pyrimidines (cytosine and thymine). Adenine, guanine, and cytosine also occur in RNA, but thymine is replaced in RNA by a closely related pyrimidine base called uracil. The structures of the four deoxyribonucleotides and the four ribonucleotides are shown in Figure 2.12.

Purine **Pyrimidine**

Though similar chemically, DNA and RNA differ dramatically in size. Molecules of DNA are enormous, with molecular weights up to several billion. Molecules of RNA, by contrast, are much smaller, containing as few as 60 nucleotides and having molecular weights as low as 22,000 for tRNA.

Nucleotides are linked together in DNA and RNA by phosphodiester bonds between the 5′-phosphate group on one nucleotide and the 3′-hydroxyl group on the sugar of another nucleotide. One end of the nucleic acid polymer thus has a free hydroxyl at C3′ (the **3′ end**), and the other end has a phosphate at C5′ (the **5′ end**). The sequence of nucleotides in a chain is described by starting at the 5′ end and identifying the bases in order of occurrence, using the abbreviations G, C, A, T (or U for RNA). Thus, a typical DNA sequence might be written as TAGGCT.

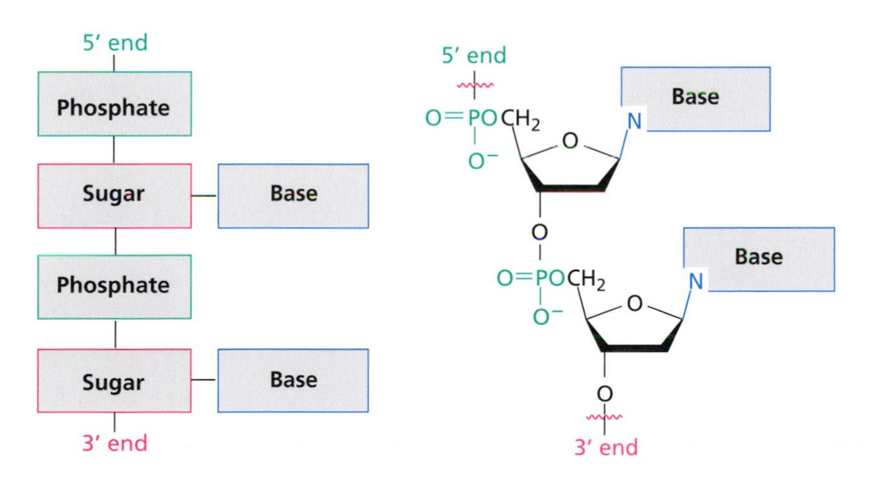

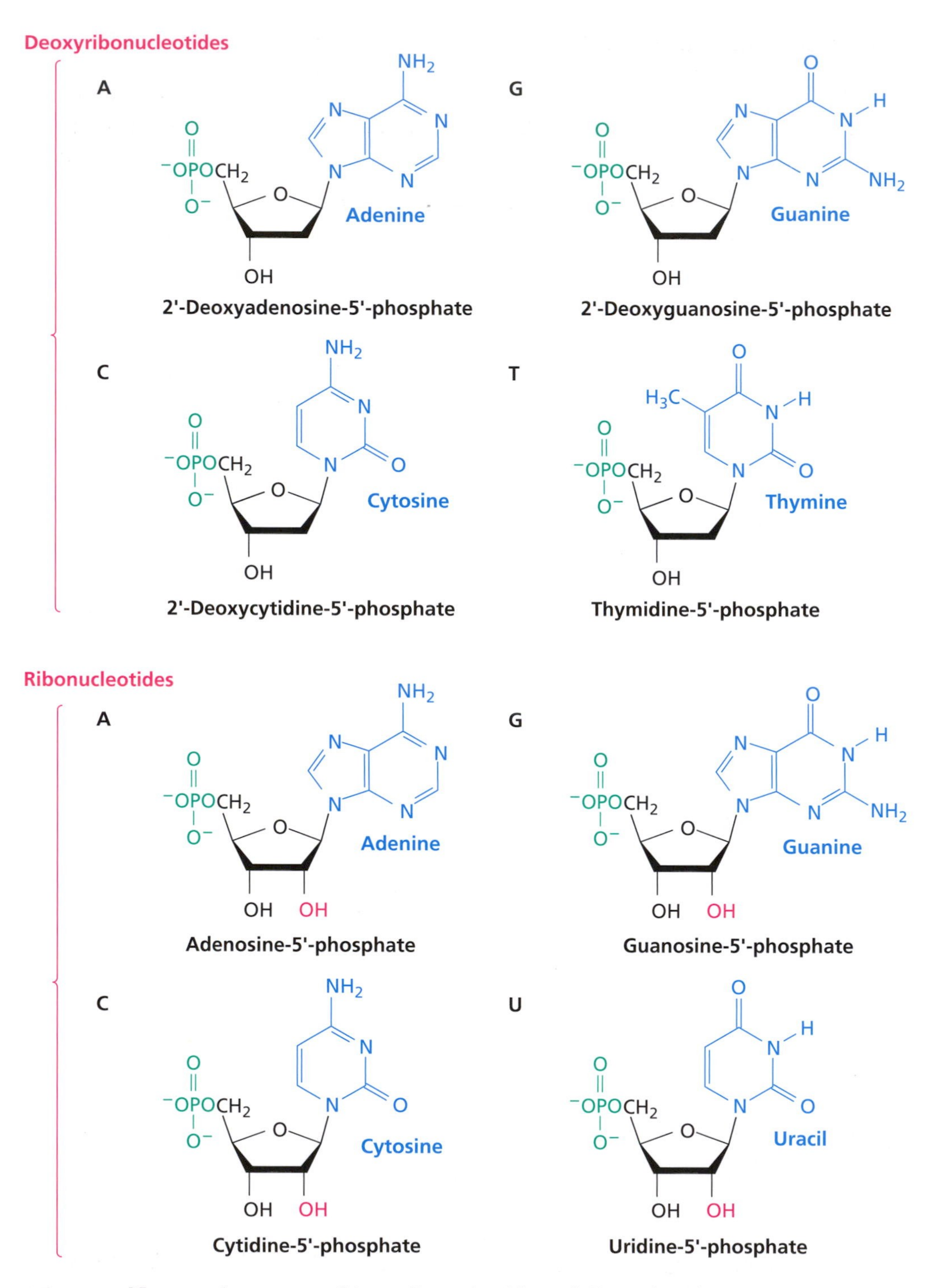

FIGURE 2.12 Names and structures of deoxyribonucleotides and ribonucleotides.

DNA: Deoxyribonucleic Acid

Under physiological conditions, DNA consists of two complementary strands, running in opposite directions and coiled around each other in a **double helix** like the handrails on a spiral staircase. One of the two strands is called the **coding strand**, or **sense strand**, and its complement is the **template strand**, or **antisense strand**. As described in all introductory textbooks, the template strand is transcribed to give messenger RNA, which is thus an RNA copy of the DNA coding strand, with U in place of T. This messenger RNA then directs protein synthesis during the translation process.

DNA strands are held together by hydrogen bonds between complementary pairs of bases, A with T and C with G. That is, whenever an A base occurs in one strand, a T base occurs opposite it in the other strand; when a C base occurs in one, a G occurs in the other (Figure 2.13).

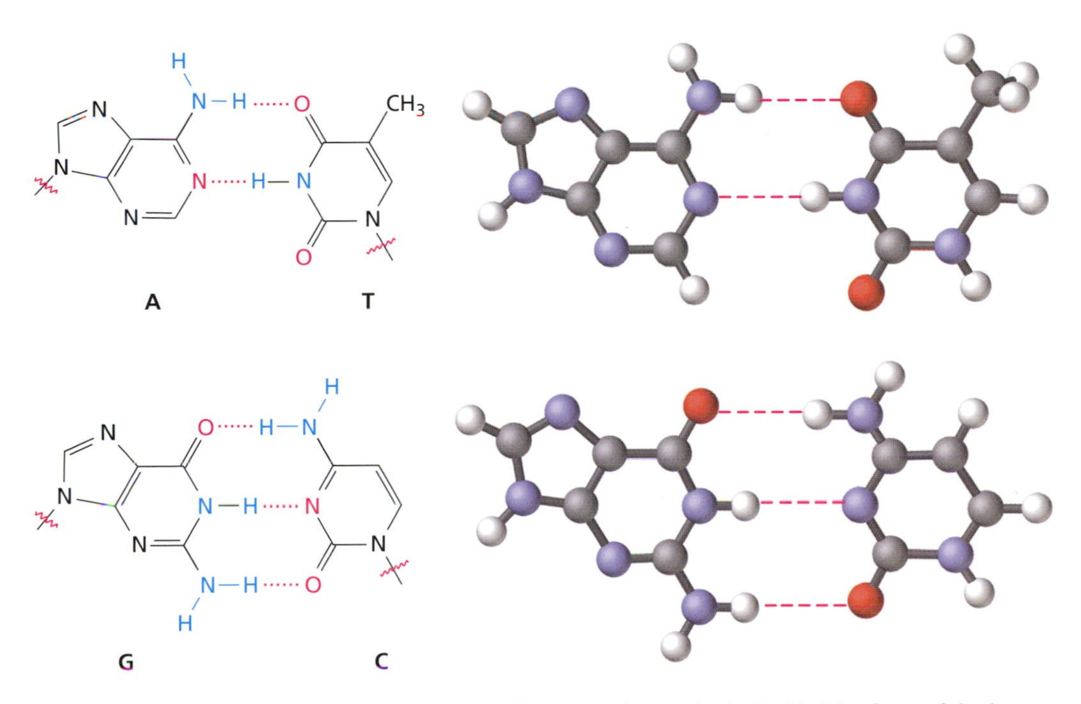

FIGURE 2.13 Hydrogen bonding between complementary base pairs in DNA. The faces of the bases are relatively neutral, while the edges have positively and negatively polarized regions. Base A is aligned for hydrogen bonding with T, and G is aligned with C.

A turn of the DNA double helix is shown in Figure 2.14. The helix is 20 Å wide, there are 10 base pairs per turn, and each turn is 34 Å in length. Note that the

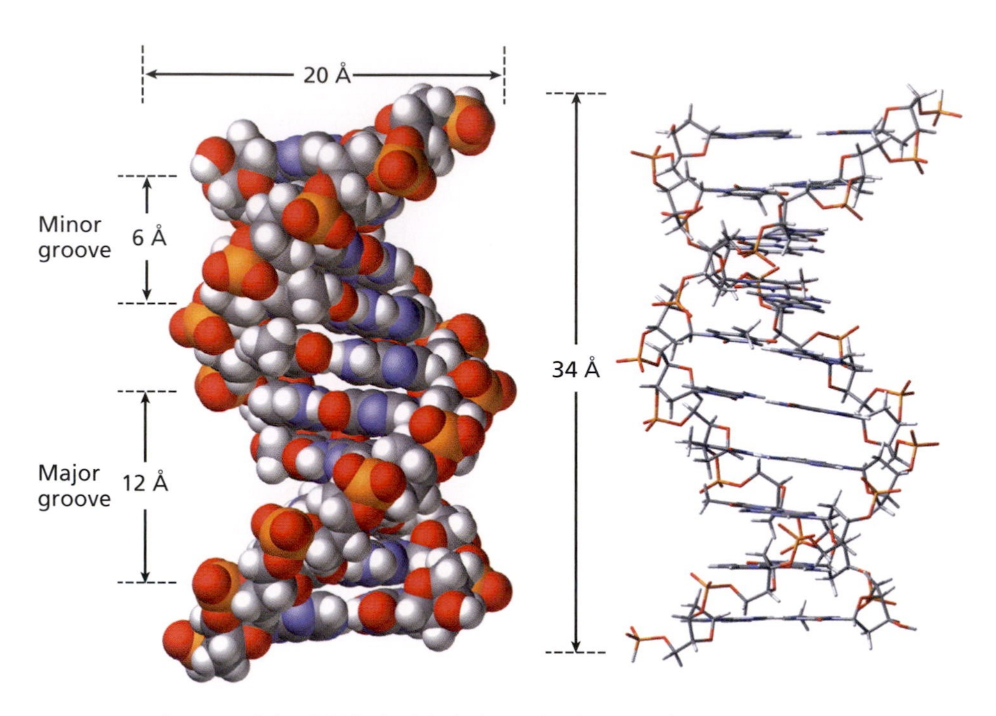

FIGURE 2.14 A turn of the DNA double helix in both space-filling and wire-frame formats. The sugar–phosphate backbone runs along the outside of the helix, and the amine bases hydrogen bond to one another on the inside.

double helix has two kinds of "grooves." The so-called **major groove** is 12 Å in height, and the **minor groove** is 6 Å in height. The major groove exposes the edges of the various bases, making them subject to chemical manipulation.

RNA: Ribonucleic Acid

RNA, unlike DNA, is single-stranded rather than double-stranded, and has several functions. **Messenger RNA (mRNA)** is transcribed from DNA and functions as the template for protein synthesis in ribosomes, small granular particles in the cytoplasm of a cell. **Ribosomal RNA (rRNA)** complexed with protein provides the physical makeup of the ribosomes. **Transfer RNA (tRNA)** transports specific amino acids to the ribosomes where they are joined together to make proteins. Figure 2.15 shows a model of yeast serine tRNA. The anticodon loop part of the structure contains the recognition sequence complementary to mRNA, and a molecule of serine bonds to the acceptor stem. Note that the molecule has several loops where the chain doubles back on itself to give short helical segments.

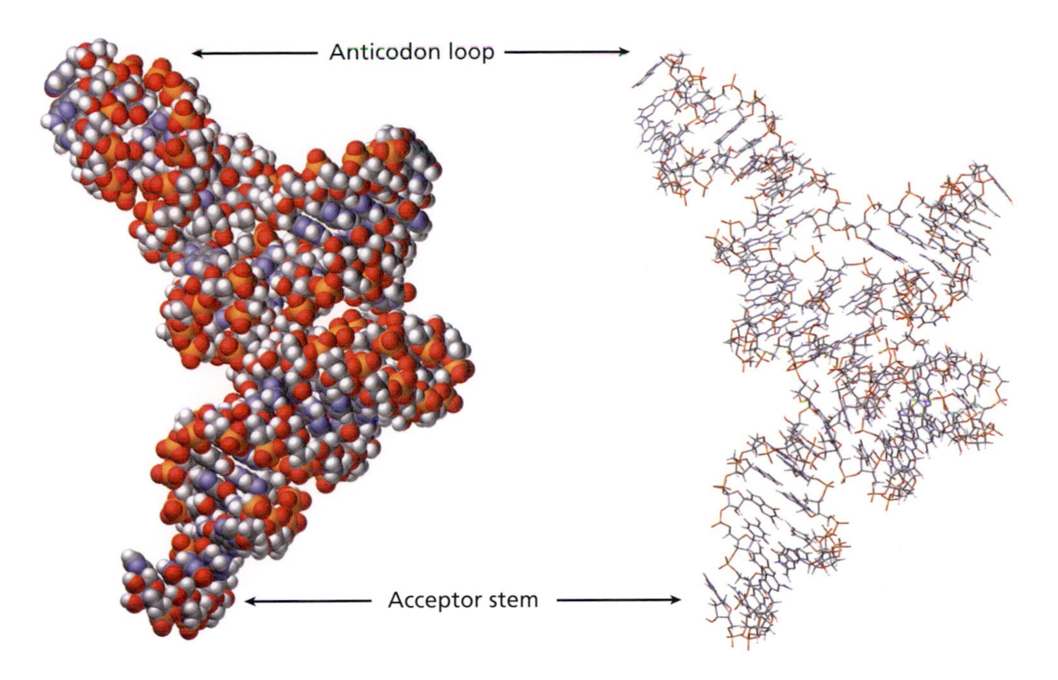

FIGURE 2.15 The structure of yeast serine transfer RNA in both space-filling and wire-frame formats. Although single-stranded, the molecule has several loops where the chain doubles back on itself to give short helical segments.

2.6 Biomolecules: Enzymes, Coenzymes, and Coupled Reactions

Enzymes

An **enzyme** is a biological catalyst—usually a protein. Like all catalysts, an enzyme speeds up the rate of a given reaction but does not affect the reaction's equilibrium constant. The enzyme does this by lowering the activation energy of the slow step or by making available a different reaction pathway, usually with multiple steps, whose rate-limiting step is lower in energy than that of the uncatalyzed pathway. Millionfold rate enhancements are common for enzyme-catalyzed processes, and a rate enhancement of 2×10^{23} has been reported for orotidine monophosphate decarboxylase (Section 6.2).

Enzymes function through a pathway that involves (1) the initial formation of an enzyme–substrate complex E·S, (2) a multistep conversion of the enzyme-bound substrate into enzyme-bound product E·P, and (3) the release of product.

$$\mathrm{E} + \mathrm{S} \underset{k_{-1}}{\overset{k_1}{\rightleftharpoons}} \mathrm{E{\cdot}S} \overset{k_2}{\rightleftharpoons} \mathrm{E{\cdot}P} \overset{k_3}{\rightleftharpoons} \mathrm{E} + \mathrm{P} \qquad (k_{cat}: \mathrm{E{\cdot}S} \rightarrow \mathrm{E} + \mathrm{P})$$

Once formed, the E·S complex can either revert back to reactants or proceed on to products. If, as is often the case, the rate constant for formation of the E·S complex (k_1) and the rate constant for release of product from the E·P complex (k_3) are large compared to the rate constant for reaction (k_2), then the conversion of bound substrate to bound product is the rate-limiting step. The overall rate constant for conversion of the E·S complex to products E + P is referred to as the *catalytic rate constant*, k_{cat}, also called the **turnover number** because it represents the number of substrate molecules the enzyme turns over into product per unit time. Catalase, for example, which catalyzes the decomposition of H_2O_2, has an extraordinarily high turnover number of $1.0 \times 10^7\ s^{-1}$. A value of about 10^3 is more typical, however—as with acetylcholinesterase, for example, which catalyzes the hydrolysis of acetylcholine at nerve synapses.

The rate acceleration achieved by enzymes is due to a combination of factors. Of primary importance is the ability of the enzyme to stabilize, and thus lower the energy of, the transition state(s) for product formation. That is, it's not the enzyme's ability to bind the *substrate* that is most critical but rather its ability to bind and thereby stabilize the *transition state.* Often, in fact, the enzyme binds the transition structure as much as 10^{12} times more tightly than it binds the substrate or products. The reactants E + S and the E·S complex are generally similar in energy, but the transition state between E·S and E·P is substantially lower in energy than the transition state for the uncatalyzed reaction. An energy diagram for an enzyme-catalyzed process might look like that in Figure 2.16.

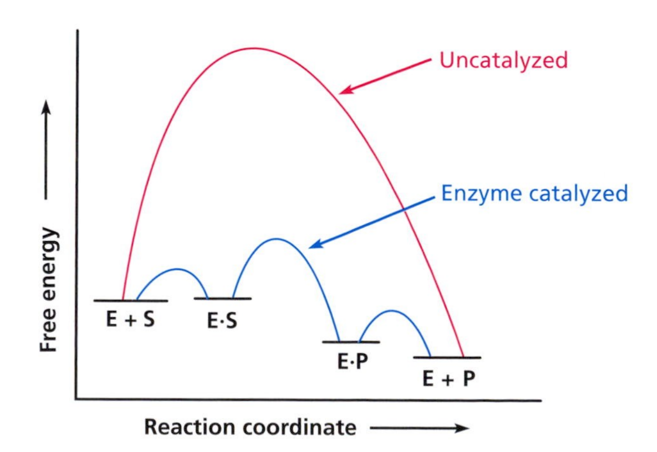

FIGURE 2.16 Energy diagrams for uncatalyzed (red) and enzyme-catalyzed (blue) processes. The enzyme makes available an alternative, lower-energy pathway in which conversion of bound substrate (E·S) to bound product (E·P) is the rate-limiting step. Rate enhancement is due to the ability of the enzyme to bind to the transition state for product formation, thereby lowering its energy.

Substrate binding and reaction catalysis take place within a cleft on the enzyme surface that is referred to as an **active site**. Since enzymes are made of chiral amino acids with a variety of different substituents, the active site is also chiral and can adopt a shape complementary to that of the substrate, much as the shape of a glove is complementary to that of a hand. The active site is lined by amino acids with the appropriate polarity, acidity, or basicity necessary to catalyze a specific reaction. In addition, the active site contains any **cofactors**—metal ions or small organic molecules—needed for the reaction. Figure 2.17 shows a molecular model of hexokinase, an enzyme that catalyzes the phosphorylation of glucose as the first step in the glycolysis pathway. The deep cleft containing the active site is clearly visible.

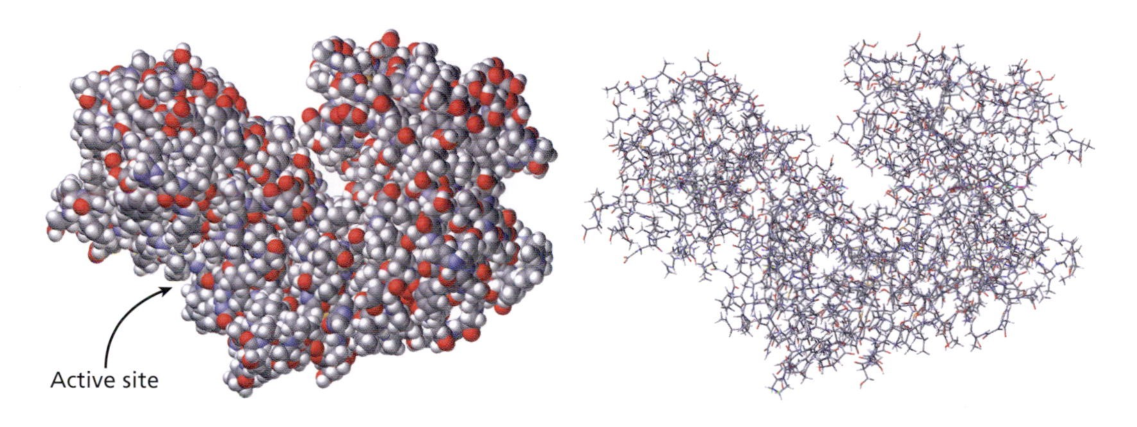

FIGURE 2.17 Models of hexokinase in space-filling and wire-frame formats, showing the deep cleft that contains the active site.

Enzymes are classified into six groups depending on the kind of reaction they catalyze, as shown in Table 2.4: (1) *Oxidoreductases* catalyze oxidations and reductions; (2) *transferases* catalyze the transfer of a group from one substrate to another; (3) *hydrolases* catalyze hydrolysis reactions of esters, amides, and related substrates; (4) *lyases* catalyze the elimination or addition of a small molecule such as H_2O from or to a substrate; (5) *isomerases* catalyze isomerizations; and (6) *ligases* catalyze the bonding together of two molecules, often coupled with the hydrolysis of ATP. The systematic name of an enzyme has two parts, ending with *-ase*. The first part identifies the enzyme's substrate, and the second part identifies its class. For example, the *hexose kinase* shown in Figure 2.17 is a transferase that catalyzes the transfer of a phosphate group from ATP to glucose.

Table 2.4 Classification of Enzymes

Class	Some subclasses	Function
1. Oxidoreductases	Dehydrogenases	Introduction of double bond
	Oxidases	Oxidation
	Reductases	Reduction
2. Transferases	Kinases	Transfer of phosphate group
	Transaminases	Transfer of amino group
3. Hydrolases	Lipases	Hydrolysis of ester
	Nucleases	Hydrolysis of phosphate
	Proteases	Hydrolysis of amide
4. Lyases	Decarboxylases	Loss of CO_2
	Dehydrases	Loss of H_2O
5. Isomerases	Epimerases	Isomerization of chirality center
6. Ligases	Carboxylases	Addition of CO_2
	Synthetases	Formation of new bond

Coenzymes

Many enzyme-catalyzed reactions, particularly those that involve oxidations or reductions, require the presence of a **coenzyme**—a small organic cofactor that takes part in the reaction. A coenzyme is not a catalyst; rather, it is chemically changed during the reaction and requires an additional step to return to its initial state and complete the catalytic cycle. The structures of some common coenzymes are shown in Table 2.5. We'll discuss their chemistry and the mechanisms of their reactions at appropriate points later in the text.

Table 2.5 Structures of Some Common Coenzymes

Adenosine triphosphate, ATP (phosphorylation)

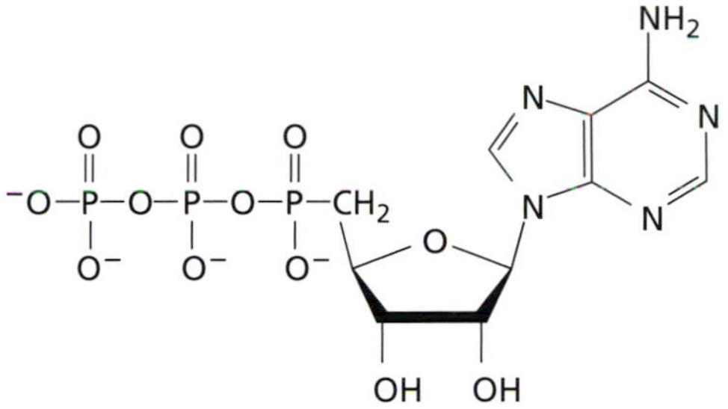

Coenzyme A (acyl transfer)

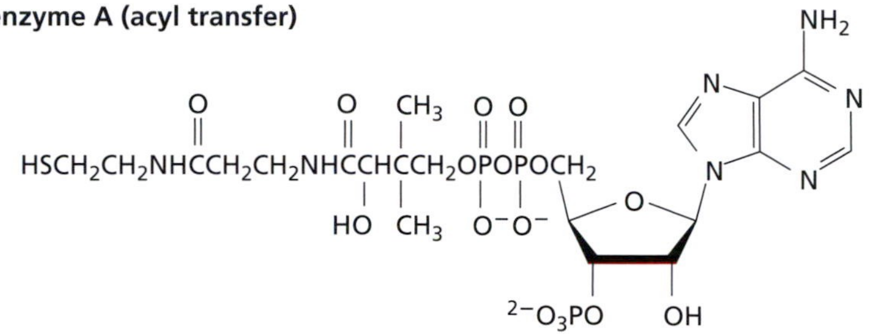

Nicotinamide adenine dinucleotide, NAD^+ (oxidation–reduction)

($NADP^+$)

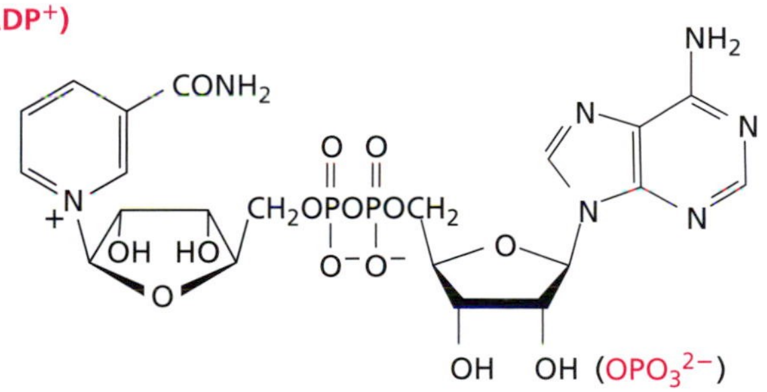

Flavin adenine dinucleotide, FAD (oxidation–reduction)

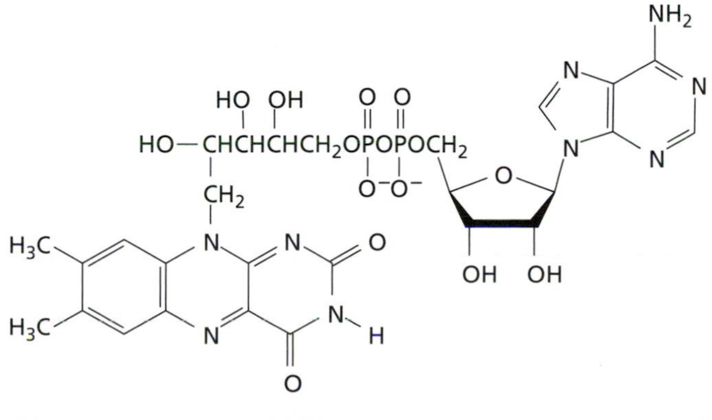

continues

Table 2.5 Structures of Some Common Coenzymes (*continued*)

Tetrahydrofolate (transfer of C_1 units)

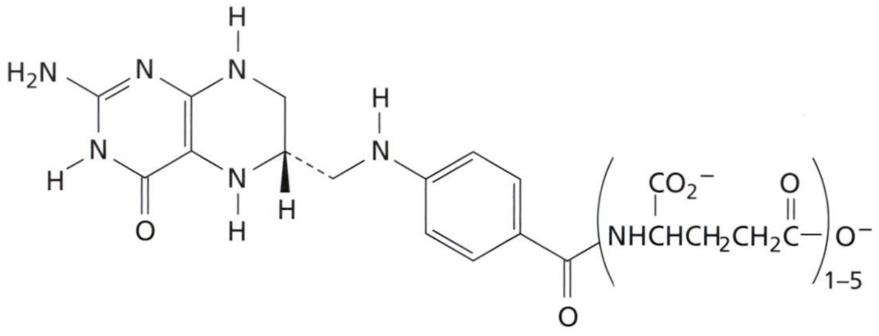

H_2N N N H H N H N H H O H N H CO_2^- O $NHCHCH_2CH_2C-$ O^- 1–5 O

Lipoic acid (acyl transfer)

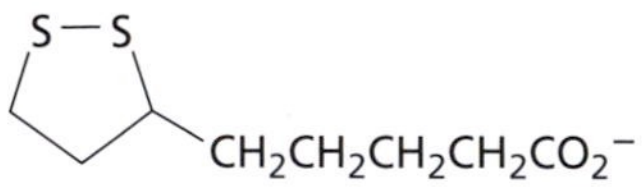

S—S $CH_2CH_2CH_2CH_2CO_2^-$

Pyridoxal phosphate (amino acid metabolism)

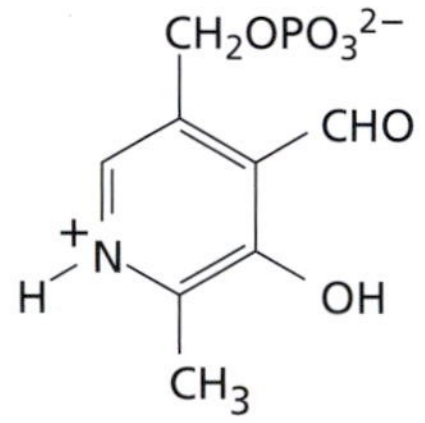

$CH_2OPO_3^{2-}$ CHO $^+$N H OH CH_3

Thiamine diphosphate (decarboxylation)

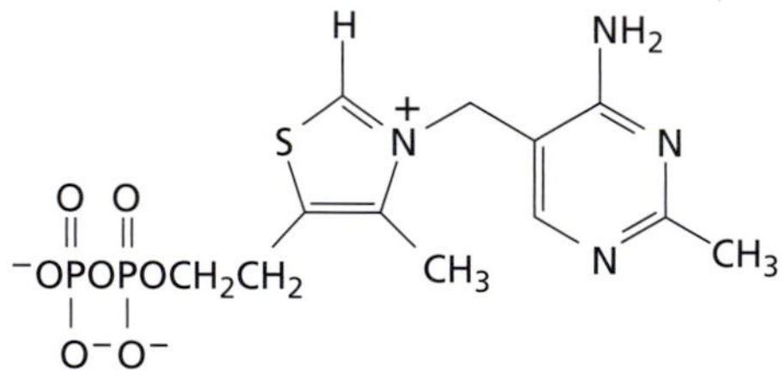

H NH_2 S $^+$N N O O $^-OPOPOCH_2CH_2$ O^-O^- CH_3 N CH_3

Biotin (carboxylation)

O H—N N—H H H H S $CH_2CH_2CH_2CH_2CO_2^-$

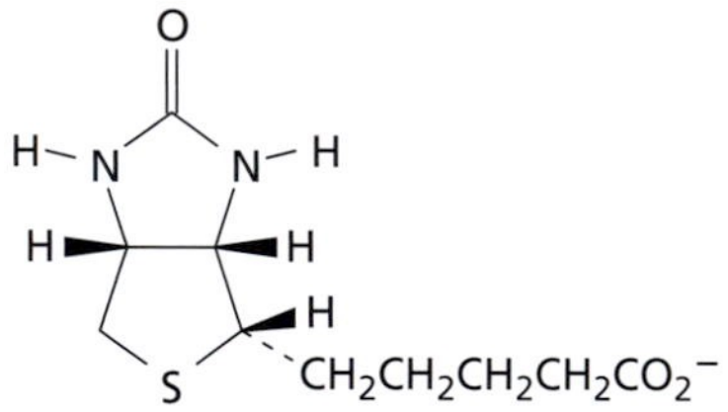

***S*-Adenosylmethionine (methyl transfer)**

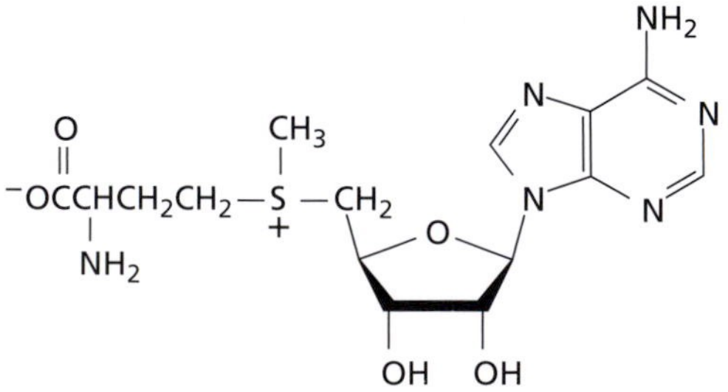

NH_2 N N O CH_3 $^-OCCHCH_2CH_2-\overset{+}{S}-CH_2$ N N NH_2 O OH OH

Many coenzymes are derived from **vitamins**—substances that an organism requires for growth yet is unable to synthesize and must therefore receive in its diet. Among those listed in Table 2.5 are coenzyme A from pantothenate (vitamin B_3), NAD^+ from niacin, FAD from riboflavin (vitamin B_2), tetrahydrofolate from folic acid, pyridoxal phosphate from pyridoxine (vitamin B_6), thiamin diphosphate from thiamin (vitamin B_1), and biotin.

Coupled Reactions and High-Energy Compounds

Every field has its jargon—words and phrases that have a special meaning not always apparent to newcomers. Biochemists, for instance, often speak about reactions as being "coupled," or they describe certain substances as being "high-energy" compounds or as having high-energy bonds. The words sound straightforward but can be misleading if not clearly defined.

To understand what it means for reactions to be coupled, imagine that reaction 1 does not occur to any reasonable extent because it has a small equilibrium constant and is energetically unfavorable; that is, the reaction has $\Delta G > 0$. (Recall that a reaction with a negative value of ΔG is *exergonic*, thermodynamically favorable, and can occur spontaneously. A reaction with a positive value of ΔG is *endergonic*, thermodynamically unfavorable, and cannot occur spontaneously.)

$$(1) \quad \mathbf{A} + m \rightleftharpoons \mathbf{B} + n \qquad \Delta G > 0$$

where **A** and **B** are the biochemically "interesting" substances undergoing transformation while *m* and *n* are enzyme cofactors, H_2O, or various other substances.

Imagine also that product n can react with substance o to yield p and q in a second, strongly favorable reaction that has a large equilibrium constant and $\Delta G << 0$:

$$(2) \quad n + o \rightleftharpoons p + q \qquad \Delta G << 0$$

Considering the two reactions together, they share, or are coupled by, the common intermediate n. When even a tiny amount of n is formed in reaction 1, it undergoes essentially complete conversion in reaction 2, thereby removing it from the first equilibrium and forcing reaction 1 to continually replenish n until the reactants **A** and m are gone. That is, the two reactions added together have a favorable $\Delta G < 0$, and we say that the favorable reaction 2 "drives" the unfavorable reaction 1. Because the two reactions are coupled through n, the transformation of **A** to **B** becomes possible.

$$
\begin{array}{llll}
(1) & \mathbf{A} + m \rightleftharpoons & \mathbf{B} + n & \Delta G > 0 \\
(2) & n + o \rightleftharpoons & p + q & \Delta G << 0 \\
\hline
\text{Net:} & \mathbf{A} + m + o \rightleftharpoons & \mathbf{B} + p + q & \Delta G < 0
\end{array}
$$

An example of coupled reactions occurs in a pathway for the biosynthesis of fatty acids from acetic acid (Section 3.4). The first step in that pathway is the reaction between acetate and coenzyme A (CoASH) to give the thioester acetyl coenzyme A (acetyl CoA). The direct reaction of acetate with coenzyme A does not occur to any extent because it is energetically unfavorable, with $\Delta G^{\circ\prime} \approx +32$ kJ/mol. (The standard free-energy change for a biological reaction is denoted $\Delta G^{\circ\prime}$ and refers to a process in which reactants and products have a concentration of 1.0 M in an aqueous solution at pH = 7.)

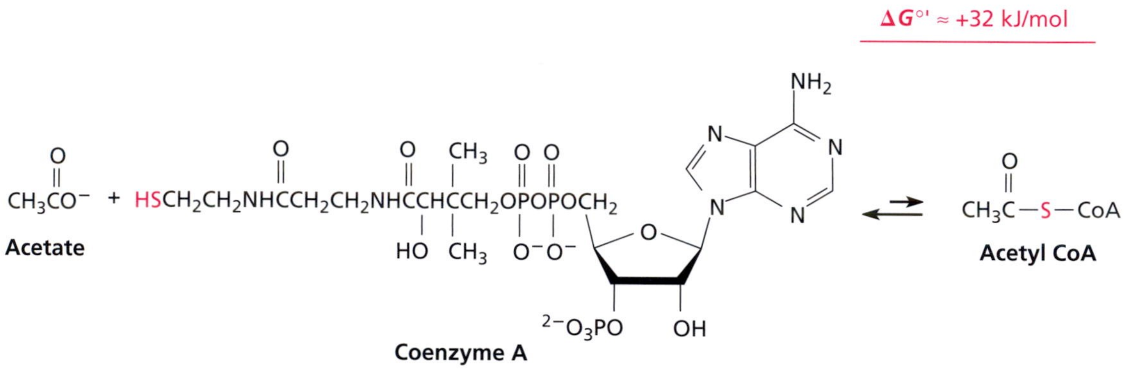

$\Delta G^{\circ\prime} \approx +32$ kJ/mol

$CH_3CO_2^-$ (Acetate) + $HSCH_2CH_2NHCOCH_2CH_2NHCOCH(OH)C(CH_3)_2CH_2OPO_2^-OPO_2^-OCH_2$–(adenosine 3'-phosphate) (Coenzyme A) $\rightleftharpoons$ $CH_3CO{-}S{-}CoA$ (Acetyl CoA)

To get around the unfavorable nature of the direct reaction, two coupled reactions are used. Acetate reacts first with guanosine triphosphate (GTP) to give acetyl phosphate and guanosine diphosphate (GDP) in a somewhat unfavorable process with $\Delta G^{\circ\prime} \approx +11$ kJ/mol:

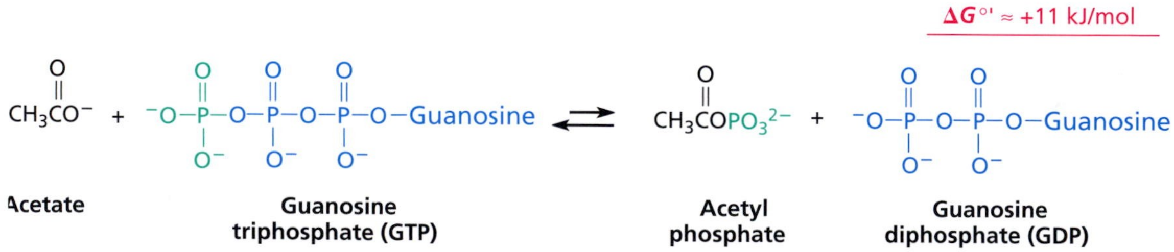

$\Delta G^{\circ\prime} \approx +11$ kJ/mol

$CH_3CO_2^-$ (Acetate) + $^-O{-}PO(O^-){-}O{-}PO(O^-){-}O{-}PO(O^-){-}O{-}$Guanosine (Guanosine triphosphate (GTP)) $\rightleftharpoons$ $CH_3COOPO_3^{2-}$ (Acetyl phosphate) + $^-O{-}PO(O^-){-}O{-}PO(O^-){-}O{-}$Guanosine (Guanosine diphosphate (GDP))

The acetyl phosphate intermediate then reacts with coenzyme A to give acetyl CoA in a favorable second step with $\Delta G°' \approx -10$ kJ/mol:

$\Delta G°' \approx -10$ kJ/mol

$$CH_3\overset{O}{\overset{\|}{C}}OPO_3^{2-} + CoASH \rightleftharpoons CH_3\overset{O}{\overset{\|}{C}}{-}SCoA + HOPO_3^{2-}$$

Acetyl phosphate **Acetyl CoA**

As a result of coupling the two reactions, the thermodynamics of the conversion of acetate to acetyl CoA changes from an unfavorable $\Delta G°' \approx +32$ kJ/mol for the direct reaction to an essentially neutral $\Delta G°' \approx +1$ kJ/mol and an equilibrium constant near unity. Subsequent reactions then remove acetyl CoA and further drive the equilibrium.

(1) Acetate + GTP $\rightleftharpoons$ ~~Acetyl phosphate~~ + GDP $\Delta G°' \approx +11$ kJ/mol

(2) ~~Acetyl phosphate~~ + CoASH $\rightleftharpoons$ Acetyl CoA + $HOPO_3^{2-}$ $\Delta G°' \approx -10$ kJ/mol

Net: Acetate + GTP + CoASH $\rightleftharpoons$ Acetyl CoA + GDP + $HOPO_3^{2-}$ $\Delta G°' \approx +1$ kJ/mol

Guanosine triphosphate and, more commonly, adenosine triphosphate (ATP, Table 2.5) are examples of what biochemists call "high-energy" compounds. Other examples are various acyl phosphates and phosphoenolpyruvate.

Acyl phosphate **Phosphoenolpyruvate**

Calling ATP, GTP, and phosphoenolpyruvate high-energy compounds doesn't mean that they are different from other compounds in any fundamental way, it means only that their reactions are usually exergonic—they generally release a large amount of energy when they react because they are high in energy to begin with. The energy released is frequently coupled with and used to drive reactions that would otherwise not occur.

Problems

2.1 Identify the chirality centers in each of the following molecules:

(a) HO, CH_3O, N, N

Quinine

(b) HO, O, NCH_3, HO

Morphine

(c) OH, HO, O, HO, O, CH_2OH

Ascorbic acid

(d) $CH_3C(=O)-SCH_2CH_2NHC(=O)CH_2CH_2NHC(=O)CH(OH)C(CH_3)_2CH_2OP(=O)(O^-)OP(=O)(O^-)OCH_2$–[ribose: O, $^{2-}O_3PO$, OH]–adenine (NH_2, N, N, N, N)

Acetyl Coenzyme A

2.2 Convert the following molecular model of the anti-inflammatory steroid prednisolone into a line drawing, and identify the chirality centers. As noted previously, the standard color convention for molecular models is that H = ivory, C = black, O = red, N = blue, and S = yellow.

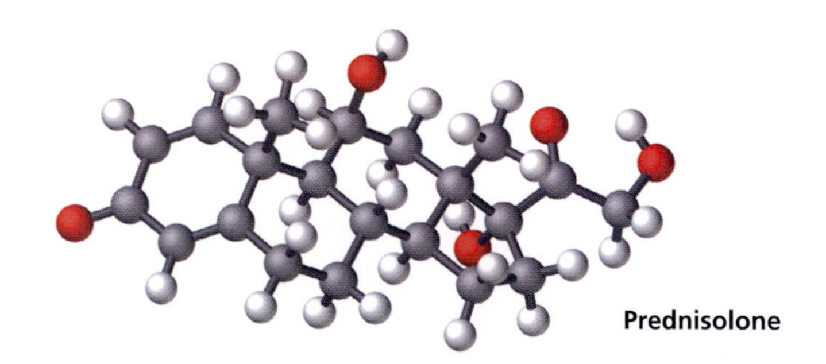

Prednisolone

2.3 Tell whether the following pairs of compounds are enantiomers, diastereomers, epimers, or meso compounds:

(a) H_3C H CH_3 H and H_3C H H CH_3

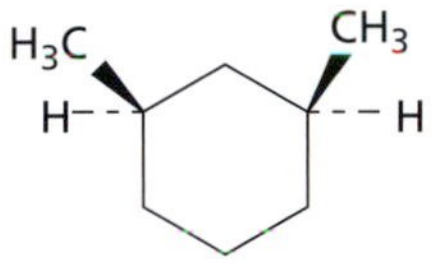

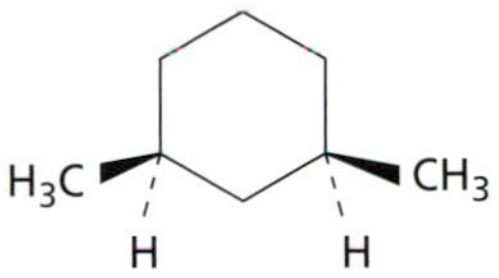

(b) CH_2OH O HO OH OH OH and CH_2OH O HO OH HO OH

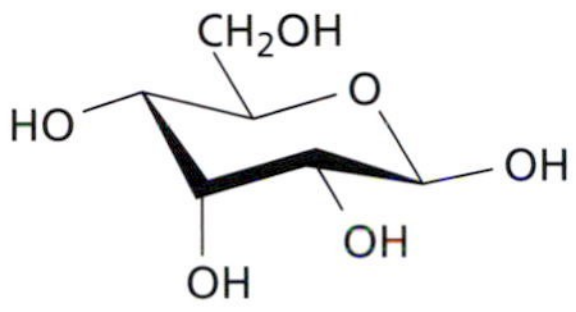

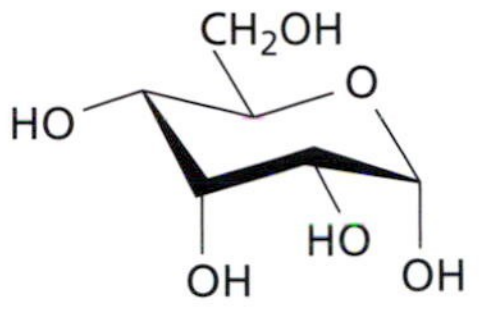

(c) $H_3\overset{+}{N}$ C H CH_3 O C N H H_3C H C C O^- O and ^-O C O C H CH_3 N H O C C H_3C H $\overset{+}{N}H_3$

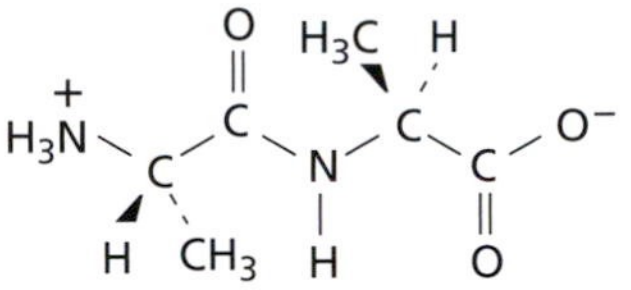

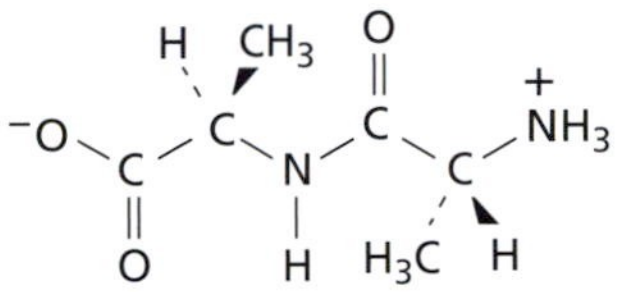

2.4 Assign *R,S* configurations to the chirality centers in each of the following molecules:

(a) H H H_3C OH H $CH(CH_3)_2$

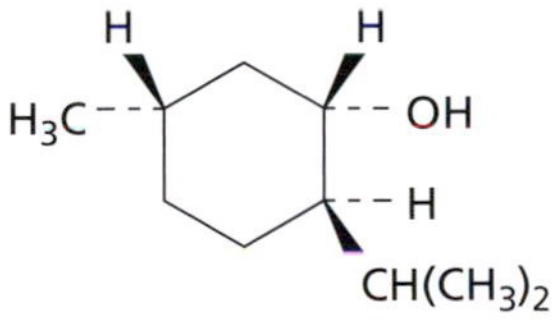

Menthol

(b) O H—N N—H H H H S $CH_2CH_2CH_2CH_2CO_2^-$

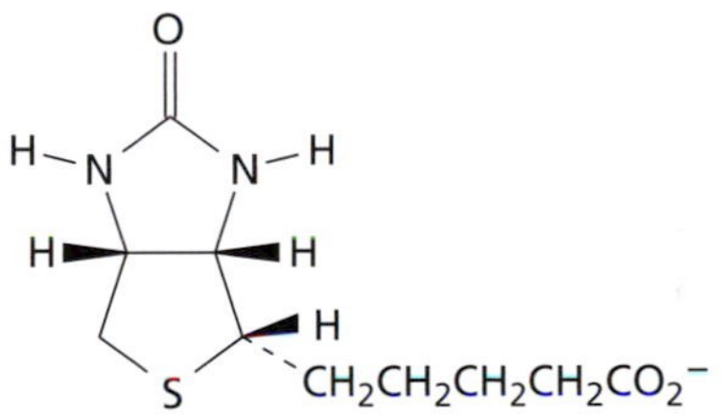

Biotin

(c) HO CH_2OH O HO OH NH_2

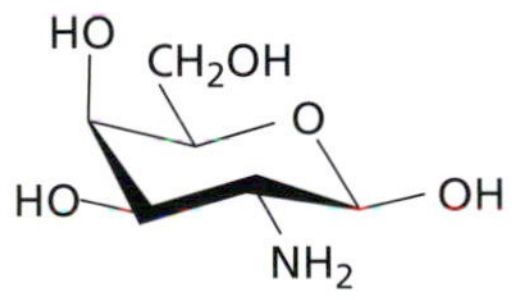

Galactosamine

(d) O H CO_2H HO H H HO H

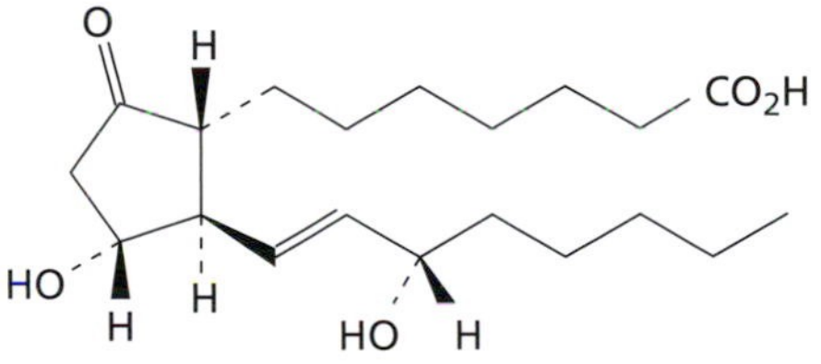

Prostaglandin E_1

2.5 Assign *R,S* configurations to the chirality centers in the following molecular model of pseudoephedrine:

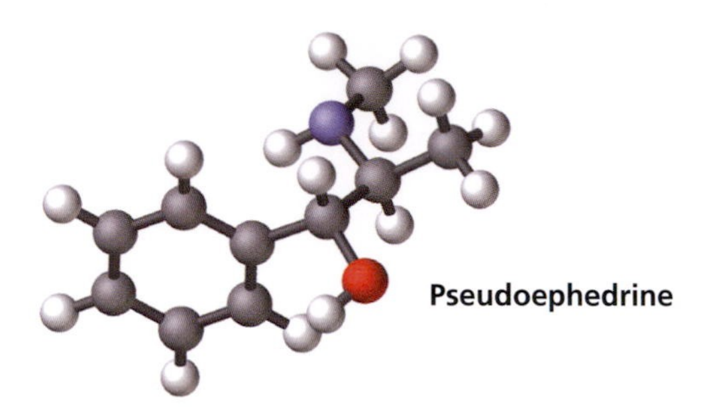

2.6 Identify the indicated hydrogens in the following molecules as *pro-R* or *pro-S*:

(a) H H HO_2C CO_2H HO H

Malic acid

(b) H H CH_3S CO_2^- H H $H_3\overset{+}{N}$ H

Methionine

(c) H H HS CO_2^- $H_3\overset{+}{N}$ H

Cysteine

2.7 Identify the *re* face and *si* face with respect to the indicated carbon atoms in each of the following molecules:

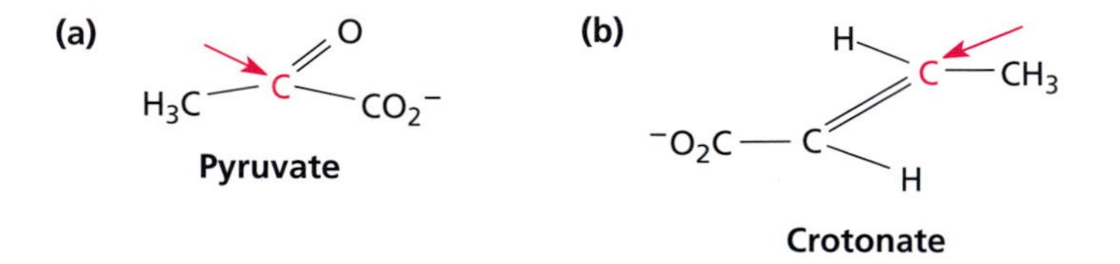

2.8 Draw the products of the following reactions, and assign *R* or *S* stereochemistry to each new chirality center (see Problem 2.7):

(a) Reduction of the ketone carbonyl group in pyruvate from the *re* face.

(b) Addition of —OH to the *si* face of crotonate at C3, followed by protonation at C2, also from the *si* face.

2.9 The dehydration of citrate to *cis*-aconitate, a step in the citric acid cycle, involves the *pro-R* carboxymethyl "arm" of citrate rather than the *pro-S* arm. Which of the following two products is formed?

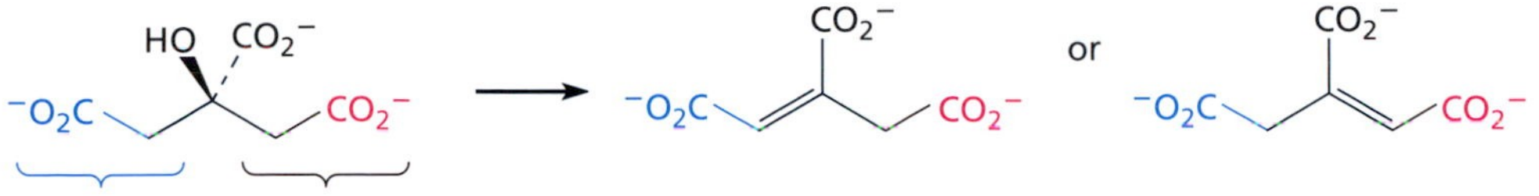

Citrate *cis*-**Aconitate**

2.10 Hydration of *cis*-aconitate yields (2*R*,3*S*)-isocitrate. Show the stereochemistry of the product, tell whether the initial attack of H_2O takes place on the *re* or *si* face of *cis*-aconitate, and also whether the addition is *syn* or *anti*—that is, whether the —OH and —H add from the same face or from opposite faces of the double bond.

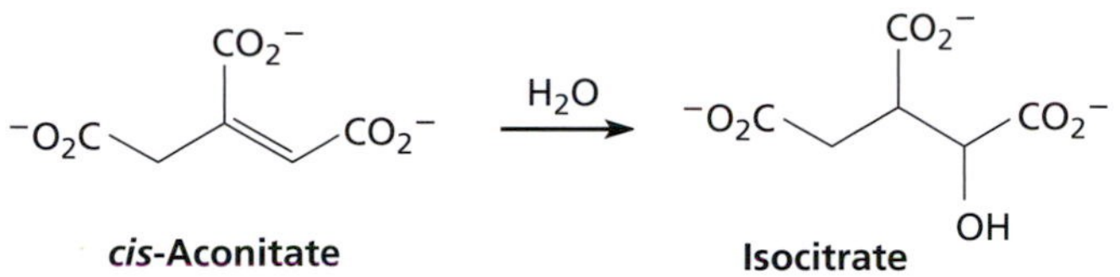

cis-**Aconitate** **Isocitrate**

2.11 We said in Section 2.1 that cholestanol and coprostanol are epimeric at C5. Draw three-dimensional structures of both using chair conformations for the 6-membered rings, and tell whether the —OH group in each is axial or equatorial.

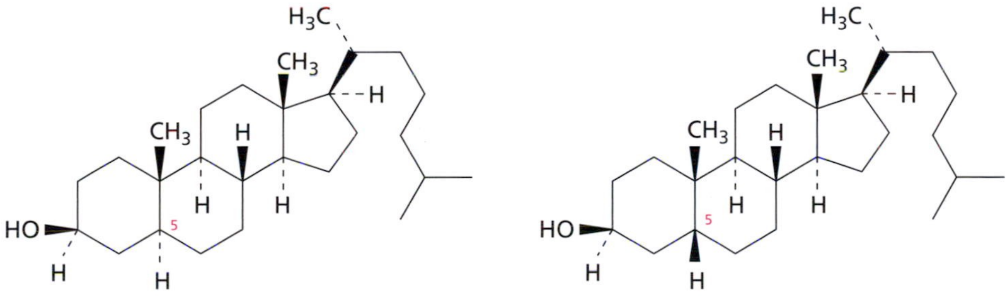

Cholestanol **Coprostanol**

2.12 On treatment with an acid catalyst, the naturally occurring terpenoid ψ-ionone is transformed into β-ionone. Suggest a mechanism for the reaction.

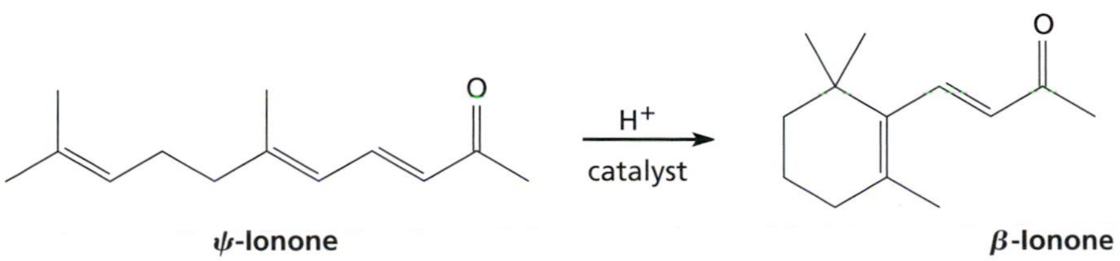

ψ-**Ionone** β-**Ionone**

2.13 Use Figure 2.7 to identify the following aldoses. Draw a Fischer projection of each in its open-chain form, and tell whether each is a D sugar or an L sugar.

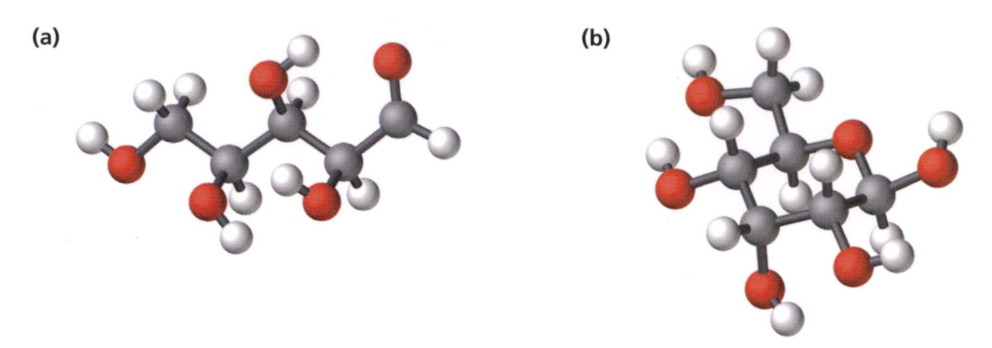

2.14 The following model is that of a naturally occurring aldohexose found in ivory.

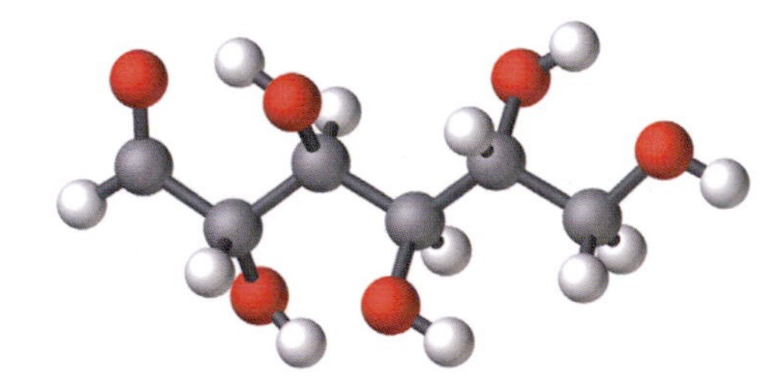

(a) Draw Fischer projections of the sugar, its enantiomer, and an epimer.

(b) Draw the β anomer of the sugar in its pyranose form.

2.15 Draw Fischer projections of the following sugars in their open-chain forms:

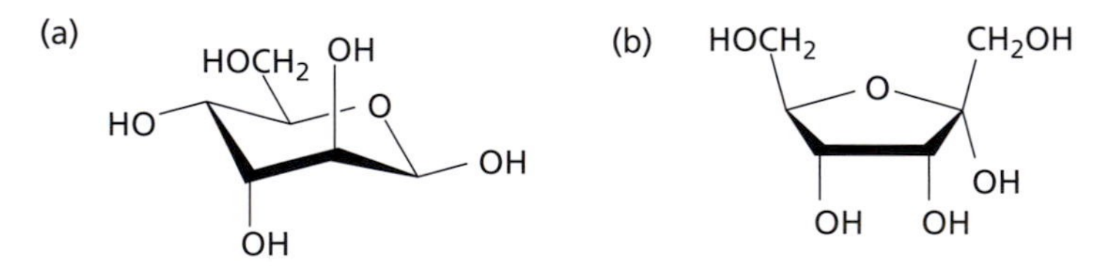

2.16 D-Ribulose has the systematic name 1,3R,4R,5-tetrahydroxy-2-pentanone. Draw the cyclic β-D-ribulofuranose.

2.17 Give the sequence of the following tetrapeptide:

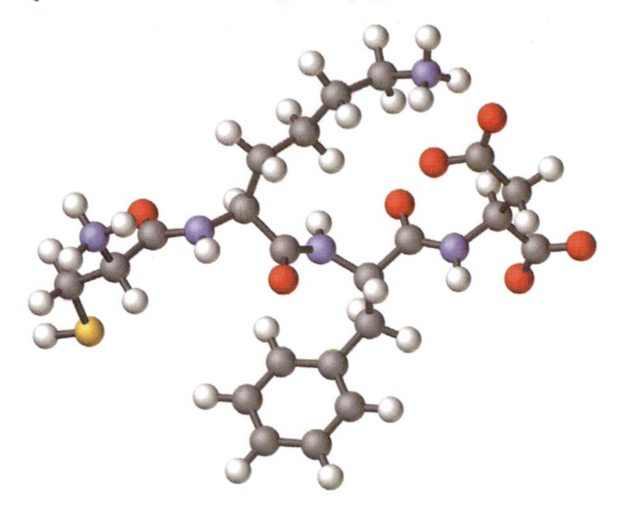

2.18 Draw the structure of the following peptide:

Val-Glu-Pro-Ala-Cys

2.19 The α-helical segments of proteins stop whenever a proline is encountered in the chain. Suggest a reason why proline is not present in an α helix.

2.20 An imidazole ring like that in histidine has two nitrogen atoms. Draw structures for the two protonated forms of 5-methylimidazole, and tell which nitrogen atom is more basic.

N, N, H, H_3C

5-Methylimidazole

2.21 Write the complete structure of the DNA dinucleotide G-C.

2.22 What class of enzyme catalyzes each of the following reactions?

(a) $CH_3\overset{OH}{\overset{|}{C}}HCO_2^- \longrightarrow CH_3\overset{O}{\overset{\|}{C}}CO_2^-$

(b) $CH_3\overset{OH}{\overset{|}{C}}HCH_2\overset{O}{\overset{\|}{C}}SR \longrightarrow CH_3CH{=}CH\overset{O}{\overset{\|}{C}}SR$

(c) $CH_3\overset{O}{\overset{\|}{C}}SCoA \longrightarrow {}^-O_2CCH_2\overset{O}{\overset{\|}{C}}SCoA$

3 Lipid Metabolism

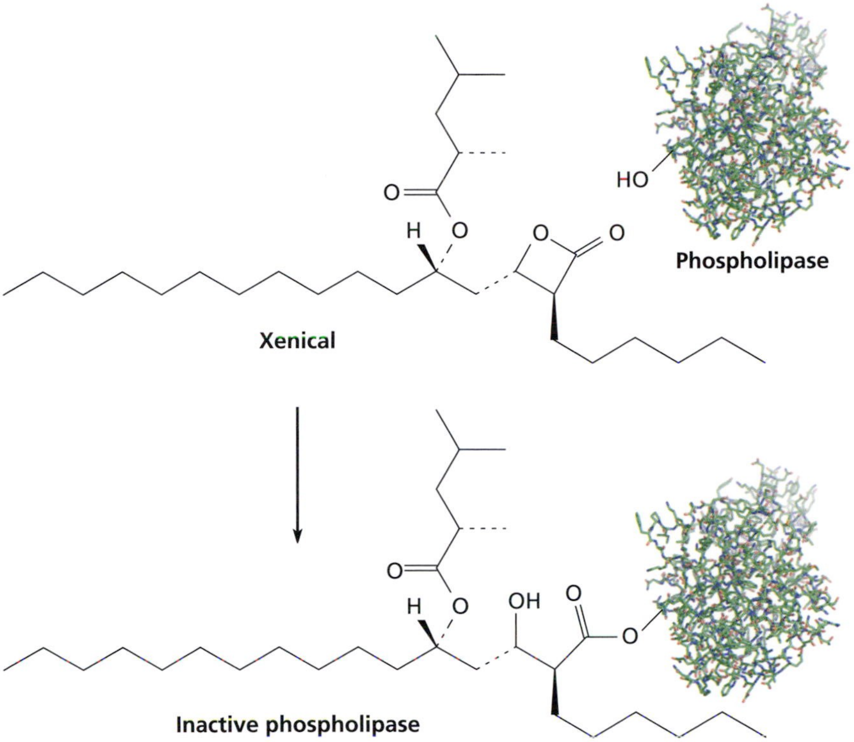

Xenical is a widely used anti-obesity drug. It functions by irreversible acylation of an active-site serine in phospholipase by the strained four-membered-ring lactone. The resulting inactive enzyme is no longer able to hydrolyze triacylglycerol, thus blocking fat uptake by the adipocyte.

Dietary lipids are composed of approximately 90% triacylglycerols along with small amounts of phospholipids and cholesterol. Actually, the definition of the term "lipid" is not universally agreed on; some workers restrict the term to triacylglycerols and related compounds while other workers adopt a broader definition that includes all naturally occurring substances soluble in nonpolar solvents. Although we'll focus on triacylglycerols in this chapter, we'll also look at two other widely occurring classes of compounds—terpenoids and steroids—because all three are biosynthesized from acetyl CoA. Acetyl CoA, a key molecule in numerous biological pathways, contains an acetyl group linked by a thioester bond to the sulfur atom of phosphopantetheine. The phosphopantetheine is in turn linked to adenosine 3′,5′-bisphosphate (the prefix *bis-* means two).

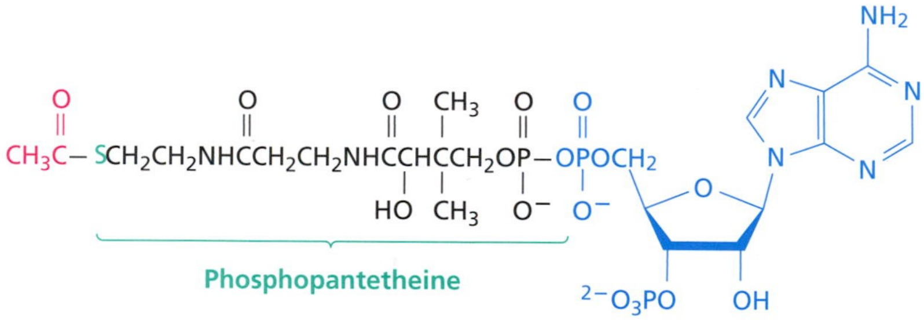

$CH_3C(=O)-SCH_2CH_2NHC(=O)CH_2CH_2NHC(=O)CH(OH)C(CH_3)_2CH_2OP(=O)(O^-)-OP(=O)(O^-)OCH_2$–(ribose, $3'$-OPO_3^{2-}, OH)–adenine (NH_2)

Phosphopantetheine

Acetyl CoA — a thioester

3.1 Digestion and Transport of Triacylglycerols

The metabolic breakdown of dietary fats begins with their digestion in the stomach and small intestine. Because they are insoluble in water, triacylglycerols must first be emulsified into tiny droplets by the action of bile acid salts. These salts, of which taurocholate and glycocholate are the most abundant, act much like detergents in that they have both nonpolar, fat-soluble parts and polar, water-soluble parts in their structures. The nonpolar parts are attracted to fats, and the polar parts are attracted to water, thereby bringing the two phases into contact.

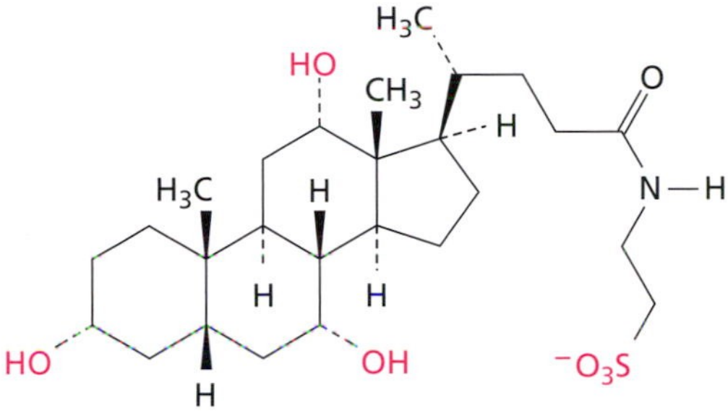

Taurocholate

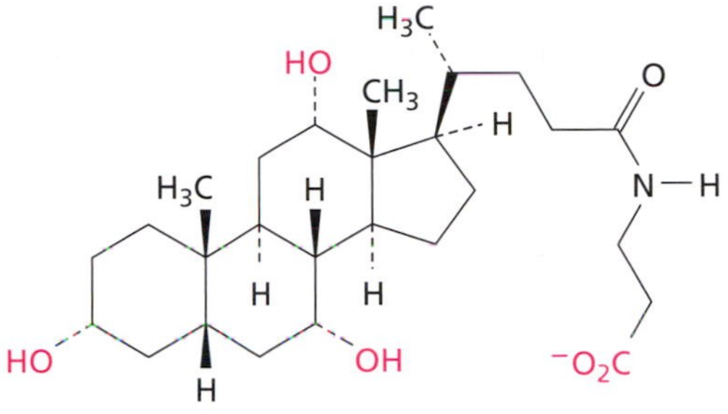

Glycocholate

Triacylglycerol Hydrolysis

Once emulsified, triacylglycerols are hydrolyzed at C1 and C3 to yield 2-monoacylglycerols plus fatty acids.

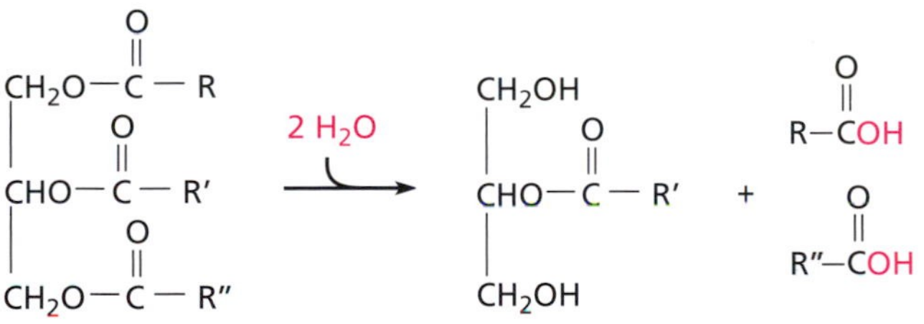

A triacylglycerol $\xrightarrow{2\,H_2O}$ **A 2-acyl glycerol** + **Fatty acids**

The hydrolysis is catalyzed by human pancreatic lipase, a globular protein of 449 amino acids, whose mechanism of action[1, 2] is summarized in Figure 3.1. The active site contains a catalytic triad of aspartic acid, histidine, and serine residues, which act cooperatively to provide the necessary acid and base catalysis for the individual steps. Hydrolysis is accomplished by two sequential nucleophilic acyl substitution reactions, one that covalently binds an acyl group to the side chain —OH of a serine residue on the enzyme to give an acyl enzyme intermediate and a second that frees the fatty acid from the enzyme.

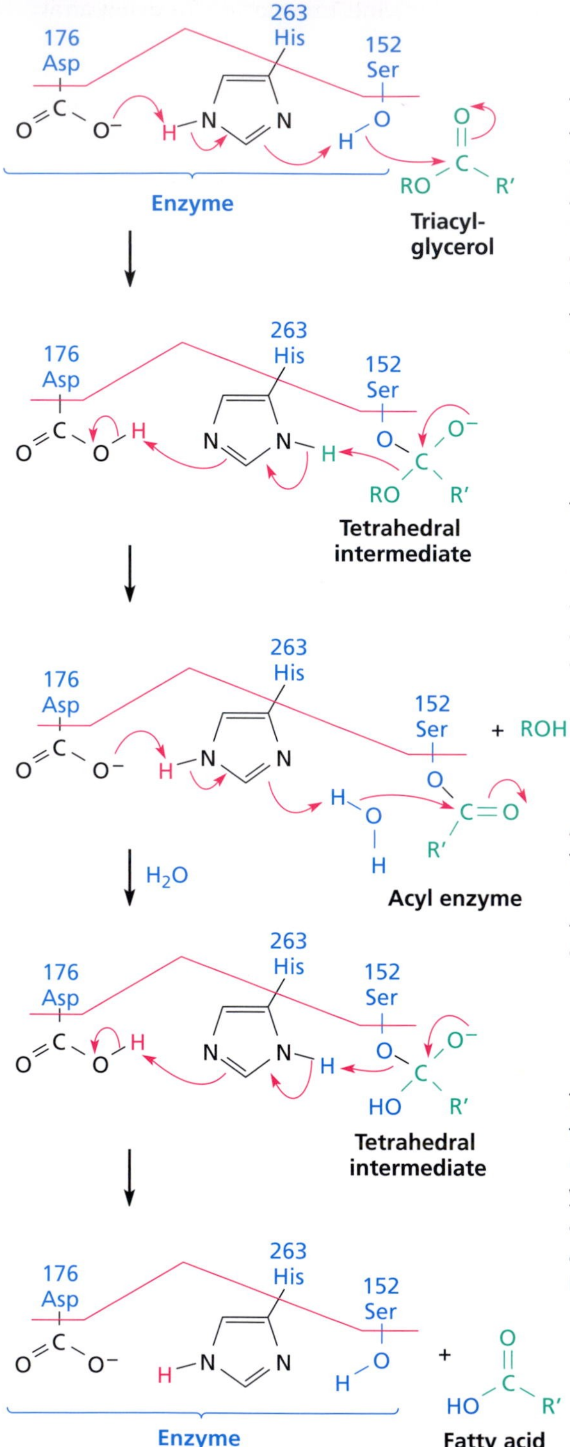

The triacylglycerol is held at the active site of the pancreatic lipase with Asp-176, His-263, and Ser-152 nearby. The side-chain carboxylate of Asp-176 acts as a base to remove a proton from His-263, which thus becomes a stronger base capable of deprotonating the —OH group of Ser-152. Ser-152 adds to the carbonyl group of the triacylglycerol, yielding a tetrahedral alkoxide intermediate.

This intermediate expels an acylglycerol as the leaving group in a nucleophilic acyl substitution reaction, giving an acyl enzyme. The glycerol accepts a proton from His-263, which in turn accepts a proton from the side-chain carboxyl of Asp-176.

Asp-176 and His-263 react cooperatively to deprotonate a water molecule, giving a hydroxide ion that adds to the acyl group. A tetrahedral alkoxide intermediate is again formed.

The tetrahedral alkoxide intermediate expels the Ser-152 as the leaving group in a second nucleophilic acyl substitution reaction, yielding a free fatty acid. The Ser-152 accepts a proton from His-263, which in turn accepts a proton from Asp-176. The enzyme has now returned to its starting structure.

FIGURE 3.1 Mechanism of action of human pancreatic lipase. A catalytic triad of Asp-176, His-263, and Ser-152 is involved, and two nucleophilic acyl substitution reactions occur.

The presence of the catalytic triad and the acyl enzyme has been established by X-ray crystallography of an analog in which the Ser-152 residue is covalently bonded to a methyl alkylphosphonate [$RPO(OR)_2$] serving as a stable mimic for the first tetrahedral intermediate (Figure 3.2).

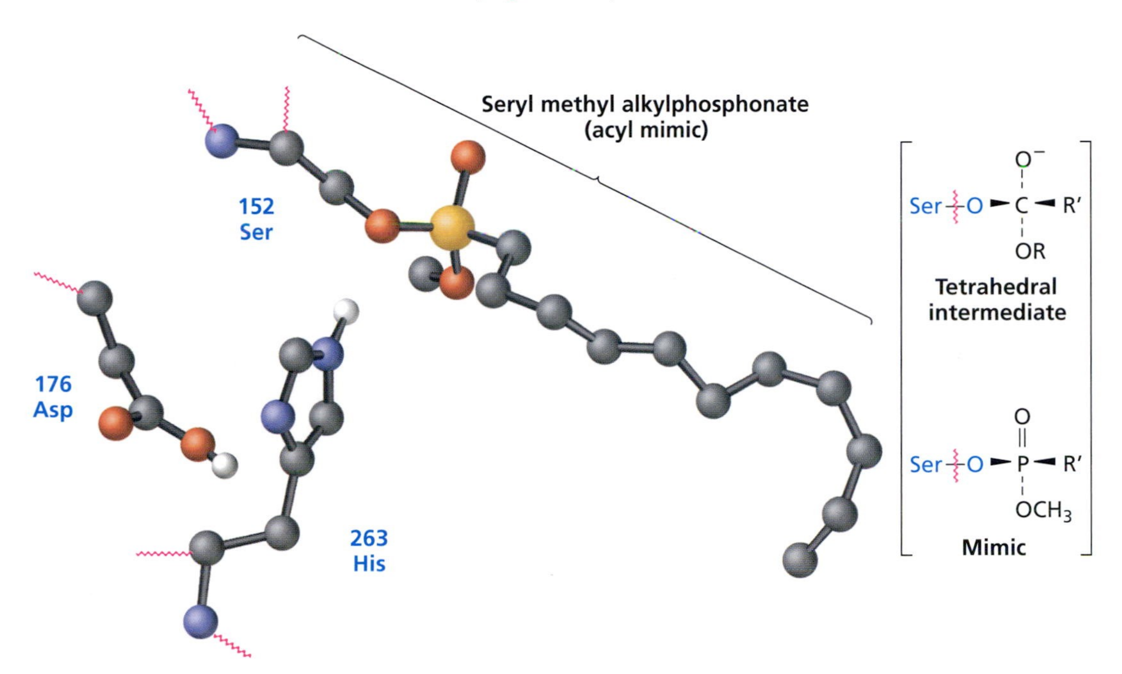

FIGURE 3.2 An X-ray crystal structure of an enzyme-bound methyl alkylphosphonate in the active site of human pancreatic lipase shows the catalytic triad of amino acid residues. The methyl alkylphosphonate group bonded to Ser-152 acts as a stable analog of an acyl enzyme intermediate.

A Brief Note about Visualizing Enzyme Structures

Handheld molecular models are extremely useful in organic chemistry for examining and studying the three-dimensionality of small molecules. Enzymes and other proteins, however, are far too large for making handheld models, so computer-based visualization is essential. As with the human pancreatic lipase shown in Figure 3.2, it's even possible to focus in on the active site of enzymes and identify the residues nearby. Several excellent visualization programs are freely available, including the Swiss PDB Viewer whose use is described in Appendix A. Understanding how to use this viewer will give you access to the three-dimensional structures of more than 25,000 proteins and other substances currently available online from the Protein Data Bank (PDB).

Operated by Rutgers University and funded by the U.S. National Science Foundation, the Protein Data Bank is a worldwide repository for processing and distributing three-dimensional structural data for biological macromolecules. To access a file of atomic coordinates for a protein, go to the PDB website at *http://www.rcsb.org/pdb/*, use the search function to find the protein you're interested in, and download the PDB coordinate file. Then use the Swiss PDB Viewer to open the coordinate file and display the protein structure on your computer. Once the structure is up, a set of tools allows you to rotate, color, zoom in, or otherwise manipulate the view, even finding and viewing the active site of an enzyme.

The first nucleophilic acyl substitution step—reaction of the side-chain hydroxyl of Ser-152 with the triacylglycerol to give an acyl enzyme—could in principle occur directly without the participation of other residues in the enzyme. In fact, though, a neutral alcohol is only weakly nucleophilic, so the addition step must be catalyzed by first converting the free alcohol into a more strongly nucleophilic alkoxide ion. This proton transfer is accomplished cooperatively when the side-chain carboxylate anion of Asp-176 abstracts an N—H hydrogen from the side-chain imidazole ring of His-263, thereby making the histidine more basic and enabling it to deprotonate the side chain —OH of Ser-152. The deprotonated serine then adds to a carbonyl group of a triacylglycerol to give a tetrahedral intermediate.

176 Asp 263 His 152 Ser → 176 Asp 263 His 152 Ser

Free alcohol — poor nucleophile

Alkoxide ion — good nucleophile

The tetrahedral intermediate next expels a mono- or diacylglycerol as the leaving group and produces an acyl enzyme. This step, too, could in principle occur directly, but a negatively charged alkoxide ion is a poor leaving group. Thus, the step is catalyzed by a proton transfer from His-263 that makes the leaving group a neutral alcohol. Asp-176 then transfers its side-chain carboxyl proton to histidine.

176 Asp 263 His 152 Ser RO R′ → 176 Asp 263 His 152 Ser ROH + C=O R′

Neutral alcohol – good leaving group

The second nucleophilic acyl substitution step hydrolyzes the acyl enzyme and gives the free fatty acid by a mechanism essentially identical to that just discussed. Water, a relatively poor nucleophile, is deprotonated to yield a much better nucleophile that adds to the enzyme-bound acyl group. The tetrahedral intermediate then expels the neutral serine residue as the leaving group, freeing the fatty acid and returning the enzyme to its active form.

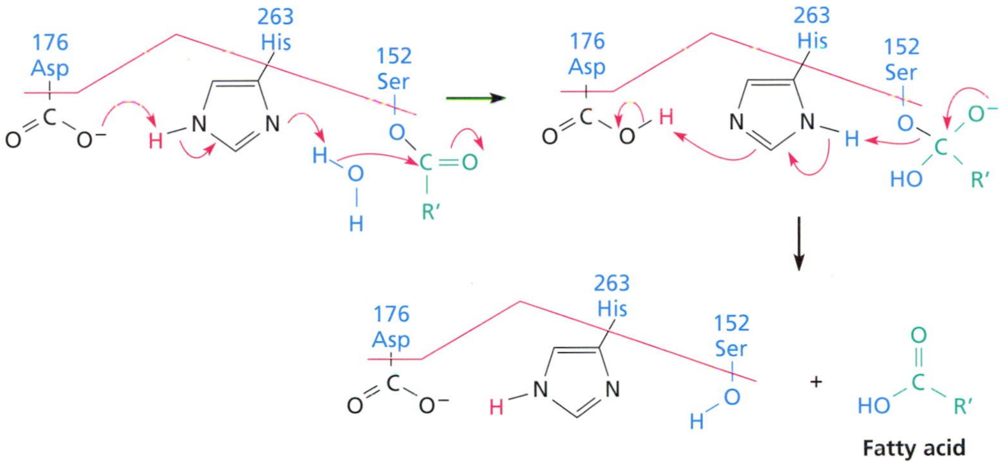

Triacylglycerol Resynthesis

The 2-monoacylglycerols and free fatty acids produced by lipase-catalyzed hydrolysis are transported by bile salts into the cells lining the intestinal wall, where the fatty acids are converted into fatty acyl CoA's, which react with monoacylglycerols to regenerate triacylglycerols. Thus, we're now back where we started: Triacylglycerols in the diet undergo hydrolysis in the small intestine, and the hydrolysis products then recombine to give triacylglycerols in the intestinal mucosa. The only net change is one of transport. Triacylglycerol molecules are too large and insoluble for direct absorption from the aqueous environment of the digestive system through cell walls, but their hydrolysis yields smaller, more polar molecules that are transported more easily.

Resynthesis of triacylglycerols from fatty acids and 2-monoacylglycerols takes place in three steps: Acyl-CoA synthetase first activates the fatty acid by converting it into a thioester with coenzyme A; monoacylglycerol acyltransferase next couples the fatty acyl CoA with a monoacylglycerol to give a diacylglycerol; and diacylglycerol acyltransferase then catalyzes a further coupling to give the final product. Note that we'll use the abbreviation —SCoA when drawing chemical structures of acyl CoA's to underscore the point that thioester bonds are present. Note also that there is a difference between a *synthetase* and a *synthase*. A synthetase catalyzes a reaction that is coupled to the hydrolysis of ATP or similar high-energy compound, while a synthase catalyzes a reaction that does not require ATP.

Fatty acid (RCO_2^-) —[HSCoA, ATP → AMP, PP_i]→ Fatty acyl CoA (RCOSCoA) —[$HOCH_2CH(OCOR)CH_2OH$ → HSCoA]→ Diacylglycerol ($RCOOCH_2$–CH(OCOR)–CH_2OH) —[RCOSCoA → HSCoA]→ Triacylglycerol ($RCOOCH_2$–CH(OCOR)–CH_2OCOR)

Free carboxylates are relatively unreactive toward attack by nucleophiles because they are negatively charged. Thus, the fatty acid is first activated by converting it into a more reactive acid derivative. In the chemical laboratory, we might activate a carboxylic acid by conversion to an acid chloride, but living organisms generally use either a thioester, an acyl phosphate, or the related acyl adenosyl phosphate (Figure 3.3) for activation. As noted in Section 1.6, thioesters and acyl phosphates are substantially more reactive in nucleophilic acyl substitution reactions than carboxylic acids, esters, or amides.

Conversion of a fatty acid into the thioester of coenzyme A is catalyzed by an acyl-CoA synthetase, with ATP as cofactor.[3] As shown in Figure 3.3, ATP first coordinates to an Mg^{2+} ion to help neutralize its negative charges and make it more electrophilic, and nucleophilic addition of fatty-acid carboxylate to phosphorus then occurs. The resultant pentacoordinate phosphorus intermediate expels diphosphate ion (abbreviated PP_i) as the leaving group and generates an acyl adenosyl phosphate in a process analogous to a nucleophilic acyl substitution reaction. The diphosphate ion undergoes hydrolysis to give phosphate ion (abbreviated P_i), and the fatty acyl group undergoes a nucleophilic acyl substitution reaction with coenzyme A to give the thioester product. A basic site in the enzyme helps catalyze the process by deprotonating the —SH group of coenzyme A.

ATP

ATP is first activated by coordination to magnesium ion, and nucleophilic addition of a fatty-acid carboxylate to phosphorus then yields a pentacoordinate intermediate...

(Pentacoordinate phosphorus intermediate)

...which expels diphosphate ion (PP_i) as the leaving group and gives an acyl adenosyl phosphate in a process analogous to a nucleophilic acyl substitution reaction.

Acyl adenosyl phosphate

A basic site in the enzyme catalyzes addition of the —SH group of coenzyme A to the acyl adenosyl phosphate, giving a tetrahedral alkoxide intermediate...

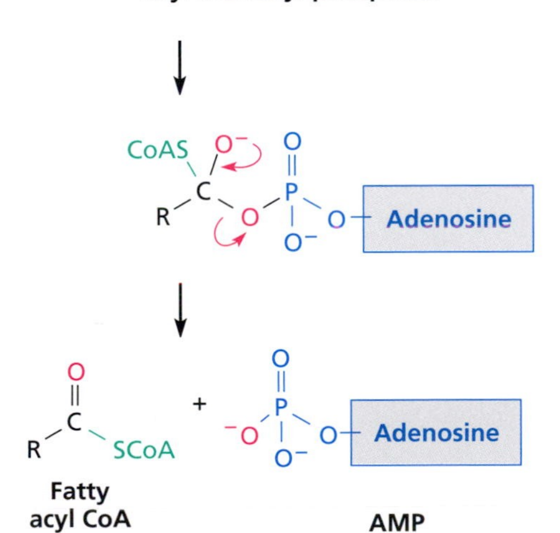

...which expels adenosine monophosphate (AMP) as the leaving group and yields the fatty acyl CoA.

FIGURE 3.3 Mechanism of fatty acyl CoA formation from a fatty acid. The fatty acid is activated by reaction with ATP to form an acyl adenosyl phosphate.

Reaction of the fatty acyl CoA with a 2-monoacylglycerol to give a 1,2-diacylglycerol is catalyzed by monoacylglycerol acyltransferase. Fatty acyl CoA first reacts with a cysteine —SH group on the enzyme to give an enzyme-bound fatty acyl intermediate, which reacts with 2-monoacylglycerol. The two steps are then repeated with diacylglycerol acyltransferase to give the final triacylglycerol product. All steps are yet further examples of nucleophilic acyl substitution reactions and occur through tetrahedral intermediates.

Cys–SH + R–C(=O)–SCoA → Cys–S–C(=O)–R + HSCoA

Cys–S–C(=O)–R + $HOCH_2$–CH(OCOR)–CH_2OH → Cys–SH + $RCOOCH_2$–CH(OCOR)–CH_2OH

Enzyme **Fatty acyl CoA** **Acyl enzyme** **Enzyme** **Diacylglycerol**

Triacylglycerols synthesized in the intestinal mucosa are too insoluble for transport in the bloodstream. Instead, they combine with phospholipids and proteins to give large, globular, lipoprotein assemblies called *chylomicrons.* The triacylglycerols cluster in the hydrophobic core of the chylomicron, while phospholipids and charged proteins provide a hydrophilic coating. The chylomicrons are then released into the bloodstream and delivered to muscles for energy and to adipose tissue for storage.

3.2 Triacylglycerol Catabolism: The Fate of Glycerol

After their transport by chylomicrons, metabolic breakdown of triacylglycerols begins with hydrolysis to yield glycerol plus fatty acids. The reaction is catalyzed by a lipoprotein lipase, whose mechanism of action is essentially identical to that of the pancreatic lipase shown previously in Figure 3.1. The fatty acids released on hydrolysis are transported to mitochondria and oxidized to provide energy, while the glycerol is carried to the liver for further metabolism. In the liver, glycerol is first phosphorylated on the *pro R* —CH_2OH group (Section 2.1) by reaction with ATP. Oxidation by NAD^+ then yields dihydroxyacetone phosphate (DHAP), which enters the carbohydrate metabolic pathway (Section 4.2).

pro S
pro R
ATP ADP
NAD^+ $NADH/H^+$

Glycerol

***sn*-Glycerol 3-phosphate**

Dihydroxyacetone phosphate (DHAP)

The phosphorylation of glycerol is catalyzed by glycerol kinase. (As noted in Table 2.4, a kinase is an enzyme that catalyzes a phosphorylation reaction, usually with ATP.) As in the formation of acyl adenosyl phosphate shown in Figure 3.3, phosphorylation of glycerol begins with activation of ATP with Mg^{2+}, followed by nucleophilic attack of glycerol on phosphorus and subsequent expulsion of ADP (Figure 3.4). A basic site in the enzyme helps catalyze the process by deprotonating the glycerol —OH group. Note that the phosphorylation product is named *sn*-glycerol 3-phosphate, where the *sn*- prefix means "stereospecific numbering." The molecule is drawn in Fischer projection with the —OH group at C2 pointing to the left, and the glycerol carbon atoms are numbered beginning at the top.

It has been shown using an isotopically substituted chiral phosphate group that phosphorylations take place with inversion of configuration at phosphorus[4] and occur through a pentacoordinate oxyphosphorane,[5] but the exact mechanism is not known. Either a direct backside displacement of the leaving group by an S_N2-like pathway (giving an oxyphosphorane-like transition state) or an addition–elimination mechanism by a nucleophilic acyl substitution pathway (giving an oxyphosphorane as an intermediate) would account for the observed results.

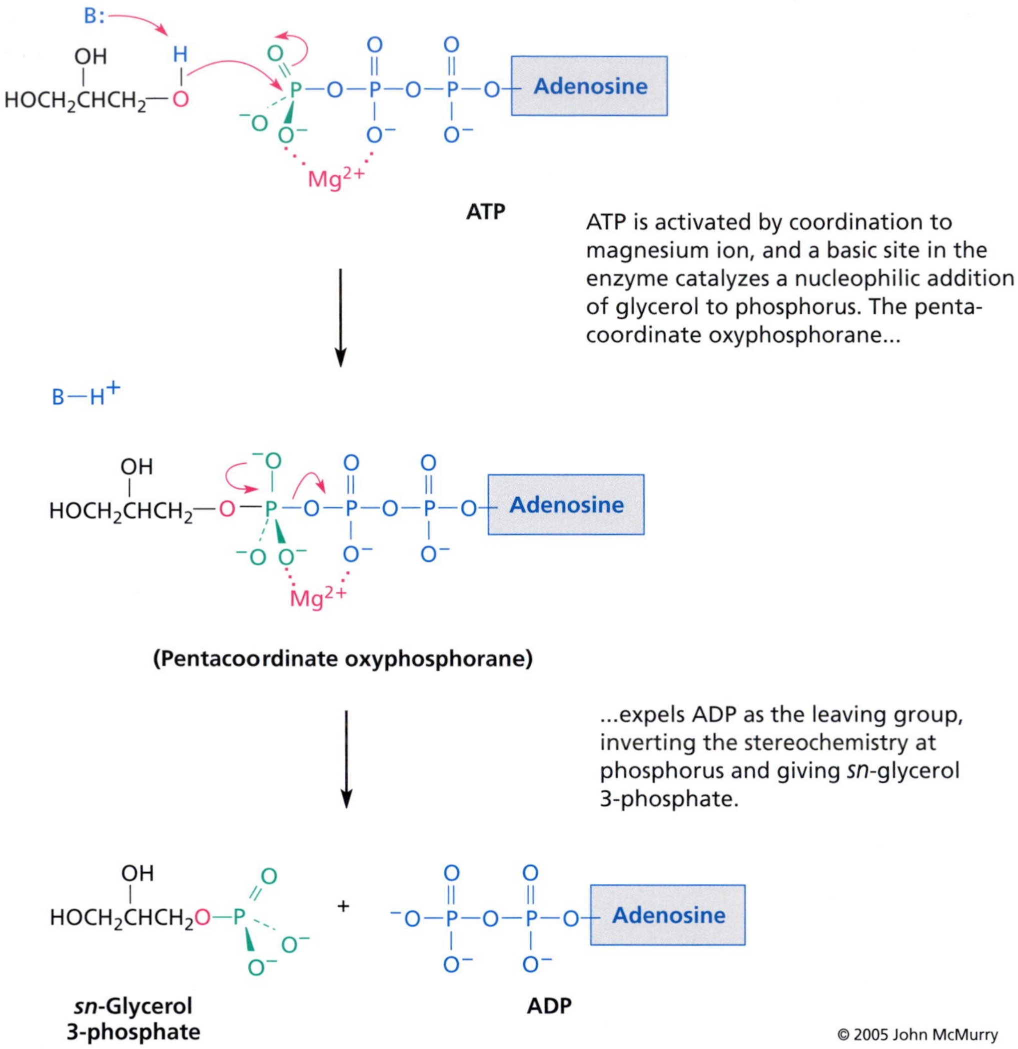
B:
OH
H
HOCH$_2$CHCH$_2$—O
Adenosine
Mg^{2+}
ATP
ATP is activated by coordination to magnesium ion, and a basic site in the enzyme catalyzes a nucleophilic addition of glycerol to phosphorus. The penta-coordinate oxyphosphorane...
B—H$^+$
Adenosine
(Pentacoordinate oxyphosphorane)
...expels ADP as the leaving group, inverting the stereochemistry at phosphorus and giving *sn*-glycerol 3-phosphate.
Adenosine
sn-Glycerol 3-phosphate
ADP
© 2005 John McMurry

FIGURE 3.4 Mechanism of the phosphorylation of glycerol by glycerol kinase.

Oxidation of *sn*-glycerol 3-phosphate to give dihydroxyacetone phosphate is catalyzed by *sn*-glycerol-3-phosphate dehydrogenase. As noted in Section 1.9, this oxidation of an alcohol to a ketone involves nicotinamide adenine dinucleotide (NAD^+) as cofactor and involves transfer of a hydride ion from the hydroxyl-bearing carbon of the alcohol to the C═C—C═N^+ of the nicotinamide ring in a conjugate nucleophilic addition reaction (Section 1.5). Zinc ion acts as a cofactor for the reaction, coordinating to the alcohol and thereby increasing its acidity (Figure 3.5).

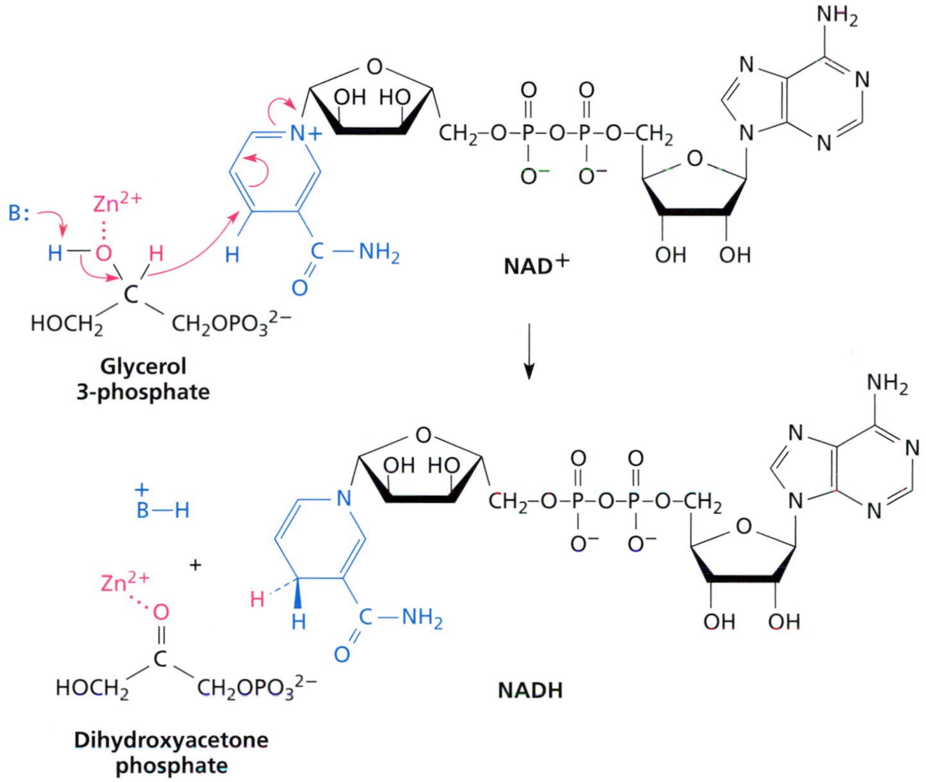

FIGURE 3.5 Mechanism of the oxidation of glycerol 3-phosphate by NAD^+.

The hydride-ion addition is stereospecific, occurring exclusively on the *re* face (Section 2.1) of the nicotinamide ring and adding a hydrogen with *pro-R* stereochemistry. All alcohol dehydrogenases are stereospecific, although the specificity is different depending on the enzyme.

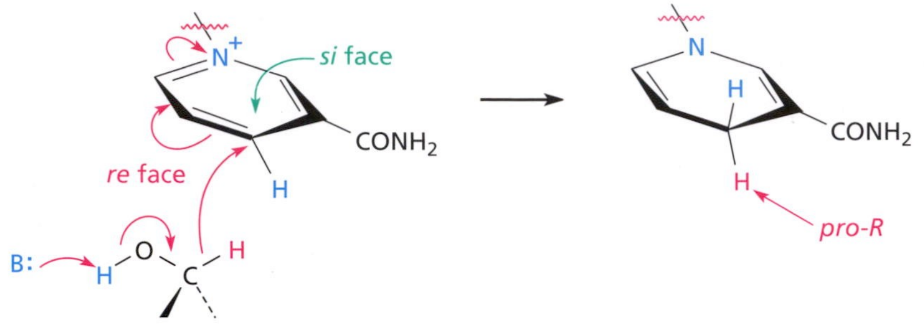

You might note when thinking about oxidations and reductions, that a hydrogen *atom* is equivalent to a hydrogen *ion*, H^+, plus an electron, e^-. Thus, for the two hydrogen atoms removed in the oxidation of an alcohol, 2 H atoms = $2\ H^+ + 2\ e^-$. When NAD^+ is involved as the oxidant, both electrons accompany one H^+, in effect adding a hydride ion, $H:^-$, to NAD^+ to give NADH. The second hydrogen removed from the oxidized substrate enters the solution as H^+.

3.3 Triacylglycerol Catabolism: Fatty-Acid Oxidation

Fatty acids are catabolized in the mitochondria of cells by a repetitive four-step sequence of enzyme-catalyzed reactions called the ***β*-oxidation pathway**, shown in Figure 3.6. Each passage through the pathway results in the cleavage of an acetyl group from the carboxyl end of the fatty-acid chain, until the entire molecule is ultimately degraded. As each acetyl group is produced, it enters the citric acid cycle and is further degraded to CO_2, as we'll see in Section 4.4.

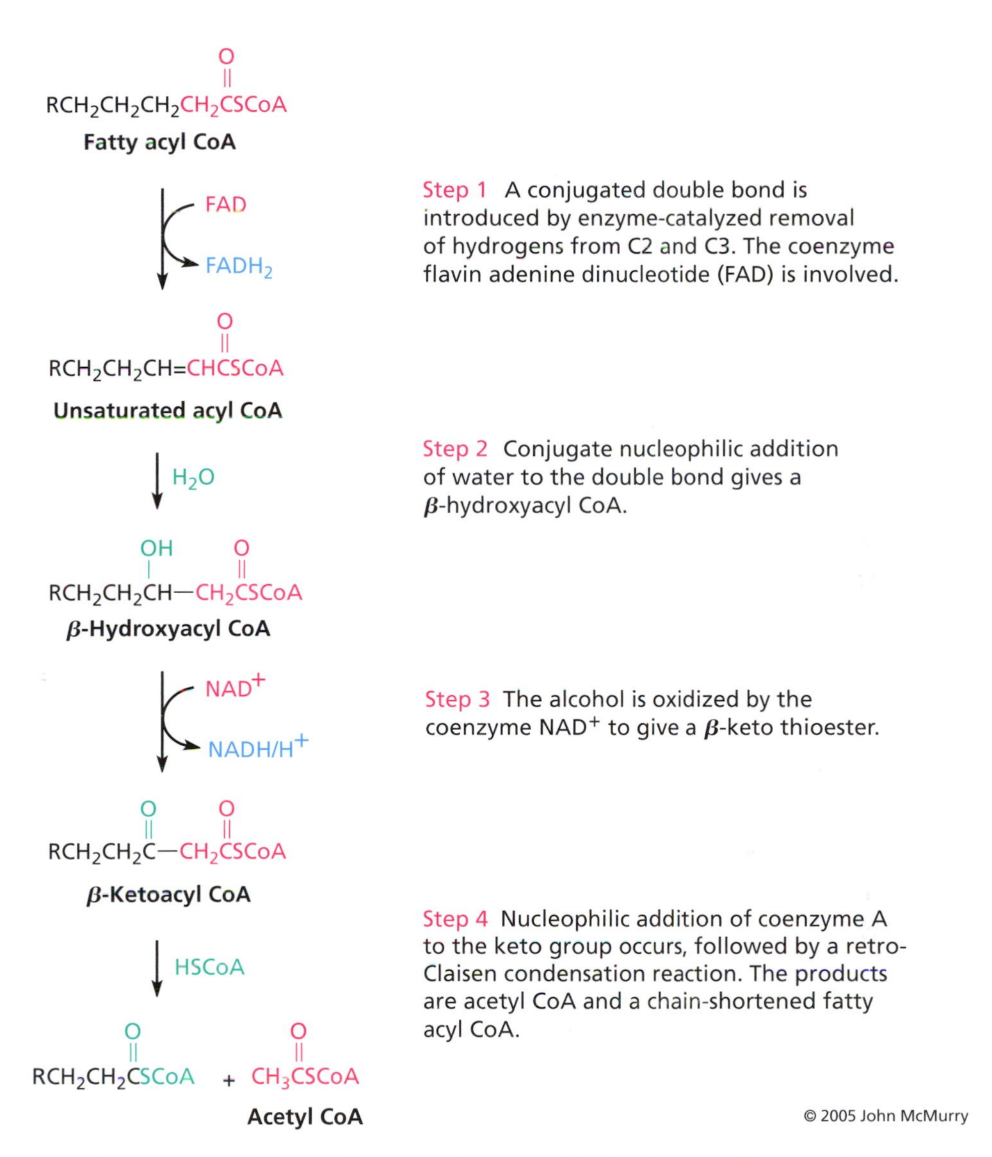

$RCH_2CH_2CH_2CH_2CSCoA$ (C=O)
Fatty acyl CoA
FAD
$FADH_2$
Step 1 A conjugated double bond is introduced by enzyme-catalyzed removal of hydrogens from C2 and C3. The coenzyme flavin adenine dinucleotide (FAD) is involved.
$RCH_2CH_2CH{=}CHCSCoA$ (C=O)
Unsaturated acyl CoA
H_2O
Step 2 Conjugate nucleophilic addition of water to the double bond gives a β-hydroxyacyl CoA.
OH
$RCH_2CH_2CH{-}CH_2CSCoA$ (C=O)
β-Hydroxyacyl CoA
NAD^+
$NADH/H^+$
Step 3 The alcohol is oxidized by the coenzyme NAD^+ to give a β-keto thioester.
$RCH_2CH_2C{-}CH_2CSCoA$ (two C=O)
β-Ketoacyl CoA
HSCoA
Step 4 Nucleophilic addition of coenzyme A to the keto group occurs, followed by a retro-Claisen condensation reaction. The products are acetyl CoA and a chain-shortened fatty acyl CoA.
RCH_2CH_2CSCoA + CH_3CSCoA
Acetyl CoA
© 2005 John McMurry

FIGURE 3.6 The four steps of the β-oxidation pathway, resulting in cleavage of an acetyl group from the end of the fatty-acid chain. The key chain-shortening step is a retro-Claisen condensation reaction of a β-keto thioester.

Step 1. Introduction of a double bond The β-oxidation pathway begins when two hydrogen atoms are removed from carbons 2 and 3 by one of a family of acyl-CoA dehydrogenases[6] to yield an α,β-unsaturated acyl CoA. This kind of oxidation—the introduction of a conjugated double bond into a carbonyl compound—occurs frequently in biochemical pathways and is usually catalyzed by the coenzyme *flavin adenine dinucleotide* (*FAD*). Reduced flavin ($FADH_2$) is the by-product.

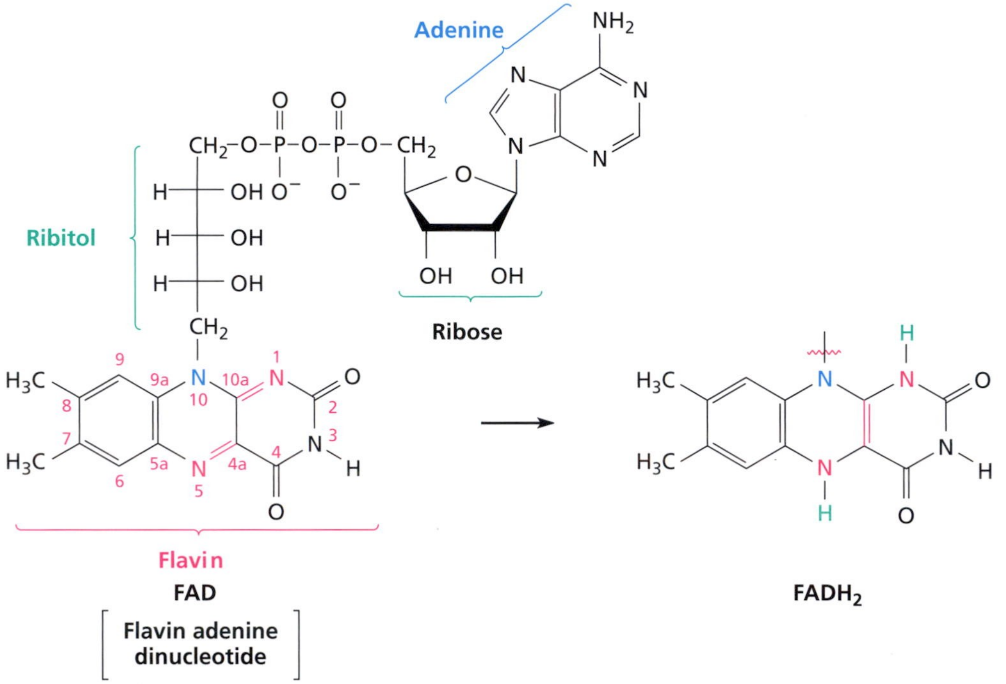

Adenine
Ribitol
Ribose
Flavin
FAD
[**Flavin adenine dinucleotide**]
$FADH_2$

An X-ray crystal structure of the active site in an enzyme–substrate complex of acyl-CoA dehydrogenase shows the flavin ring and the disposition of the two hydrogens that are removed (Figure 3.7). An α hydrogen of the acyl CoA is removed by Glu-376, and a β hydrogen is donated to the N5 nitrogen atom of FAD.

The mechanisms of FAD-catalyzed reactions are often difficult to establish because flavin coenzymes can operate by both two-electron (polar) and one-electron (radical) pathways. As a result, extensive studies of the family of acyl-CoA dehydrogenases have not yet provided a clear mechanistic picture of how

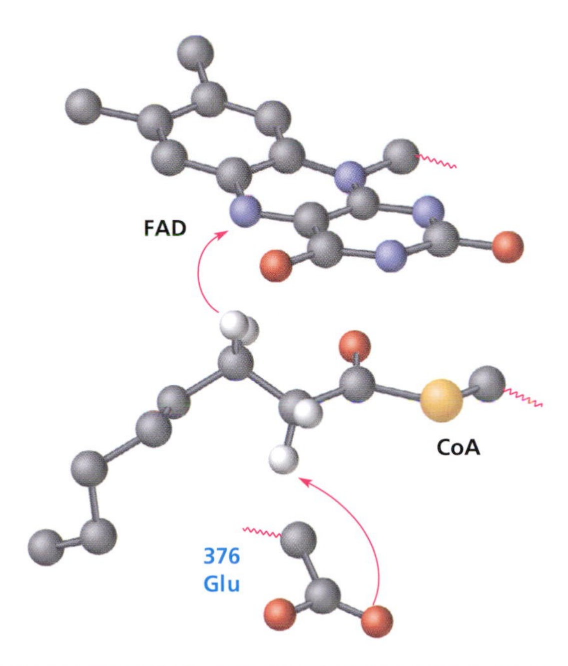

FIGURE 3.7 X-ray crystal structure of the active site in an enzyme–substrate complex of acyl-CoA dehydrogenase. Glutamate-376 deprotonates the α carbon, and a β hydrogen is donated to FAD.

these enzymes function. What is known[6, 7] is that: (1) The first step is abstraction of the *pro-R* hydrogen from the acidic α position of the acyl CoA to give a thioester enolate ion. Hydrogen bonding between the acyl carbonyl group and the ribitol hydroxyls of FAD greatly increases the acidity of the substrate, and a glutamate residue acts as the base. (2) The *pro-R* hydrogen at the β position is transferred to FAD; and (3) the α,β-unsaturated acyl CoA that results has a trans double bond.

Ribitol
pro-R
pro-R
trans

Thioester enolate

The details of the reaction are not clear, although several possibilities have been suggested. One suggestion is that reaction takes place by a conjugate hydride-transfer mechanism,[8] similar to what occurs during alcohol oxidations with NAD^+. Electrons on the enolate ion might expel a β hydride ion, which could add to the doubly bonded N5 nitrogen on FAD. Protonation of the intermediate at N1 would give the final product.

Hydride-transfer mechanism

Alternatively, an electron-transfer mechanism has been suggested. Donation of one electron from the thioester enolate ion to the unsaturated π-electron system of FAD would yield a substrate radical and a resonance-stabilized flavin anion radical. Protonation of the anion radical would give a *semiquinone*, and abstraction of the β hydrogen atom from the acyl-CoA radical by the flavin semiquinone would then give the product (Figure 3.8).

Step 2. Conjugate addition of water The α,β-unsaturated acyl CoA produced in step 1 reacts with water by a conjugate addition pathway (Section 1.5) to yield a β-hydroxyacyl CoA in a process catalyzed by enoyl-CoA hydratase.[9, 10] Water as nucleophile adds to the β carbon of the double bond, yielding a thioester enolate ion intermediate that is then protonated.

In the enzyme, the nucleophilic water molecule is deprotonated by Glu-144 to make it a more reactive donor, and addition occurs from the *si* face of the molecule. At the same time, the carbonyl oxygen atom is hydrogen-bonded to the amide NH of alanine and glycine residues to increase its reactivity as an acceptor. Protonation of the enolate ion is carried out by the acid form of Glu-164,

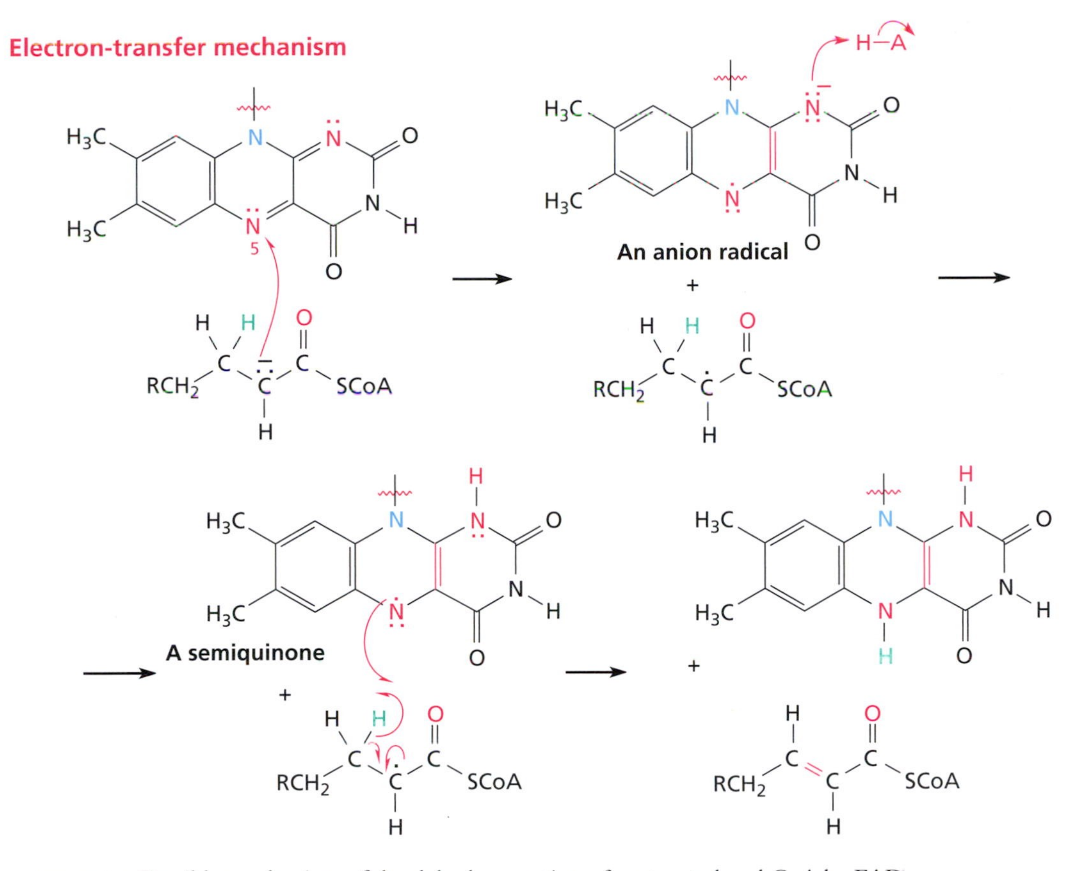

FIGURE 3.8 Possible mechanism of the dehydrogenation of a saturated acyl CoA by FAD.

which adds a *pro-R* hydrogen to the α position. Thus, both —OH and —H add from the same face of the double bond—a so-called *syn* addition.

144 Glu
98 Ala
141 Gly
164 Glu
CoAS
S
pro-R

(3*S*)-Hydroxyacyl CoA

Step 3. Alcohol oxidation The β-hydroxyacyl CoA from step 2 is oxidized to a β-ketoacyl CoA in a reaction catalyzed by one of a family of L-3-hydroxyacyl-CoA dehydrogenases,[11] which differ in substrate specificity according to the chain length of the acyl group. As in the oxidation of *sn*-glycerol 3-phosphate to dihydroxyacetone phosphate (Section 3.2, Figure 3.5), this alcohol oxidation requires NAD^+ as a coenzyme and yields reduced $NADH/H^+$ as by-product. Deprotonation of the hydroxyl group is carried out by the His-158 residue at the active site, and addition of hydride ion occurs stereospecifically on the *si* face of the nicotinamide ring. Note that this stereochemistry differs from that observed during oxidation of glycerol 3-phosphate.

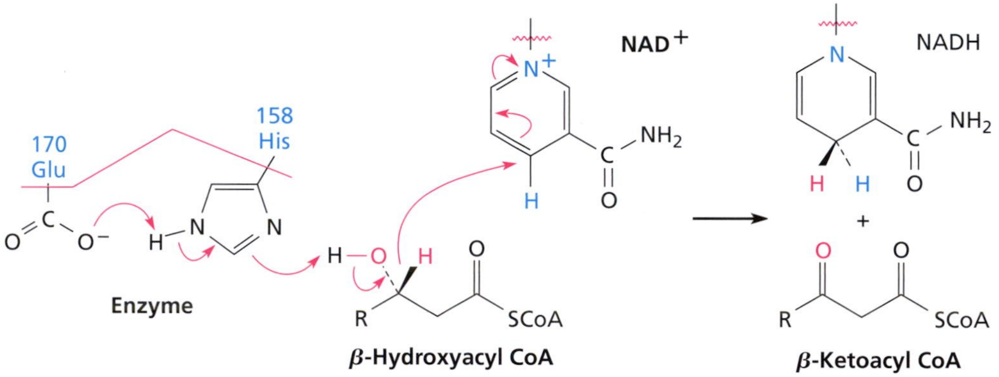

An X-ray crystal structure of the enzyme–substrate complex from hydroxyacyl CoA dehydrogenase (Figure 3.9) shows how His-158 deprotonates the hydroxyl group of the substrate and the adjacent hydride ion on the β carbon is transferred to NAD^+.

Step 4. Chain cleavage Acetyl CoA is split off from the acyl chain in the final step of β-oxidation, leaving an acyl CoA that is two carbon atoms shorter than the original. The reaction is catalyzed by β-ketoacyl-CoA thiolase[12, 13] and is mechanistically the reverse of a Claisen condensation reaction (Section 1.7, Figure 1.14). Remember that in the *forward* direction, a Claisen condensation joins two esters together to form a β-keto ester product. In the *reverse* direction, a

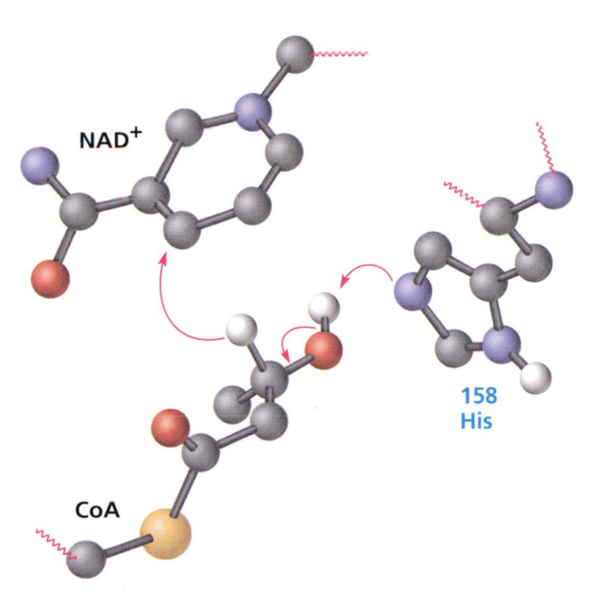

FIGURE 3.9 X-ray crystal structure of the enzyme–substrate complex from hydroxyacyl CoA dehydrogenase. His-158 deprotonates the hydroxyl group, and the adjacent hydride ion is transferred to NAD^+.

retro-Claisen reaction splits a β-keto ester (or β-keto thioester) apart to form two esters (or two thioesters).

$$\underset{\text{H H}}{\diagdown\text{C}}\text{–C(=O)–OR (SR)} + \text{H–}\underset{\text{H}}{\text{C}}\text{–C(=O)–OR (SR)} \underset{\text{Retro-Claisen}}{\overset{\text{Claisen}}{\rightleftharpoons}} \underset{\text{H H}}{\diagdown\text{C}}\text{–C(=O)–}\underset{\text{H}}{\text{C}}\text{–C(=O)–OR (SR)} + \text{H–OR (H–SR)}$$

As shown in Figure 3.10, the chain-cleavage reaction is thought to occur by initial nucleophilic addition of the Cys-125 —SH group on the enzyme to the keto group of the β-ketoacyl CoA to yield an alkoxide-ion intermediate. The keto carbonyl group is activated for this addition by hydrogen bonding to the NH of Gly-405, and the nucleophilic cysteine —SH is activated through deprotonation by His-375. Cleavage of the C2–C3 bond then follows, with expulsion of an acetyl CoA enolate ion in a retro-Claisen reaction. Note that the

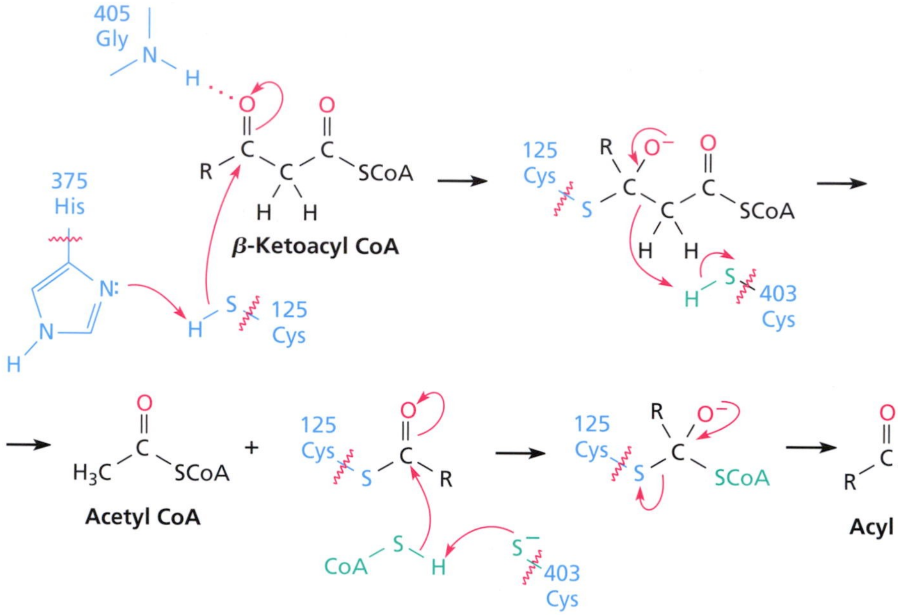

FIGURE 3.10 Mechanism of the retro-Claisen reaction of a β-ketoacyl CoA to yield acetyl CoA and a chain-shortened acyl CoA.

chain-shortened acyl group is now bound to the enzyme by a thioester bond. Protonation of the enolate ion by Cys-403 gives acetyl CoA, and the bound acyl group undergoes nucleophilic acyl substitution by reaction with a molecule of coenzyme A. The chain-shortened acyl CoA that results then enters another round of the β-oxidation pathway for further degradation.

Look at the catabolism of myristic acid shown in Figure 3.11 to see the overall results of the β-oxidation pathway. The first passage along the pathway converts the 14-carbon myristyl CoA into the 12-carbon lauryl CoA plus acetyl CoA; the second passage converts lauryl CoA into the 10-carbon capryl CoA plus acetyl CoA; the third passage converts capryl CoA into the 8-carbon caprylyl CoA; and so on. Note that the final passage produces *two* molecules of acetyl CoA because the precursor has four carbons. The number of passages is always one less than the number of acetyl CoA molecules produced.

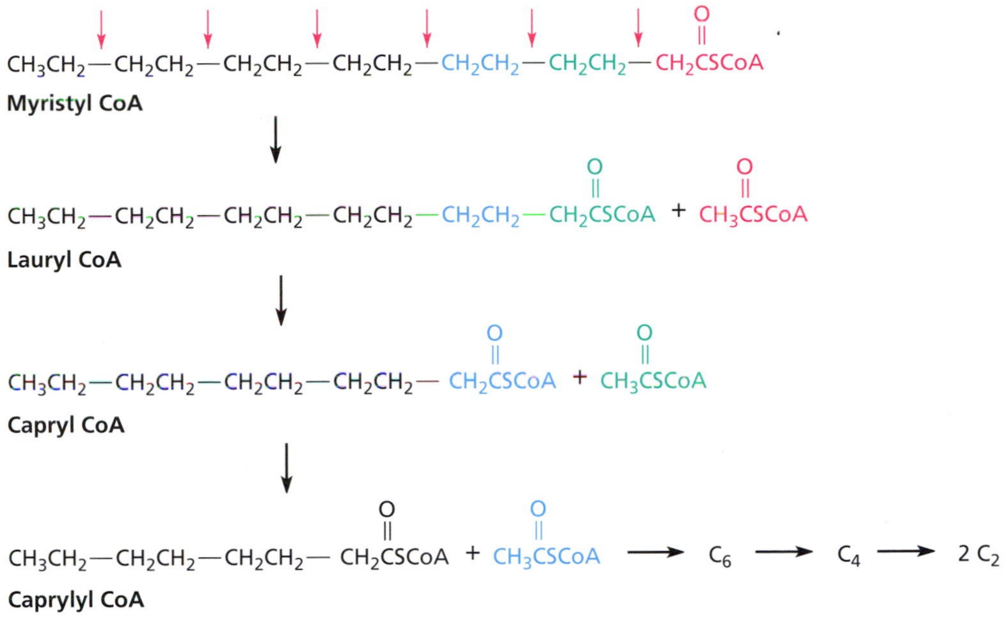

$$CH_3CH_2{-}CH_2CH_2{-}CH_2CH_2{-}CH_2CH_2{-}CH_2CH_2{-}CH_2CH_2{-}CH_2\overset{O}{\overset{\|}{C}}SCoA$$

Myristyl CoA

↓

$$CH_3CH_2{-}CH_2CH_2{-}CH_2CH_2{-}CH_2CH_2{-}CH_2CH_2{-}CH_2\overset{O}{\overset{\|}{C}}SCoA + CH_3\overset{O}{\overset{\|}{C}}SCoA$$

Lauryl CoA

↓

$$CH_3CH_2{-}CH_2CH_2{-}CH_2CH_2{-}CH_2CH_2{-}CH_2\overset{O}{\overset{\|}{C}}SCoA + CH_3\overset{O}{\overset{\|}{C}}SCoA$$

Capryl CoA

↓

$$CH_3CH_2{-}CH_2CH_2{-}CH_2CH_2{-}CH_2\overset{O}{\overset{\|}{C}}SCoA + CH_3\overset{O}{\overset{\|}{C}}SCoA \longrightarrow C_6 \longrightarrow C_4 \longrightarrow 2\ C_2$$

Caprylyl CoA

FIGURE 3.11 Catabolism of the 14-carbon myristic acid by the β-oxidation pathway yields seven molecules of acetyl CoA after six passages. The final passage yields two molecules of acetyl CoA because a 4-carbon acid is cleaved to two 2-carbon fragments.

Most fatty acids have an even number of carbon atoms, so that none are left over after β-oxidation. Those fatty acids with an odd number of carbon atoms yield the three-carbon propionyl CoA in the final β-oxidation. Propionyl CoA is then converted to succinate by a complex, multistep pathway whose mechanistic details are not fully understood. (The same conversion occurs during threonine metabolism, Section 5.3.)

3.4 Fatty-Acid Biosynthesis

One of the more striking features of the common fatty acids is that all have an even number of carbon atoms (Table 2.2). This even number results because all fatty acids are derived biosynthetically from the two-carbon precursor acetyl CoA by sequential addition of two-carbon units to a growing chain. Acetyl CoA, in turn, arises primarily from the metabolic breakdown of carbohydrates in the glycolysis pathway, as we'll see in Chapter 4. Thus, dietary carbohydrates are effectively turned into fats for storage.

The anabolic scheme by which organisms synthesize fatty acids *from* acetyl CoA is closely related to the β-oxidation pathway by which organisms convert fatty acids *to* acetyl CoA. The schemes are not, however, the exact reverse of one another; they *must* differ in at least some details for both to be energetically favorable and for independent regulatory mechanisms to operate. One difference, for example, is that fatty-acid oxidation takes place in cellular mitochondria, while fatty-acid synthesis occurs in the cytosol. Other differences include the identity of the acyl-group carrier, the stereochemistry of the β-hydroxyacyl reaction intermediate, and the identity of the redox coenzyme. FAD is used for introduction of a double bond in β-oxidation, while NADPH is used for reduction of the double bond in fatty-acid biosynthesis.

In bacteria, each step in fatty-acid synthesis is catalyzed by discrete enzymes that catalyze specific steps. In vertebrates, however, fatty acid-synthesis is catalyzed by a large, multienzyme complex that contains two identical subunits of 2505 amino acids each and is able to catalyze all steps in the pathway. An overview of fatty-acid biosynthesis is shown in Figure 3.12.

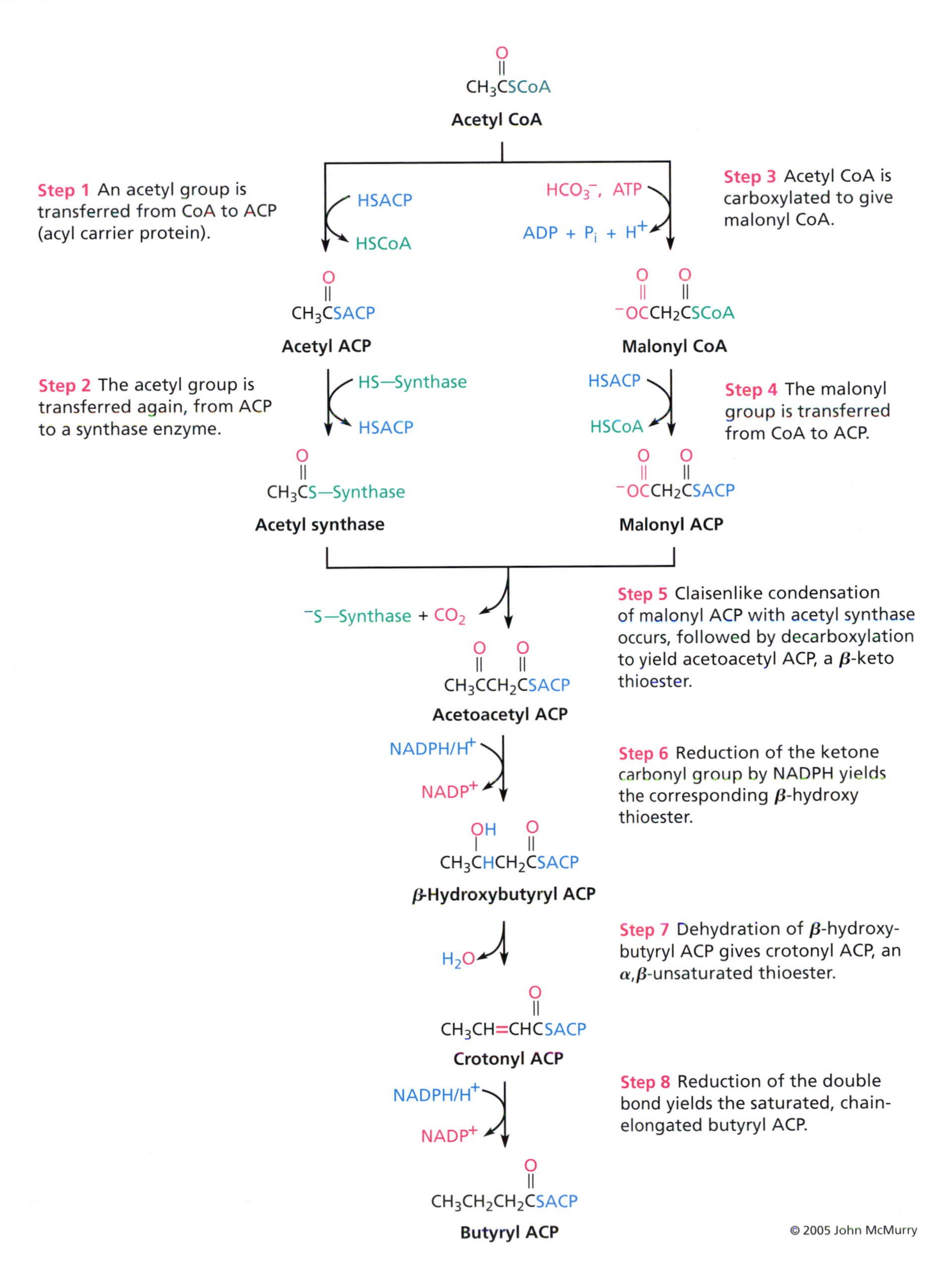

FIGURE 3.12 The biological pathway for fatty-acid synthesis from the two-carbon precursor, acetyl CoA. Individual steps are explained in more detail in the text.

Steps 1–2. Acyl transfers The starting material for fatty-acid synthesis is the thioester acetyl CoA, the ultimate product of carbohydrate breakdown in the glycolysis pathway. Fatty-acid synthesis begins with several "priming reactions," which transport and convert acetyl CoA into more reactive species. The first such reaction is a typical nucleophilic acyl substitution reaction that converts acetyl CoA into acetyl ACP (acyl carrier protein). The reaction is catalyzed by ACP transacylase.

In bacteria, ACP is a small protein of 77 residues that transports an acyl group from enzyme to enzyme. In vertebrates, however, ACP appears to be a long arm on a multienzyme synthase complex, whose apparent function is to shepherd an acyl group from site to site within the complex. As in acetyl CoA, the acyl group in acetyl ACP is linked by a thioester bond to the sulfur atom of a phosphopantetheine group. The phosphopantetheine is in turn linked to ACP through the side-chain —OH group of a serine residue.

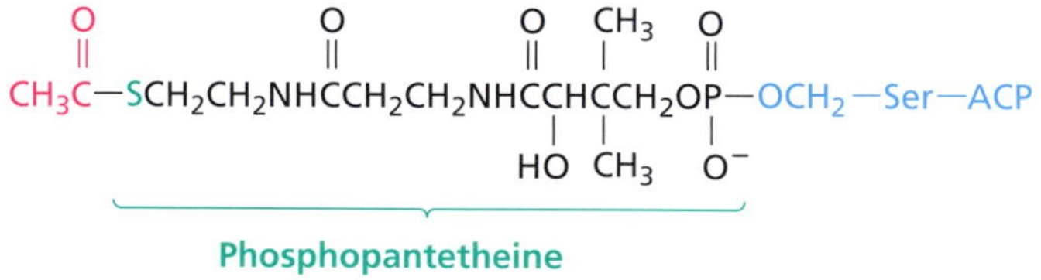

$$CH_3\overset{O}{\overset{\|}{C}}-SCH_2CH_2NH\overset{O}{\overset{\|}{C}}CH_2CH_2NH\overset{O}{\overset{\|}{C}}\underset{HO}{\underset{|}{C}}H\overset{CH_3}{\overset{|}{\underset{CH_3}{\underset{|}{C}}}}CH_2O\overset{O}{\overset{\|}{\underset{O^-}{\underset{|}{P}}}}-OCH_2-Ser-ACP$$

Phosphopantetheine

Acetyl ACP

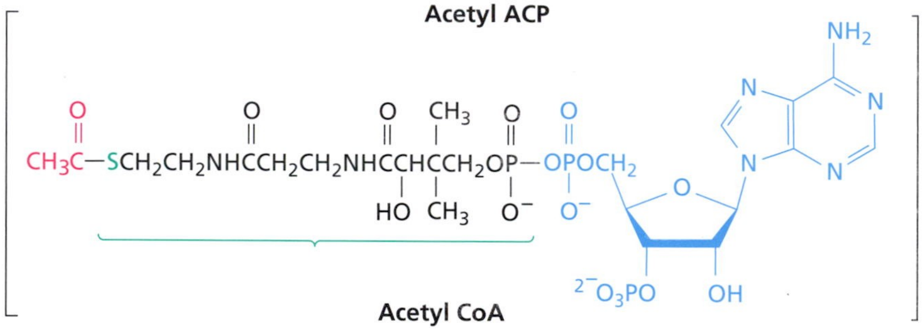

$$CH_3\overset{O}{\overset{\|}{C}}-SCH_2CH_2NH\overset{O}{\overset{\|}{C}}CH_2CH_2NH\overset{O}{\overset{\|}{C}}\underset{HO}{\underset{|}{C}}H\overset{CH_3}{\overset{|}{\underset{CH_3}{\underset{|}{C}}}}CH_2O\overset{O}{\overset{\|}{\underset{O^-}{\underset{|}{P}}}}-O\overset{O}{\overset{\|}{\underset{O^-}{\underset{|}{P}}}}OCH_2\text{-(adenosine 3'-}{}^{2-}O_3PO\text{)}$$

Acetyl CoA

Step 2, another priming reaction, involves a further exchange of thioester linkages by another nucleophilic acyl substitution, and results in covalent bonding of the acetyl group to a cysteine residue in the synthase complex that will catalyze the upcoming condensation step.

Steps 3-4. Carboxylation and acyl transfer The third step is a "loading reaction" in which acetyl CoA is carboxylated by reaction with HCO_3^- and ATP to yield malonyl CoA plus ADP. This step requires the coenzyme biotin, which is bonded to the terminal —NH_2 group of a lysine residue in the acetyl-CoA carboxylase. Biotin first reacts with bicarbonate ion to give *N*-carboxybiotin, which transfers CO_2 to acetyl CoA. Thus, biotin acts as a carrier of CO_2, covalently binding it in one step and releasing it in another.

ATP, HCO_3^- → ADP, P_i

Biotin → ***N*-Carboxybiotin**

Acetyl CoA + ***N*-Carboxybiotin** → **Malonyl CoA** + **Biotin**

The mechanism of this and related biotin-dependent carboxylation reactions[14] has been the subject of numerous studies, but no definitive conclusions have yet been reached. One likely possibility, shown in Figure 3.13, involves a preliminary reaction of bicarbonate ion with ATP and Mg^{2+} to transfer a phosphate group and generate a reactive carboxyphosphate intermediate. Decomposition of carboxyphosphate at the active site then yields phosphate ion (P_i) plus noncovalently bound CO_2 as the active carboxylating species. Simultaneously, an N—H group on biotin undergoes deprotonation by a thiolate group on the enzyme to yield a nucleophilic anion, which immediately reacts with the CO_2 by nucleophilic addition to a C═O bond in an aldol-like condensation reaction. Note in Figure 3.13 that the initial phosphorylation reaction of bicarbonate ion with ATP is shown in an abbreviated form without drawing the pentacoordinate phosphorus intermediate as in Figure 3.4. We'll continue to do this from now on to save space.

Once *N*-carboxybiotin has been formed, its reaction with acetyl CoA to give malonyl CoA may again involve noncovalently bound CO_2 as the reactive species. One proposal[15] is that loss of CO_2 is favored by hydrogen-bond formation between the *N*-carboxybiotin carbonyl group and a nearby acidic site in the

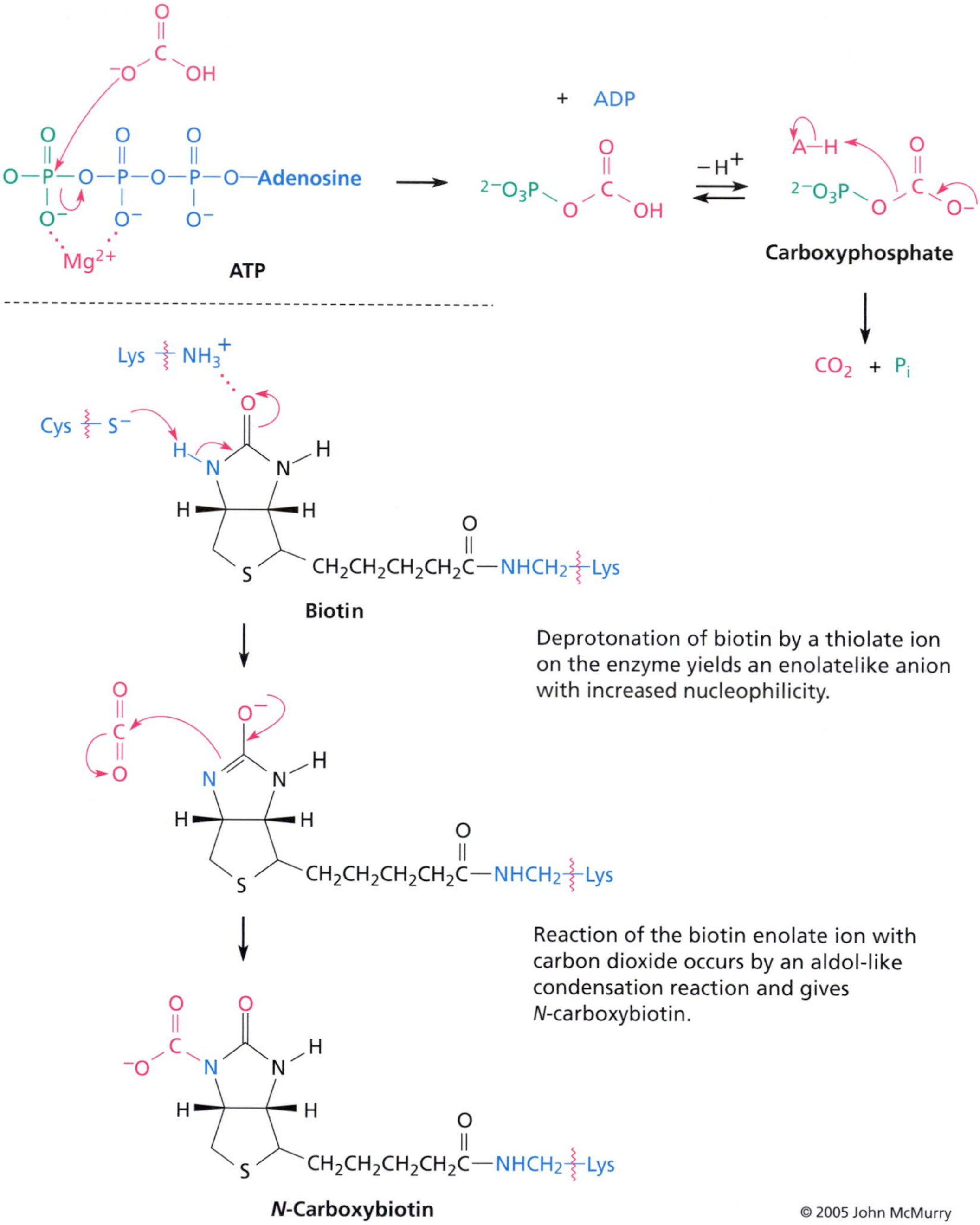

ATP + ADP

Carboxyphosphate

CO_2 + P_i

Biotin

Deprotonation of biotin by a thiolate ion on the enzyme yields an enolatelike anion with increased nucleophilicity.

Reaction of the biotin enolate ion with carbon dioxide occurs by an aldol-like condensation reaction and gives *N*-carboxybiotin.

***N*-Carboxybiotin**

FIGURE 3.13 A possible mechanism for the formation of *N*-carboxybiotin. Reaction of bicarbonate ion with ATP gives carboxyphosphate, which then gives carbon dioxide as the active carboxylating agent. The initial phosphorylation occurs through a pentacoordinate phosphorus intermediate but is shown here in an abbreviated form.

enzyme. Deprotonation of acetyl CoA by the resultant biotin anion gives a thioester enolate ion, which can immediately react with CO_2 (Figure 3.14).

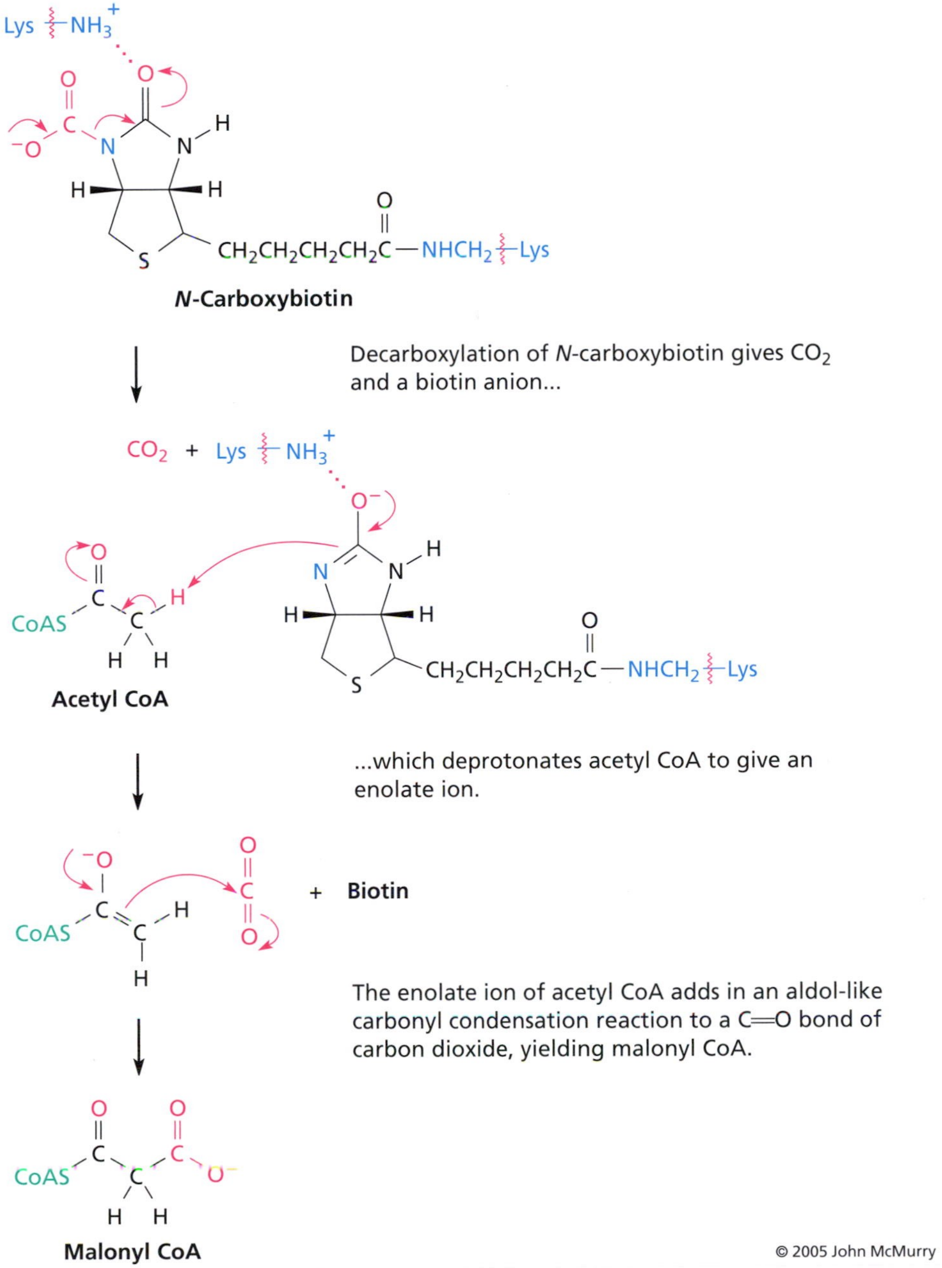

***N*-Carboxybiotin**

Decarboxylation of *N*-carboxybiotin gives CO_2 and a biotin anion...

CO_2 + Lys–NH_3^+

Acetyl CoA

...which deprotonates acetyl CoA to give an enolate ion.

+ **Biotin**

The enolate ion of acetyl CoA adds in an aldol-like carbonyl condensation reaction to a C═O bond of carbon dioxide, yielding malonyl CoA.

Malonyl CoA

FIGURE 3.14 A possible mechanism of the carboxylation of acetyl CoA by reaction with *N*-carboxybiotin. CO_2 is thought to be the reactive carboxylating species.

Following the formation of malonyl CoA, another nucleophilic acyl substitution reaction converts it into malonyl ACP, thus binding the malonyl group to an ACP arm of the multienzyme synthase complex. At this point, both acetyl and malonyl groups are bound to the enzyme, and the stage is set for their condensation.

Step 5. Condensation The key carbon–carbon bond-forming reaction that builds the fatty-acid chain occurs in step 5. This step is simply a Claisen condensation (Section 1.7) between acetyl synthase as the electrophilic acceptor and malonyl ACP as the nucleophilic donor. The mechanism of the condensation is thought to involve decarboxylation of malonyl ACP to give an enolate ion and immediate addition of the enolate ion to the carbonyl group of acetyl synthase.[16] Breakdown of the tetrahedral intermediate gives the four-carbon condensation product acetoacetyl ACP and frees the synthase binding site for attachment of the chain-elongated acyl group at the end of the sequence.

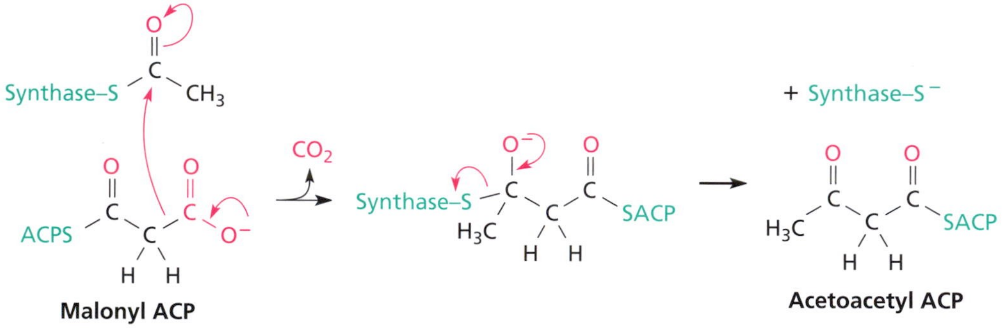

Synthase–S–C(=O)–CH_3 + Malonyl ACP → (CO_2) → + Synthase–S^- + Acetoacetyl ACP

Note that the decarboxylation of malonyl ACP is not a general reaction of carboxylic acids. Most carboxylic acids are stable and do not lose CO_2 under any but the most extreme laboratory conditions. Malonic acid is special, however, as are all carboxylic acids that have an additional carbonyl group two atoms away at their β carbon. Such compounds undergo an unusually easy decarboxylation reaction because the β carbonyl group can act as an electron acceptor in what is essentially a retro-aldol reaction. Without this acceptor carbonyl group at the β carbon to stabilize the developing negative charge, no loss of CO_2 occurs.

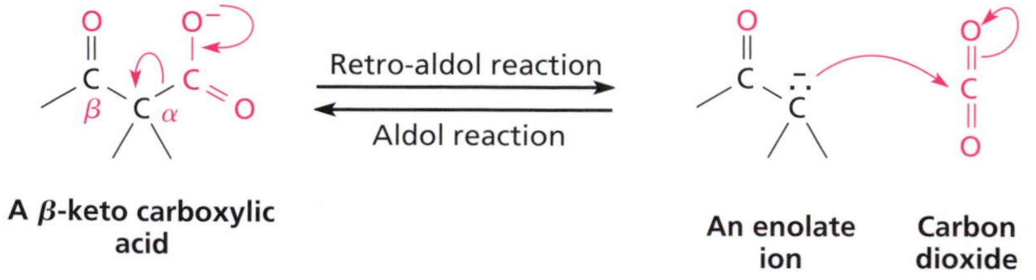

A β-keto carboxylic acid ⇌ (Retro-aldol reaction / Aldol reaction) An enolate ion + Carbon dioxide

One further question about the condensation step: Why does the route for fatty-acid synthesis proceed through malonyl ACP, first adding CO_2 to an acetyl group in step 3 and then immediately removing it in step 5? Why not simply effect a Claisen condensation between two molecules of acetyl ACP? The reason is that it's necessary to provide a strong thermodynamic driving force for the condensation step. Direct Claisen condensation reactions are thermodynamically unfavorable, but the ATP-driven carboxylation of biotin in step 3 is exergonic, as is the subsequent decarboxylation and condensation, which yields a stable CO_2 molecule. Thus, the coupling of the condensation step with ATP hydrolysis makes the overall scheme possible.

Step 6. Reduction The ketone carbonyl group in acetoacetyl ACP is next reduced to the alcohol β-hydroxybutyryl ACP by β-keto thioester reductase and NADPH (nicotinamide adenine dinucleotide phosphate), a reducing coenzyme closely related to NADH. The *pro-R* hydrogen of NADPH is transferred to the *si* face of the ketoacyl group, giving *R* stereochemistry at the newly formed chirality center in the β-hydroxy thioester product (Figure 3.15).

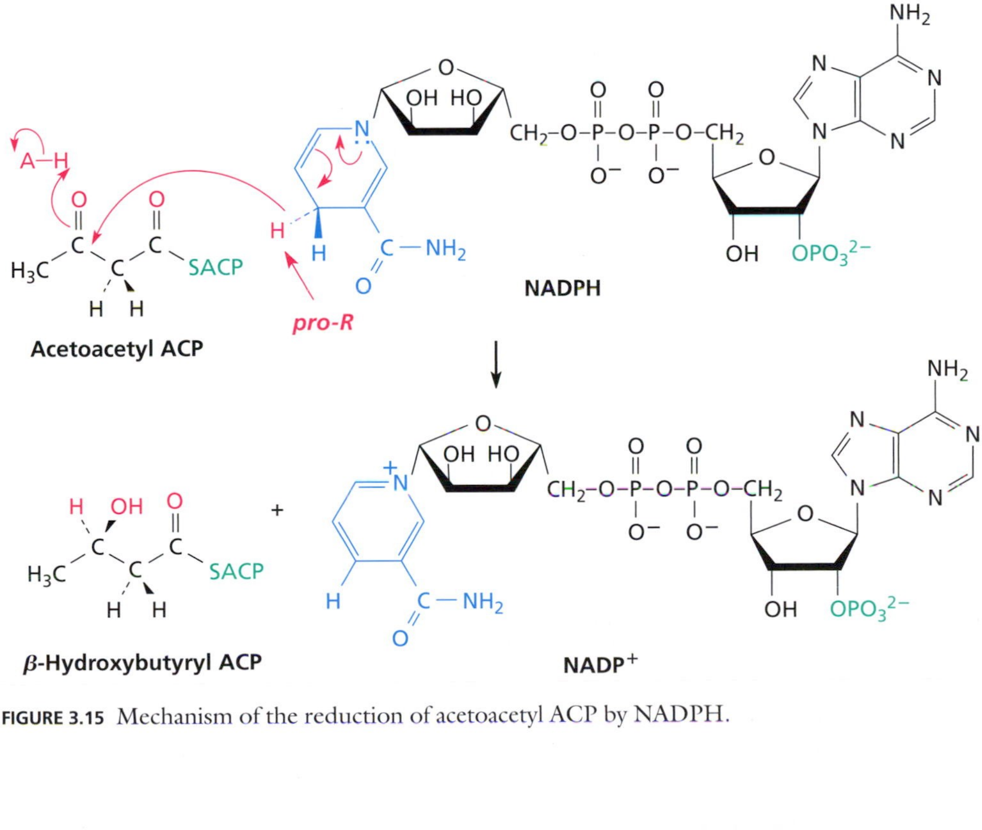

FIGURE 3.15 Mechanism of the reduction of acetoacetyl ACP by NADPH.

Steps 7–8. Dehydration and reduction Subsequent dehydration of the β-hydroxy thioester in step 7 yields *trans*-crotonyl ACP, and reduction of the carbon–carbon double bond of crotonyl ACP by NADPH in step 8 gives butyryl ACP. The dehydration step occurs with stereospecific removal of the *pro-R* hydrogen from C2 by an elimination mechanism that is thought to proceed through an intermediate thioester enol. A basic residue in the enzyme (probably histidine) acts first as a base for deprotonation and then as a conjugate acid for protonation of the —OH leaving group.

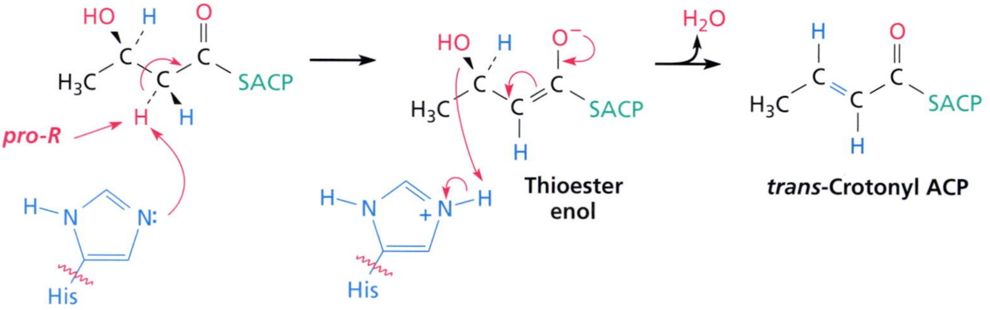

The double-bond reduction in step 8 occurs by conjugate addition of a hydride ion from NADPH to the β carbon of *trans*-crotonyl ACP. In vertebrates, the *pro-R* hydrogen of NADPH adds to C3 of the crotonyl group from the *re* face, and protonation on C2 occurs on the *si* face in an overall *syn* addition. Other organisms carry out similar chemistry, but with different stereochemistry.

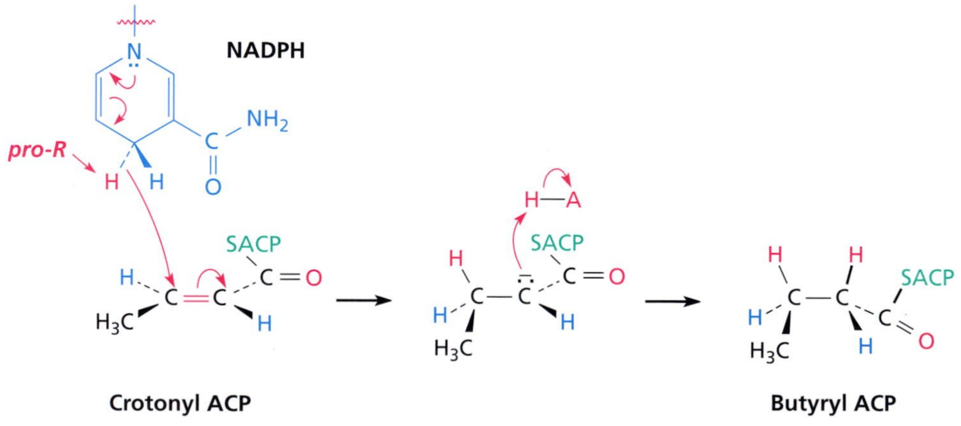

The net effect of the eight steps in the fatty-acid synthesis pathway is to take two 2-carbon groups and combine them into one 4-carbon saturated acyl group. The butyryl ACP thus produced is then transferred to the cysteine —SH at the acyl binding site of the synthase enzyme by a nucleophilic acyl substitution reaction, where it undergoes condensation with another malonyl ACP to yield a 6-carbon unit. Still further repetitions of the pathway add two more carbon atoms to the chain each time until the 16-carbon palmitoyl ACP is reached.

$$CH_3CH_2CH_2\overset{O}{\overset{\|}{C}}SACP \xrightarrow[\text{HSACP}]{\text{HS—Synthase}} CH_3CH_2CH_2\overset{O}{\overset{\|}{C}}S\text{—Synthase} \longrightarrow CH_3CH_2CH_2CH_2CH_2\overset{O}{\overset{\|}{C}}SACP$$

$$\rightleftharpoons CH_3CH_2CH_2CH_2CH_2CH_2CH_2CH_2CH_2CH_2CH_2CH_2CH_2CH_2CH_2\overset{O}{\overset{\|}{C}}SACP$$

Palmitoyl ACP

Further chain elongation of palmitic acid occurs by reactions similar to those just described, but CoA rather than ACP is the carrier group, and separate enzymes are needed for each step rather than a multienzyme synthase complex.

3.5 Terpenoid Biosynthesis

As noted in Section 2.2, *terpenoids* are a large group of substances found primarily in plants, bacteria, archaea, and fungi, as well as in animals. Some examples were shown in Figure 2.3. Their exact biological functions within specific organisms are not generally known, but many are thought to play some sort of defensive role to protect an organism from predation.

More than 22,000 terpenoids with a vast diversity of structure are known. Although they appear structurally unrelated, all contain a multiple of 5 carbon atoms. Monoterpenes contain 10 carbons, sesquiterpenes contain 15 carbons, diterpenes contain 20 carbons, sesterterpenes contain 25 carbons, triterpenes contain 30 carbons, and so on. Terpenoids have multiples of 5 carbon atoms because they arise biosynthetically from the 5-carbon precursor isopentenyl diphosphate, formerly called isopentenyl pyrophosphate and therefore abbreviated IPP.

Isopentenyl diphosphate is biosynthesized by two different pathways depending on the organism and the structure of the final terpenoid. In animals and higher plants, sesquiterpenes and triterpenes arise primarily from mevalonate, while monoterpenes, diterpenes, and tetraterpenes are biosynthesized from 1-deoxy-D-xylulose 5-phosphate. In bacteria, both pathways are used.[17]

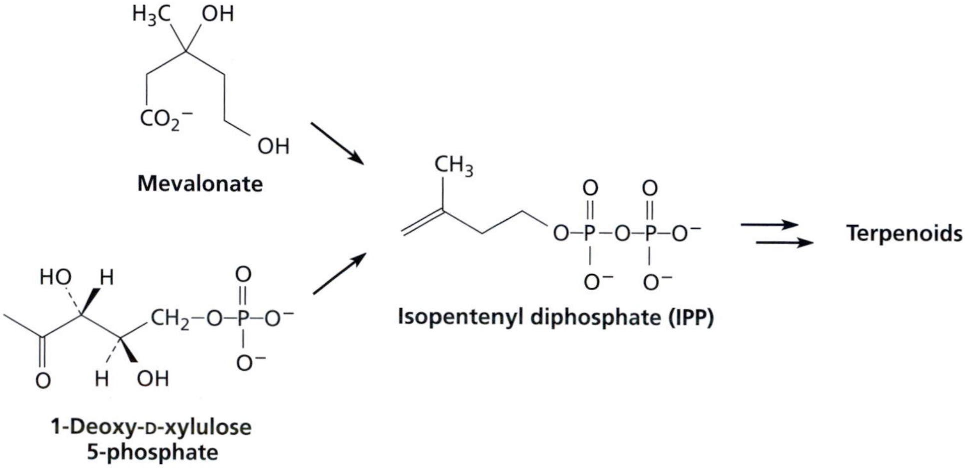

H_3C OH

CO_2^- OH

Mevalonate

HO H

CH_2–O–P–O$^-$

O H OH O$^-$

1-Deoxy-D-xylulose 5-phosphate

CH_3

O O

O–P–O–P–O$^-$

O$^-$ O$^-$

Isopentenyl diphosphate (IPP)

Terpenoids

The Mevalonate Pathway to Isopentenyl Diphosphate

The mevalonate pathway for terpenoid biosynthesis,[18] summarized in Figure 3.16, begins with the conversion of acetate to acetyl CoA, followed by Claisen condensation to yield acetoacetyl CoA. A second carbonyl condensation reaction with a third molecule of acetyl CoA, this one an aldol-like process, then yields the six-carbon compound 3-hydroxy-3-methylglutaryl CoA, which is reduced to give mevalonate. Phosphorylation, followed by simultaneous loss of CO_2 and phosphate ion, completes the process.

Acetyl CoA

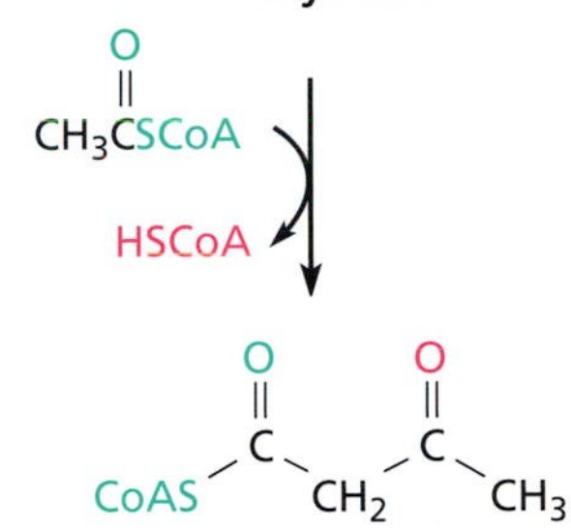

Step 1 Claisen condensation of two molecules of acetyl CoA gives acetoacetyl CoA.

Acetoacetyl CoA

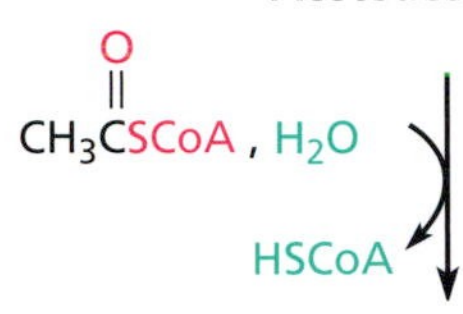

Step 2 Aldol-like condensation of acetoacetyl CoA with a third molecule of acetyl CoA, followed by hydrolysis, gives (3*S*)-3-hydroxy-3-methylglutaryl CoA.

(3*S*)-3-Hydroxy-3-methylglutaryl CoA

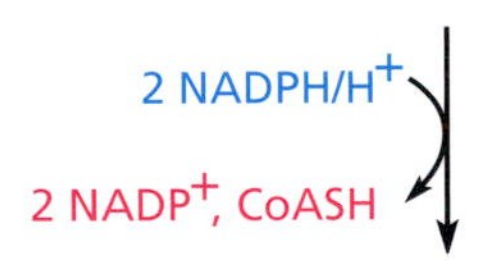

Step 3 Reduction of the thioester group by two equivalents of NADPH gives (*R*)-mevalonate, a dihydroxy acid.

(*R*)-Mevalonate

3 ATP

3 ADP, P_i, CO_2

Step 4 Phosphorylation of the tertiary hydroxyl and diphosphorylation of the primary hydroxyl, followed by decarboxylation and simultaneous expulsion of phosphate, gives isopentenyl diphosphate, the precursor of terpenoids.

Isopentenyl diphosphate

FIGURE 3.16 The mevalonate pathway for the biosynthesis of isopentenyl diphosphate from three molecules of acetate.

Step 1. Claisen condensation The first step in mevalonate biosynthesis is a Claisen condensation to yield acetoacetyl CoA. The reaction is catalyzed by acetoacetyl-CoA acetyltransferase, the same kind of enzyme that catalyzes (in the opposite direction) the retro-Claisen reaction in the final step of β-oxidation (Figure 3.6). Note that the condensation reaction occurs directly and does not involve an intermediate carboxylation to give malonyl CoA, as occurs in fatty-acid synthesis. As shown in Figure 3.17, an acetyl group is first bound to the enzyme by a nucleophilic acyl substitution reaction with a cysteine —SH group. Formation of an enolate ion from a second molecule of acetyl CoA, followed by Claisen condensation, yields the product.

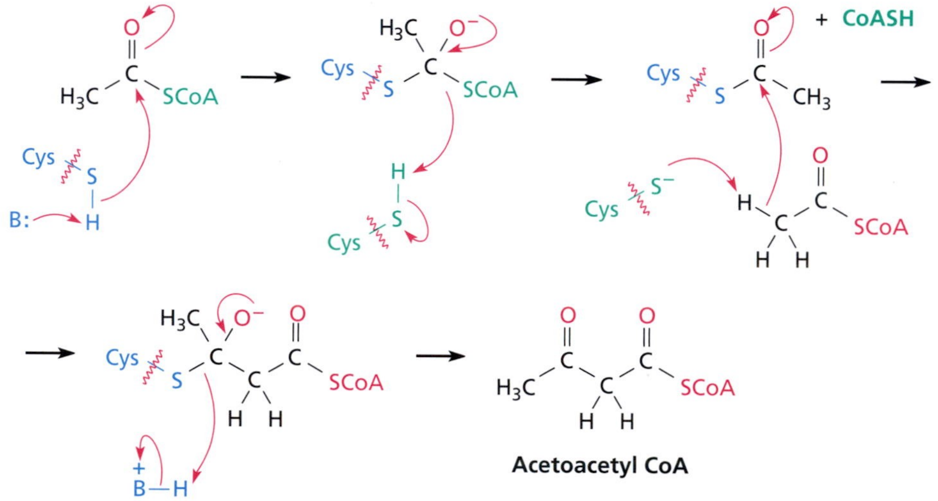

FIGURE 3.17 Mechanism of the Claisen condensation of acetyl CoA to give acetoacetyl CoA. The reaction is catalyzed by the same kind of thiolase enzyme involved in β-oxidation of fatty acids.

Step 2. Aldol condensation Acetoacetyl CoA next undergoes an aldol-like addition of an acetyl CoA enolate ion in a reaction catalyzed by 3-hydroxy-3-methylglutaryl-CoA synthase. The reaction occurs by initial binding of the substrate to a cysteine —SH group in the synthase, followed by enolate-ion addition and subsequent hydrolysis (Figure 3.18). The aldol addition step occurs on the *re* face of the ketone carbonyl group and gives (3*S*)-3-hydroxy-3-methylglutaryl CoA.

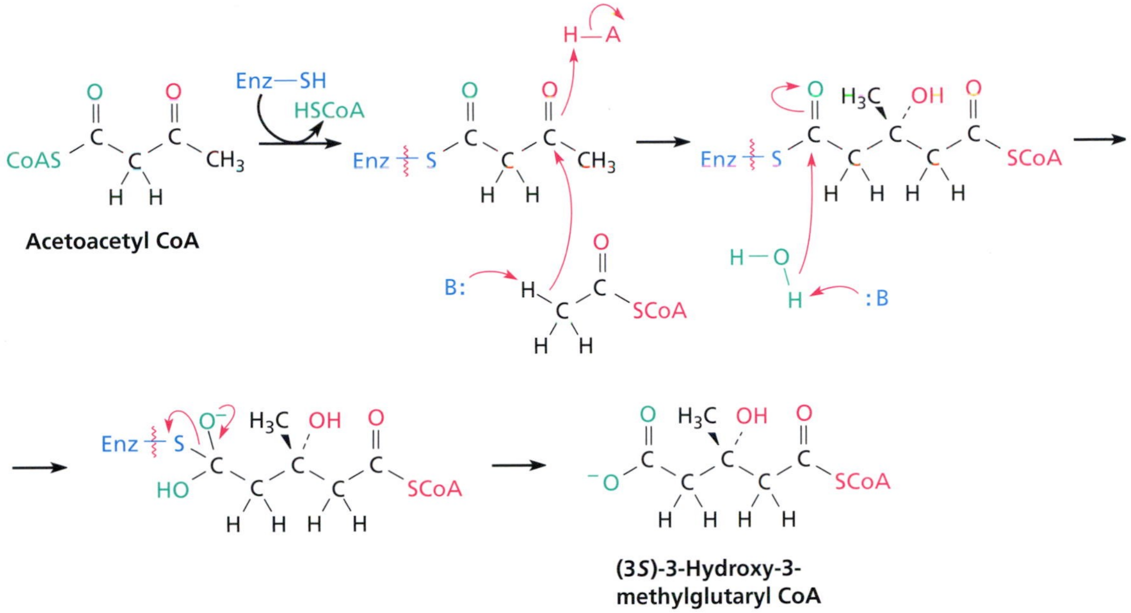

Acetoacetyl CoA

(3*S*)-3-Hydroxy-3-methylglutaryl CoA

FIGURE 3.18 Mechanism of the aldol-like addition of acetyl CoA to acetoacetyl CoA. The reaction occurs from the *re* face and gives (3*S*)-3-hydroxy-3-methylglutaryl CoA.

Step 3. Reduction Reduction of (3*S*)-3-hydroxy-3-methylglutaryl CoA (HMG-CoA) to give (*R*)-mevalonate requires two equivalents of NADPH and is catalyzed by 3-hydroxy-3-methylglutaryl-CoA reductase.[19, 20] As shown in Figure 3.19, the reaction requires several steps and proceeds through an aldehyde intermediate. The first step is a nucleophilic acyl substitution reaction involving hydride transfer from NADPH to the thioester carbonyl group of HMG-CoA. The step requires an acid catalyst to activate the carbonyl group toward addition—possibly the protonated side chain of a lysine residue—and gives a hemithioacetal intermediate. Following expulsion and protonation of coenzyme A as the leaving group, the aldehyde intermediate undergoes a second hydride-transfer addition to give mevalonate.

Step 4. Phosphorylation and decarboxylation Three additional reactions are required to convert mevalonate to isopentenyl diphosphate. The first two are straightforward phosphorylations that occur by nucleophilic substitution reactions on the terminal phosphorus of ATP: Mevalonate is first converted to

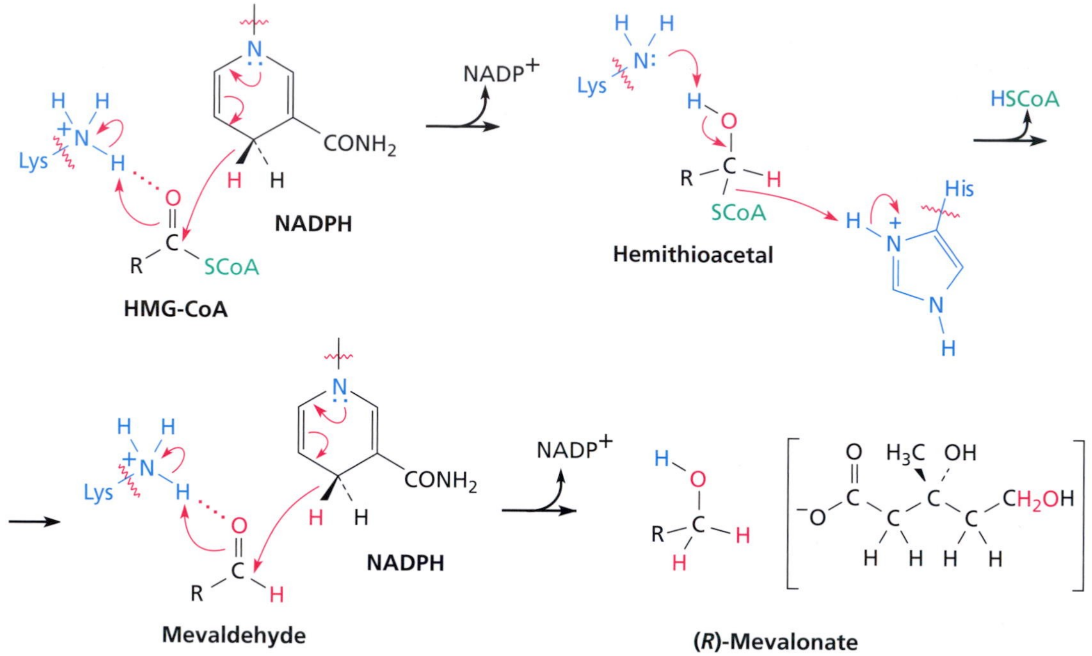

FIGURE 3.19 Mechanism of the reduction of (3*S*)-3-hydroxy-3-methylglutaryl CoA (HMG-CoA) to give (*R*)-mevalonate.

mevalonate 5-phosphate (phosphomevalonate) by reaction with ATP in a process catalyzed by mevalonate kinase. Mevalonate 5-phosphate then reacts with a second ATP to give mevalonate 5-diphosphate (diphosphomevalonate). The third reaction results in phosphorylation of the tertiary hydroxyl group, followed by decarboxylation and loss of phosphate ion.[21, 22]

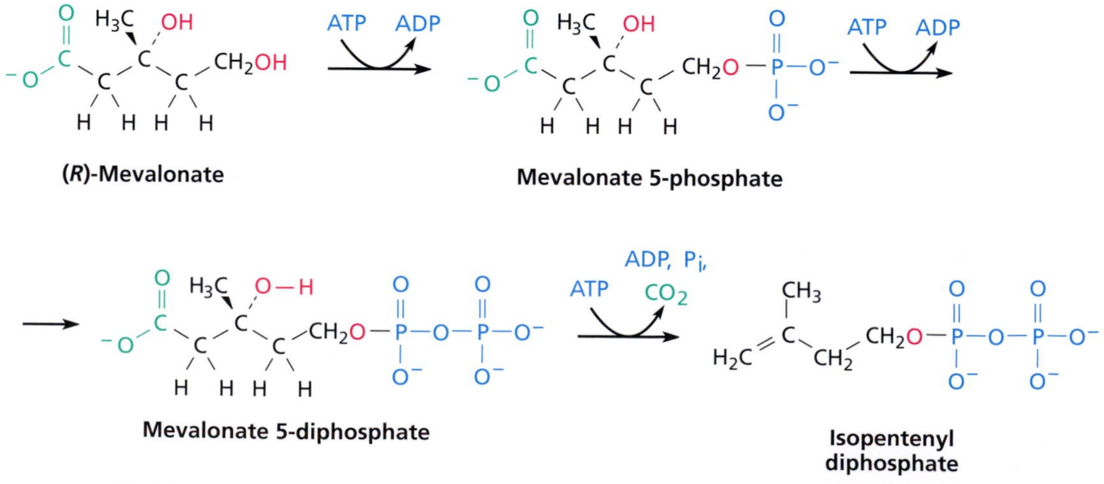

The phosphorylation/decarboxylation of mevalonate 5-diphosphate to give isopentenyl diphosphate appears unusual because, as noted previously, decarboxylations don't typically occur except in β-keto acids and malonic acids; that is, in compounds where the carboxylate group is two atoms away from an additional carbonyl group at the β carbon. The function of this second carbonyl group is to act as an electron acceptor and stabilize the charge resulting from loss of CO_2 by forming an enolate ion. In fact, however, the decarboxylation of a β-keto acid and the decarboxylation of mevalonate 5-diphosphate are closely related.

Catalyzed by mevalonate-5-diphosphate decarboxylase,[22] the substrate is first phosphorylated on the free —OH group by reaction with ATP to give a tertiary phosphate, which is believed to undergo spontaneous dissociation to give a tertiary carbocation (Section 1.8). The positive charge then acts as an electron acceptor to facilitate decarboxylation in exactly the same way a β-carbonyl group does, giving isopentenyl diphosphate as product. The decarboxylation occurs with *anti* stereochemistry so that the *pro-R* hydrogen at C2 ends up trans to the methyl group (Figure 3.20).

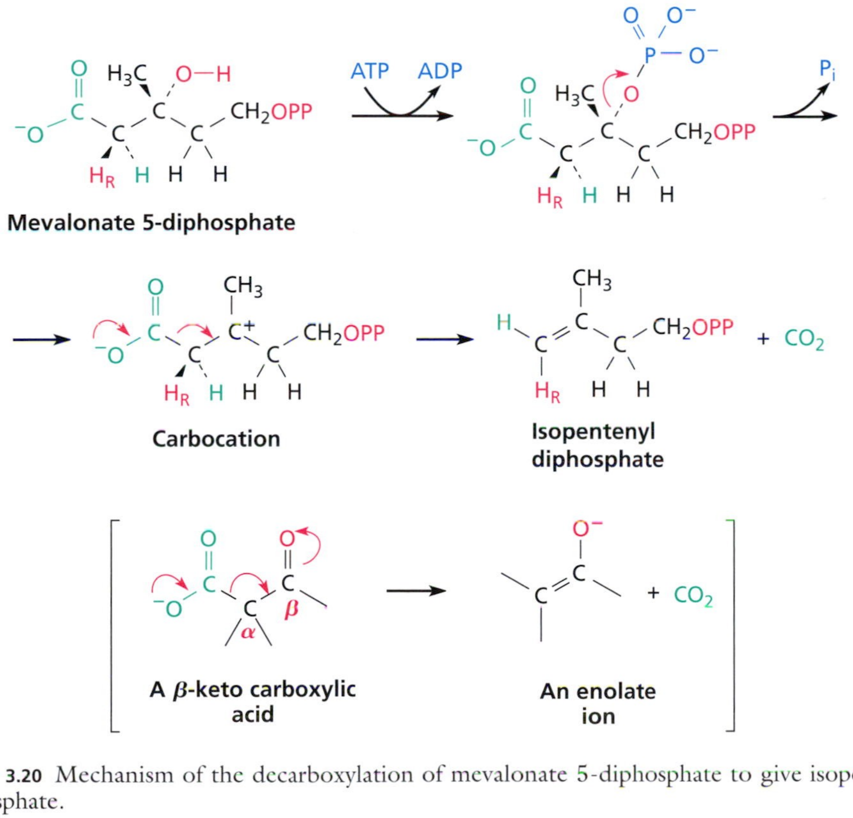

FIGURE 3.20 Mechanism of the decarboxylation of mevalonate 5-diphosphate to give isopentenyl diphosphate.

The Deoxyxylulose Phosphate Pathway to Isopentenyl Diphosphate

The mevalonate pathway for terpenoid biosynthesis has been known for many years, and its details have been extensively studied. By contrast, discovery of the deoxyxylulose phosphate (DXP) pathway occurred only in 1993, and not all its details are yet understood.[23, 24] What is known is summarized in Figure 3.21.

Isopentenyl diphosphate arises in the DXP pathway by an initial coupling of two 3-carbon units, pyruvate and glyceraldehyde 3-phosphate, with concurrent loss of CO_2. The reaction is catalyzed by 1-deoxy-D-xylulose-5-phosphate synthase and requires thiamin diphosphate, a derivative of vitamin B_1, as coenzyme. As we'll see on several future occasions, thiamin diphosphate is often involved when an α-keto acid, such as pyruvate, undergoes decarboxylation. (Thiamin diphosphate, formerly called thiamin pyrophosphate, is usually abbreviated as TPP. The spelling *thiamine* is also correct and is frequently used.)

The key structural feature in thiamin diphosphate is the presence of a thiazolium ring—a five-membered, unsaturated heterocycle containing a sulfur atom and a positively charged nitrogen atom. The thiazolium ring is weakly acidic, with a pK_a of approximately 18 for the ring hydrogen between N and S. Bases can therefore deprotonate thiamin diphosphate, leading to formation of an *ylid*—a neutral species with adjacent + and − charges. In enzymes, this deprotonation is thought to involve the attached aminopyrimidine ring, which undergoes tautomerization by base attack on the —NH_2 group followed by protonation on a ring nitrogen. The tautomer is then deprotonated on the pyrimidine ring, followed by an internal proton transfer from the thiazolium ring to the amino group.

Thiamin diphosphate (TPP)

(Tautomer)

Thiamin diphosphate ylid

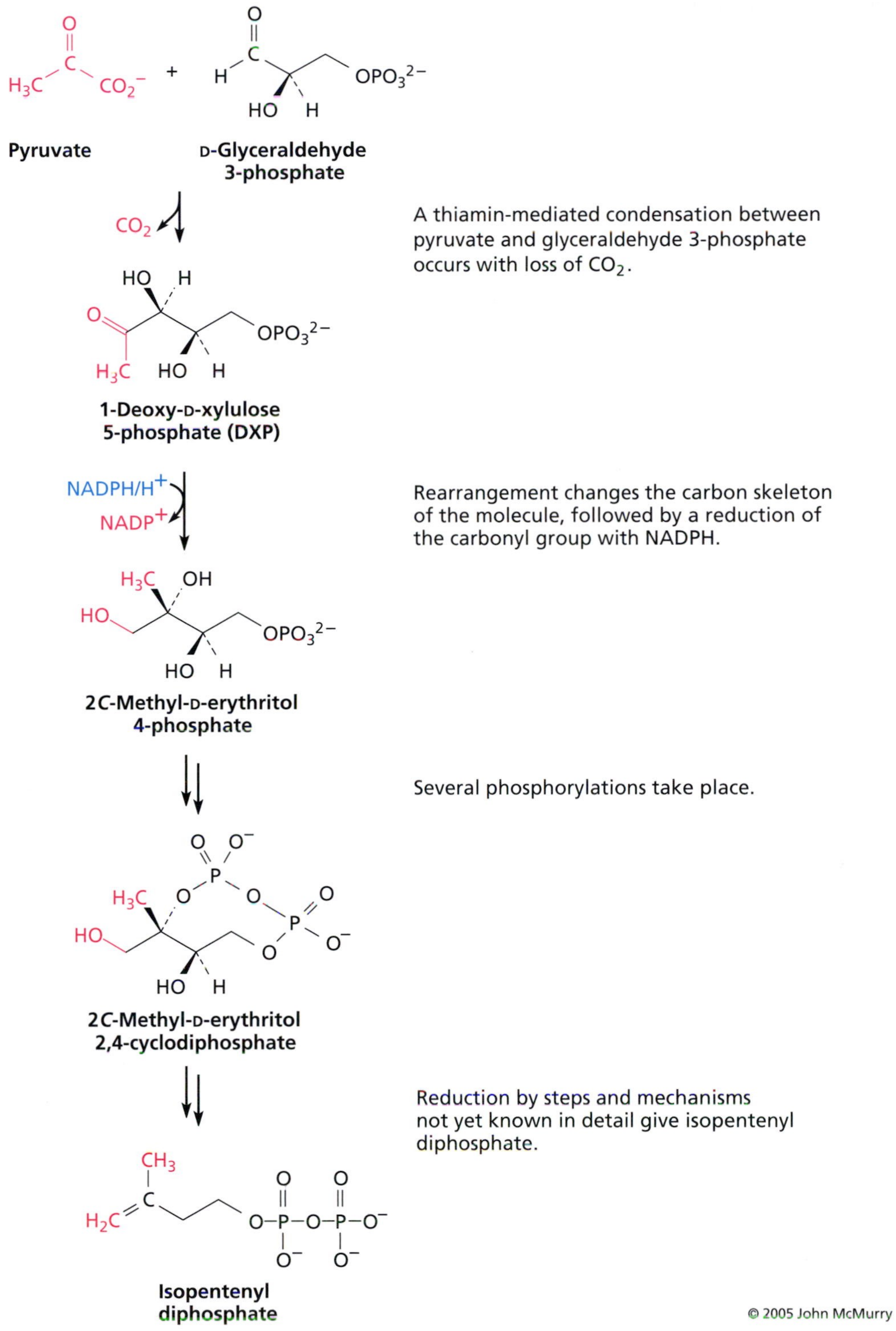

Pyruvate + **D-Glyceraldehyde 3-phosphate**

CO_2

A thiamin-mediated condensation between pyruvate and glyceraldehyde 3-phosphate occurs with loss of CO_2.

1-Deoxy-D-xylulose 5-phosphate (DXP)

NADPH/H^+

$NADP^+$

Rearrangement changes the carbon skeleton of the molecule, followed by a reduction of the carbonyl group with NADPH.

2C-Methyl-D-erythritol 4-phosphate

Several phosphorylations take place.

2C-Methyl-D-erythritol 2,4-cyclodiphosphate

Reduction by steps and mechanisms not yet known in detail give isopentenyl diphosphate.

Isopentenyl diphosphate

FIGURE 3.21 The deoxyxylulose phosphate pathway for the biosynthesis of isopentenyl diphosphate from pyruvate and glyceraldehyde 3-phosphate. The details of some steps are not yet known.

A likely mechanism of the thiamin-mediated coupling reaction between pyruvate and glyceraldehyde 3-phosphate to give 1-deoxy-D-xylulose 5-phosphate is shown in Figure 3.22. Four steps are needed.

Step 1. Nucleophilic addition Because of its electron lone pair and formal negative charge, the ylid carbon on the thiazolium ring is nucleophilic. It therefore adds to the ketone carbonyl group of pyruvate to give an alcohol in a typical nucleophilic addition reaction. The product alcohol is an α-hydroxy *iminium ion*, containing the C=N$^+$ functional group.

TPP ylid **Pyruvate** **An α-hydroxy iminium ion**

Step 2. Decarboxylation The alcohol addition product from step 1 has an iminium ion (C=N$^+$ group) two atoms away from the carboxylate, making it structurally similar to a β-keto acid. As in a β-keto acid (Section 3.4), the C=N$^+$ group can accept electrons, allowing the thiamin addition product to undergo loss of CO_2. The product is an *enamine*, a compound that has an amino group as a substituent on a C=C bond.

Thiamin adduct **An enamine**

A β-keto acid **An enolate ion**

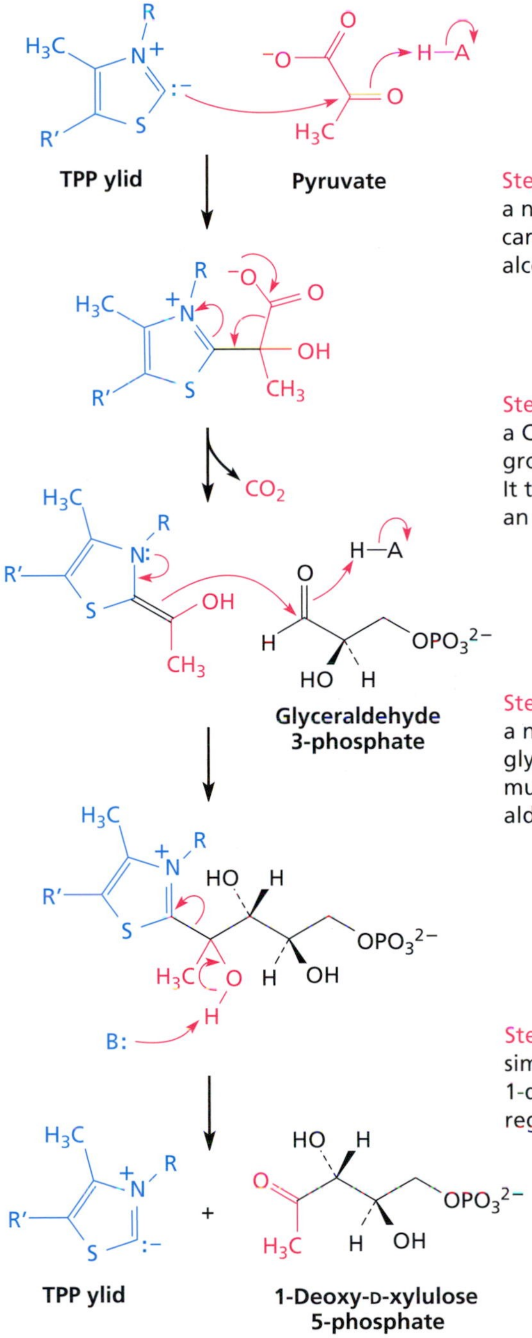

Step 1 Thiamin diphosphate ylid does a nucleophilic addition to the ketone carbonyl group of pyruvate to yield an alcohol addition product.

Step 2 The addition product, which contains a C═N two carbons away from the carboxylate group, is structurally similar to a β-keto acid. It therefore undergoes decarboxylation, giving an enamine.

Step 3 The enamine double bond does a nucleophilic addition to the *si* face of the glyceraldehyde 3-phosphate carbonyl group, much as an enolate ion might add in an aldol reaction.

Step 4 Cleavage of the adduct in a process similar to a retro-aldol reaction gives 1-deoxy-D-xylulose 5-phosphate and regenerates the thiamin diphosphate ylid.

FIGURE 3.22 Mechanism of the thiamin-dependent coupling of pyruvate and glyceraldehyde 3-phosphate to yield 1-deoxy-D-xylulose 5-phosphate.

Step 3. Nucleophilic addition The chemical reactivity of enamines is similar in many respects to that of enolate ions. Both have a carbon–carbon double bond with a strongly electron-donating substituent (—O^- or —NR_2), and both can therefore react with carbonyl compounds in aldol-like condensation reactions. When an enolate ion adds to an aldehyde or ketone, the product is a β-hydroxy carbonyl compound. When an enamine adds, the product is a β-hydroxy iminium ion. Addition occurs from the *si* face of the glyceraldehyde 3-phosphate carbonyl group, leading to *R* stereochemistry at the new chirality center.

An enamine **Glyceraldehyde 3-phosphate** → **A β-hydroxy iminium ion**

An enolate ion → **A β-hydroxy carbonyl**

Step 4. Cleavage The final step is a reverse of the first. In step 1, thiamin ylid adds to a carbonyl group to give an α-hydroxy iminium ion product; in step 4, an α-hydroxy iminium ion decomposes to give a carbonyl compound and thiamin ylid.

An α-hydroxy iminium ion → **TPP ylid** + **1-Deoxy-D-xylulose 5-phosphate**

After 1-deoxy-D-xylulose 5-phosphate has been biosynthesized by the thiamin-mediated coupling of pyruvate and glyceraldehyde 3-phosphate, rearrangement and reduction occur to give 2*C*-methyl-D-erythritol 4-phosphate. The process is catalyzed by deoxyxylulose-5-phosphate reductoisomerase and requires NADPH as coenzyme, along with Mg^{2+}. A likely mechanism[25, 26] is shown in Figure 3.23.

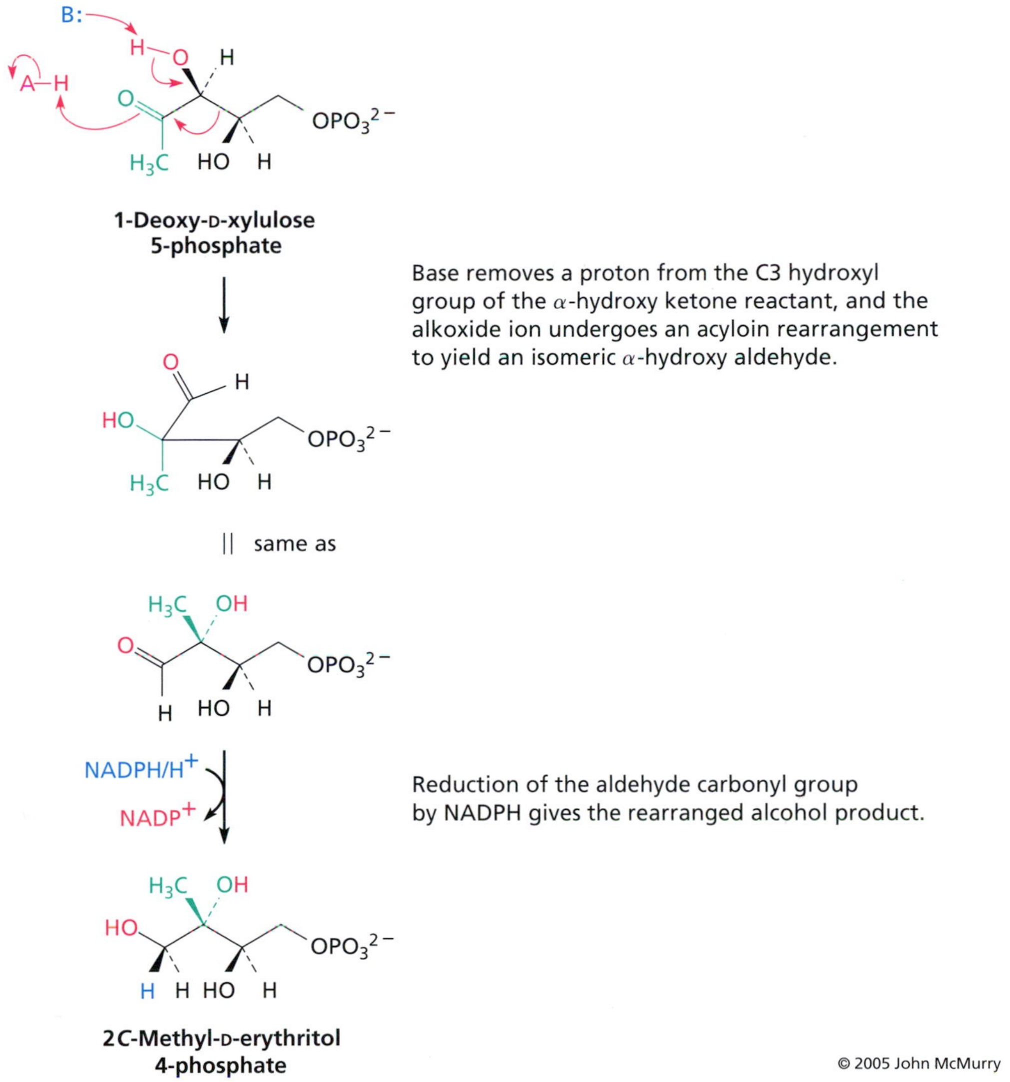

B:
A–H
H–O H
O
OPO_3^{2-}
H_3C HO H

1-Deoxy-D-xylulose 5-phosphate

Base removes a proton from the C3 hydroxyl group of the α-hydroxy ketone reactant, and the alkoxide ion undergoes an acyloin rearrangement to yield an isomeric α-hydroxy aldehyde.

O
H
HO
OPO_3^{2-}
H_3C HO H

|| same as

H_3C OH
O
OPO_3^{2-}
H HO H

$NADPH/H^+$
$NADP^+$

Reduction of the aldehyde carbonyl group by NADPH gives the rearranged alcohol product.

H_3C OH
HO
OPO_3^{2-}
H H HO H

2*C*-Methyl-D-erythritol 4-phosphate

FIGURE 3.23 Probable mechanism of the reductive rearrangement of 1-deoxy-D-xylulose 5-phosphate to 2*C*-methyl-D-erythritol 4-phosphate in the deoxyxylulose phosphate pathway. The key step is an acyloin rearrangement.

The key step in the conversion of 1-deoxy-D-xylulose 5-phosphate to 2*C*-methylerythritol 4-phosphate is an *acyloin rearrangement.* This process, well known in laboratory chemistry, is simply the base-catalyzed isomerization of one α-hydroxy carbonyl compound to another. The isomerization occurs by initial conversion to an alkoxide ion, followed by stereospecific migration of a substituent and its bonding pair of electrons from one carbon to a neighboring carbon.

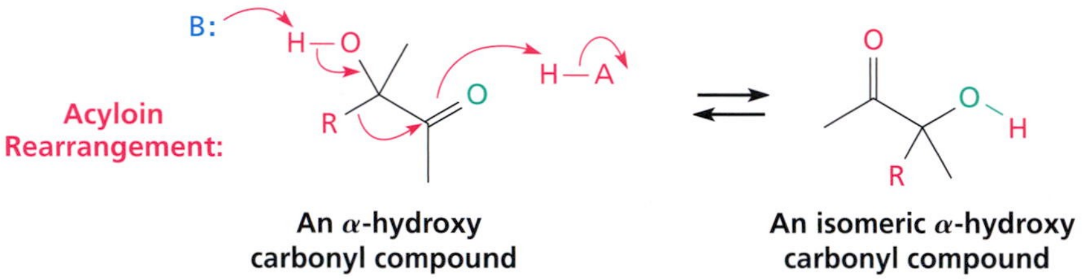

Once the rearrangement has occurred, reduction of the aldehyde carbonyl group by NADPH takes place, with the *pro-S* hydrogen on NADPH adding to the *re* face of the aldehyde substrate.

Next, 2*C*-methyl-D-erythritol 4-phosphate reacts with cytidine triphosphate (CTP), and the diphosphocytidylerythritol that results is further phosphorylated by reaction with ATP. Cyclization then occurs, with loss of cytidine monophosphate and formation of 2*C*-methyl-D-erythritol 2,4-cyclodiphosphate.[27, 28]

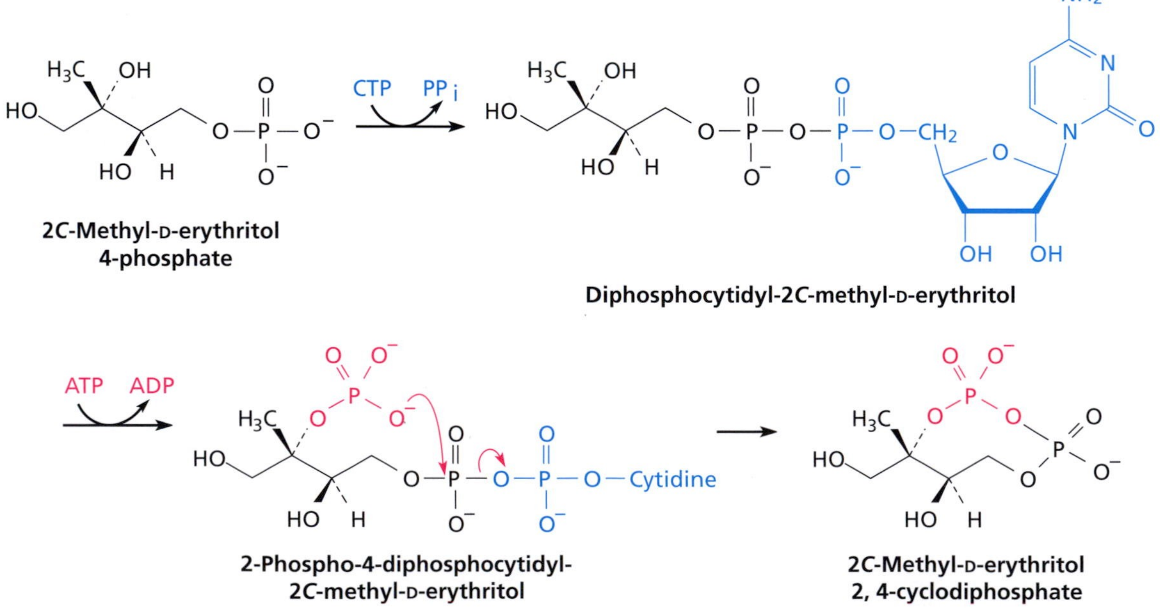

Little is known of the intermediates or mechanisms involved in the further conversion of 2*C*-methyl-D-erythritol 2,4-cyclodiphosphate to isopentenyl diphosphate, although it appears that radical reactions, perhaps mediated by iron–sulfur clusters, are involved. One possibility is that the cyclodiphosphate is converted to an epoxide, which then undergoes a deoxygenation and subsequent reduction of the allylic alcohol.

2C-Methyl-D-erythritol 2,4-cyclodiphosphate

Isopentenyl diphosphate

Conversion of Isopentenyl Diphosphate to Terpenoids

The conversion of isopentenyl diphosphate (IPP) to terpenoids begins with its isomerization to dimethylallyl diphosphate, formerly called dimethylallyl pyrophosphate and abbreviated DMAPP. These two C_5 building blocks then combine to give the C_{10} unit geranyl diphosphate (GPP), which combines with another IPP to give the C_{15} unit farnesyl diphosphate (FPP), and so on up to C_{25}. Terpenoids with more than 25 carbons—that is, triterpenes (C_{30}) and tetraterpenes (C_{40})—are synthesized by dimerization of C_{15} and C_{20} units, respectively. Triterpenes and steroids, in particular, arise from reductive dimerization of farnesyl diphosphate to give squalene (Figure 3.24).

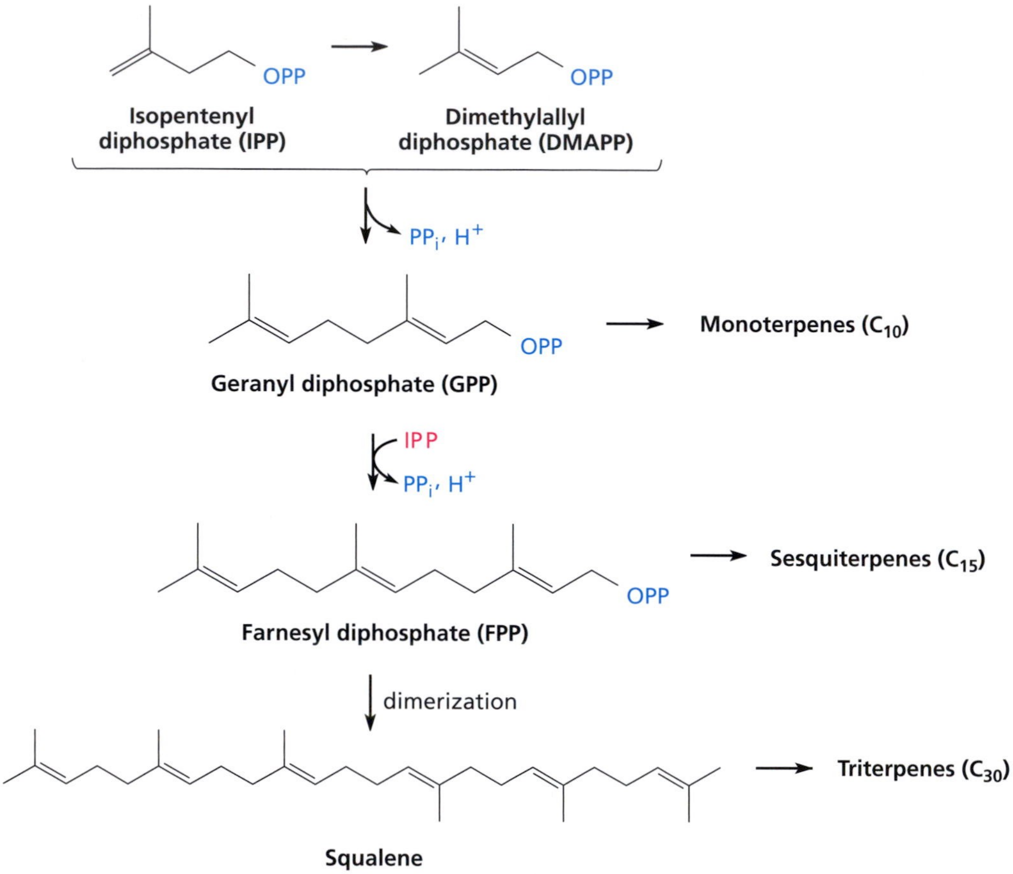

Isopentenyl diphosphate (IPP) → Dimethylallyl diphosphate (DMAPP)

↓ PP_i, H^+

Geranyl diphosphate (GPP) → Monoterpenes (C_{10})

↓ IPP; PP_i, H^+

Farnesyl diphosphate (FPP) → Sesquiterpenes (C_{15})

↓ dimerization

Squalene → Triterpenes (C_{30})

FIGURE 3.24 An overview of terpenoid synthesis from isopentenyl diphosphate.

The isomerization of isopentenyl diphosphate to dimethylallyl diphosphate is catalyzed by IPP isomerase.[29, 30] The enzyme requires Mg^{2+} and Zn^{2+} ions, and studies indicate that reaction occurs through a carbocation pathway. As shown in Figure 3.25, protonation of the IPP double bond by a hydrogen-bonded cysteine residue gives a tertiary carbocation intermediate. Deprotonation by glutamate as base then removes the *pro-R* hydrogen from C2 and yields DMAPP. X-ray structural studies on the enzyme show that it holds the substrate in an unusually deep, well-protected pocket, presumably to shield the highly reactive carbocation from reaction with solvent or other external substances.

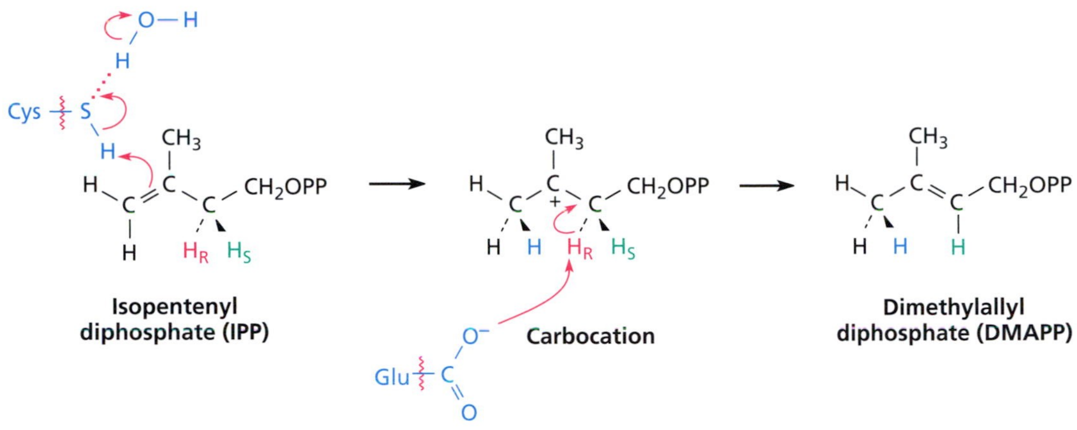

FIGURE 3.25 Mechanism of the isomerization of isopentenyl diphosphate to dimethylallyl diphosphate. The reaction takes place through a carbocation intermediate.

Both the initial coupling of DMAPP with IPP to give geranyl diphosphate and the subsequent coupling of GPP with a second molecule of IPP to give farnesyl diphosphate are catalyzed by farnesyl diphosphate synthase. The process requires Mg^{2+} ion, and the key step is a nucleophilic substitution reaction in which the double bond of IPP behaves as a nucleophile in displacing the diphosphate-ion leaving group (PP_i) from an allylic diphosphate substrate. The intermediate tertiary carbocation that results is then deprotonated with loss of the *pro-R* hydrogen.[30] The exact mechanism of the nucleophilic substitution step—whether S_N1 or S_N2—is difficult to establish conclusively. Available evidence[31] suggests, however, that the substrate develops considerable cationic character and that dissociation of the allylic diphosphate ion in an S_N1-like pathway probably occurs (Figure 3.26).

The further conversion of geranyl diphosphate into monoterpenes typically involves carbocation intermediates and multistep reaction pathways catalyzed by a terpene cyclase.[31] Monoterpene cyclases function by first isomerizing geranyl diphosphate to its allylic isomer linalyl diphosphate (LPP), a process that occurs

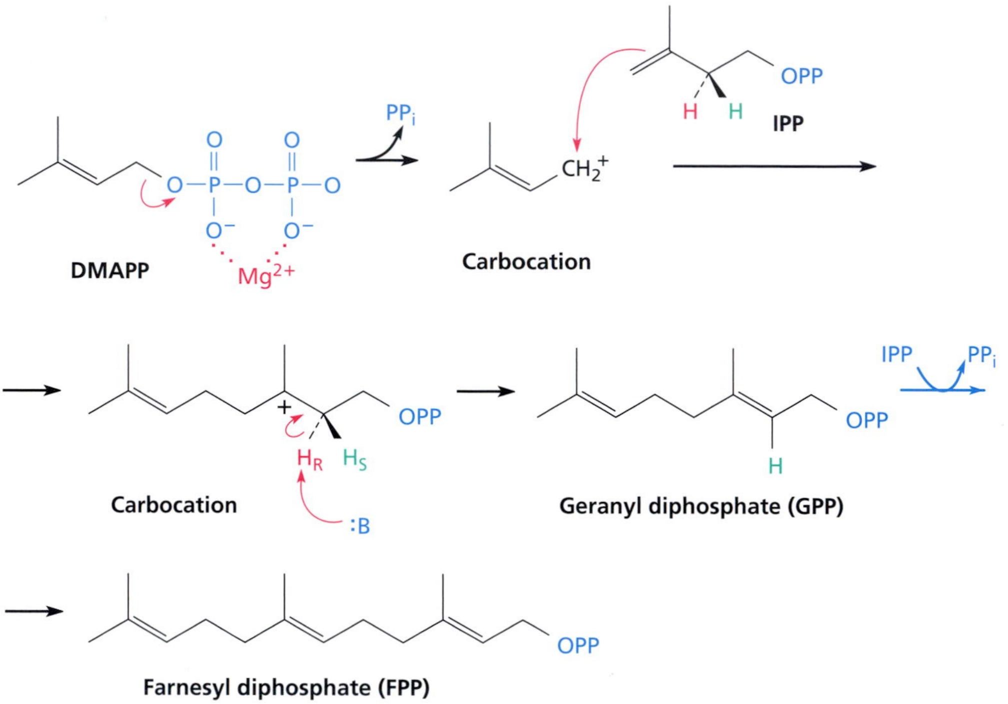

DMAPP Mg^{2+} PP$_i$ Carbocation IPP OPP

Carbocation H$_R$ H$_S$:B OPP Geranyl diphosphate (GPP) IPP PP$_i$ OPP

Farnesyl diphosphate (FPP) OPP

FIGURE 3.26 Mechanism of the coupling reaction of the two 5-carbon molecules, dimethylallyl diphosphate (DMAPP) and isopentenyl diphosphate (IPP), to give geranyl diphosphate (GPP). Further coupling of GPP with another molecule of IPP gives farnesyl diphosphate (FPP).

by spontaneous S_N1-like dissociation to an allylic carbocation, followed by recombination. The effect of this isomerization is to convert the C2–C3 double bond of GPP into a single bond, thereby making possible its *E/Z* isomerization and allowing subsequent cyclization. Further dissociation and cyclization by electrophilic addition of the cationic carbon to the terminal double bond then gives a cyclic cation, which might either rearrange, undergo a hydride shift, be captured by a nucleophile, or be deprotonated to give any of the several hundred known monoterpenes. As just one example, the monoterpene limonene arises by the biosynthetic pathway shown in Figure 3.27.

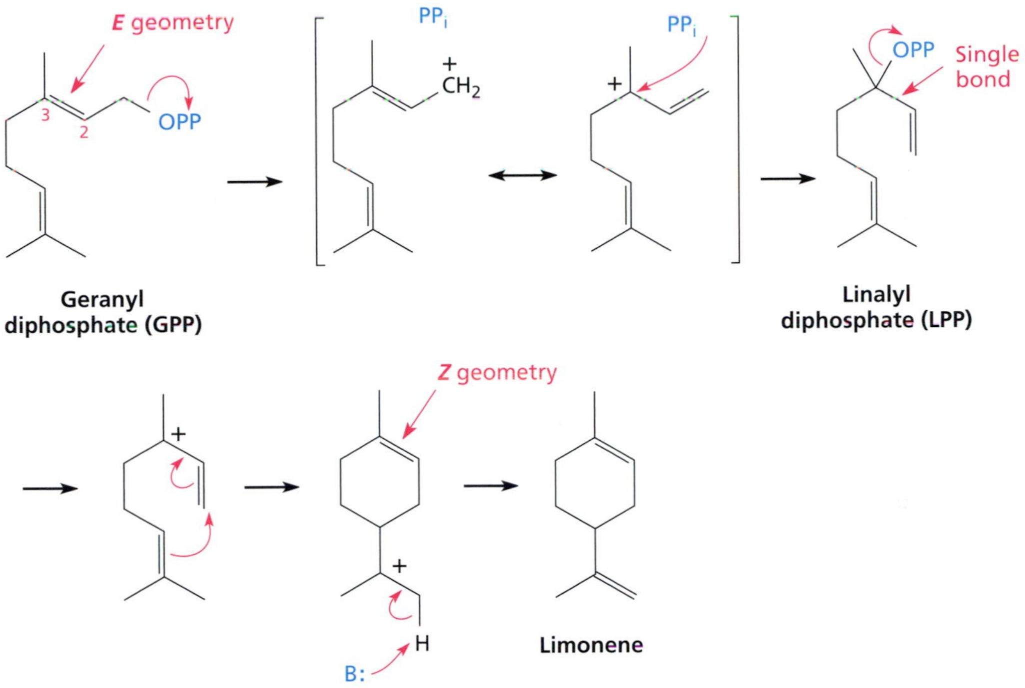

FIGURE 3.27 Mechanism of the formation of the monoterpene limonene from geranyl diphosphate.

Just as monoterpenes arise from cationic cyclizations of geranyl diphosphate, sesquiterpenes arise by cationic cyclizations of farnesyl diphosphate. The sesquiterpene epi-aristolochene, for instance, has been the subject of particularly detailed studies,[32] which have revealed the biosynthetic pathway shown in Figure 3.28. Loss of diphosphate ion from farnesyl diphosphate gives an allylic cation (step 1), which undergoes cationic cyclization to yield a macrocyclic, tertiary carbocation (step 2). Loss of a proton next gives a neutral triene (step 3), which is reprotonated on C6 to give an isomeric cation (step 4). Further cyclization yields a bicyclic cation (step 5), which undergoes a 1,2-hydride migration (step 6), a methyl shift (step 7), and a deprotonation (step 8) before giving the final product.

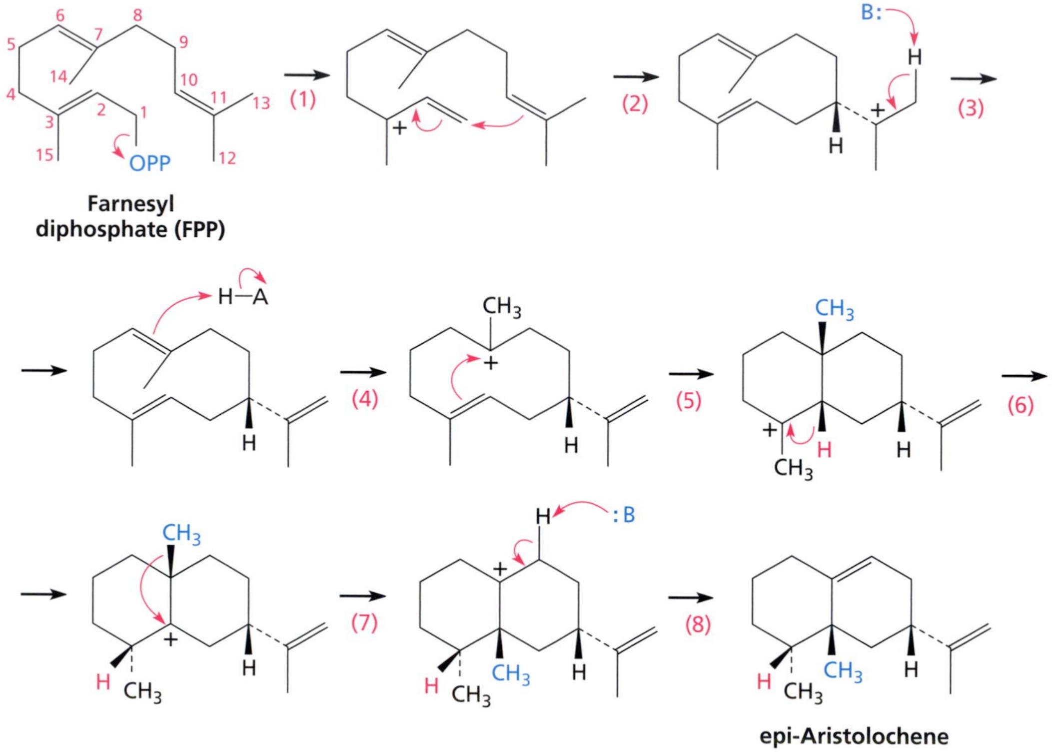

FIGURE 3.28 Mechanism of the biosynthesis of the sesquiterpene epi-aristolochene from farnesyl diphosphate.

3.6 Steroid Biosynthesis

Although generally considered as a separate class of compounds, steroids are actually modified and degraded triterpenoids. As such, they are derived from farnesyl diphosphate (C_{15}) via reductive dimerization to squalene (C_{30}) and subsequent cyclization. The initial product of squalene cyclization, and thus the biosynthetic precursor of all steroids, is the triterpene lanosterol (Figure 3.29).

Conversion of Farnesyl Diphosphate to Squalene

The biosynthetic dimerization of farnesyl diphosphate to squalene occurs in two steps, both catalyzed by squalene synthase.[33, 34] The first step yields the

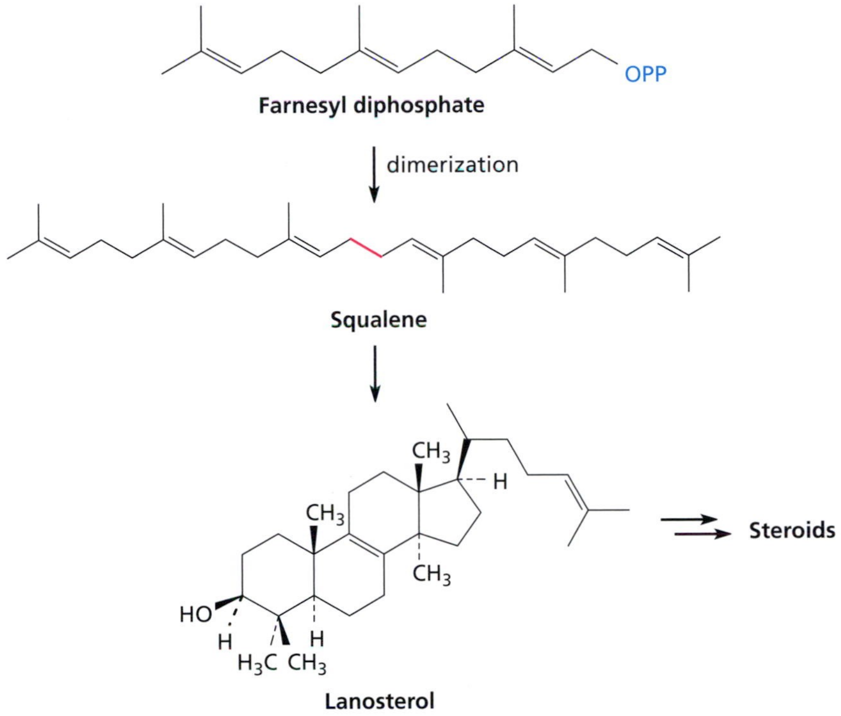

Farnesyl diphosphate

dimerization

Squalene

Lanosterol

Steroids

FIGURE 3.29 A general overview of steroid biosynthesis from farnesol diphosphate.

cyclopropane-containing intermediate presqualene diphosphate, which undergoes reductive rearrangement to squalene.

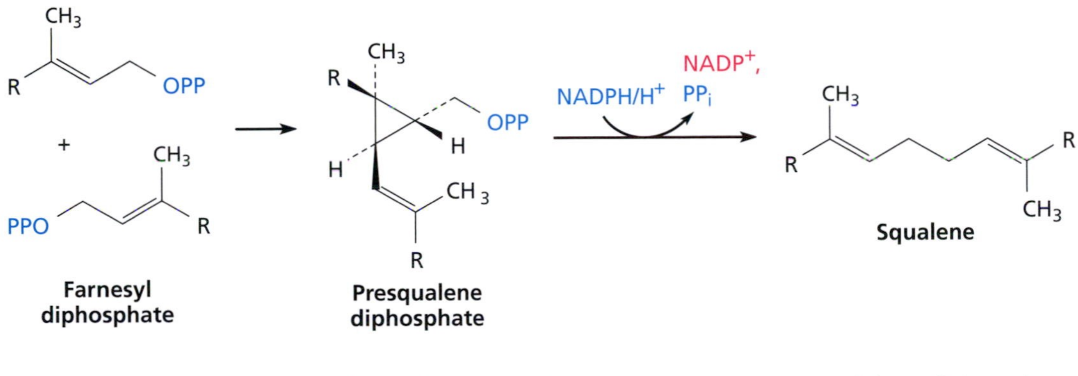

Farnesyl diphosphate

Presqualene diphosphate

NADPH/H^+ NADP$^+$, PP_i

Squalene

Both steps are thought to involve carbocation intermediates, although in neither case are the details clear. The first step, whose proposed mechanism is shown in Figure 3.30, occurs by an S_N1-like dissociation of farnesyl diphosphate to give an allylic carbocation intermediate. The dissociation is thought to be facilitated by an active-site tyrosine residue, which stabilizes the cation by electrostatic interaction of its π electrons with the positive carbon. Nucleophilic addition of the cation to the double bond of a second FPP molecule then gives a tertiary carbocation, which undergoes loss of the *pro-S* hydrogen from the position two carbons away to yield presqualene diphosphate.

Tyr
O
H
R
CH_3
OPP
H
H_R
H_S
Farnesyl diphosphate
Tyr
O^-
PP_i
R
CH_3
+
H_R
H
H_S
PPO
CH_3
H
R
Farnesyl diphosphate
Tyr
O^-
R
CH_3
H_S
H_R
H
+
CH_3
R
PPO
H
R
CH_3
H
H
PPO
CH_3
H
R
Presqualene diphosphate
[R =]

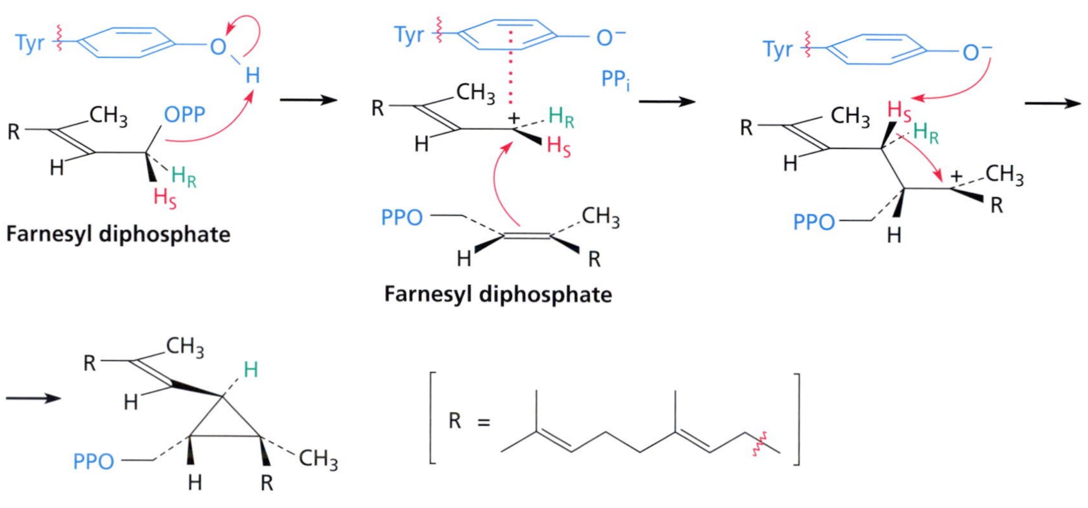

FIGURE 3.30 Proposed mechanism for the formation of presqualene diphosphate from farnesyl diphosphate.

Although it appears unusual, this 1,3 elimination of H^+ from a carbocation to give a cyclopropane has been observed previously on numerous occasions in laboratory chemistry. It is, in fact, simply the reverse of the well-known and much-studied acid-catalyzed ring-opening reaction of cyclopropanes.

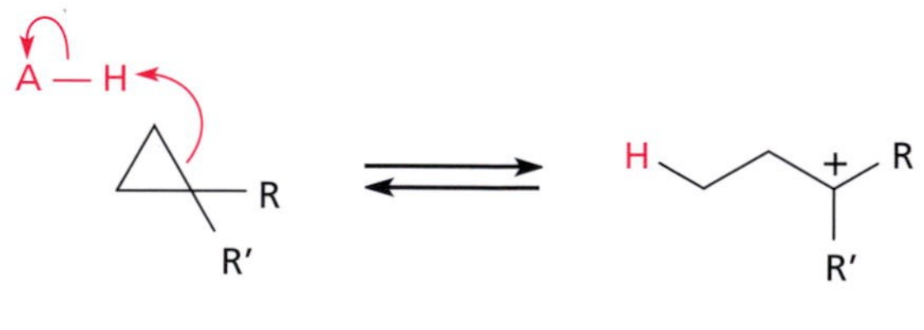

A — H
R
R'
H
+
R
R'

The second step in the formation of squalene from farnesol also involves carbocation intermediates[35] and is thought to proceed by the mechanism outlined in Figure 3.31. S_N1-like dissociation of presqualene diphosphate gives what is

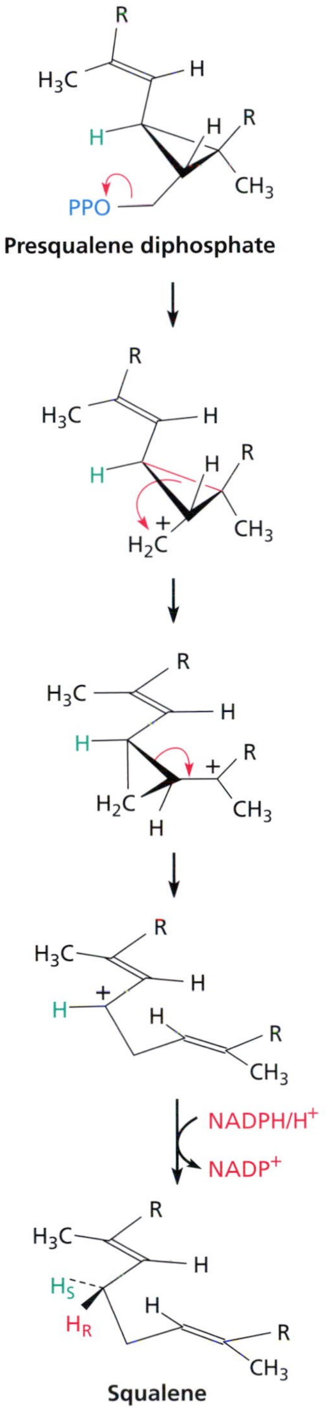

Presqualene diphosphate

Spontaneous dissociation by loss of diphosphate ion yields a primary cyclopropylcarbinyl cation.

Rearrangement by 1,3-migration of the nonadjacent cyclopropane bond gives an isomeric tertiary cyclopropylcarbinyl cation...

...which undergoes further rearrangement by opening of the cyclopropane ring to give an allylic carbocation.

NADPH/H^+

$NADP^+$

Reduction of the cation by NADPH occurs with addition of a *pro-R* hydrogen and gives squalene.

Squalene

FIGURE 3.31 Proposed mechanism for the conversion of presqualene diphosphate to squalene.

called a *cyclopropylcarbinyl cation*, in which the positively charged carbon atom is adjacent to the cyclopropane ring. Cyclopropylcarbinyl cations have been extensively studied in the laboratory and have been found to be unusually stable because of an electronic interaction between the adjacent bent cyclopropane bonds and the vacant *p* orbital on the positive carbon. Furthermore, cyclopropylcarbinyl cations are well known to undergo rapid rearrangement to isomeric cyclopropylcarbinyl products. In the present instance, 1,3-migration of a cyclopropane bond yields a rearranged tertiary cation, which undergoes ring-opening of the cyclopropane. The allylic carbocation that results is then reduced by NADPH with addition of the *pro-R* hydrogen to give squalene.

Conversion of Squalene to Lanosterol

The conversion of squalene to lanosterol is among the most intensively studied of all biosynthetic transformations. Decades of effort have gone into the work, and several Nobel Prizes have been awarded for the results obtained. Starting from an achiral, open-chain polyene, the entire process requires only two enzymes—squalene epoxidase and oxidosqualene:lanosterol cyclase[36]—and results in the formation of a tetracyclic triterpene with seven chirality centers.

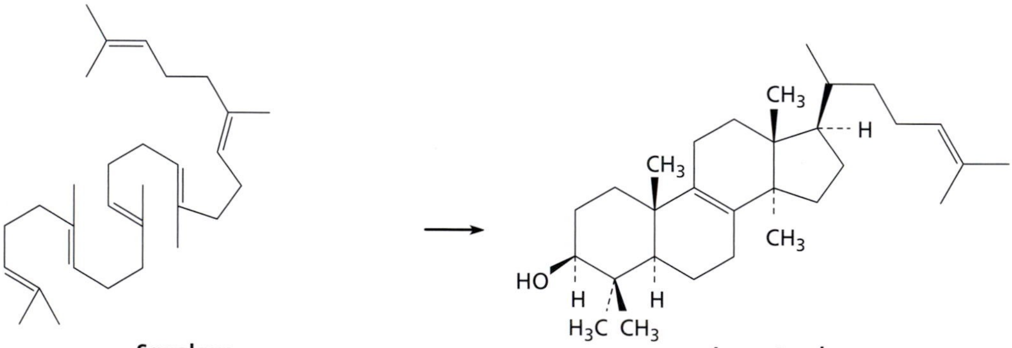

CH_3 CH_3 H CH_3 HO H H H_3C CH_3

Squalene **Lanosterol**

The first step in lanosterol biosynthesis is the conversion of squalene to its epoxide, (3*S*)-2,3-oxidosqualene, catalyzed by squalene epoxidase. Molecular O_2 provides the source of the epoxide oxygen atom, and NADPH is required along with a flavin coenzyme. Because only one of the two oxygen atoms from the O_2 molecule is incorporated into the product, squalene epoxidase is called a **monooxygenase**. The proposed mechanism involves reaction of $FADH_2$ with O_2 to produce a flavin hydroperoxide intermediate (ROOH), which transfers an oxygen to squalene in a pathway initiated by nucleophilic attack of the squalene double bond on the terminal hydroperoxide oxygen (Figure 3.32). The flavin alcohol formed as a by-product loses H_2O to give FAD, which is reduced back to $FADH_2$ by NADPH. Such an epoxidation mechanism is analogous to that by which peroxyacids (RCO_3H) react with alkenes to give epoxides in the laboratory.

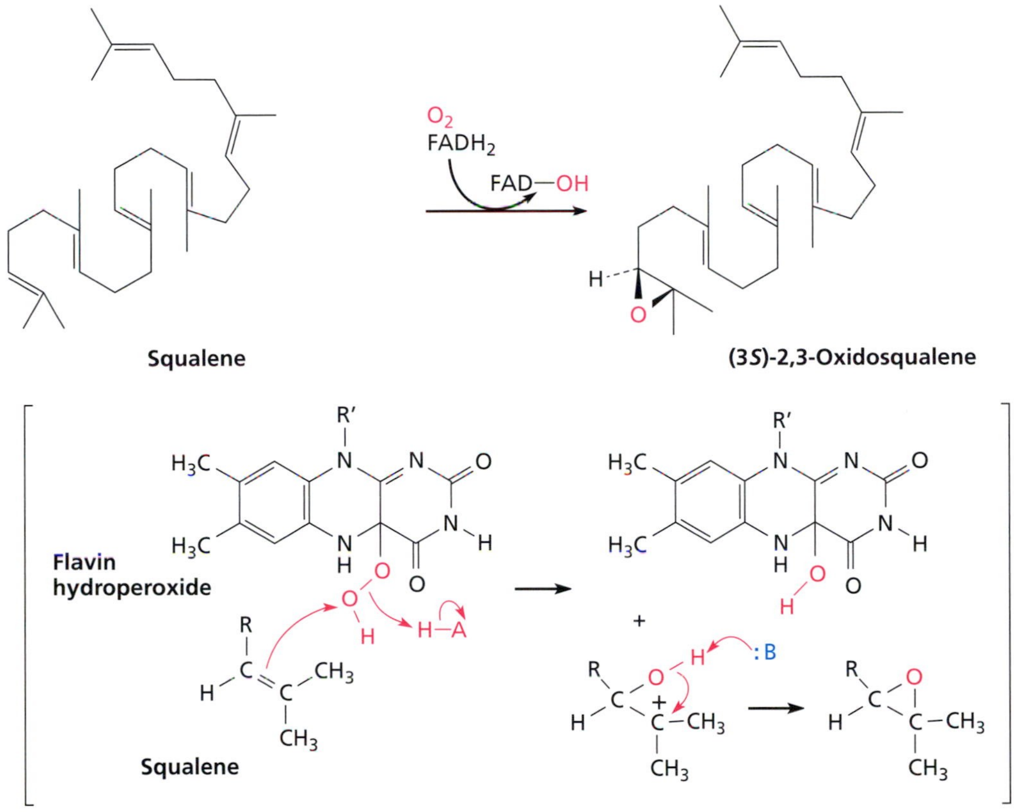

FIGURE 3.32 Proposed mechanism of the oxidation of squalene by flavin hydroperoxide.

The second part of lanosterol biosynthesis is catalyzed by oxidosqualene:lanosterol cyclase, which accomplishes the extraordinary feat of forming four rings, six carbon–carbon bonds, and seven chirality centers in one reaction. As shown in Figure 3.33, squalene is folded by the enzyme into a conformation that aligns the various double bonds for undergoing a cascade of successive intramolecular electrophilic additions, followed by a series of hydride and methyl migrations. Except for the initial epoxide protonation/cyclization, the process is probably stepwise rather than concerted and appears to involve discrete carbocation intermediates that are stabilized by electrostatic interactions with electron-rich aromatic amino acids in the enzyme.

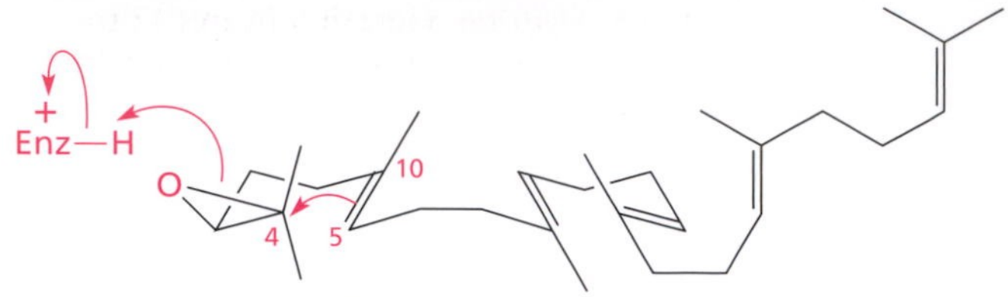

2,3-Oxidosqualene

Step 1 Protonation on oxygen opens the epoxide ring and gives a tertiary carbocation at C4. Intramolecular electrophilic addition of C4 to the 5,10 double bond then yields a tertiary monocyclic carbocation at C10.

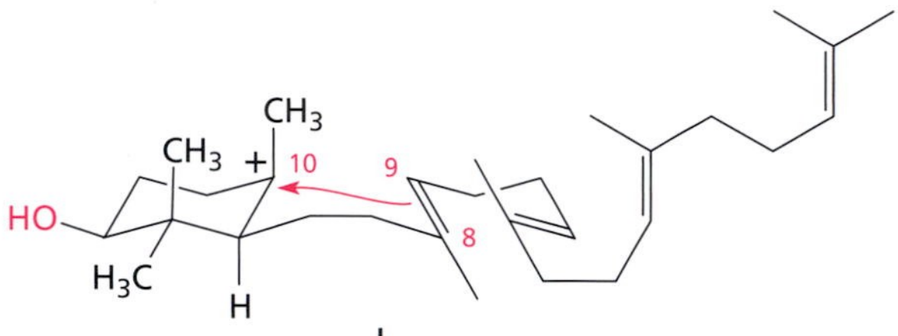

Step 2 The C10 carbocation adds to the 8,9 double bond, giving a C8 tertiary bicyclic carbocation with a boatlike B ring.

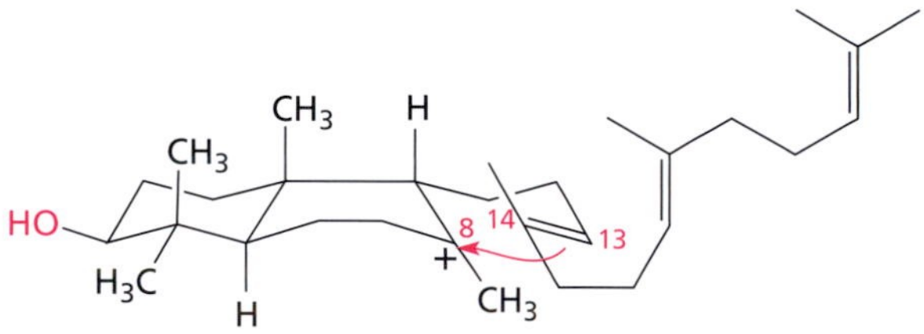

Step 3 Further intramolecular addition of the C8 carbocation to the 13,14 double bond occurs with non-Markovnikov regiochemistry and gives a tricyclic *secondary* carbocation at C13.

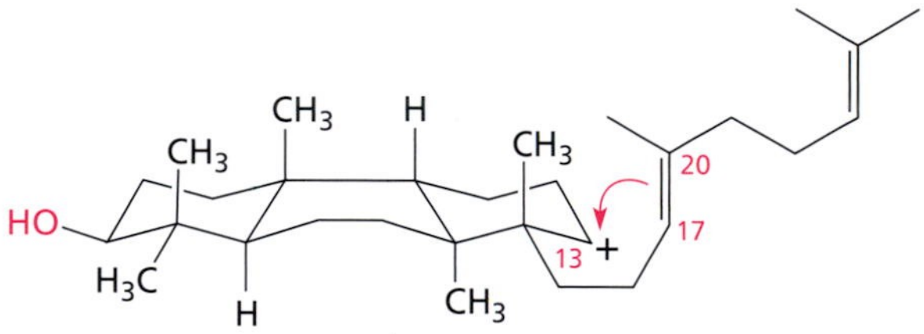

Step 4 The fourth and final cyclization occurs by addition of the C13 cation to the 17,20 double bond, giving the protosteryl cation with 17β stereochemistry.

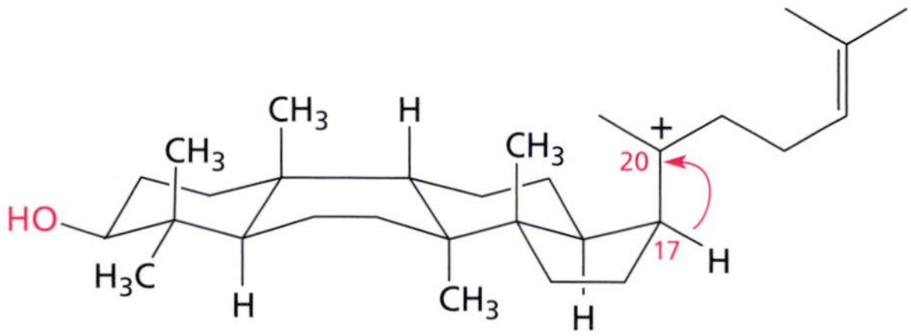

Protosteryl cation

continues

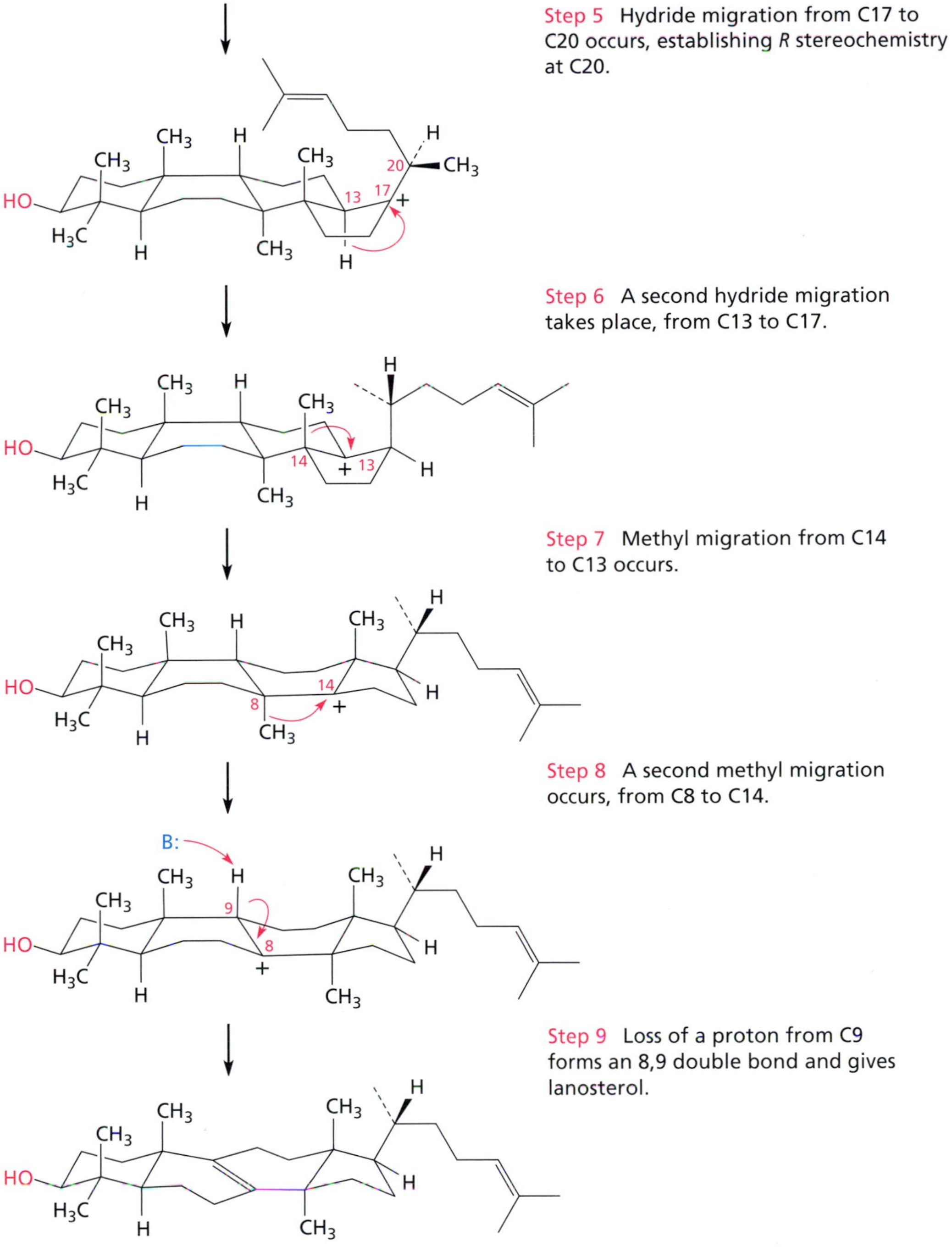

Step 5 Hydride migration from C17 to C20 occurs, establishing *R* stereochemistry at C20.

Step 6 A second hydride migration takes place, from C13 to C17.

Step 7 Methyl migration from C14 to C13 occurs.

Step 8 A second methyl migration occurs, from C8 to C14.

Step 9 Loss of a proton from C9 forms an 8,9 double bond and gives lanosterol.

Lanosterol

FIGURE 3.33 Mechanism of the conversion of 2,3-oxidosqualene to lanosterol. Four cationic cyclizations are followed by four rearrangements and a final loss of H^+ from C9. The steroid numbering system is used for referring to specific positions in the intermediates. (See Section 2.2.)

Steps 1–2. Epoxide opening and initial cyclizations Cyclization is initiated in step 1 by protonation of the epoxide ring by an aspartic acid residue in the enzyme. Nucleophilic opening of the protonated epoxide by the nearby 5,10 double bond (steroid numbering; Section 2.2) then yields a tertiary cation at C10. Further addition of C10 to the 8,9 double bond in step 2 then gives a bicyclic tertiary cation at C8. Note that this second cyclization must form a boatlike B ring to account for the observed stereochemistry of the final product. That is, for the methyl group at C14 in lanosterol to have an α (down) orientation, that same methyl group must have an α orientation at C13 in the migration precursor (step 8). This, in turn, implies a boatlike conformation for the B ring.

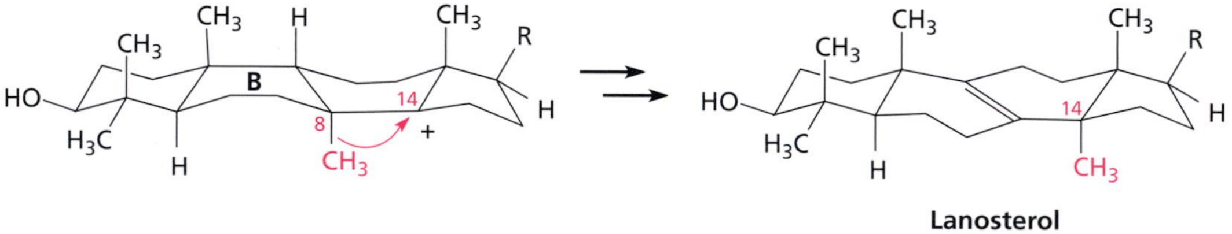

Lanosterol

Step 3. Third cyclization The third cationic cyclization, in step 3, is somewhat unusual because it occurs with non-Markovnikov regiochemistry and gives a secondary cation at C13 rather than the alternative tertiary cation. There is growing evidence, however, that the tertiary carbocation may in fact be formed initially and that the secondary cation arises by subsequent rearrangement. The secondary cation is probably stabilized in the enzyme pocket by the proximity of an electron-rich aromatic ring.

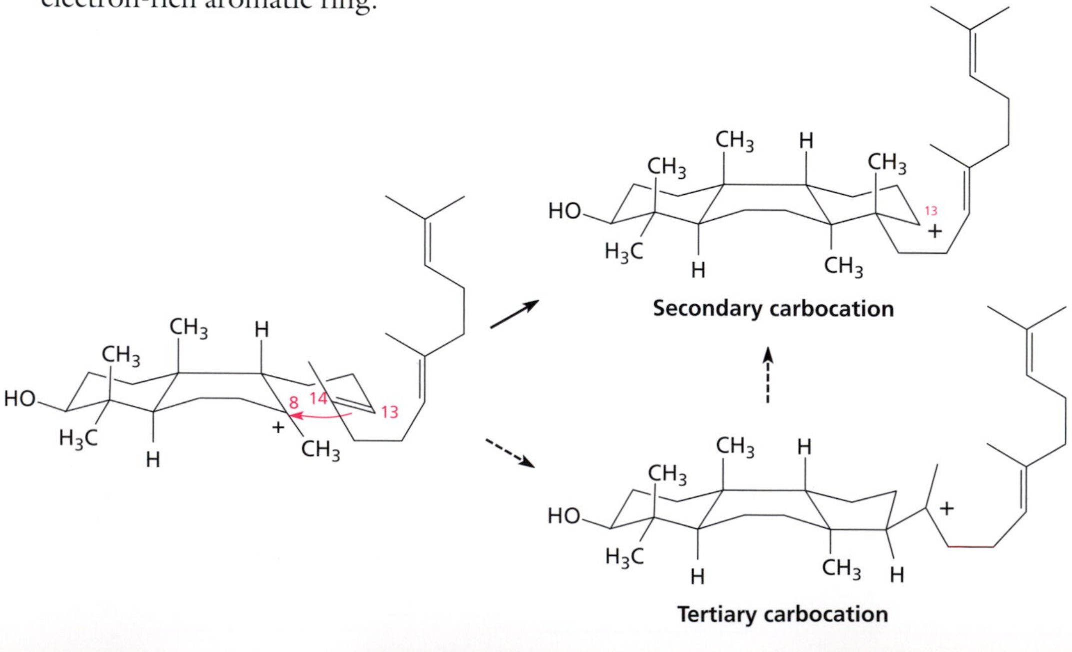

Secondary carbocation

Tertiary carbocation

Step 4. Final cyclization The fourth and last cyclization occurs in step 4 by addition of the cationic center at C13 to the 17,20 double bond, yielding what is known as the *protosteryl* cation. As indicated in Figure 3.33, the side-chain alkyl group at C17 has β (up) stereochemistry, although this stereochemistry is lost in step 5 and then reset in step 6.

Steps 5–9. Carbocation rearrangements Once the tetracyclic carbon skeleton of lanosterol has been formed, a series of carbocation rearrangements take place. The first rearrangement, hydride migration from C17 to C20, occurs in step 5 and results in establishment of *R* stereochemistry at C20 in the side chain. A second hydride migration then occurs from C13 to C17 on the α (bottom) face of the ring in step 6 and reestablishes the 17β orientation of the side chain. Finally, two methyl group migrations, the first from C14 to C13 and the second from C8 to C14, place the positive charge at C8, where it is ultimately quenched in step 9 by loss of the neighboring proton from C9.

From lanosterol, the pathway for steroid biosynthesis continues on to yield cholesterol.[37] A number of changes take place in the process: Three methyl groups are removed, one double bond is reduced, and another double bond is relocated. Cholesterol then becomes a branch point, serving as the common precursor from which other steroids are derived.

Lanosterol

Cholesterol

References

1. Winkler, F. K.; D'Arcy, A.; Hunziker, W., "Structure of Human Pancreatic Lipase," *Nature*, **1990**, *343*, 771–774.
2. Winkler, F. K.; Gubernator, K., "Structure and Mechanism of Human Pancreatic Lipase," in *Pharma Research-New Technologies*, Editors: Woolley, P.; Petersen, S. B., *Lipases*, **1994**, 139–157, Cambridge University Press, Cambridge, UK.

3. Lehner, R.; Kuksis, A., "Triacylglycerol Synthesis by Purified Triacylglycerol Synthetase of Rat Intestinal Mucosa," *J. Biol. Chem.,* **1995**, *270*, 13630–13636.
4. Mueller, E. G.; Crowder, M. W.; Averil, B. A.; Knowles, J. R., "Purple Acid Phosphatase: A Diiron Enzyme that Catalyzes a Direct Phospho Group Transfer to Water," *J. Am. Chem. Soc.,* **1993**, *115*, 2914–2915.
5. Lahiri, S. D.; Zhang, G.; Dunaway-Mariano, D.; Allen, K. N., "The Pentacovalent Phosphorus Intermediate of a Phosphoryl Transfer Reaction," *Science*, **2003**, *299*, 2067–2071.
6. Engst, S.; Vock, P.; Wang, M.; Kim, J.-J. P.; Ghisla, S., "Mechanism of Activation of Acyl-CoA Substrates by Medium Chain Acyl-CoA Dehydrogenase: Interaction of the Thioester Carbonyl with the Flavin Adenine Dinucleotide Ribityl Side Chain," *Biochemistry*, **1999**, *38*, 257–267.
7. Thorpe, C.; Kim, J.-J., "Structure and Mechanism of Action of the Acyl-CoA Dehydrogenases," *FASEB Journal*, **1995**, *9*, 718–725.
8. Umhau, S.; Pollegioni, L.; Molla, G.; Diederichs, K.; Welte, W.; Pilone, M. S.; Ghisla, S., "The X-Ray Structure of D-Amino Acid Oxidase at Very High Resolution Identifies the Chemical Mechanism of Flavin-Dependent Substrate Dehydrogenation," *Proc. Natl. Acad. Sci. USA*, **2000**, *97*, 12463–12468.
9. Wu, W.-J.; Feng, Y.; He, X.; Hofstein, H. A.; Raleigh, D. P.; Tonge, P. J., "Stereospecificity of the Reaction Catalyzed by Enoyl-CoA Hydratase," *J. Am. Chem. Soc.*, **2000**, *122*, 3987–3994.
10. Holden, H. M.; Benning, M. M.; Haller, T.; Gerlt, J. A., "The Crotonase Superfamily: Divergently Related Enzymes that Catalyze Different Reactions Involving Acyl Coenzyme A Thioesters," *Acc. Chem. Res.*, **2001**, *34*, 145–157.
11. Barycki, J. J.; O'Brien, L. K.; Bratt, J. M.; Zhang, R.; Sanishvili, R.; Strauss, A. W.; Banaszak, L. J., "Biochemical Characterization and Crystal Structure Determination of Human Heart Short Chain L-3-Hydroxyacyl-CoA Dehydrogenase Provide Insights into Catalytic Mechanism," *Biochemistry*, **1999**, *38*, 5786–5798.
12. Thompson, S.; Mayerl, F.; Peoples, O. P.; Masamune, S.; Sinskey, A. J.; Walsh, C. T., "Mechanistic Studies on β-Ketoacyl Thiolase from *Zoogloea ramigera*: Identification of the Active-Site Nucleophile as Cys89, Its Mutation to Ser89, and Kinetic and Thermodynamic Characterization of Wild-Type and Mutant Enzymes," *Biochemistry*, **1989**, *28*, 5735–5742.
13. Mathieu, M.; Modis, Y.; Zeelen, J. P.; Engel, C. K.; Abagyan, R. A.; Ahlberg, A.; Rasmussen, B.; Lamzin, V. S.; Kunau, W. H.; Wierenga, R. K., "The 1.8 Å Crystal Structure of the Dimeric Peroxisomal 3-Ketoacyl-CoA Thiolase of *Saccharomyces cerevisiae*: Implications for Substrate Binding and Reaction Mechanism," *J. Mol. Bio.*, **1997**, *273*, 714–728.
14. Attwood, P. V., "The Structure and the Mechanism of Action of Pyruvate Carboxylase," *Internat. J. Biochem. Cell Bio.*, **1995**, *27*, 231–249.
15. Zhang, H.; Yang, Z.; Shen, Y.; Tong, L., "Crystal Structure of the Carboxyltransferase Domain of Acetyl–Coenzyme A Carboxylase," *Science*, **2003**, *299*, 2064–2067.

16. Abbadi, A.; Brummel, M.; Schutt, B. S.; Slabaugh, M. B.; Schuch, R.; Spener, F., "Reaction Mechanism of Recombinant 3-Oxoacyl-(acyl-carrier-protein) Synthase III from *Cuphea wrightii* Embryo, a Fatty Acid Synthase Type II Condensing Enzyme," *Biochem. J.*, **2000**, *345*, 153–160.
17. Dewick, P. M., "The Biosynthesis of C_5–C_{25} Terpenoid Compounds," *Nat. Prod. Rept.*, **2002**, *19*, 181–222.
18. Bochar, D. A.; Friesen, J. A.; Stauffacher, C. V.; Rodwell, V. W., "Biosynthesis of Mevalonic Acid from Acetyl CoA," *Compr. Nat. Prod. Chem. Vol. 2,* **1999**, 15–44, Elsevier Science, Oxford, UK.
19. Tabernero, L.; Bochar, D. A.; Rodwell, V. W.; Stauffacher, C. V., "Substrate-Induced Closure of the Flap Domain in the Ternary Complex Structures Provides Insights into the Mechanism of Catalysis by 3-Hydroxy-3-methylglutaryl-CoA Reductase," *Proc. Natl. Acad. Sci. USA*, **1999**, *96*, 7167–7177.
20. Istvan, E. S.; Deisenhofer, J., "The Structure of the Catalytic Portion of Human HMG-CoA Reductase," *Biochim. Biophys. Acta*, **2000**, *1529*, 9–18.
21. Jabalquinto, A. M.; Alvear, M; Cardemil, E., "Physiological Aspects and Mechanism of Action of Mevalonate 5-Diphosphate Decarboxylase," *Comp. Biochem. Physiol., B: Comp. Biochem.*, **1988**, *90B*, 671–677.
22. Dhe-Paganon, S.; Magrath, J.; Abeles, R. H., "Mechanism of Mevalonate Pyrophosphate Decarboxylase: Evidence for a Carbocationic Transition State," *Biochemistry*, **1994**, *33*, 13355–13362.
23. Eisenreich, W.; Schwarz, M.; Cartayrade, A.; Arigoni, D.; Zenk, M. H.; Bacher, A., "The Deoxyxylulose Phosphate Pathway of Terpenoid Biosynthesis in Plants and Microorganisms," *Chem. Biol.*, **1998**, *5*, R221–R233.
24. Rohmer, M., "A Mevalonate-Independent Route to Isopentenyl Diphosphate," *Compr. Nat. Prod. Chem. Vol. 2,* **1999**, 45–68, Elsevier Science, Oxford, UK.
25. Radykewicz, T.; Rohdich, F.; Wungsintaweekul, J.; Herz, S.; Kis, K.; Eisenreich, W.; Bacher, A.; Zenk, M. H.; Arigoni, D., "Biosynthesis of Terpenoids: 1-Deoxy-D-xylulose 5-Phosphate Reductoisomerase from *Escherichia coli* Is a Class B Dehydrogenase," *FEBS Lett.*, **2000**, *465*, 157–160.
26. Kuzuyama, T.; Takahashi, S.; Takagi, M.; Seto, H., "Characterization of 1-Deoxy-D-xylulose 5-Phosphate Reductoisomerase, an Enzyme Involved in Isopentenyl Diphosphate Biosynthesis, and Identification of Its Catalytic Amino Acid Residues," *J. Biol. Chem.*, **2000**, *275*, 19928–19932.
27. Herz, S.; Wungsintaweekul, J.; Schuhr, C. A.; Hecht, S.; Luttgen, H.; Sagner, S.; Fellermeier, M.; Eisenreich, W.; Zenk, M. H.; Bacher, A.; Rohdich, F., "Biosynthesis of Terpenoids: ygbB Protein Converts 4-Diphosphocytidyl-2*C*-methyl-D-erythritol 2-Phosphate to 2*C*-Methyl-D-erythritol 2,4-Cyclodiphosphate," *Proc. Natl. Acad. Sci. USA,* **2000**, *97*, 2486–2490.
28. Rohdich, F.; Zepeck, F.; Adam, P.; Hecht, S.; Kaiser, J.; Laupitz, R.; Gräwert, T.; Amslinger, S.; Eisenreich, W.; Bacher, A.; Arigoni, D., "The Deoxyxylulose Phosphate Pathway of Isoprenoid Biosynthesis: Studies on the Mechanisms of the Reactions Catalyzed by IspG and IspH Protein," *Proc. Nat'l. Acad. Sci., USA*, **2003**, *100*, 1586–1591.

29. Durbecq, V.; Sainz, G.; Oudjama, Y.; Clantin, B.; Bompard-Gilles, C.; Tricot, C.; Caillet, J.; Stalon, V.; Droogmans, L.; Villeret, V., "Crystal Structure of Isopentenyl Diphosphate:Dimethylallyl Diphosphate Isomerase," *EMBO J.*, **2001**, *20*, 1530–1537.
30. Leyes, A. E.; Baker, J. A.; Poulter, C. D., "Biosynthesis of Isoprenoids in *Escherichia coli*: Stereochemistry of the Reaction Catalyzed by Farnesyl Diphosphate Synthase," *Org. Lett.*, **1999**, *1*, 1071–1073.
31. Dolence, J. M.; Poulter, C. D., "Electrophilic Alkylations, Isomerizations, and Rearrangements," in *Compr. Nat. Prod. Chem. Vol. 5,* **1999**, 315–341, Elsevier Science, Oxford, UK.
32. Starks, C. M.; Back, K.; Chappell, J.; Noel, J. P., "Structural Basis for Cyclic Terpene Biosynthesis by Tobacco 5-epi-Aristolochene Synthase," *Science*, **1997**, *277*, 1815–1820.
33. Gu, P.; Ishii, Y.; Spencer, T. A.; Shechter, I., "Function–Structure Studies and Identification of Three Enzyme Domains Involved in the Catalytic Activity in Rat Hepatic Squalene Synthase," *J. Biol. Chem.*, **1998**, *273*, 12515–12525.
34. Radisky, E. S.; Poulter, C. D., "Squalene Synthase: Steady-State, Pre-Steady-State, and Isotope-Trapping Studies," *Biochemistry*, **2000**, *39*, 1748–1760.
35. Jarstfer, M. B.; Blagg, B. S. J.; Rogers, D. H.; Poulter, C. D., "Biosynthesis of Squalene. Evidence for a Tertiary Cyclopropylcarbinyl Cationic Intermediate in the Rearrangement of Presqualene Diphosphate to Squalene," *J. Am. Chem. Soc.*, **1996**, *118*, 13089–13090.
36. Abe, I.; Prestwich, G. D., "Squalene Epoxidase and Oxidosqualene:Lanosterol Cyclase—Key Enzymes in Cholesterol Biosynthesis," *Compr. Nat. Prod. Chem., Vol. 2,* **1999**, 267–298, Elsevier Science, Oxford, UK.
37. Risley, J. M., "Cholesterol Biosynthesis: Lanosterol to Cholesterol," *J. Chem. Educ.*, **2002**, *79*, 377–384.

Problems

3.1 Show the product of each of the following reactions

(a) $CH_3CH_2CH_2CH_2CH_2C(=O)SCoA \xrightarrow[\text{Acyl-CoA dehydrogenase}]{FAD \rightarrow FADH_2}$

(b) Product of (a) + $H_2O \xrightarrow{\text{Enoyl-CoA hydratase}}$

(c) Product of (b) $\xrightarrow[\beta\text{-Hydroxyacyl-CoA dehydrogenase}]{NAD^+ \rightarrow NADH/H^+}$

3.2 Digitoxigenin is a heart stimulant obtained from the purple foxglove *Digitalis purpurea* and used in the treatment of heart disease. Draw the three-dimensional conformation of digitoxigenin, and identify the two $-$OH groups as axial or equatorial.

Digitoxigenin

3.3 Draw a Fischer projection of *sn*-glycerol 1-phosphate, and assign *R* or *S* configuration to the chiral center. Do the same for *sn*-glycerol 2,3-diacetate.

3.4 The following model is that of glycocholate, a constituent of human bile. Identify each of the three hydroxyl groups as axial or equatorial. Is cholic acid an A–B trans steroid or an A–B cis steroid?

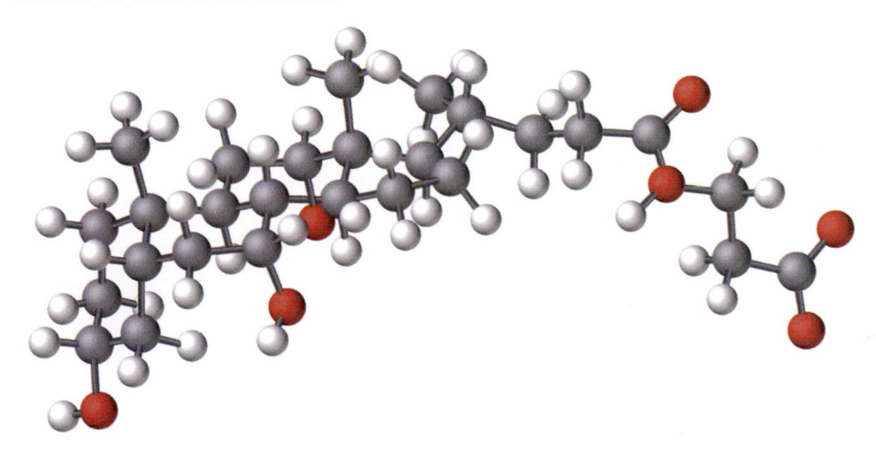

3.5 Assuming that acetate containing a ^{14}C isotopic label in the carboxyl carbon is used as the starting material and that the mevalonate pathway is followed, identify the positions in each of the following molecules where the label would appear.

(a) **Dimethylallyl diphosphate**

(b) **Limonene**

(c) **Caryophyllene**

3.6 Propose a biosynthetic pathway for the sesquiterpene helminthogermacrene from farnesyl diphosphate.

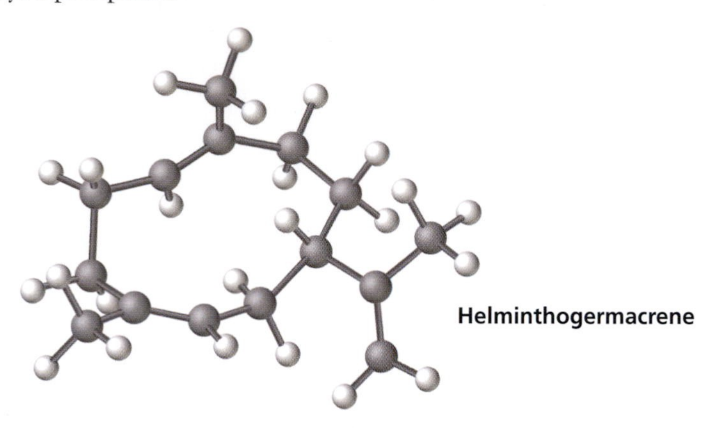

3.7 Propose a mechanistic pathway for the biosynthesis of borneol.

H_3C CH_3 CH_3 H OH

Borneol

3.8 Isoborneol is converted into camphene on treatment with dilute sulfuric acid. Propose a mechanism for the reaction, which involves a carbocation rearrangement.

H_3C CH_3 CH_3 OH H $\xrightarrow{H_2SO_4}$ H_2C H_3C CH_3

Isoborneol **Camphene**

3.9 When farnesyl diphosphate is treated with squalene synthase in the absence of NADPH, the following two compounds are formed. Propose a mechanism for the reaction (see Figure 3.31).

CH_3 R OPP $\longrightarrow$ CH_3 OH R R CH_3 + R CH_3 CH_3 R

Farnesyl diphosphate

3.10 Assuming that acetate containing a ^{14}C isotopic label in the carboxyl carbon is used as the starting material and that the mevalonate pathway is followed, identify the positions in lanosterol where the label would appear.

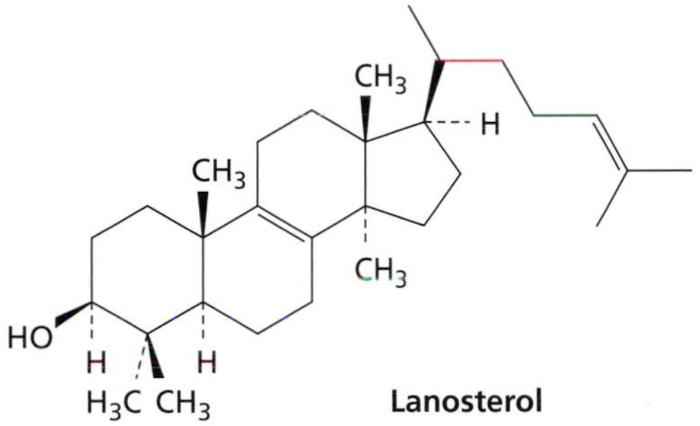

CH_3 H CH_3 CH_3 HO H H H_3C CH_3

Lanosterol

3.11 Propose a mechanism for the biosynthesis of the sesquiterpene trichodiene from farnesyl diphosphate. The process involves cyclization to give an intermediate secondary carbocation, followed by several migrations.

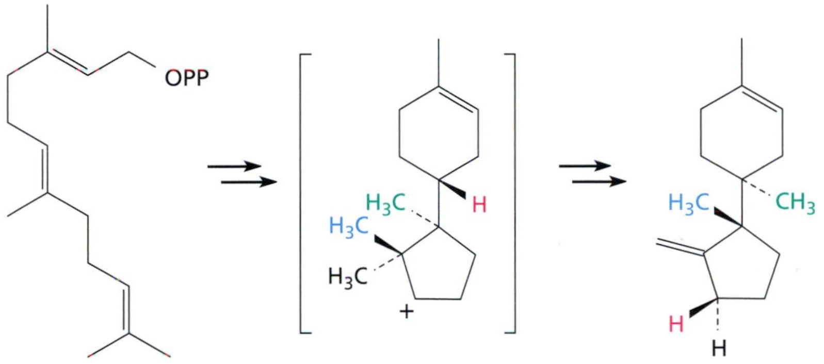

OPP

H_3C H H_3C H_3C +

H_3C CH_3 H H

Farnesyl diphosphate (FPP)

Trichodiene

3.12 In many plants, monoterpenes are biosynthesized through the DXP pathway. Predict the location of the ^{13}C label in limonene starting from pyruvate specifically labeled with ^{13}C on C1. Do the same using pyruvate labeled on C2 and on C3.

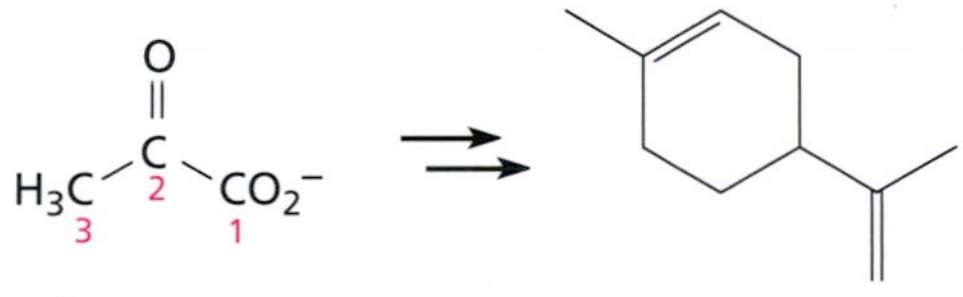

O ‖ H_3C C CO_2^- 3 2 1

Pyruvate

Limonene

3.13 Propose a mechanism for the following rearrangement:

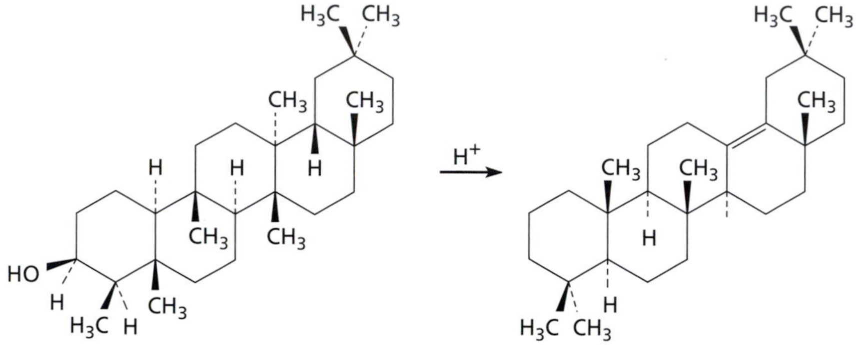

3.14 Retrieve the PDB coordinate file for human pancreatic lipase (Figure 3.2), display the structure using the Swiss PDB viewer, and answer the following questions: (The PDB code is 1LPB.)

(a) What is the distance between the basic nitrogen of His-263 and the oxygen of Ser-152?

(b) The P═O bond of the inhibitor is polarized for addition of serine by hydrogen bonding to the amide NH bonds of two residues, a situation called the *oxyanion hole*, which occurs frequently in the catalysis of addition reactions to carbonyl and phosphate groups. Which two residues in the active site are involved? (As a general guide, two electron-rich atoms separated by approximately 3.0 Å can participate in a hydrogen bond.)

(c) Asp-176 is thought to activate His-263 by proton transfer. What is the distance between the carboxylate oxygen of aspartate and the basic nitrogen of histidine?

(d) His-151 is also in the active site. Why is His-263 rather than His-151 the active-site base?

3.15 Retrieve the PDB coordinate file for acyl-CoA dehydrogenase (Figure 3.7), display the structure using the Swiss PDB viewer, and answer the following questions: (The PDB code is 3MDE.)

(a) What is the distance between the β carbon of the substrate (the hydride donor) and N5 of the flavin? Is this distance sufficiently short for a direct hydride transfer?

(b) Glu-376 is the proposed active-site base involved in deprotonation of the thioester. Is it sufficiently close to the α hydrogen of the thioester?

(c) Enolate ions are usually stabilized at enzyme active sites by hydrogen bonding or coordination to a metal ion. Identify the enolate-stabilizing interaction at the active site of acyl-CoA dehydrogenase.

4 Carbohydrate Metabolism

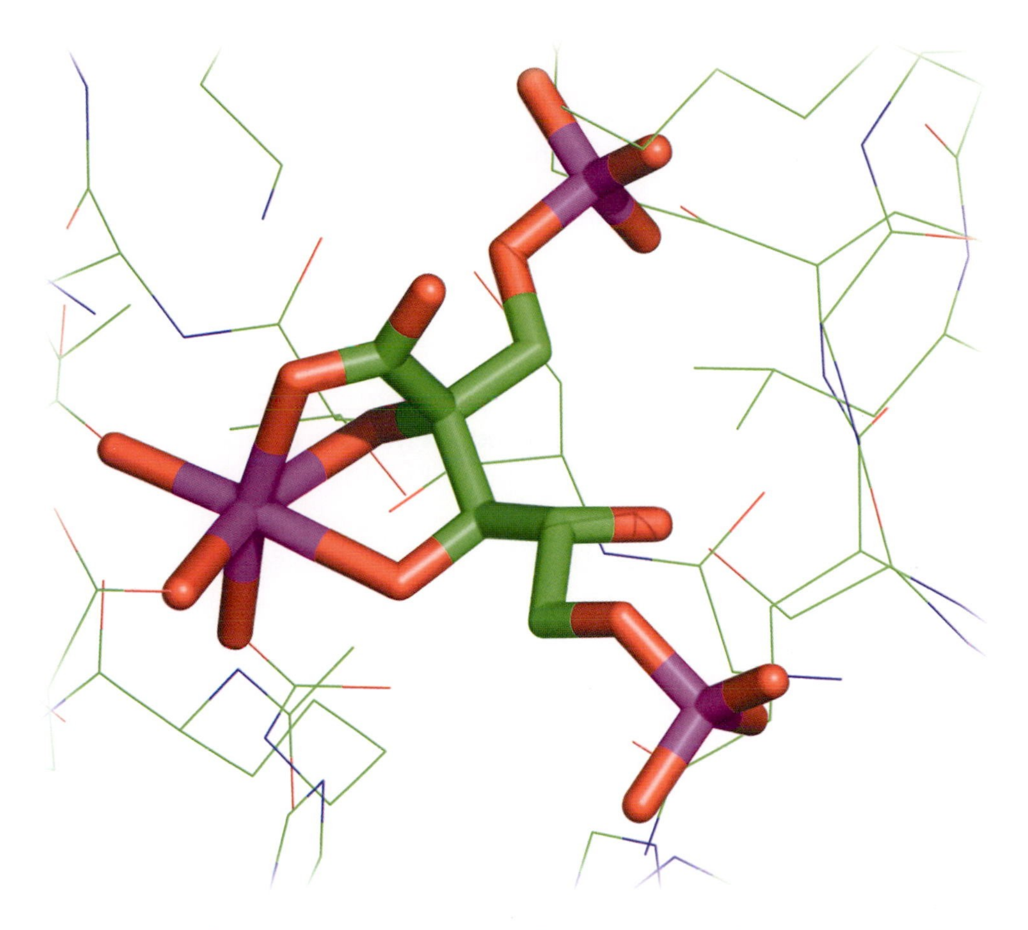

The combination of photosynthesis and the citric acid cycle is responsible for the circulation of 100 billion tons of carbon through the biosphere each year. Ribulose bisphosphate carboxylase, or "Rubisco," catalyzes the first step in carbon dioxide fixation and is the most abundant protein on earth. The structure shows the active-site containing an analog of a reaction intermediate. In this chapter, we'll examine the chemistry of this and other important reactions in carbohydrate metabolism.

Carbohydrates are vastly more abundant than any other class of biomolecules. It has been estimated, in fact, that more than 50% of the dry weight of the earth's biomass—all plants and animals—consists of glucose polymers. Carbohydrates are synthesized by green plants during photosynthesis, and when eaten and metabolized they provide the major source of energy used by organisms. Thus, carbohydrates act as the chemical intermediaries by which solar energy is stored and used to support life. Recall from Section 2.3 that simple carbohydrates are polyhydroxy aldehydes or ketones and that glucose, the most abundant simple carbohydrate, exists in aqueous solution as an equilibrium mixture of open-chain form and cyclic α and β pyranose anomers.

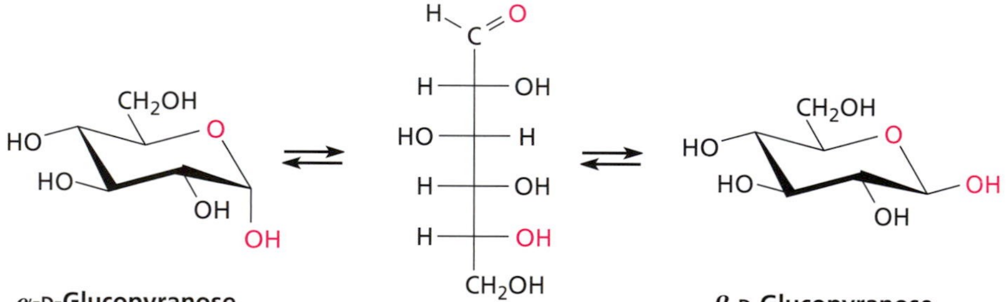

α-D-Glucopyranose (37.3%) ⇄ **(0.002%)** ⇄ **β-D-Glucopyranose (62.6%)**

4.1 Digestion and Hydrolysis of Complex Carbohydrates

Dietary carbohydrate is largely starch, a glucose polymer in which the monosaccharide units are linked by α-($1\rightarrow4'$) glycoside bonds (Section 2.3). Starch consists of two main fractions: amylose, which is insoluble in cold water, and amylopectin, which is soluble in cold water. Amylose makes up about 20% of starch by mass and is a linear polymer of several hundred α-(1→4′)-linked glucose units. Amylopectin accounts for the remaining 80% of starch by mass and is a branched polymer of up to 5000 glucose units with α-(1→6′) branches every 25 or so units (Figure 4.1).

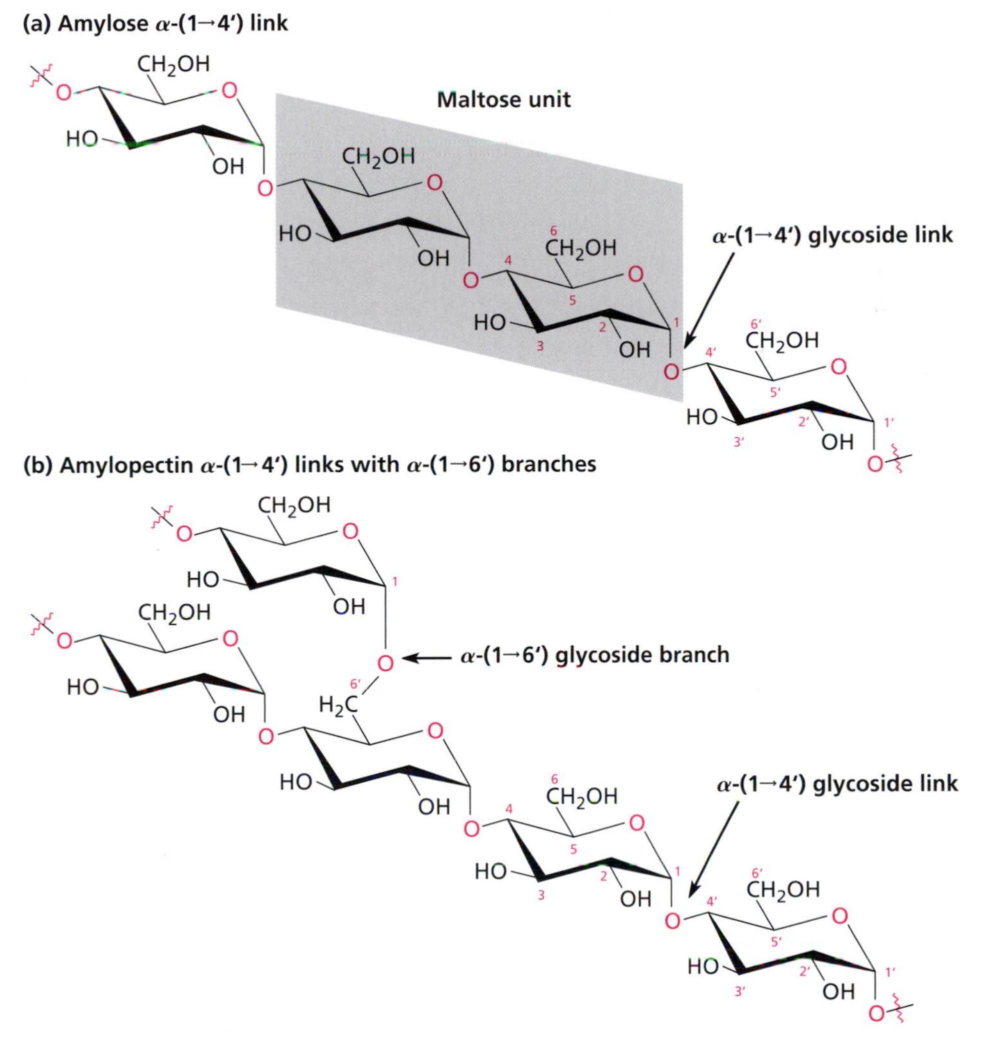

FIGURE 4.1 The structures of (a) amylose and (b) amylopectin, complex carbohydrates found in starch. Amylose is a linear polymer of glucose units connected by α-(1→4′) links. Amylopectin has additional α-(1→6′) branches.

Digestion of starch begins in the mouth, where many of the internal (1→4′) glycoside links, but not the (1→6′) links or terminal (1→4′) links, are randomly hydrolyzed by a glycosidase called α-amylase. Further digestion continues in the small intestine to give a mixture of the disaccharide maltose (Figure 4.1a), the trisaccharide maltotriose, and small oligosaccharides called *limit dextrins*, which contain the (1→6′) branches. Final processing in the intestinal mucosa by additional glycosidases then yields glucose, which is absorbed by the intestine and transported through the bloodstream.

Glycosidase-catalyzed hydrolysis of a glycoside bond in a polysaccharide can occur with either inversion or retention of stereochemistry at the anomeric center.[1–3] Both mechanisms probably proceed through a short-lived oxonium ion in which one face is effectively shielded by the leaving group, leaving the other side open to nucleophilic attack (Figure 4.2a). Inverting glycosidases operate through a single S_N2-like inversion in which a carboxylate residue, either aspartate or glutamate, acts as a base to deprotonate water, which then adds to the oxonium ion from the side opposite the leaving group (Figure 4.2b). Retaining glycosidases operate through two inversions. In the first, a carboxylate ion adds to the oxonium ion from the side opposite the leaving group, giving a covalently bonded, glycosylated enzyme. In the second, water displaces the carboxylate (Figure 4.2c).

4.2 Glucose Catabolism: Glycolysis

Glucose, whether obtained directly from dietary starch, indirectly from the body's store of glycogen, or by synthesis in the liver (gluconeogenesis), is the body's primary energy source for short-term needs. Its catabolism begins with **glycolysis**, a series of 10 enzyme-catalyzed reactions that break down glucose into two equivalents of pyruvate, $CH_3COCO_2^-$. The steps of glycolysis, also called the *Embden–Meyerhoff pathway* after its discoverers, are summarized in Figure 4.3.

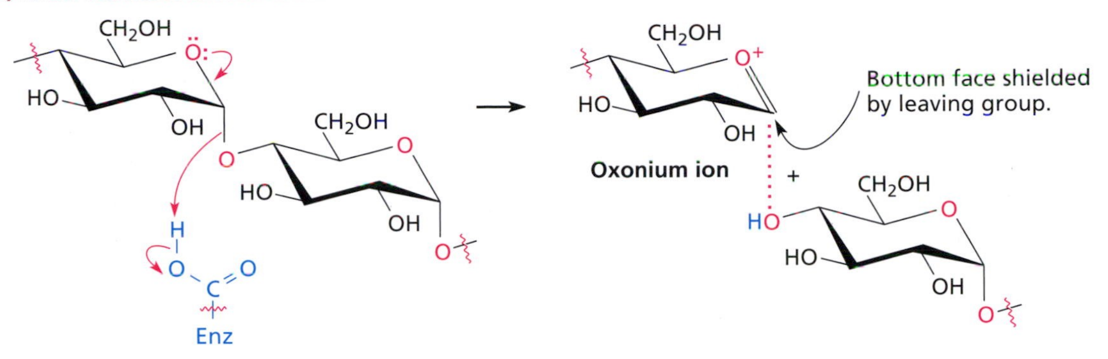

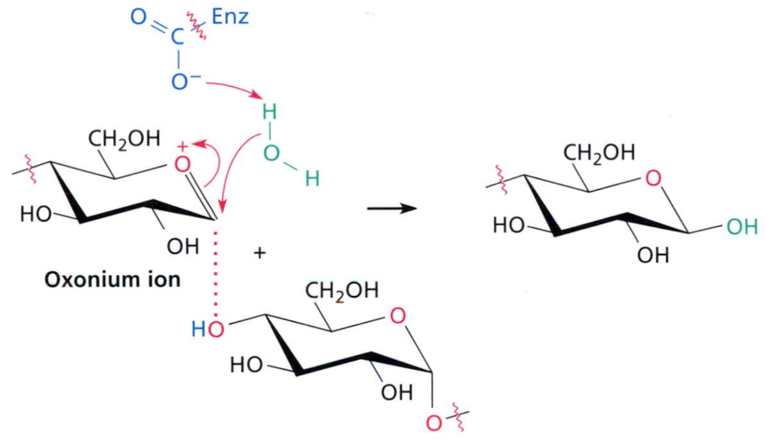

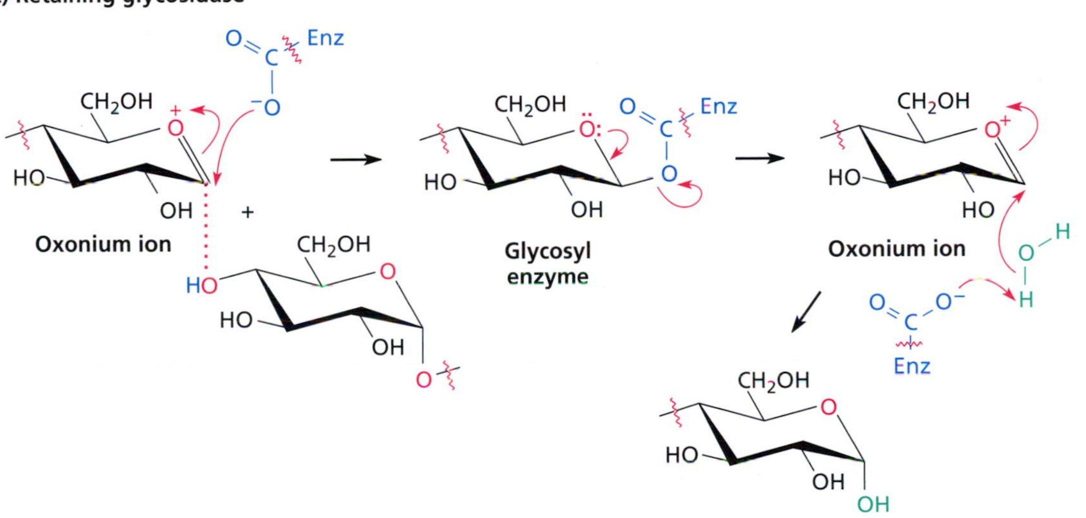

FIGURE 4.2 Generalized mechanism of glycosidases. (**a**) Initial formation of an oxonium ion is followed by either one inversion or two. (**b**) Inverting glycosidases use a single inversion by nucleophilic attack of water. (**c**) Retaining glycosidases use two inversions, the first by a carboxylate ion to give a glycosylated enzyme intermediate and the second by nucleophilic attack of water.

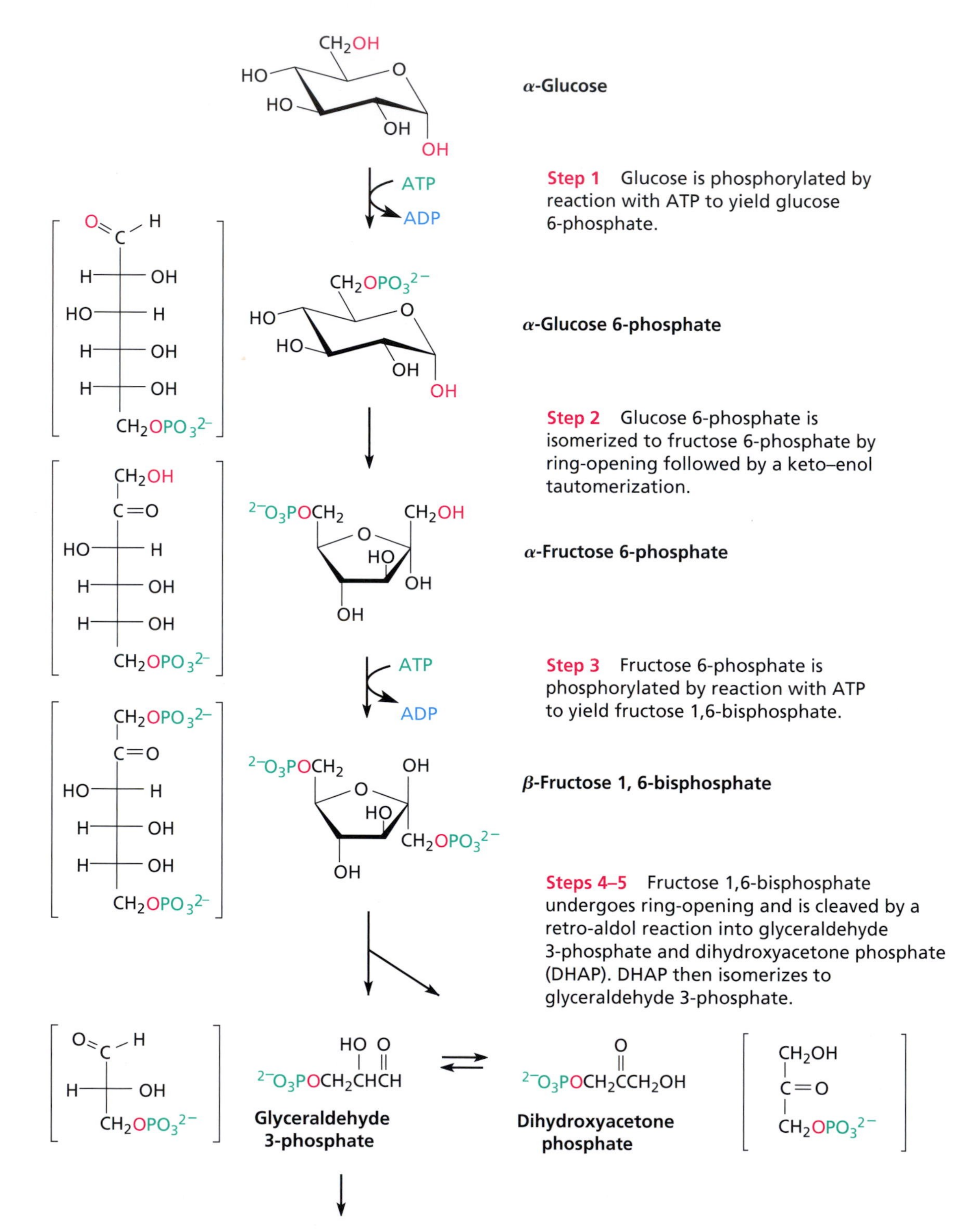

α-Glucose
ATP
ADP
Step 1 Glucose is phosphorylated by reaction with ATP to yield glucose 6-phosphate.
α-Glucose 6-phosphate
Step 2 Glucose 6-phosphate is isomerized to fructose 6-phosphate by ring-opening followed by a keto–enol tautomerization.
α-Fructose 6-phosphate
ATP
ADP
Step 3 Fructose 6-phosphate is phosphorylated by reaction with ATP to yield fructose 1,6-bisphosphate.
β-Fructose 1, 6-bisphosphate
Steps 4–5 Fructose 1,6-bisphosphate undergoes ring-opening and is cleaved by a retro-aldol reaction into glyceraldehyde 3-phosphate and dihydroxyacetone phosphate (DHAP). DHAP then isomerizes to glyceraldehyde 3-phosphate.
Glyceraldehyde 3-phosphate
Dihydroxyacetone phosphate

continues

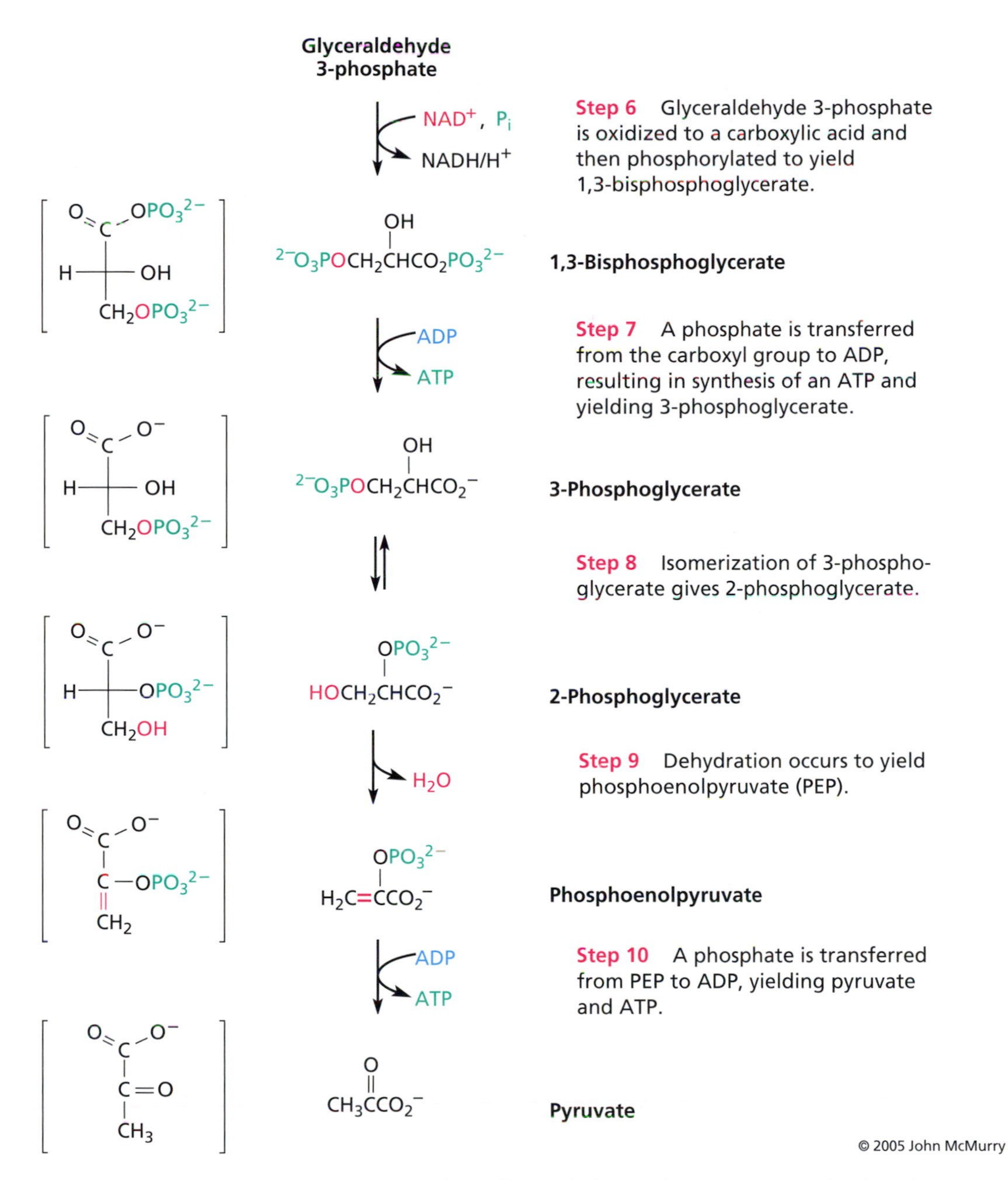

Glyceraldehyde 3-phosphate
NAD+, Pi
NADH/H+
Step 6 Glyceraldehyde 3-phosphate is oxidized to a carboxylic acid and then phosphorylated to yield 1,3-bisphosphoglycerate.
1,3-Bisphosphoglycerate
ADP
ATP
Step 7 A phosphate is transferred from the carboxyl group to ADP, resulting in synthesis of an ATP and yielding 3-phosphoglycerate.
3-Phosphoglycerate
Step 8 Isomerization of 3-phosphoglycerate gives 2-phosphoglycerate.
2-Phosphoglycerate
H2O
Step 9 Dehydration occurs to yield phosphoenolpyruvate (PEP).
Phosphoenolpyruvate
ADP
ATP
Step 10 A phosphate is transferred from PEP to ADP, yielding pyruvate and ATP.
Pyruvate

FIGURE 4.3 The 10-step glycolysis pathway for catabolizing glucose to two molecules of pyruvate. The individual steps are described in more detail in the text.

Step 1. Phosphorylation Glucose, produced by the digestion of dietary carbohydrates, is first phosphorylated at the C6 hydroxyl group by reaction with ATP in a process catalyzed by hexokinase.[4] As in the phosphorylation of glycerol shown in Section 3.2, Figure 3.4, the reaction requires Mg^{2+} as a cofactor to complex with the negatively charged ATP. Glucose is tightly bound in the enzyme active site, with an aspartate residue acting as base to deprotonate the C6 hydroxyl and with several asparagine and glutamate residues forming hydrogen bonds to other hydroxyls. (As noted previously, the phosphorylation is shown in an abbreviated form without drawing the pentacoordinate phosphorus intermediate to save space.)

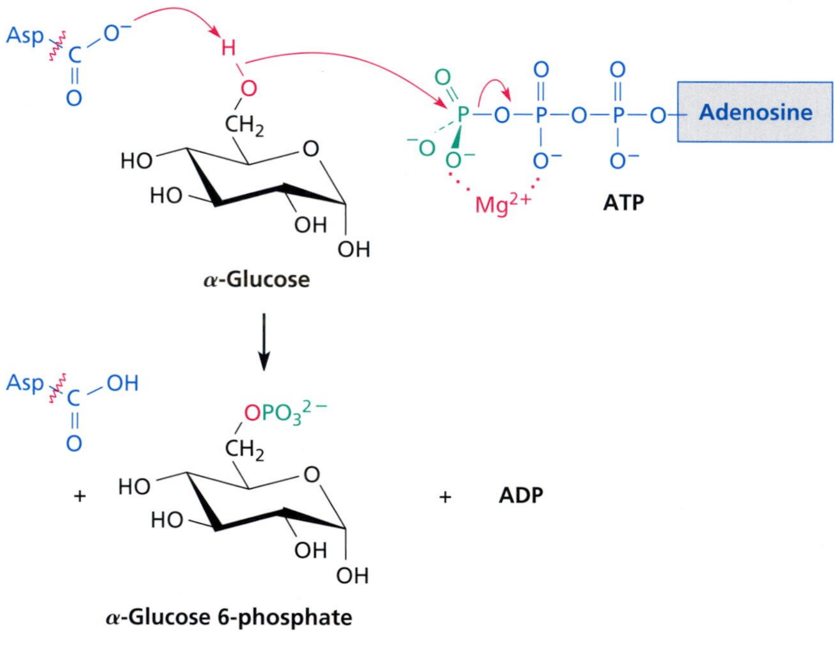

α-Glucose + ATP

α-Glucose 6-phosphate + ADP

Step 2. Isomerization The glucose 6-phosphate formed in step 1 is isomerized by glucose 6-phosphate isomerase (phosphoglucose isomerase) to give fructose 6-phosphate. The isomerization reaction takes place through *tautomerization*, the rapid interconversion of keto and enol forms of a carbonyl compound. Keto–enol tautomerism is catalyzed by both acid and base, and the equilibrium normally favors the keto form. Acetone, for instance, contains only about 0.000 000 1% enol form and 99.999 999 9% keto form at room temperature.

H_3O^+ or OH^-

Keto form **Enol form**

The glucose–fructose interconversion is a multistep process whose details are not yet fully understood.[5–7] The process begins by opening of the hemiacetal ring to an open-chain aldehyde, which undergoes keto–enol tautomerization to its enol form (actually, its ene*di*ol: HO—C═C—OH). But because glucose and fructose share a common enediol, further tautomerization to a different keto form produces open-chain fructose, and cyclization completes the process (Figure 4.4).

The initial ring-opening is thought to be catalyzed by protonation of the ring oxygen, either by the side-chain ammonium group of a lysine residue or by the imidazolium group of a histidine residue. The open-chain glucose that results is then deprotonated by removal of the acidic hydrogen at C2 by glutamate, followed by reprotonation on the C1 oxygen atom to yield the enediol. Deprotonation of the enediol and reprotonation by glutamate with addition of a *pro-R* hydrogen to C1 yields fructose, which then undergoes cyclization catalyzed by a lysine (or histidine) residue (Figure 4.4). When the reaction is carried out using glucose labeled with deuterium at C2, some of the deuterium is lost to solvent but some is added back to C1. This result indicates that the glutamate residue is probably positioned in a hydrophobic pocket of the enzyme active site so that the deuterated glutamate undergoes only slow proton exchange with solvent. The entire process is reversible, with an equilibrium constant near unity.

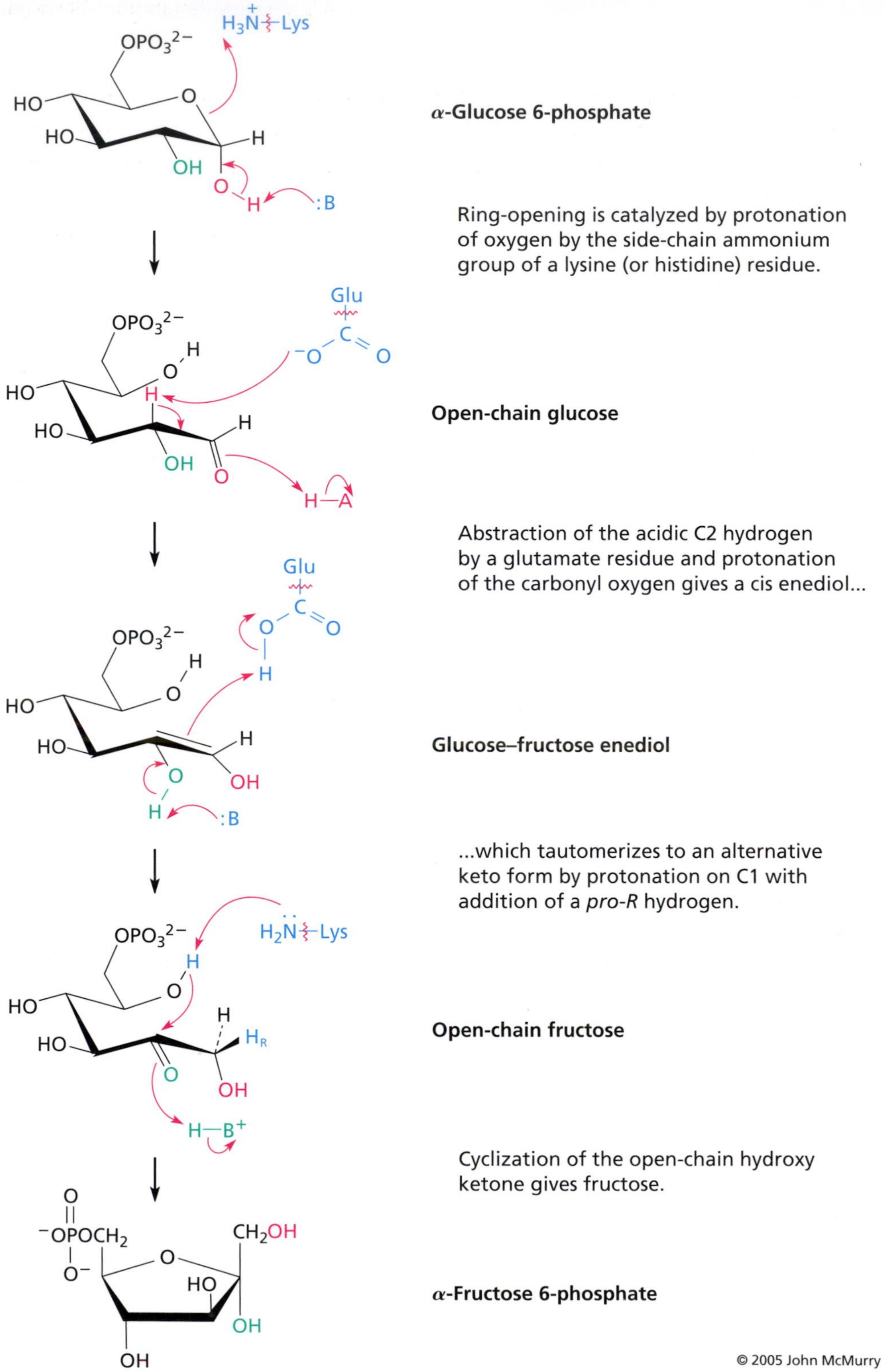

α-Glucose 6-phosphate

Ring-opening is catalyzed by protonation of oxygen by the side-chain ammonium group of a lysine (or histidine) residue.

Open-chain glucose

Abstraction of the acidic C2 hydrogen by a glutamate residue and protonation of the carbonyl oxygen gives a cis enediol...

Glucose–fructose enediol

...which tautomerizes to an alternative keto form by protonation on C1 with addition of a *pro-R* hydrogen.

Open-chain fructose

Cyclization of the open-chain hydroxy ketone gives fructose.

α-Fructose 6-phosphate

FIGURE 4.4 Mechanism of the isomerization of glucose 6-phosphate to fructose 6-phosphate, catalyzed by phosphoglucose isomerase.

Step 3. Phosphorylation Fructose 6-phosphate is next converted to fructose 1,6-bisphosphate (FBP) by a phosphofructokinase-catalyzed reaction with ATP (recall that the prefix *bis-* means two). The mechanism[8] is similar to that in step 1, with Mg^{2+} ion again required as cofactor. Interestingly, the product of step 2 is the α anomer of fructose 6-phosphate but it is the β anomer that is phosphorylated in step 3, implying that the two anomers equilibrate rapidly through the open-chain form.

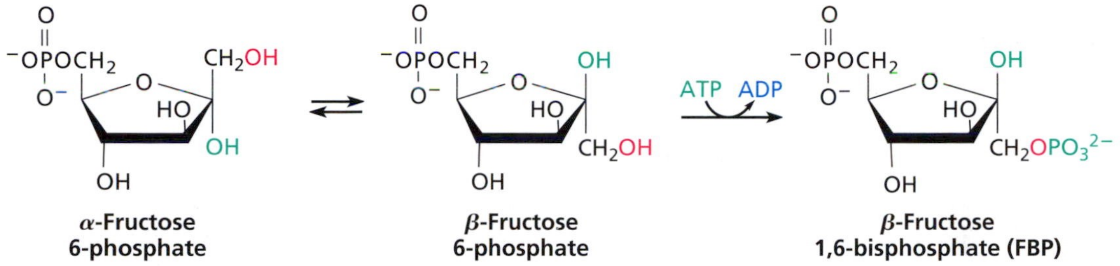

α-Fructose 6-phosphate ⇌ **β-Fructose 6-phosphate** —(ATP → ADP)→ **β-Fructose 1,6-bisphosphate (FBP)**

Steps 4–5. Cleavage and isomerization Fructose 1,6-bisphosphate is cleaved in step 4 into the two 3-carbon pieces, dihydroxyacetone phosphate (DHAP) and glyceraldehyde 3-phosphate (GAP). The bond between C3 and C4 of fructose 1,6-bisphosphate breaks, and a C=O group is formed at C4. The cleavage is a retro-aldol reaction (Section 1.7) and is catalyzed by an aldolase.

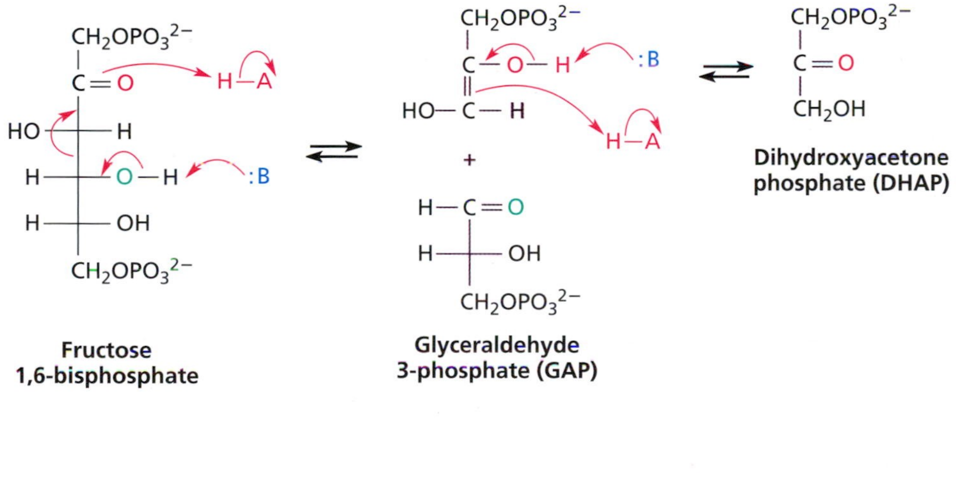

Fructose 1,6-bisphosphate ⇌ **Glyceraldehyde 3-phosphate (GAP)** + enediol ⇌ **Dihydroxyacetone phosphate (DHAP)**

Two classes of aldolases are used by organisms for catalysis of the retro-aldol reaction. In fungi, algae, and some bacteria, the retro-aldol reaction is catalyzed by class II aldolases, which function by coordination of the fructose carbonyl group with Zn^{2+} as the Lewis acid. In plants and animals, however, the reaction is catalyzed by class I aldolases[9] and does not take place on the free ketone. Instead, fructose 1,6-bisphosphate undergoes reaction with the side-chain —NH_2 group of a lysine residue (Lys-229) on the aldolase to yield an enzyme-bound iminium ion (Section 1.5), or protonated Schiff base. Because of its positive charge, the iminium ion is more reactive as an electron acceptor than a ketone carbonyl group. Retro-aldol reaction ensues, giving glyceraldehyde 3-phosphate and an enamine ($R_2N—C═C$), which is protonated to give another iminium ion. Nucleophilic addition of water then hydrolyzes the iminium ion to yield dihydroxyacetone phosphate (Figure 4.5).

Dihydroxyacetone phosphate is next isomerized to glyceraldehyde 3-phosphate by triose phosphate isomerase. As in the conversion of glucose 6-phosphate to fructose 6-phosphate in step 2, the isomerization takes place by keto–enol tautomerization through a common enediol. Glutamate-165 anion is thought to be the base that effects deprotonation of C1 by removal of the *pro-R* hydrogen,[10] and the same glutamate is the acid that reprotonates on C2 using the same hydrogen (Figure 4.6). The net result of steps 4 and 5 is the production of two glyceraldehyde 3-phosphate molecules, both of which pass down the rest of the pathway. Thus, each of the remaining five steps of glycolysis takes place twice for every glucose molecule that enters at step 1.

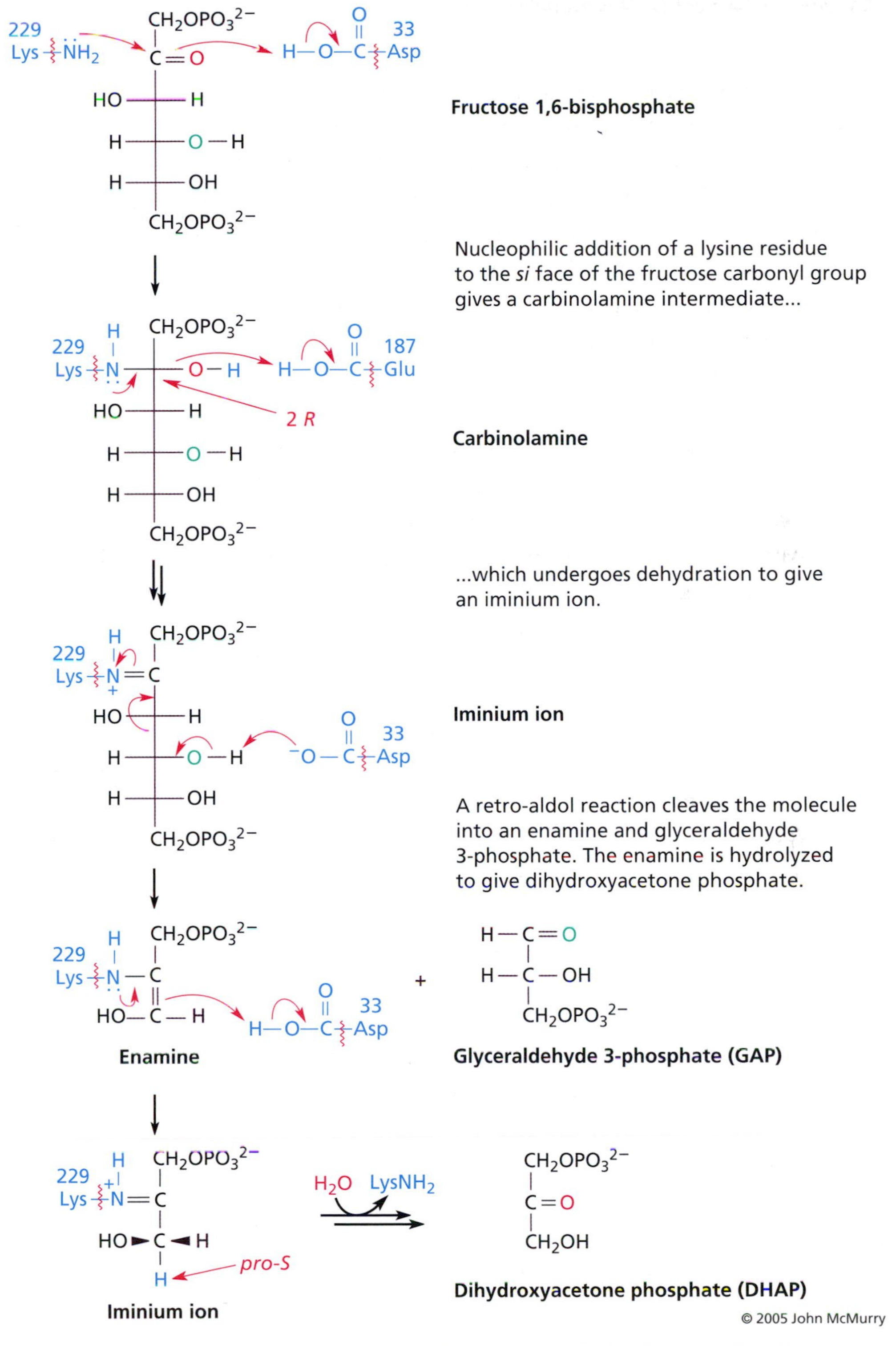

229
Lys
NH_2
$CH_2OPO_3^{2-}$
C=O
33
H—O—C—Asp
HO—H
H—O—H
H—OH
Fructose 1,6-bisphosphate
Nucleophilic addition of a lysine residue to the *si* face of the fructose carbonyl group gives a carbinolamine intermediate...
187
Glu
2 *R*
Carbinolamine
...which undergoes dehydration to give an iminium ion.
Iminium ion
A retro-aldol reaction cleaves the molecule into an enamine and glyceraldehyde 3-phosphate. The enamine is hydrolyzed to give dihydroxyacetone phosphate.
Enamine
Glyceraldehyde 3-phosphate (GAP)
H_2O LysNH$_2$
pro-S
Iminium ion
Dihydroxyacetone phosphate (DHAP)

FIGURE 4.5 Mechanism of the cleavage reaction of fructose 1,6-bisphosphate, catalyzed by a class I aldolase.

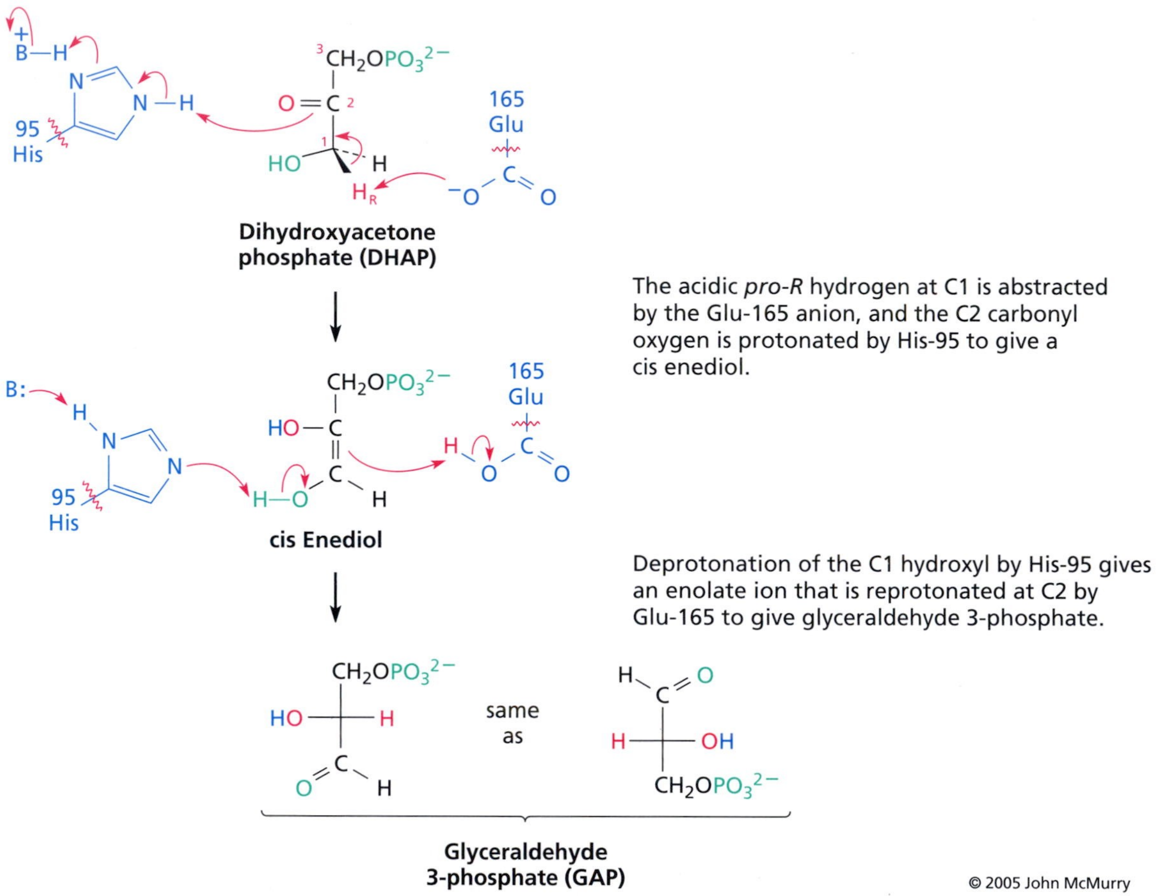

95 His
165 Glu
Dihydroxyacetone phosphate (DHAP)
The acidic *pro-R* hydrogen at C1 is abstracted by the Glu-165 anion, and the C2 carbonyl oxygen is protonated by His-95 to give a cis enediol.
cis Enediol
Deprotonation of the C1 hydroxyl by His-95 gives an enolate ion that is reprotonated at C2 by Glu-165 to give glyceraldehyde 3-phosphate.
same as
Glyceraldehyde 3-phosphate (GAP)

FIGURE 4.6 Mechanism of the keto–enol tautomerization of dihydroxyacetone phosphate to glyceraldehyde 3-phosphate. The same *pro-R* hydrogen removed in the first step is added in the second step.

Steps 6–8. Oxidation and phosphorylation Glyceraldehyde 3-phosphate is oxidized and phosphorylated to give 1,3-bisphosphoglycerate (Figure 4.7). The reaction is catalyzed by glyceraldehyde 3-phosphate dehydrogenase and begins by nucleophilic addition of the —SH group of a cysteine residue in the enzyme to the aldehyde carbonyl group to yield a *hemithioacetal*, the sulfur analog of a hemiacetal. Oxidation of the hemithioacetal —OH group by NAD^+ then yields a thioester, which reacts with phosphate ion in a nucleophilic acyl substitution step to yield 1,3-bisphosphoglycerate, a mixed anhydride between a carboxylic acid and phosphoric acid. The NAD^+ oxidation is stereospecific, with addition of hydride ion to the *si* face of the nicotinamide ring.

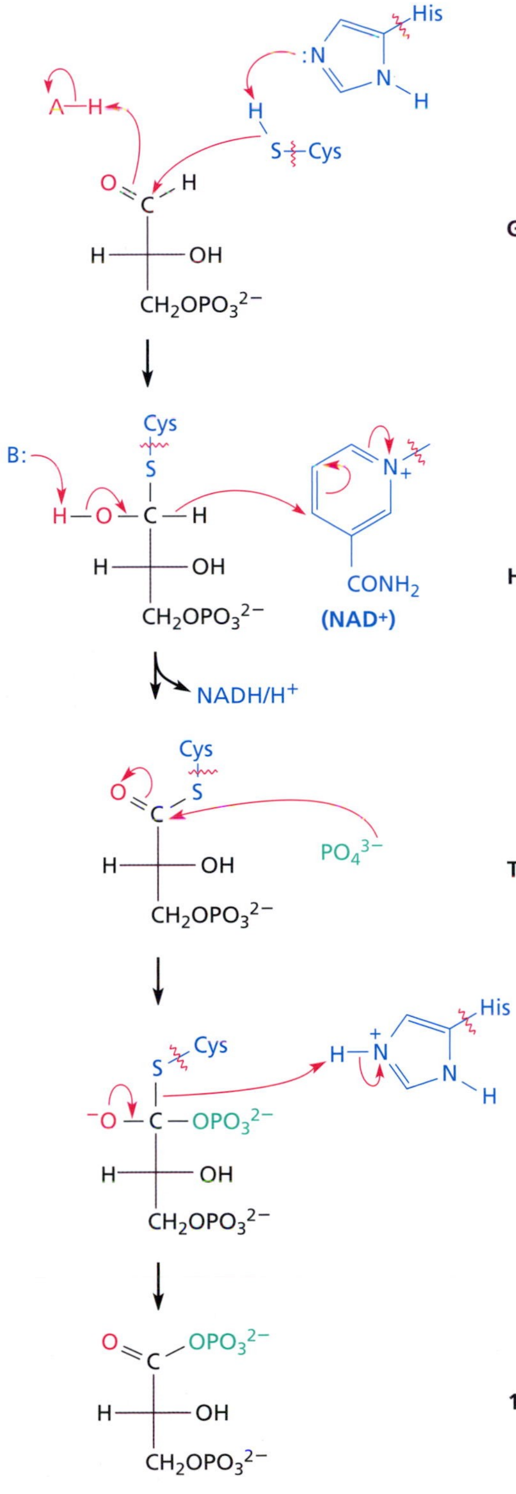

Glyceraldehyde 3-phosphate

Nucleophilic addition of the —SH group in a cysteine residue yields an enzyme-bound hemithioacetal...

Hemithioacetal

(NAD^+)

...which is oxidized by NAD^+ by the usual hydride-transfer mechanism to give a thioester.

$NADH/H^+$

Thioester

Nucleophilic addition of phosphate ion to the thioester yields a tetrahedral intermediate...

...which eliminates the cysteine in an acyl substitution reaction to give 1,3-bisphosphoglycerate.

1,3-Bisphosphoglycerate

FIGURE 4.7 Mechanism of the oxidation and phosphorylation of glyceraldehyde 3-phosphate to 1,3-bisphosphoglycerate.

Like all anhydrides, the mixed carboxylic–phosphoric anhydride formed in step 6 is very reactive in nucleophilic acyl (phosphoryl) substitution reactions—a so-called "high-energy compound." Reaction of 1,3-bisphosphoglycerate with ADP occurs with nucleophilic substitution on phosphorus and results in transfer of a phosphoryl group to yield ATP and 3-phosphoglycerate. The process is catalyzed by phosphoglycerate kinase and requires Mg^{2+} as cofactor. Note that steps 6 and 7 together accomplish the conversion of an aldehyde to a carboxylic acid.

1,3-Bisphosphoglycerate + **ADP** ⟶ **3-Phosphoglycerate** + **ATP**

Isomerization of 3-phosphoglycerate next gives 2-phosphoglycerate in a step catalyzed by phosphoglycerate mutase. In plants, the substrate transfers its phosphoryl group from the C3 oxygen to a histidine residue on the enzyme in one step and then accepts the same phosphoryl group back onto the C2 oxygen in a second step. In animals and yeast, however, the enzyme functions differently: The active enzyme contains a phosphorylated histidine (His-8), which transfers its phosphoryl group to the C2 oxygen of 3-phosphoglycerate and forms 2,3-bisphosphoglycerate as intermediate.[11] The same histidine then accepts a phosphoryl group from the C3 oxygen to yield the isomerized product and regenerated active enzyme. Thus, the phosphoryl group removed from one molecule of substrate is moved into position on the enzyme and added to the next molecule. A glutamic acid residue (Glu-86) is thought to be the active-site base.

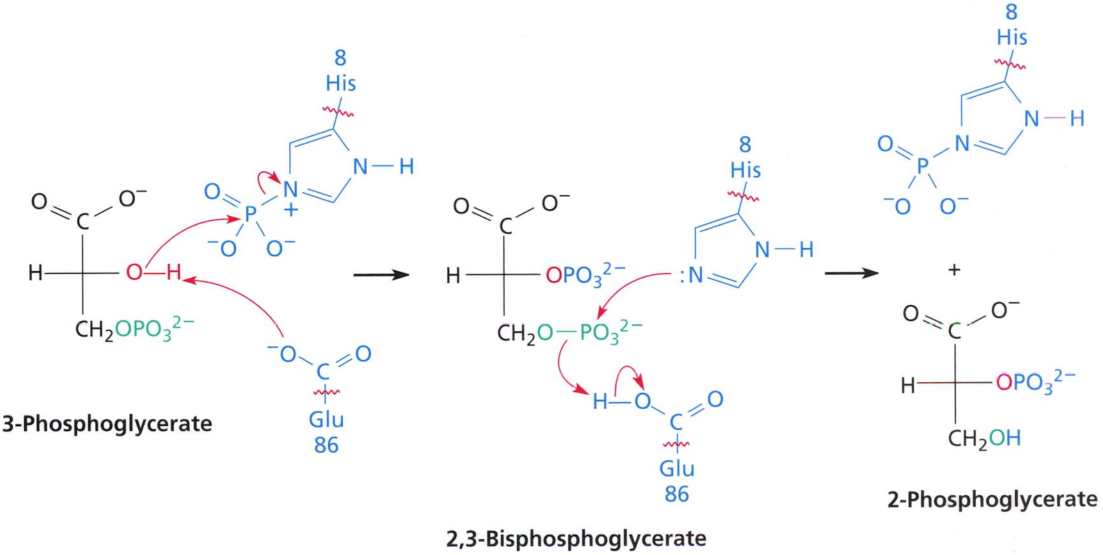

3-Phosphoglycerate

2,3-Bisphosphoglycerate

2-Phosphoglycerate

Steps 9–10. Dehydration and dephosphorylation Like most β-hydroxy carbonyl compounds produced in aldol reactions, 2-phosphoglycerate undergoes a ready dehydration (Section 1.8). The process is catalyzed by enolase,[12, 13] and the product is phosphoenolpyruvate, abbreviated PEP. Two Mg^{2+} ions are associated with the 2-phosphoglycerate to neutralize the negative charges, and a lysine residue in the active site of enolase (Lys-345) is the base that abstracts the acidic α hydrogen. Labeling studies have shown that the α hydrogen exchanges with the medium faster than elimination occurs, so the mechanism is probably an E1cB process. Glu-211 is thought to be the acid catalyst that protonates the leaving group.

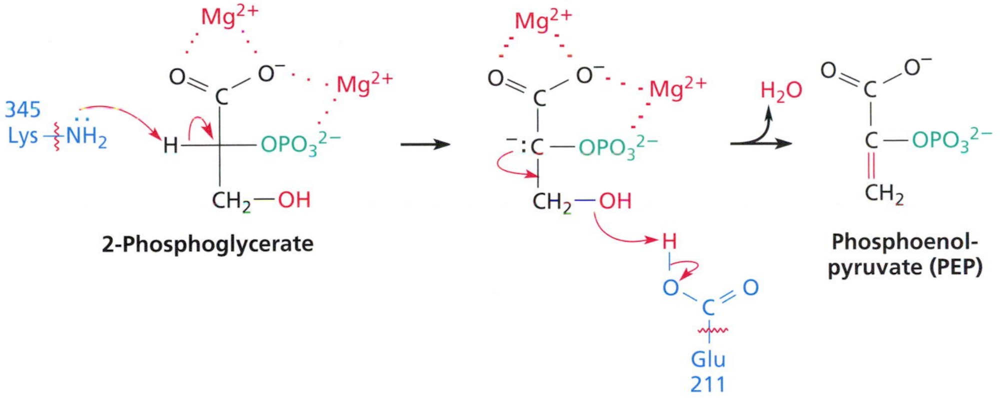

2-Phosphoglycerate

Phosphoenolpyruvate (PEP)

Transfer of the phosphoryl group to ADP in a reaction catalyzed by pyruvate kinase[14] then generates ATP and gives enolpyruvate, which undergoes tautomerization to pyruvate. Full enzyme activity requires that a molecule of fructose 1,6-bisphosphate also be present, as well as two equivalents of Mg^{2+}. One Mg^{2+} ion coordinates to ADP, and the other acts to increase the acidity of a water molecule necessary for protonation of the enolate ion.

ADP
ATP

Phosphoenol-pyruvate **Enolpyruvate** **Pyruvate**

The overall result of glycolysis can be summarized by the following equation:

$$\text{Glucose} + 2\,NAD^+ + 2\,P_i + 2\,ADP \longrightarrow 2\ \text{Pyruvate} + 2\,NADH + 2\,ATP + 2\,H_2O$$

Glucose **Pyruvate**

4.3 Transformations of Pyruvate

Pyruvate, produced in the catabolism of glucose, can undergo several further transformations depending on the conditions and on the organism. In the absence of oxygen, pyruvate can be either reduced by NADH to yield lactate [$CH_3CH(OH)CO_2^-$] or, in yeast, fermented to give ethanol. Under typical aerobic conditions in mammals, however, pyruvate is converted by oxidative decarboxylation to acetyl CoA plus CO_2.

Conversion of Pyruvate to Lactate

Pyruvate is converted in muscles during intense activity to (*S*)-lactate in a reaction catalyzed by lactate dehydrogenase. NADH is the reducing coenzyme, and the reaction occurs by transfer of the *pro-R* hydrogen from NADH to the *re* face of pyruvate, with concurrent protonation of the alkoxide by the imidazolium ring of His-195. The NAD^+ produced by the reaction cycles back into step 6 of glycolysis, the oxidation and phosphorylation of glyceraldehyde 3-phosphate to yield 1,3-bisphosphoglycerate.

NADH + Pyruvate (His 195) → *S*-Lactate (L-Lactate] + NAD^+

Conversion of Pyruvate to Ethanol

Pyruvate is converted by yeast under anaerobic conditions to ethanol and CO_2. This fermentation process has been known for more than 2500 years and is, of course, the fundamental step by which alcoholic beverages are made. Two reactions are needed: The first converts pyruvate to acetaldehyde by a decarboxylation, and the second reduces acetaldehyde to ethanol using NADH as cofactor. As shown in Figure 4.8, the first step is catalyzed by yeast pyruvate decarboxylase[15] and requires thiamin diphosphate (TPP) cofactor.

Step 1. Addition of thiamin diphosphate The conversion of pyruvate to acetaldehyde begins by reaction of pyruvate with thiamin diphosphate (TPP) ylid. As discussed in Section 3.5 (Figure 3.22), the hydrogen on the thiazolium ring of thiamin diphosphate is weakly acidic and can be removed by reaction with base to yield a nucleophilic ylid. This nucleophilic ylid adds to the ketone carbonyl group of pyruvate to yield an alcohol addition product.

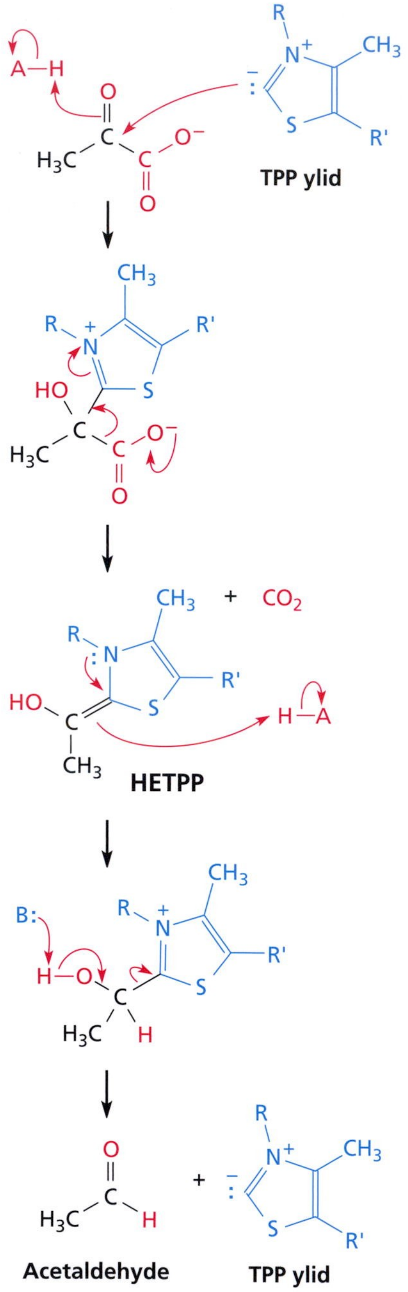

Step 1 Thiamin diphosphate (TPP) ylid does a nucleophilic addition to the ketone carbonyl group of pyruvate to yield an alcohol addition product.

Step 2 The addition product undergoes decarboxylation, giving an enamine, hydroxyethylthiamin diphosphate (HETPP).

Step 3 The enamine double bond is protonated on carbon to give a tetrahedral intermediate...

Step 4 ... which eliminates TPP ylid as the leaving group and gives acetaldehyde.

FIGURE 4.8 Mechanism of the conversion of pyruvate to acetaldehyde, catalyzed by the thiamin-dependent enzyme pyruvate decarboxylase. Individual steps are explained in the text.

Step 2. Decarboxylation Decarboxylation of the pyruvate–thiamin addition product occurs just as in the 1-deoxy-D-xylulose 5-phosphate biosynthesis (Section 3.5, Figure 3.22) to give hydroxyethylthiamin diphosphate (HETPP). The C=N$^+$ double bond of the pyruvate addition product accepts the electrons as CO_2 leaves.

Steps 3–4. Protonation and TPP elimination The double bond of the HETPP decarboxylation product is protonated to give a tetrahedral intermediate, which expels TPP ylid by a mechanism that is the exact reverse of the ketone addition in step 1.

The acetaldehyde resulting from decarboxylation of pyruvate is then reduced by alcohol dehydrogenase to give ethanol. The reaction requires NADH as cofactor and proceeds by transfer of the *pro-R* hydrogen in NADH to the *re* face of acetaldehyde.

R
N
H–A
H_2N
C
O
H H
O
C
H_3C H
H OH
C
H_3C H
+ NAD^+
NADH **Acetaldehyde** **Ethanol**

Conversion of Pyruvate to Acetyl CoA

The third transformation of pyruvate—aerobic conversion to acetyl CoA plus CO_2—occurs through a multistep sequence of reactions catalyzed by a complex of three enzymes and their cofactors called the *pyruvate dehydrogenase complex.*[15, 16] The process occurs in three stages, each catalyzed by one of the enzymes in the complex, as outlined in Figure 4.9. Acetyl CoA, the ultimate product, then acts as "fuel" for the final stage of catabolism, the citric acid cycle.

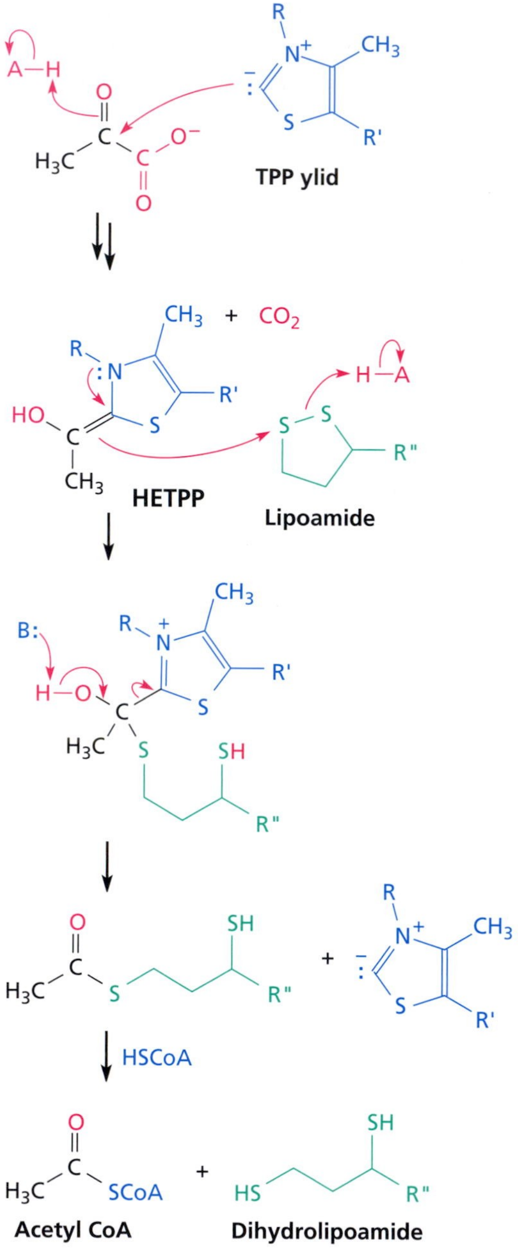

TPP ylid

HETPP **Lipoamide**

Acetyl CoA **Dihydrolipoamide**

Steps 1–2 Thiamin diphosphate (TPP) ylid reacts with pyruvate to form an addition product that undergoes decarboxylation to give HETPP.

Step 3 The enamine double bond attacks a sulfur atom of lipoamide and carries out an S_N2-like displacement of the second sulfur to yield a hemithioacetal.

Step 4 Elimination of thiamin diphosphate ylid from the hemithioacetal intermediate yields acetyl dihydrolipoamide...

Step 5 ...which reacts with coenzyme A in a nucleophilic acyl substitution reaction to exchange one thioester for another and give acetyl CoA plus dihydrolipoamide.

FIGURE 4.9 Mechanism of the conversion of pyruvate to acetyl CoA through a multistep sequence of reactions that requires three different enzymes and five different coenzymes. The individual steps are explained in more detail in the text.

Steps 1–2. Addition of thiamin diphosphate and decarboxylation The conversion of pyruvate to acetyl CoA begins by reaction of pyruvate with TPP ylid, followed by decarboxylation of the pyruvate–thiamin addition product to give HETPP, as seen previously in acetaldehyde formation (Figure 4.8).

Step 3. Reaction with lipoamide The HETPP decarboxylation product is a nucleophilic enamine ($R_2N—C=C$), which ultimately transfers its acetyl group to dihydrolipoamide, forming a thioester. The details of the process are not well understood, but a likely route involves reaction with the enzyme-bound disulfide lipoamide by nucleophilic attack on a sulfur atom, displacing the second sulfur in an S_N2-like process.

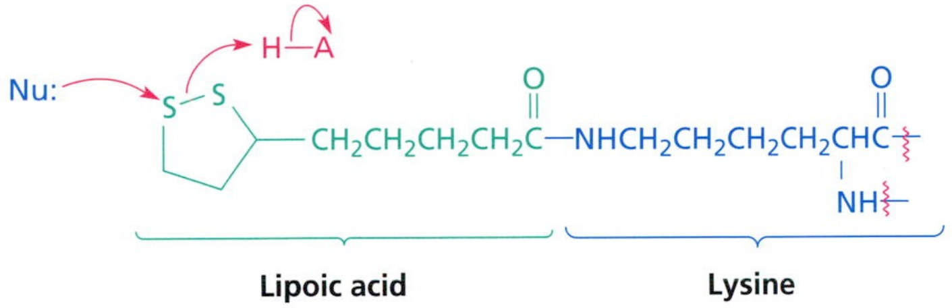

Nu: H—A S—S $CH_2CH_2CH_2CH_2C(=O)—NHCH_2CH_2CH_2CH_2CHC(=O)$ NH

Lipoic acid **Lysine**

Lipoamide: Lipoic acid is linked through an amide bond to a lysine residue in the enzyme

Step 4. Elimination of thiamin diphosphate The product of the HETPP reaction with lipoamide is a hemithioacetal, which can eliminate thiamin diphosphate ylid as the leaving group. This elimination, the reverse of the ketone addition in step 1, generates acetyl dihydrolipoamide.

Step 5. Acyl transfer Steps 1–4, which result in the formation of acetyl dihydrolipoamide, are all catalyzed by the first enzyme (E_1) in the pyruvate dehydrogenase complex.[15, 16] Step 5, reaction of acetyl dihydrolipoamide with coenzyme A to yield acetyl CoA plus dihydrolipoamide, is a typical nucleophilic acyl substitution reaction and is catalyzed by the second enzyme, E_2.[15, 16]

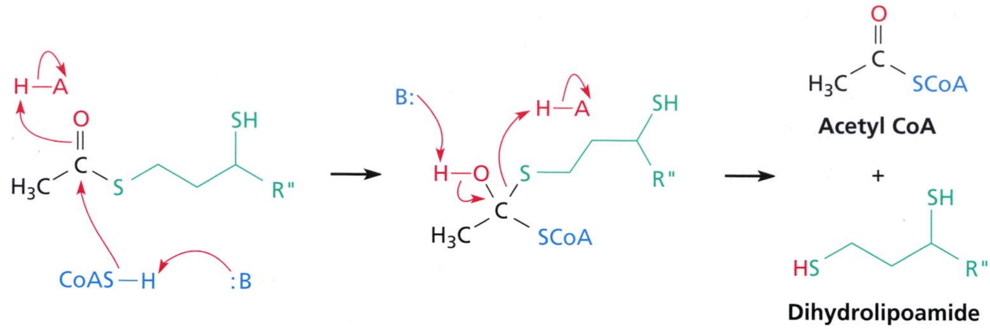

Finally, the dihydrolipoamide is oxidized back to lipoamide by FAD, and the $FADH_2$ that results is in turn oxidized back to FAD by NAD^+ to complete the catalytic cycle. The process is catalyzed by the third enzyme in the complex (E_3, lipoamide dehydrogenase) by the mechanism[15–17] shown in Figure 4.10. The

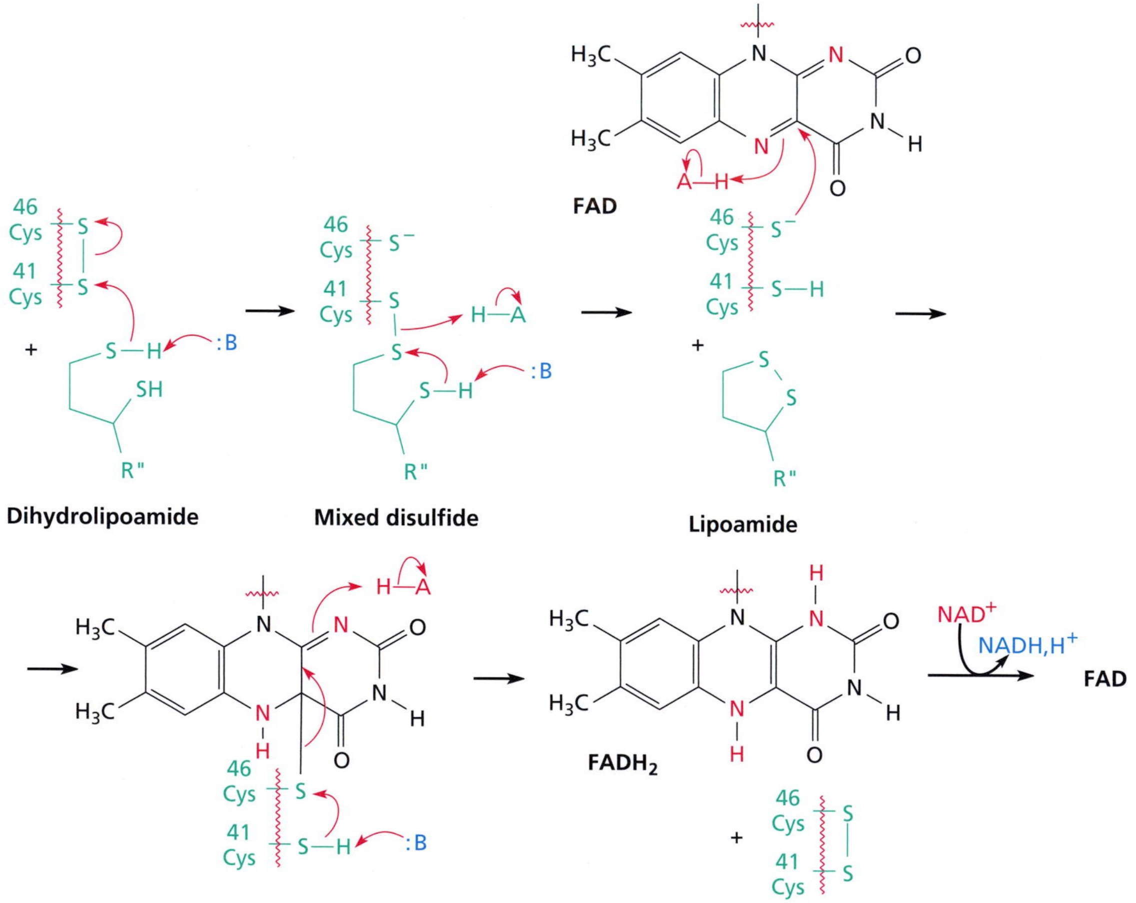

FIGURE 4.10 Mechanism of the oxidation of dihydrolipoamide to lipoamide, catalyzed by lipoamide dehydrogenase.

reaction involves a disulfide link between cysteine residues 41 and 46 and occurs by initial formation of a mixed disulfide between dihydrolipoamide and Cys-41. Closure of the lipoamide ring by nucleophilic addition of the Cys-46 thiol group to FAD, re-formation of the Cys–Cys disulfide bond, and loss of reduced $FADH_2$ complete the process.

Note the difference between the acetaldehyde-forming reaction (Figure 4.8) and the acetyl CoA-forming reaction. The conversion of pyruvate to acetaldehyde is a *nonoxidative* decarboxylation because the HETPP intermediate is simply protonated to yield an aldehyde, but the conversion of pyruvate to acetyl CoA is an *oxidative* decarboxylation because HETPP reacts with lipoamide to yield a thioester.

4.4 The Citric Acid Cycle

The initial stages of catabolism result in the conversion of fats and carbohydrates into acetyl groups that are bonded through a thioester link to coenzyme A. Acetyl CoA now enters the next stage of catabolism—the *citric acid cycle*, also called the *tricarboxylic acid (TCA) cycle*, or *Krebs cycle*, after Hans Krebs, who unraveled its complexities in 1937. The overall result of the cycle is the conversion of an acetyl group into two molecules of CO_2 plus reduced coenzymes by the eight-step sequence of reactions shown in Figure 4.11.

As its name implies, the citric acid *cycle* is a closed loop of reactions in which the product of the final step (oxaloacetate) is a reactant in the first step. The intermediates are constantly regenerated and continuously flow through the cycle, which operates as long as the oxidizing coenzymes NAD^+ and FAD are available. To meet this condition, the reduced coenzymes NADH and $FADH_2$ must be reoxidized via the electron-transport chain, which in turn relies on oxygen as the ultimate electron acceptor. Thus, the cycle is dependent on the availability of oxygen and on the operation of the electron-transport chain.

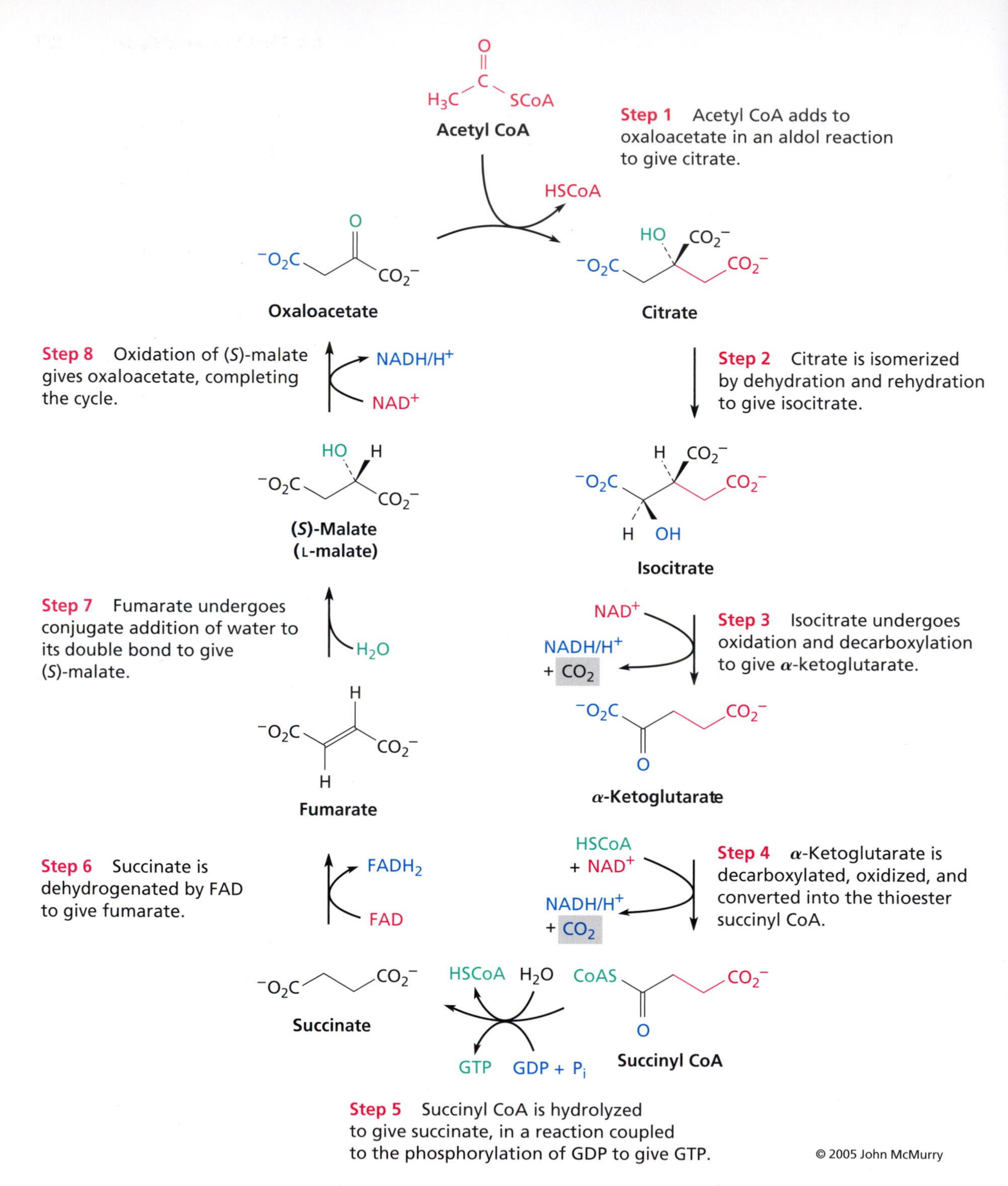

FIGURE 4.11 The citric acid cycle is an eight-step series of reactions that results in the conversion of an acetyl group into two molecules of CO_2 plus reduced coenzymes. Individual steps are explained in more detail in the text.

Step 1. Addition to oxaloacetate Acetyl CoA enters the citric acid cycle in step 1 by aldol-like addition of its enolate ion (or enol) to the *si* face of the ketone carbonyl group of oxaloacetate, giving (*S*)-citryl CoA. The reaction is catalyzed by citrate synthase and occurs by the mechanism[18] shown in Figure 4.12. Asp-375 is the base that deprotonates the acetyl CoA, and His-320 is the acid that protonates the aldol product.

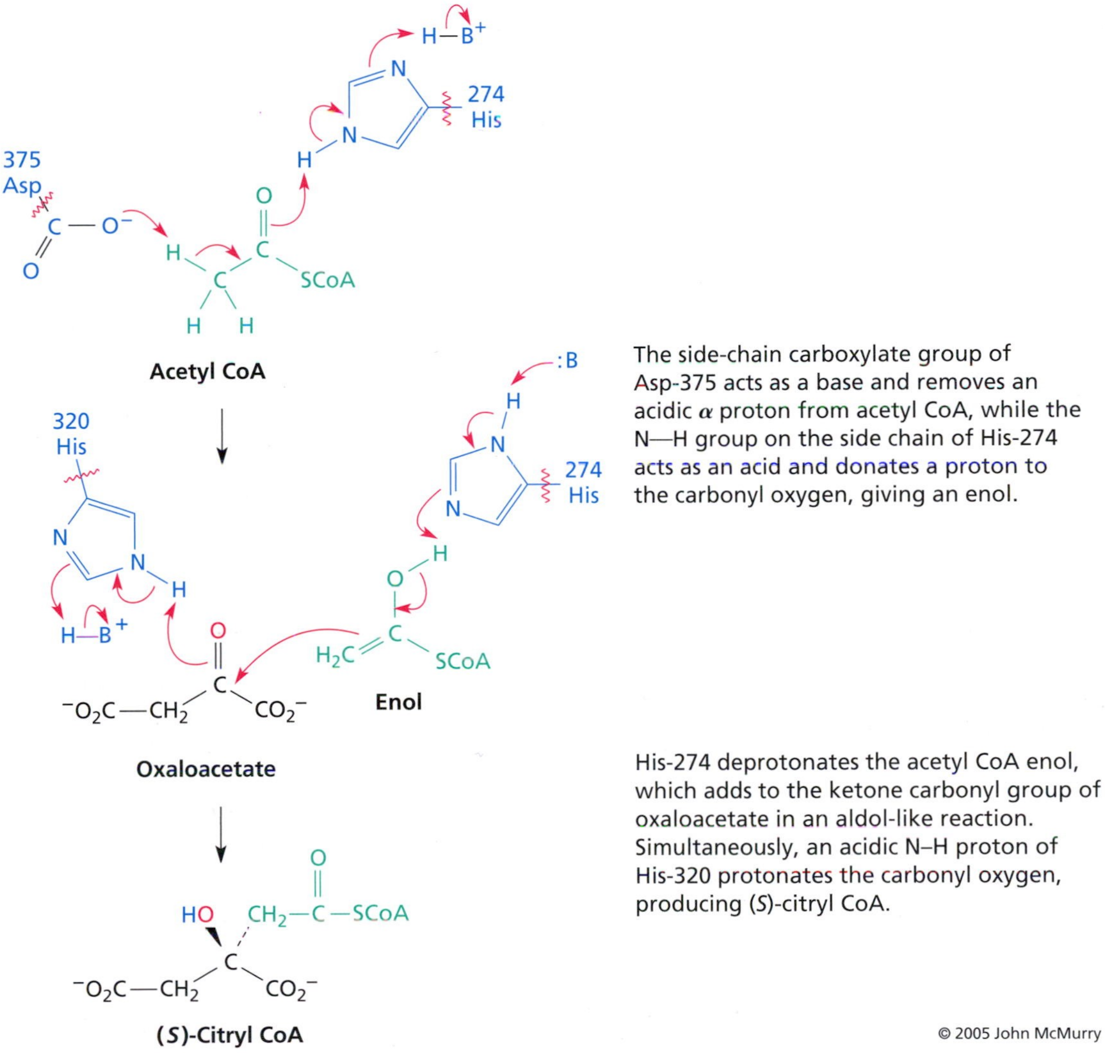

The side-chain carboxylate group of Asp-375 acts as a base and removes an acidic α proton from acetyl CoA, while the N—H group on the side chain of His-274 acts as an acid and donates a proton to the carbonyl oxygen, giving an enol.

His-274 deprotonates the acetyl CoA enol, which adds to the ketone carbonyl group of oxaloacetate in an aldol-like reaction. Simultaneously, an acidic N–H proton of His-320 protonates the carbonyl oxygen, producing (*S*)-citryl CoA.

FIGURE 4.12 Mechanism of the enzyme-catalyzed addition of acetyl CoA to oxaloacetate to give (*S*)-citryl CoA.

Following its formation, (*S*)-citryl CoA is hydrolyzed to citrate by a typical nucleophilic acyl substitution reaction, catalyzed by the same citrate synthase enzyme. Note that C3 of citrate is a prochiral center, with the *pro-S* arm of the molecule derived from acetyl CoA and the *pro-R* arm derived from oxaloacetate.

(*S*)-Citryl CoA $\xrightarrow{H_2O \quad HSCoA}$ Citrate

Step 2. Isomerization Citrate, a prochiral tertiary alcohol, is converted into its isomer, (2*R*,3*S*)-isocitrate, a chiral secondary alcohol. The isomerization occurs in two steps, both catalyzed by aconitase.[19] The initial step is dehydration of a β-hydroxy acid to give *cis*-aconitate, the same sort of reaction that occurs during dehydration of 2-phosphoglycerate in step 9 of glycolysis (Figure 4.3). The second step is a conjugate nucleophilic addition of water to the C=C bond, the same sort of reaction that occurs in step 2 of the β-oxidation pathway (Figure 3.6).

Citrate $\xrightarrow{H_2O}$ *cis*-Aconitate $\xrightarrow{H_2O}$ (2*R*,3*S*)-Isocitrate

The dehydration of citrate takes place specifically on the *pro-R* arm, away from the carbon atoms of the acetyl group that added to oxaloacetate in step 1, and the *pro-R* hydrogen on that *pro-R* arm is removed, implying *anti* geometry for the elimination. An iron–sulfur cluster (Section 1.2) coordinates to the hydroxyl oxygen and acts as a Lewis-acid catalyst for the dehydration, while a serine residue in the enzyme removes the hydrogen. Labeling studies have shown that the —OH group eliminated in the dehydration step is lost to solvent but the —H is not. That is, the —OH group added in the subsequent hydration step is different from the one previously eliminated, but the H atom added is the same. Furthermore, the hydration step has *anti* stereochemistry, with oxygen adding to the *re* face of C2 and protonation occurring from the *re* face of C3 (Figure 4.13).

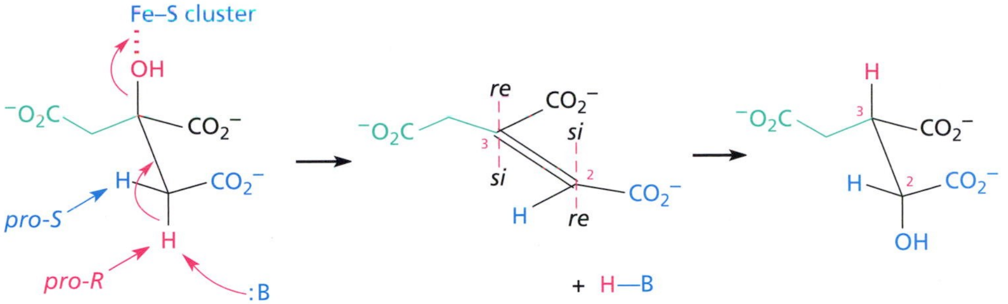

Fe–S cluster

Citrate

cis-Aconitate

(2*R*,3*S*)-Isocitrate

FIGURE 4.13 Stereochemistry of dehydration and hydration steps in the aconitase-catalyzed isomerization of citrate to (2*R*,3*S*)-isocitrate. Both steps occur with *anti* stereochemistry, and the same proton lost from one face during dehydration is added back to the other face during hydration.

The reaction stereochemistry summarized in Figure 4.13 is complex, but think about the implications: The proton removed during dehydration is lost from the *re* face of C2 but is then added in the next step to the *re* face of C3. But the *re* faces of C2 and C3 are 180° apart, on opposite sides of the molecule! Thus, the *cis*-aconitate molecule must flip 180° between steps, being released from the enzyme, rotating, and then reattaching.

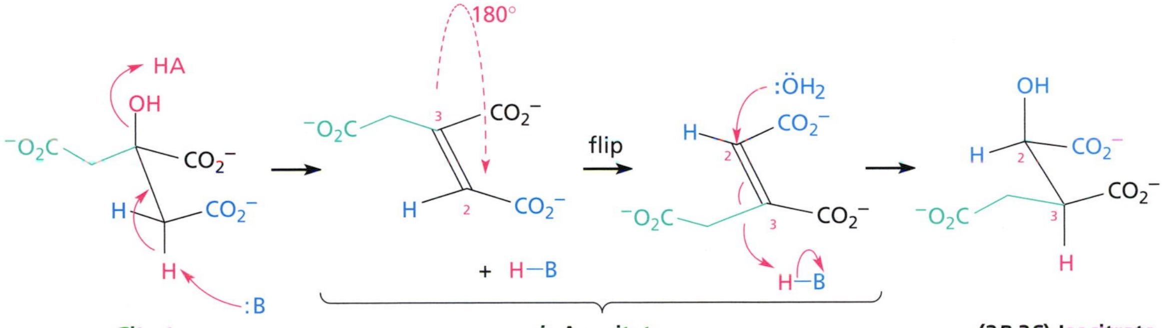

Citrate

cis-Aconitate

(2*R*,3*S*)-Isocitrate

Step 3. Oxidation and decarboxylation (2*R*,3*S*)-Isocitrate, a secondary alcohol, is oxidized by NAD^+ in step 3 to give a ketone, oxalosuccinate, which loses CO_2 to give α-ketoglutarate. Catalyzed by isocitrate dehydrogenase, the decarboxylation is a typical reaction of a β-keto acid, like that in step 5 of fatty-acid synthesis (Section 3.4). The enzyme requires a divalent cation as cofactor, presumably to polarize the ketone carbonyl group.

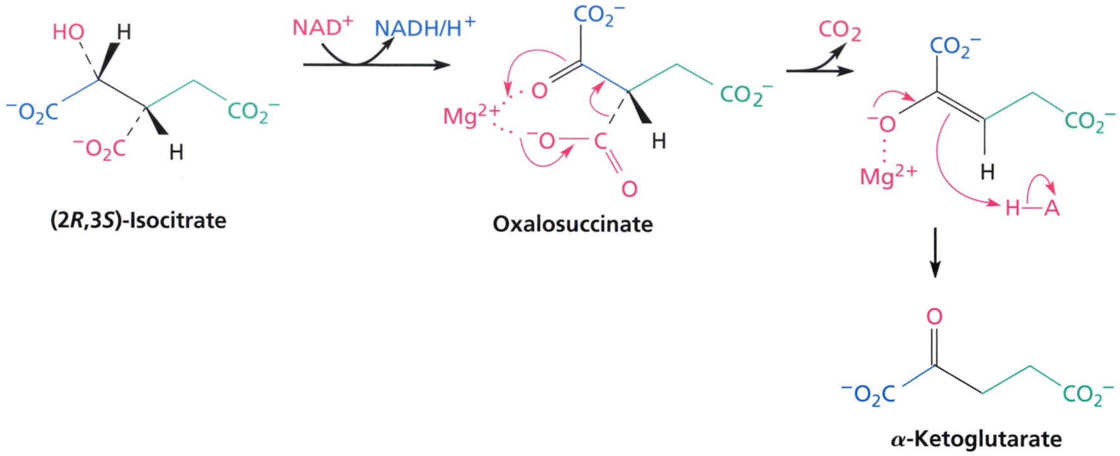

(2*R*,3*S*)-Isocitrate

Oxalosuccinate

α-Ketoglutarate

Step 4. Oxidative decarboxylation The transformation of α-ketoglutarate to succinyl CoA in step 4 is a multistep process analogous to the transformation of pyruvate to acetyl CoA that we saw in Figure 4.9. In both cases, an α-keto acid loses CO_2 and is oxidized to a thioester in a series of steps catalyzed by a multienzyme dehydrogenase complex. The reaction involves an initial nucleophilic addition reaction to α-ketoglutarate by thiamin diphosphate ylid, followed by decarboxylation, reaction with lipoamide, elimination of TPP ylid, and finally a transesterification of the dihydrolipoamide thioester with coenzyme A.

Step 5. Hydrolysis Succinyl CoA is hydrolyzed to succinate in step 5, a reaction that appears to be a straightforward nucleophilic acyl substitution. As in steps 6 and 7 of glycolysis in which an enzyme-bound thioester is hydrolyzed, however, the reaction is more complex than it appears. Catalyzed by succinyl CoA synthetase, the hydrolysis involves an acyl phosphate intermediate and is coupled with phosphorylation of guanosine diphosphate (GDP) to give guanosine triphosphate (GTP). The process is shown in Figure 4.14.

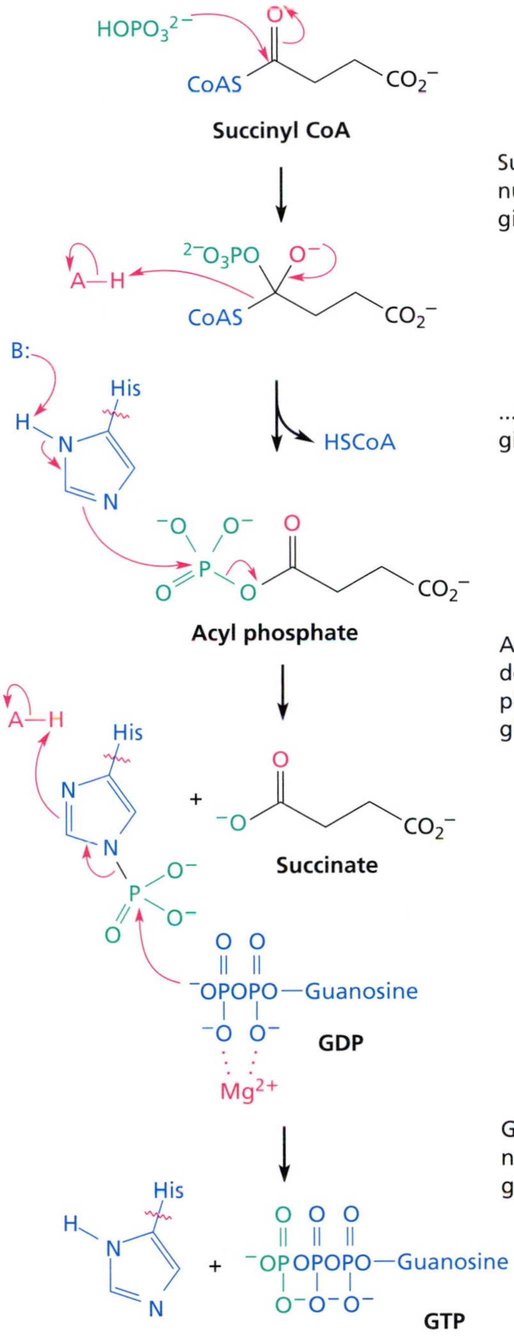

Succinyl CoA

Succinyl CoA adds phosphate ion in a nucleophilic acyl substitution reaction, giving a tetrahedral intermediate...

HSCoA

...which expels CoA as leaving group and gives an acyl phosphate.

Acyl phosphate

A histidine residue in the enzyme does a nucleophilic substitution on phosphorus, expelling succinate and giving an enzyme-bound phosphate.

Succinate

GDP

Mg^{2+}

Guanosine diphosphate (GDP) does a further nucleophilic substitution on phosphorus, giving GTP.

GTP

FIGURE 4.14 Mechanism of the hydrolysis of succinyl CoA in the citric acid cycle. The hydrolysis is coupled to the synthesis of a GTP molecule.

Step 6. Dehydrogenation Succinate is next dehydrogenated by the FAD-dependent succinate dehydrogenase to give fumarate in a process analogous to that of step 1 in the fatty-acid β-oxidation pathway (Section 3.3). The reaction is stereospecific, removing the *pro-S* hydrogen from one carbon and the *pro-R* hydrogen from the other.

Succinate → (Enz-FAD → Enz-$FADH_2$) → Fumarate

Steps 7–8. Hydration and oxidation The penultimate step in the citric acid cycle is the conjugate nucleophilic addition of water to fumarate to yield (*S*)-malate (L-malate). The addition is catalyzed by fumarase and is mechanistically similar to the addition of water to *cis*-aconitate in step 3. The reaction occurs through a carbanion intermediate that is stabilized by the adjacent carbonyl group. Protonation then takes place on the side opposite the OH, leading to a net *anti* addition.

Fumarate → (carbanion) → (*S*)-Malate

The final step is the oxidation of (S)-malate by NAD^+ to give oxaloacetate, a reaction catalyzed by malate dehydrogenase. The citric acid cycle has now returned to its starting point, ready to revolve again. The overall result of the cycle can be summarized as:

$$\underset{\text{Acetyl CoA}}{H_3C{-}C({=}O){-}SCoA} + 3\,NAD^+ + FAD + GDP + P_i + 2\,H_2O \longrightarrow 2\,CO_2 + HSCoA + 3\,NADH + 2\,H^+ + FADH_2 + GTP$$

4.5 Glucose Biosynthesis: Gluconeogenesis

As noted previously in connection with fatty-acid metabolism (Section 3.4), the pathway by which an organism makes a substance is not the reverse of the pathway by which the organism degrades the same substance. **Gluconeogenesis**, the 11-step biosynthetic pathway by which organisms make glucose from pyruvate, is related to glycolysis but is not its exact reverse. The gluconeogenesis pathway is shown in Figure 4.15, and a comparison of glycolysis with gluconeogenesis is shown in Figure 4.16.

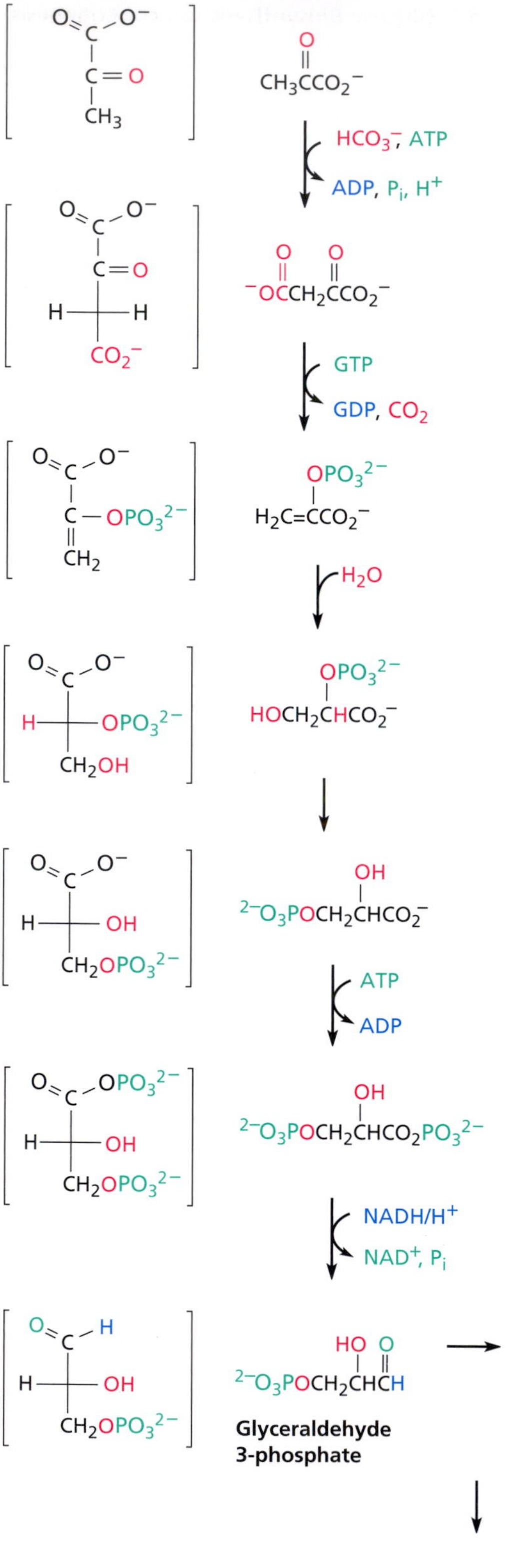

$CH_3COCO_2^-$

Pyruvate

HCO_3^-, ATP → ADP, P_i, H^+

Step 1 Pyruvate undergoes a biotin-dependent carboxylation on the methyl group to give oxaloacetate...

$^-O_2CCH_2COCO_2^-$

Oxaloacetate

GTP → GDP, CO_2

Step 2 ...which is decarboxylated and phosphorylated by GTP to give phosphoenolpyruvate.

$H_2C{=}C(OPO_3^{2-})CO_2^-$

Phosphoenolpyruvate

H_2O

Step 3 Conjugate nucleophilic addition of water to the double bond of phosphoenolpyruvate gives 2-phosphoglycerate...

$HOCH_2CH(OPO_3^{2-})CO_2^-$

2-Phosphoglycerate

Step 4 ...which is isomerized by transfer of the phosphoryl group to give 3-phosphoglycerate.

$^{2-}O_3POCH_2CH(OH)CO_2^-$

3-Phosphoglycerate

ATP → ADP

Step 5 Phosphorylation of the carboxyl group by reaction with ATP yields 1,3-bisphosphoglycerate.

$^{2-}O_3POCH_2CH(OH)CO_2PO_3^{2-}$

1,3-Bisphosphoglycerate

NADH/H^+ → NAD^+, P_i

Steps 6–7 Reduction of the acyl phosphate gives glyceraldehyde 3-phosphate, which undergoes keto–enol tautomerization to yield dihydroxyacetone phosphate.

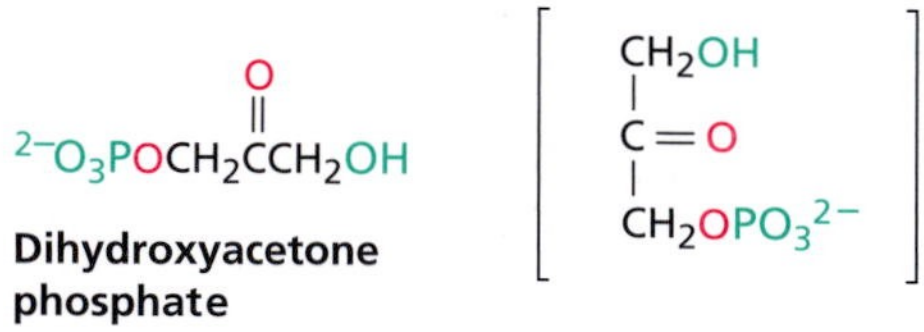

$^{2-}O_3POCH_2CH(OH)CHO$

Glyceraldehyde 3-phosphate

→

$^{2-}O_3POCH_2COCH_2OH$

Dihydroxyacetone phosphate

continues

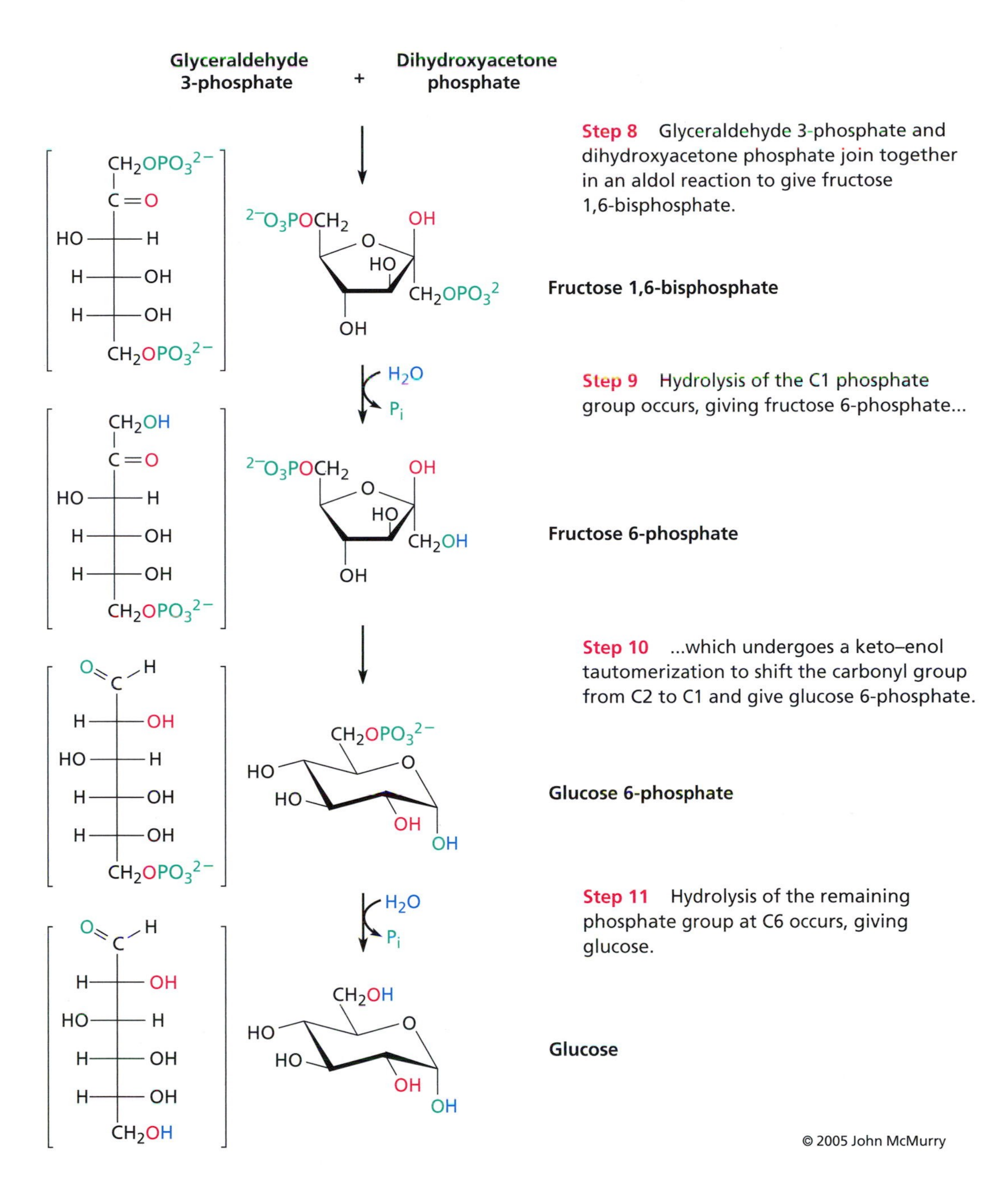

Glyceraldehyde 3-phosphate
+
Dihydroxyacetone phosphate
Step 8 Glyceraldehyde 3-phosphate and dihydroxyacetone phosphate join together in an aldol reaction to give fructose 1,6-bisphosphate.
Fructose 1,6-bisphosphate
H2O
Pi
Step 9 Hydrolysis of the C1 phosphate group occurs, giving fructose 6-phosphate...
Fructose 6-phosphate
Step 10 ...which undergoes a keto–enol tautomerization to shift the carbonyl group from C2 to C1 and give glucose 6-phosphate.
Glucose 6-phosphate
H2O
Pi
Step 11 Hydrolysis of the remaining phosphate group at C6 occurs, giving glucose.
Glucose

FIGURE 4.15 The gluconeogenesis pathway for the biosynthesis of glucose from two molecules of pyruvate. Individual steps are explained in more detail in the text.

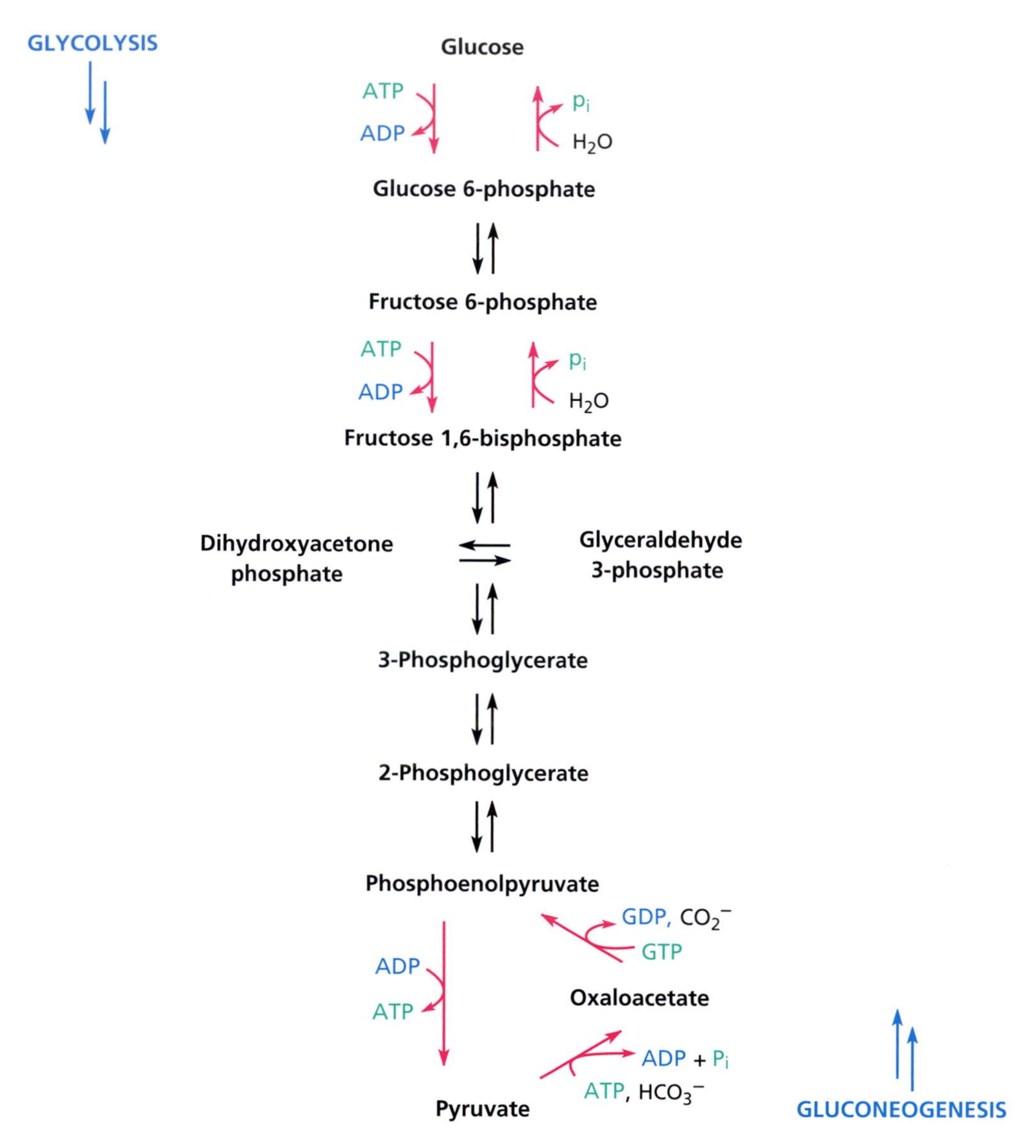

FIGURE 4.16 A comparison of glycolysis and gluconeogenesis pathways. The pathways differ at the three steps indicated by red reaction arrows.

Step 1. Carboxylation Gluconeogenesis begins with the carboxylation of pyruvate to yield oxaloacetate. The reaction is catalyzed by pyruvate carboxylase[20] and, as in the third step of fatty-acid synthesis (Section 3.4, Figures 3.13 and 3.14), requires ATP and the coenzyme biotin, acting as a carrier of CO_2. The mechanism of the carboxylation step is shown in Figure 4.17.

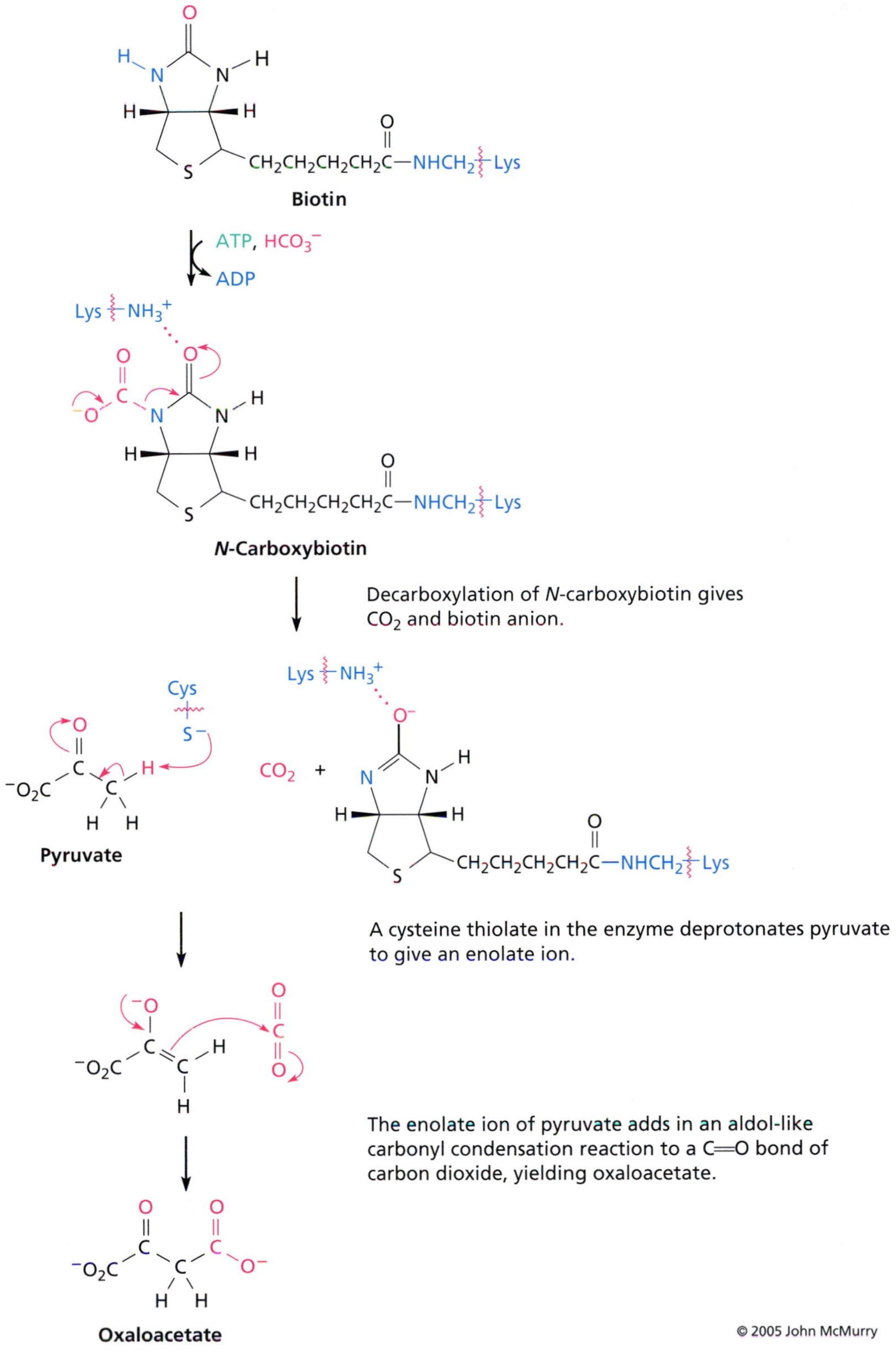

Biotin

ATP, HCO_3^-

ADP

$Lys-NH_3^+$

N-Carboxybiotin

Decarboxylation of *N*-carboxybiotin gives CO_2 and biotin anion.

Cys

Pyruvate

CO_2 +

$Lys-NH_3^+$

A cysteine thiolate in the enzyme deprotonates pyruvate to give an enolate ion.

The enolate ion of pyruvate adds in an aldol-like carbonyl condensation reaction to a C═O bond of carbon dioxide, yielding oxaloacetate.

Oxaloacetate

FIGURE 4.17 Mechanism of the biotin-dependent carboxylation reaction of pyruvate to give oxaloacetate.

Step 2. Decarboxylation and phosphorylation Decarboxylation of oxaloacetate, a β-keto acid, occurs by the usual retro-aldol-like mechanism as in step 5 of fatty-acid synthesis (Section 3.4), and phosphorylation of the resultant pyruvate enolate ion by GTP occurs concurrently to give phosphoenolpyruvate. The reaction is catalyzed by phosphoenolpyruvate carboxykinase.[21]

GTP

Oxaloacetate → Phosphoenolpyruvate + CO_2 + GDP

Why is carbon dioxide added in step 1 and then immediately removed in step 2? Why doesn't the conversion of pyruvate to phosphoenolpyruvate proceed directly in a single step by reaction of pyruvate enolate ion with GTP rather than indirectly in two steps? The answer is in the energetics of the process. Phosphoenolpyruvate is so high in energy that its direct conversion to pyruvate by reaction with ADP in step 10 of glycolysis (Section 4.2) is energetically favorable even though it generates a molecule of ATP. Thus, the reverse process, the direct conversion of pyruvate to phosphoenolpyruvate, must be energetically unfavorable even though it *consumes* a molecule of ATP. (Normally, of course, the reverse is true and reactions that consume ATP are energetically favored.) In the two-step process involving oxaloacetate, however, *two* molecules of nucleoside triphosphate are consumed (one ATP and one GTP), releasing sufficient energy to make the overall process favorable.

Steps 3–4. Hydration and isomerization Conjugate addition of water to the double bond of phosphoenolpyruvate gives 2-phosphoglycerate by a process similar to that of step 2 in the β-oxidation pathway (Section 3.3). Phosphorylation of C3 and dephosphorylation of C2 then yields 3-phosphoglycerate. Mechanistically, these steps are the exact reverse of steps 9 and 8 in glycolysis (Section 4.2), which are near-equilibrium reactions.

$$\underset{\text{Phosphoenol-pyruvate}}{H_2C{=}C(OPO_3^{2-})CO_2^-} \xrightarrow{H_2O} \underset{\text{2-Phospho-glycerate}}{HOCH_2CH(OPO_3^{2-})CO_2^-} \xrightarrow[\text{Enz}]{\text{Enz—}OPO_3^{2-}} \underset{\text{2,3-Bisphospho-glycerate}}{{}^{2-}O_3POCH_2CH(OPO_3^{2-})CO_2^-} \xrightarrow{H_2O \quad P_i} \underset{\text{3-Phospho-glycerate}}{{}^{2-}O_3POCH_2CH(OH)CO_2^-}$$

Steps 5–7. Phosphorylation, reduction, and tautomerization Reaction of 3-phosphoglycerate with ATP generates the corresponding acyl phosphate (1,3-bisphosphoglycerate), which binds to the glyceraldehyde 3-phosphate dehydrogenase by a thioester bond to a cysteine residue and is then reduced by $NADH/H^+$ to the aldehyde. Keto–enol tautomerization of the aldehyde gives dihydroxyacetone phosphate. All three steps are mechanistically the exact reverse of the corresponding steps 7, 6, and 5 of glycolysis and are near-equilibrium reactions.

$$\underset{\text{3-Phospho-glycerate}}{{}^{2-}O_3POCH_2CH(OH)CO_2^-} \xrightarrow{ATP \quad ADP} \underset{\text{1,3-Bisphosphoglycerate}}{{}^{2-}O_3POCH_2CH(OH)C(O)OPO_3^{2-}} \xrightarrow[P_i]{\text{Enz—SH}} \underset{\text{(Enzyme-bound thioester)}}{{}^{2-}O_3POCH_2CH(OH)C(O)S\text{—Cys}} \longrightarrow$$

$$\xrightarrow[\text{Enz—SH},\ NAD^+]{NADH/H^+} \underset{\text{Glyceraldehyde 3-phosphate}}{{}^{2-}O_3POCH_2CH(OH)CHO} \rightleftharpoons \underset{\text{Dihydroxyacetone phosphate}}{{}^{2-}O_3POCH_2C(O)CH_2OH}$$

Step 8. Aldol condensation Dihydroxyacetone phosphate and glyceraldehyde 3-phosphate, the two 3-carbon units produced in step 7, join by an aldol reaction to give fructose 1,6-bisphosphate, the exact reverse of step 4 in glycolysis. As in step 4 of glycolysis (Figure 4.5), this reaction takes place on an iminium ion formed by reaction of dihydroxyacetone phosphate with a side-chain lysine —NH_2 group on the enzyme and is catalyzed by a class I aldolase. Loss of a proton from the neighboring carbon then generates an enamine, an aldol-like reaction ensues, and the product is hydrolyzed.

Iminium ion

Glyceraldehyde 3-phosphate (GAP)

Fructose 1,6-bisphosphate

Steps 9–10. Hydrolysis and isomerization Hydrolysis of the phosphate group at C1 of fructose 1,6-bisphosphate gives fructose 6-phosphate. Although the result of the reaction is the exact opposite of step 3 in glycolysis, the mechanism is not the exact opposite. In glycolysis, the phosphorylation is accomplished by reaction of the fructose C1 hydroxyl group with ATP. The reverse of that process, however—the reaction of fructose 1,6-bisphosphate with ADP to give fructose 6-phosphate and ATP—is energetically unfavorable because ATP is too high in energy. Thus, an alternative pathway is used in which the C1 phosphate group is removed by a direct hydrolysis reaction, catalyzed by fructose 1,6-bisphosphatase.

The mechanism[22, 23] of the hydrolysis is not clear, particularly with respect to whether the process is *associative* or *dissociative.* In the associative pathway, water attacks phosphorus and the alcohol (fructose 6-phosphate) is expelled, forming the products in a single step. In the dissociative pathway, the substrate first dissociates to alcohol plus PO_3^-, which then reacts with water to give $HOPO_3^{2-}$. Available crystallographic data support both possibilities, and both may well operate depending on such conditions as pH and the nature of the metal cations.

Associative mechanism

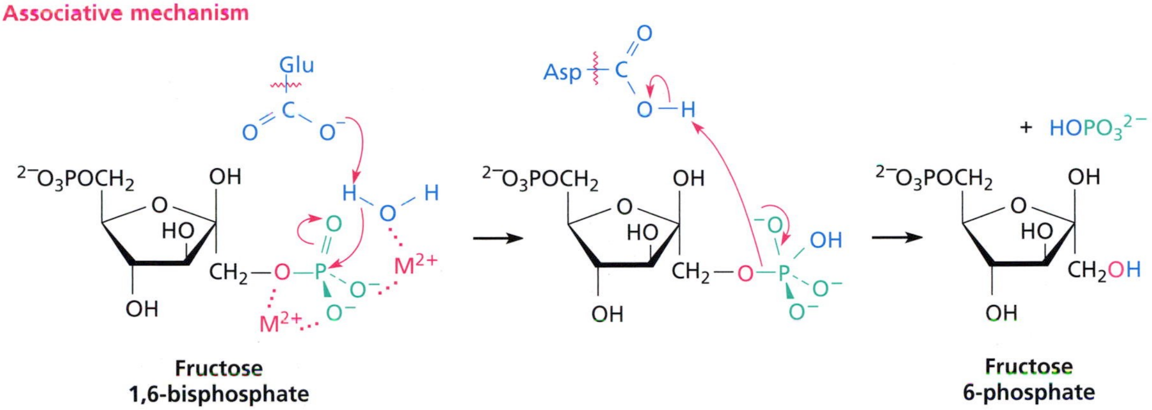

Fructose 1,6-bisphosphate

Fructose 6-phosphate

Dissociative mechanism

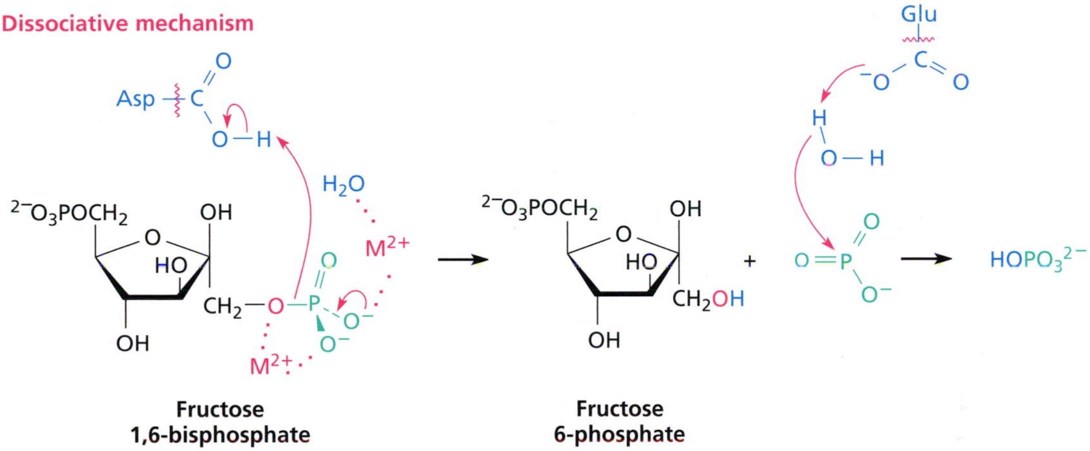

Fructose 1,6-bisphosphate

Fructose 6-phosphate

Following hydrolysis, keto–enol tautomerization of the carbonyl group from C2 to C1 gives glucose 6-phosphate. The isomerization is the exact reverse of step 2 in glycolysis (Figure 4.4).

Step 11. Hydrolysis The final step in gluconeogenesis is the conversion of glucose 6-phosphate to glucose by another phosphatase-catalyzed hydrolysis reaction. As just discussed for the hydrolysis of fructose 1,6-bisphosphate in step 9, and for the same energetic reasons, the mechanism of the glucose 6-phosphate hydrolysis is not the exact opposite of the corresponding step 1 in glycolysis. Also as in step 9, it is not yet clear whether the hydrolysis of glucose 6-phosphate is associative or dissociative.

Interestingly, the mechanisms of the two phosphate hydrolysis reactions in steps 9 and 11 are not the same. In the fructose 1,6-bisphosphate hydrolysis, water is the nucleophile, but in the glucose 6-phosphate reaction, a histidine residue on the enzyme attacks phosphorus, giving a phosphoryl enzyme intermediate that subsequently reacts with water.

A—H
His
:B
Glucose 6-phosphate
Glucose
$^{2-}O_3P$—N
H_2O
$HOPO_3^{2-}$ +

The overall result of gluconeogenesis is summarized by the following equation:

2 Pyruvate + 4 ATP + 2 GTP + 2 NADH + 2 H_2O + 2 H^+ ⟶ Glucose + 4 ADP + 2 GDP + 2 NAD^+ + 6 P_i

4.6 The Pentose Phosphate Pathway

In brain and muscle cells, where energy may be needed quickly, glucose is metabolized almost entirely by the glycolysis pathway (Section 4.2). In tissues that synthesize fatty acids and steroids, however, a second route for glucose metabolism called the **pentose phosphate pathway** is also operative. The pathway has several purposes: It allows the metabolism of five-carbon sugars, it produces the reduced coenzyme NADPH needed for other cellular functions, it produces the ribose 5-phosphate needed for ribonucleotide biosynthesis, and it synthesizes the erythrose 4-phosphate used by plants and microorganisms to make aromatic amino acids (Chapter 5).

Outlined in Figure 4.18, the pentose phosphate pathway can be divided into two stages, an initial oxidative stage (steps 1–3) where NADPH is produced and a nonoxidative stage (steps 4–7) where a series of isomerizations and carbon–carbon bond breaking and forming reactions occur. The overall result of the full pathway is the conversion of three molecules of glucose 6-phosphate into three molecules of CO_2, two of fructose 6-phosphate, and one of glyceraldehyde 3-phosphate, along with the formation of six molecules of NADPH. The products then enter the glycolysis pathway for further degradation.

$$3\ \text{Glucose 6-phosphate} + 6\ NADP^+ + 3\ H_2O \longrightarrow 2\ \text{Fructose 6-phosphate} + \text{Glyceraldehyde 3-phosphate} + 3\ CO_2 + 6\ NADPH/H^+$$

Glucose 6-phosphate   Fructose 6-phosphate   Glyceraldehyde 3-phosphate

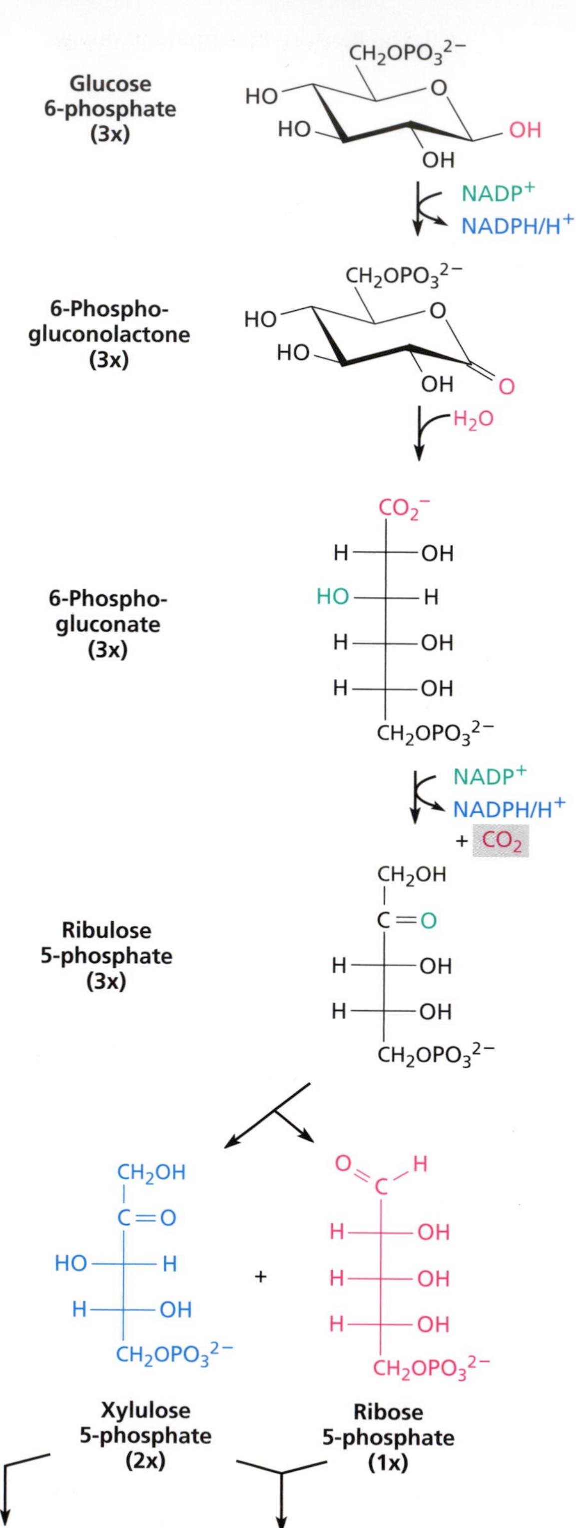

Step 1 Glucose 6-phosphate is oxidized at the anomeric center by $NADP^+$ to yield the corresponding lactone.

Step 2 Hydrolysis of the lactone gives the open-chain carboxylate.

Step 3 Oxidation of the C3 hydroxyl gives an intermediate ***β***-keto acid, which decarboxylates.

Step 4 Two different isomerizations occur by keto–enol tautomerization, giving xylulose 5-phosphate and ribose 5-phosphate.

continues

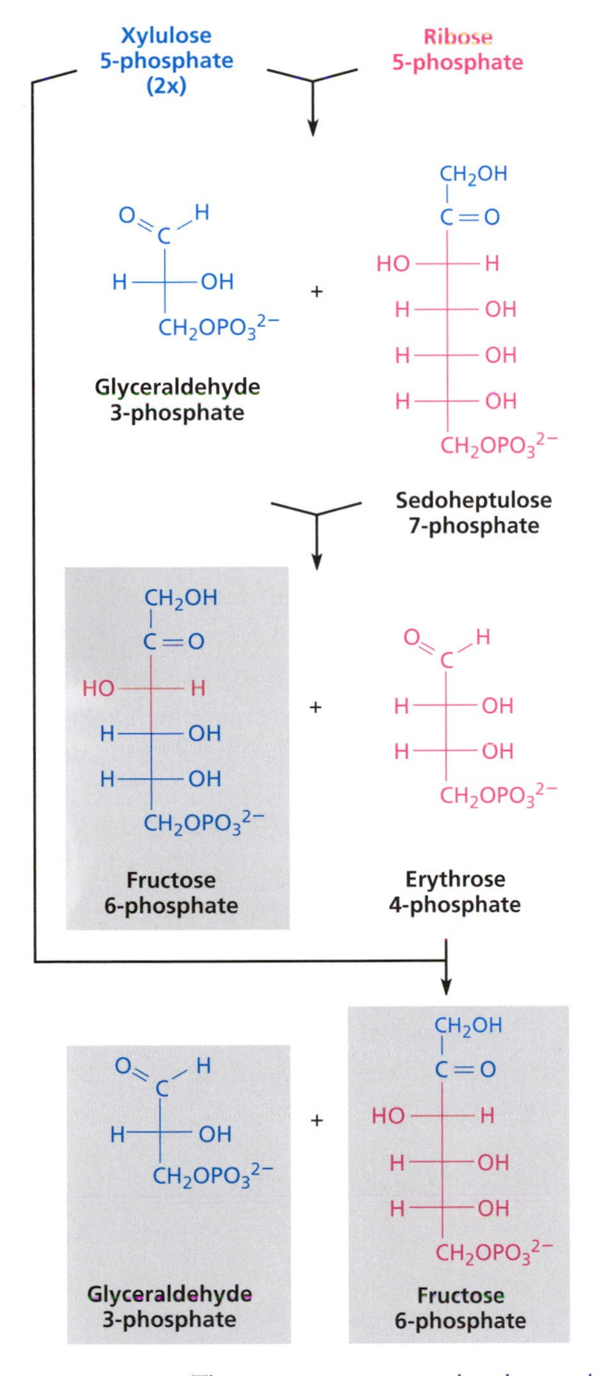

Step 5 Xylulose 5-phosphate reacts with ribose 5-phosphate, exchanging a C_2 unit and giving glyceraldehyde 3-phosphate and sedoheptulose 7-phosphate.

Step 6 Glyceraldehyde 3-phosphate and sedoheptulose 7-phosphate exchange a C_3 unit to give fructose 6-phosphate and erythrose 4-phosphate.

Step 7 Xylulose 5-phosphate and erythrose 4-phosphate exchange a C_2 unit, giving glyceraldehyde 3-phosphate and fructose 6-phosphate.

FIGURE 4.18 The seven-step pentose phosphate pathway, which results in the conversion of three molecules of glucose 6-phosphate into three molecules of CO_2, two of fructose 6-phosphate, and one of glyceraldehyde 3-phosphate. (Final products are indicated by gray screens.) Individual steps are described in more detail in the text.

Step 1. Oxidation The β anomer of glucose 6-phosphate is first oxidized by $NADP^+$ (Table 2.5) at the C1 anomeric center to give the corresponding lactone. The reaction is catalyzed by glucose 6-phosphate dehydrogenase[24] and occurs by the same hydride-transfer mechanism as NAD^+ oxidations (Section 3.2, Figure 3.5). The α glucose hydrogen is transferred to the *si* face of $NADP^+$, with assistance from a histidine residue in the enzyme.

His
Asp
$CH_2OPO_3^{2-}$
Glucose 6-phosphate
$NADP^+$
$CONH_2$
$CH_2OPO_3^{2-}$
6-Phosphogluconolactone
+ NADPH/H^+

Step 2. Hydrolysis 6-Phosphogluconolactone, a cyclic ester, is hydrolyzed to 6-phosphogluconate by nucleophilic addition of water to the carbonyl group in the usual way.

Step 3. Oxidation and decarboxylation 6-Phosphogluconate undergoes a further oxidation at C3 to give a β-keto acid intermediate that decarboxylates to form ribulose 5-phosphate. Both steps are catalyzed by phosphogluconate dehydrogenase.[25, 26] The process is similar to the conversion of isocitrate to α-ketoglutarate in step 3 of the citric acid cycle (Section 4.4) except that $NADP^+$ rather than NAD^+ is the oxidizing coenzyme and a divalent metal cation does not appear to be necessary to assist decarboxylation.

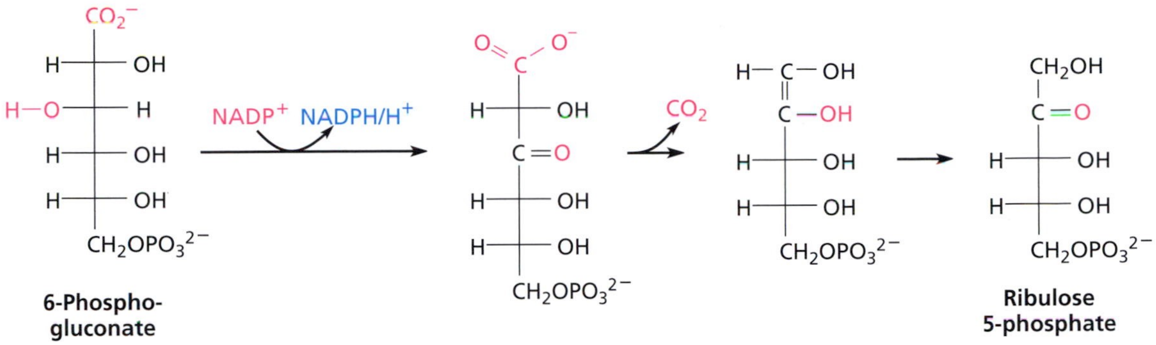

6-Phospho-gluconate

Ribulose 5-phosphate

Step 4. Isomerizations Ribulose 5-phosphate next undergoes two reversible keto–enol tautomerizations that result in formation of ribose 5-phosphate and xylulose 5-phosphate. Ribose 5-phosphate formation is catalyzed by ribulose 5-phosphate isomerase[27] and occurs through a 1,2-enediol intermediate. The acidic hydrogen at C1 is removed by base (glutamate), whose conjugate acid then reprotonates the enediol at C2.

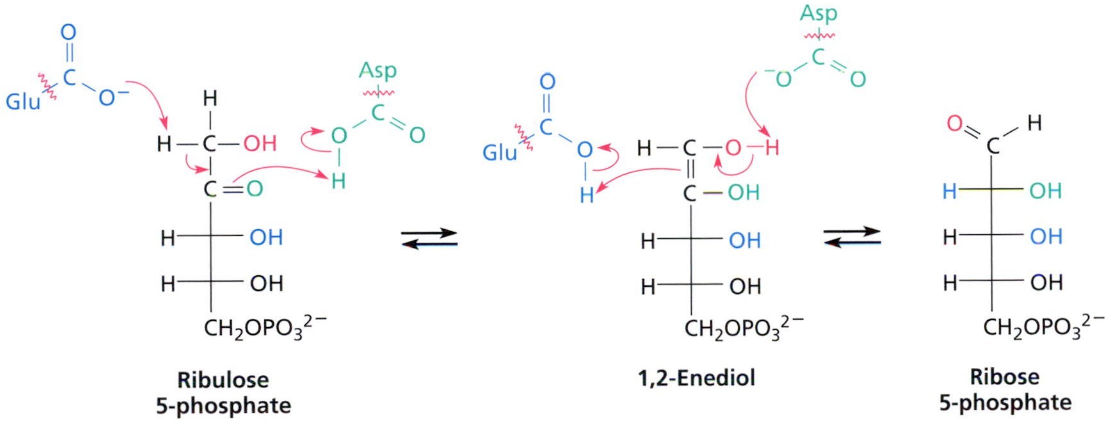

Ribulose 5-phosphate

1,2-Enediol

Ribose 5-phosphate

Xylulose 5-phosphate formation is catalyzed by ribulose 5-phosphate epimerase[28] and occurs through a 2,3-enediol or enediolate intermediate. (Recall from Section 2.1 that *epimers* are diastereomers that differ at only one chirality center and that an *epimerase* thus changes the stereochemistry at one chirality center of the substrate.) Deprotonation by Asp-38 occurs at C3 from one side of the molecule, and reprotonation by Asp-178 occurs on the same carbon from the opposite side of the molecule.

Ribulose 5-phosphate ⇌ **2,3-Enediolate** ⇌ **Xylulose 5-phosphate**

Step 5. Reaction of xylulose 5-phosphate with ribose 5-phosphate When ribose 5-phosphate is needed by dividing cells for ribonucleotide synthesis, the reversible isomerizations of step 4 produce it. When more ribose 5-phosphate than needed is available, however, the excess reacts with xylulose 5-phosphate to form intermediates that can enter the glycolysis pathway. The reaction of xylulose 5-phosphate with ribose 5-phosphate is catalyzed by a thiamin diphosphate-dependent transketolase[29] that transfers a two-carbon unit from the ketose (xylulose 5-phosphate) to the aldose (ribose 5-phosphate). As a result, the ketose is shortened by two carbons and the aldose is lengthened by two carbons.

As shown in Figure 4.19, the mechanism of the reaction is closely analogous to that found in the DXP pathway for terpenoid synthesis (Section 3.5) in which pyruvate reacts with glyceraldehyde 3-phosphate to yield 1-deoxy-D-xylulose 5-phosphate (Figure 3.22). Thiamin diphosphate (TPP) ylid first adds to the carbonyl group of xylulose 5-phosphate, the addition product cleaves in a retro-aldol-like reaction, and the resultant enamine adds to the carbonyl group of ribose 5-phosphate. Elimination of TPP ylid then gives sedoheptulose 7-phosphate.

Step 6. Reaction of sedoheptulose 7-phosphate with glyceraldehyde 3-phosphate In step 5, a trans*keto*lase transferred a two-carbon unit from a ketose to an aldose, shortening the ketose and lengthening the aldose by *two* carbons. In step 6, a trans*aldo*lase transfers a three-carbon unit from a ketose to an aldose, shortening the ketose and lengthening the aldose by *three* carbons.

As shown in Figure 4.20, the process begins by formation of a protonated Schiff base (an iminium ion) by reaction of sedoheptulose 7-phosphate with a lysine residue in the enzyme. A retro-aldol cleavage of the β-hydroxy iminium

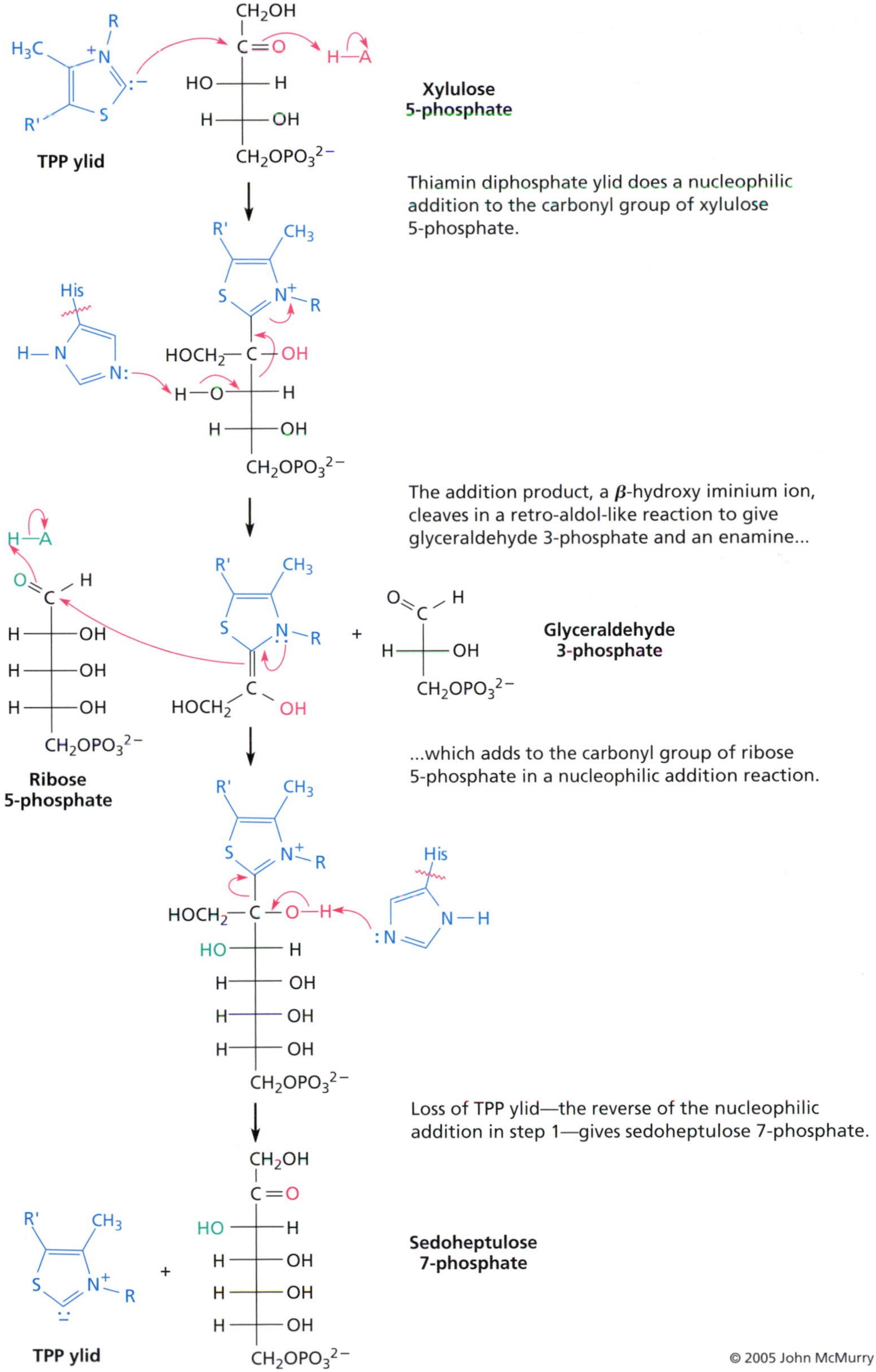
TPP ylid
Xylulose 5-phosphate
Thiamin diphosphate ylid does a nucleophilic addition to the carbonyl group of xylulose 5-phosphate.
His
The addition product, a β-hydroxy iminium ion, cleaves in a retro-aldol-like reaction to give glyceraldehyde 3-phosphate and an enamine...
Glyceraldehyde 3-phosphate
Ribose 5-phosphate
...which adds to the carbonyl group of ribose 5-phosphate in a nucleophilic addition reaction.
His
Loss of TPP ylid—the reverse of the nucleophilic addition in step 1—gives sedoheptulose 7-phosphate.
TPP ylid
Sedoheptulose 7-phosphate

FIGURE 4.19 Mechanism of the transketolase-catalyzed reaction of xylulose 5-phosphate with ribose 5-phosphate to give glyceraldehyde 3-phosphate and sedoheptulose 7-phosphate.

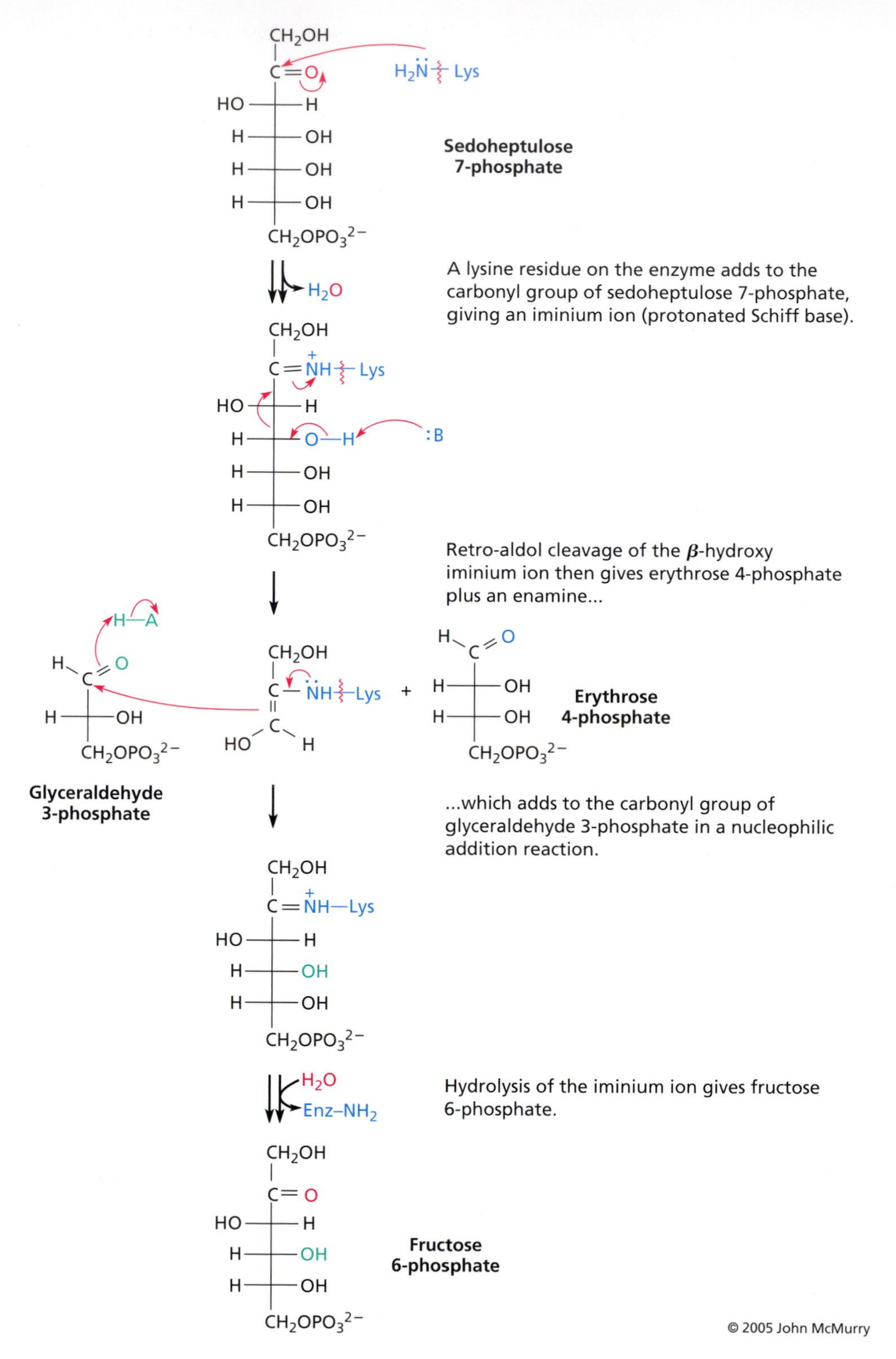

FIGURE 4.20 Mechanism of the transaldolase-catalyzed reaction of sedoheptulose 7-phosphate with glyceraldehyde 3-phosphate to give erythrose 4-phosphate and fructose 6-phosphate.

ion then occurs to give erythrose 4-phosphate plus an enamine in a process analogous to what occurs in step 4 of glycolysis (Section 4.2, Figure 4.5). The enamine then adds to the carbonyl group of glyceraldehyde 3-phosphate, a reaction identical to that in step 8 of gluconeogenesis (Section 4.5). The addition product is then hydrolyzed to give fructose 6-phosphate.

Step 7. Reaction of xylulose 5-phosphate with erythrose 4-phosphate As with step 5, the final step of the pentose phosphate pathway is again catalyzed by a transketolase. The net result is the transfer of a C_2 unit from xylulose 5-phosphate to erythrose 4-phosphate to give glyceraldehyde 3-phosphate and a second molecule of fructose 6-phosphate. The thiamin-dependent mechanism of the reaction is identical to that of step 5.

4.7 Photosynthesis: The Reductive Pentose Phosphate (Calvin) Cycle

Photosynthesis is the complex process by which plants, algae, and certain other organisms use sunlight to convert atmospheric carbon dioxide and water into molecular oxygen and carbohydrate. The process can be broken into two fundamental parts, a series of *light reactions* that split H_2O to give O_2 and the reduced coenzyme NADPH and a series of *dark reactions* that use the NADPH to reduce CO_2 and give carbohydrate, symbolized as (CH_2O).

$$H_2O + ADP + Pi + NADP^+ \xrightarrow{\text{sunlight}} O_2 + ATP + NADPH + H^+$$

$$CO_2 + ATP + NADPH + H^+ \longrightarrow (CH_2O) + ADP + Pi + NADP^+$$

$$\text{Net: } CO_2 + H_2O \longrightarrow (CH_2O) + O_2$$

Carbohydrate formation in the photosynthetic dark reactions occurs through a pathway called the **reductive pentose phosphate (RPP) cycle**, or **Calvin cycle**. In a simplified overview, the RPP cycle has three stages: (1) a *fixation* stage in which three molecules of CO_2 react with three molecules of ribulose 1,5-bisphosphate to give six molecules of 3-phosphoglycerate; (2) a *reduction* stage in which the six molecules of 3-phosphoglycerate are converted into six molecules of glyceraldehyde 3-phosphate; and (3) a *regeneration* stage in which five of the six glyceraldehyde 3-phosphate molecules are converted into three molecules of ribulose 1,5-bisphosphate, while the one remaining glyceraldehyde 3-phosphate molecule enters the gluconeogenesis pathway for conversion into a hexose sugar. Note that three carbon atoms enter the cycle as 3 CO_2 and three carbon atoms exit the cycle as glyceraldehyde 3-phosphate.

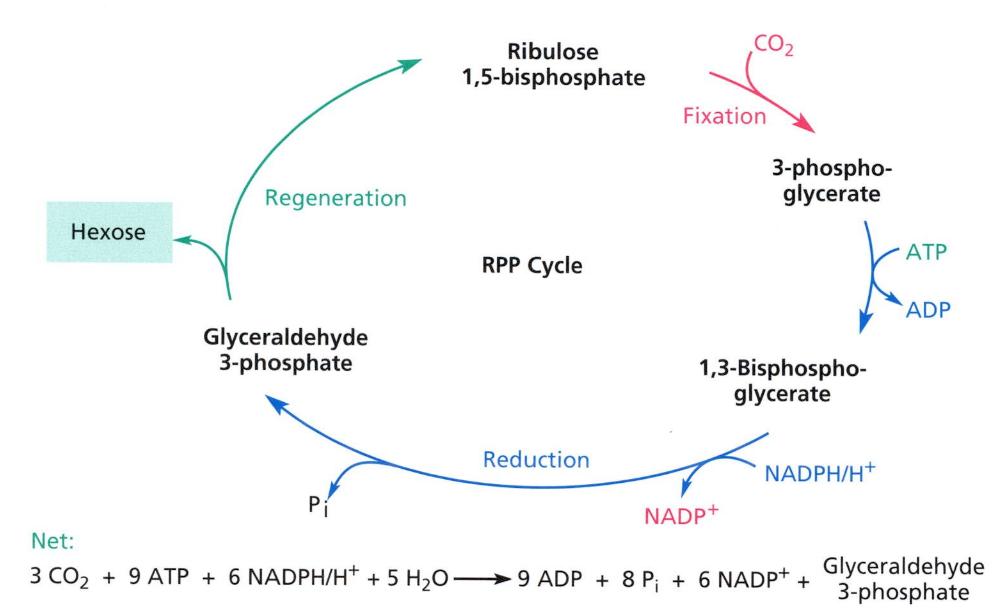

Net:

$$3\ CO_2 + 9\ ATP + 6\ NADPH/H^+ + 5\ H_2O \longrightarrow 9\ ADP + 8\ P_i + 6\ NADP^+ + \text{Glyceraldehyde 3-phosphate}$$

The 13 steps of the RPP cycle are shown in Figure 4.21. The figure is complex, but all steps except the first one are closely analogous to reactions we've already seen in other metabolic pathways, and many are identical to those of the gluconeogenesis and pentose phosphate pathways. We'll therefore look carefully only at the first step and will simply catalog the remaining ones.

Step 1. Carboxylation and cleavage of ribulose 1,5-bisphosphate The RPP cycle is initiated by reaction of CO_2 with ribulose 1,5-bisphosphate to give 3-phosphoglycerate. The reaction sequence is catalyzed by ribulose 1,5-bisphosphate carboxylase, commonly called *Rubisco*,[30, 31] and occurs by the mechanism shown in Figure 4.22.

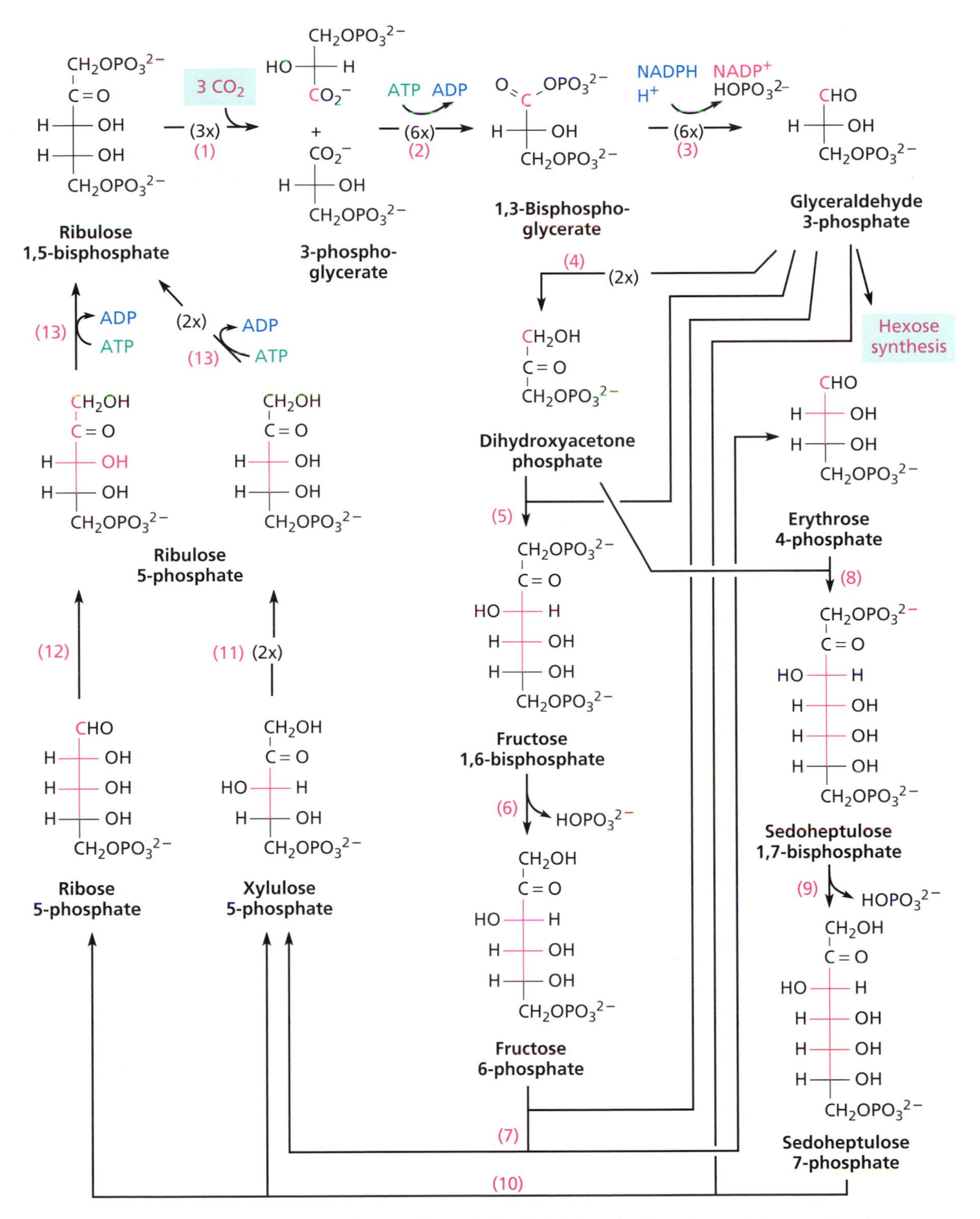

FIGURE 4.21 Reactions of the photosynthetic RPP (Calvin) cycle. Reaction of three molecules of CO_2 with three molecules of ribulose 1,5-bisphosphate leads ultimately to the entrance of one molecule of glyceraldehyde 3-phosphate into gluconeogenesis, along with regeneration of the ribulose 1,5-bisphosphate.

Ribulose 1,5-bisphosphate

Enolization occurs by abstraction of the acidic hydrogen at C3, facilitated by coordination of the oxygens at C2 and C3 to a magnesium ion.

(Enediolate)

The cis enediolate undergoes proton transfer from the oxygen at C3 to that at C2, giving an isomeric enediolate.

(Enediolate)

Nucleophilic addition of the enolate ion to carbon dioxide occurs, giving a ***β*-keto acid.**

(*β*-Keto acid)

Nucleophilic addition of water to the ketone carbonyl group gives a hydrate (a ***β*-hydroxy acid)...**

(Hydrate)

...which undergoes a retro-aldol-like reaction to give 3-phosphoglycerate and 3-phosphoglycerate anion. The anion is protonated to give a second molecule of 3-phosphoglycerate.

3-phospho-glycerate

3-phospho-glycerate

FIGURE 4.22 Mechanism of the Rubisco-catalyzed carboxylation and cleavage of ribulose 1,5-bisphosphate.

Coordination of the substrate to Mg^{2+} increases the acidity of the C3 proton, which is removed by a lysine residue in the enzyme to give an enediolate. Proton transfer from the C3 oxygen to the C2 oxygen yields an isomeric enediolate that reacts on its *si* face with CO_2. The β-keto acid that results then undergoes nucleophilic addition of water to the ketone carbonyl group, giving a hydrate that fragments by a retro-aldol-like cleavage. One of the fragments is 3-phosphoglycerate, and the other is the 3-phosphoglycerate anion, which is protonated stereospecifically by a lysine ammonium ion on the enzyme. The net result is formation of two molecules of 3-phosphoglycerate from one ribulose 1,5-bisphosphate plus CO_2.

Steps 2–3. Phosphorylation and reduction 3-Phosphoglycerate is phosphorylated by ATP to give 1,3-bisphosphoglycerate, which is reduced by NADPH to give glyceraldehyde 3-phosphate, just as occurs in steps 5 and 6 of gluconeogenesis (Figure 4.15).

Steps 4–6. Isomerization, aldol condensation, and hydrolysis Glyceraldehyde 3-phosphate isomerizes to dihydroxyacetone phosphate. The two trioses then undergo an aldol condensation to give fructose 1,6-bisphosphate, which is hydrolyzed to give fructose 6-phosphate, the same reactions as steps 7–9 of gluconeogenesis (Figure 4.15).

Step 7. Reaction of fructose 6-phosphate with glyceraldehyde 3-phosphate This reaction yields xylulose 5-phosphate and erythrose 4-phosphate, is catalyzed by a transketolase, and occurs by a mechanism analogous to that of step 5 in the pentose phosphate pathway (Figure 4.19).

Step 8. Aldol condensation of dihydroxyacetone phosphate with erythrose 4-phosphate This aldol condensation yields sedoheptulose 1,7-bisphosphate and occurs by a mechanism analogous to that of step 8 of gluconeogenesis (Figure 4.15).

Step 9. Hydrolysis Hydrolysis of sedoheptulose 1,7-bisphosphate yields sedoheptulose 7-phosphate by a mechanism analogous to that of step 9 of gluconeogenesis (Figure 4.15).

Step 10. Reaction of sedoheptulose 7-phosphate with glyceraldehyde 3-phosphate This reaction yields xylulose 5-phosphate and ribose 5-phosphate and is catalyzed by a transketolase. It is the exact reverse of step 5 in the pentose phosphate pathway (Figure 4.18).

Step 11. Isomerization Xylulose 5-phosphate isomerizes to ribulose 5-phosphate, the exact reverse of step 4 in the pentose phosphate pathway (Figure 4.18).

Step 12. Isomerization Ribose 5-phosphate isomerizes to ribulose 5-phosphate, the exact reverse of step 4 in the pentose phosphate pathway (Figure 4.18).

Step 13. Phosphorylation Ribulose 5-phosphate is phosphorylated by ATP to give ribulose 1,5-bisphosphate by a mechanism analogous to that of step 3 in glycolysis (Figure 4.3).

References

1. Rye, C. S.; Withers, S. G., "Glycosidase Mechanisms," *Curr. Opin. in Chem. Biol.*, **2002**, *6*, 573–580.
2. Vasella, A.; Davies, G. J.; Böhm, M., "Glycosidase Mechanisms," *Curr. Opin. in Chem. Biol.*, **2002**, *6*, 619–629.
3. Rydberg, E. H.; Li, C.; Maurus, R.; Overall, C. M.; Brayer, G. D.; Withers, S. G., "Mechanistic Analysis of Catalysis in Human Pancreatic α-Amylase: Detailed Kinetic and Structural Studies of Mutants of Three Conserved Carboxlic Acids," *Biochemistry*, **2002**, *41*, 4492–4502.
4. Kuser, P. R.; Krauchenco, S.; Antunes, O. A. C.; Polikarpov, I., "The High Resolution Crystal Structure of Yeast Hexokinase PII with the Correct Primary Sequence Provides New Insights into Its Mechanism of Action," J. *Biol. Chem.*, **2000**, *275*, 20814–20821.
5. Chou, C.-C.; Sun, Y.-J.; Meng, M.; Hsiao, C.-D., "The Crystal Structure of Phosphoglucose Isomerase/Autocrine Motility Factor/Neuroleukin Complexed with Its Carbohydrate Phosphate Inhibitors Suggests Its Substrate/Receptor Recognition," *J. Biol. Chem.*, **2000**, *275*, 23154–23160.
6. Lee, J. H.; Chang, K. Z.; Patel, V.; Jeffery, C. J., "Crystal Structure of Rabbit Phosphoglucose Isomerase Complexed with Its Substrate D-Fructose 6-Phosphate," *Biochemistry*, **2001**, *40*, 7799–7805.
7. Read, J.; Pearce, J.; Li, X.; Muirhead, H.; Chirgwin, J.; Davies, C., "The Crystal Structure of Human Phosphoglucose Isomerase at 1.6 Å Resolution: Implications

for Catalytic Mechanism, Cytokine Activity, and Haemolytic Anaemia," *J. Mol. Biol.*, **2001**, *309*, 447–463.

8. Kimmel, J. L.; Reinhart, G. D., "Reevaluation of the Accepted Allosteric Mechanism of Phosphofructokinase from *Bacillus stearothermophilus*," *Proc. Natl. Acad. Sci., USA,* **2000,** *97*, 3844–3849.
9. Choi, K. H.; Shi, J.; Hopkins, C. E.; T.; Dean R.; Allen, K. N., "Snapshots of Catalysis: The Structure of Fructose-1,6-bisphosphate Aldolase Covalently Bound to the Substrate Dihydroxyacetone Phosphate," *Biochemistry*, **2001**, *40*, 13868–13875.
10. Harris, T. K.; Cole, R. N.; Comer, F. I.; Mildvan, A. S., "Proton Transfer in the Mechanism of Triosephosphate Isomerase," *Biochemistry*, **1998**, *37*, 16828–16838.
11. Rigden, D. J.; Walter, R. A.; Phillips, S E. V.; Fothergill-Gilmore, L. A., "Sulfate Ions Observed in the 2.12 Å Structure of a New Crystal Form of *S. cerevisiae* Phosphoglycerate Mutase Provide Insights into Understanding the Catalytic Mechanism," *J. Mol. Biol.*, **1999**, *286*, 1507–1517.
12. Zhang, E.; Brewer, J. M.; Minor, W.; Carreira, L. A.; Lebioda, L., "Mechanism of Enolase: The Crystal Structure of Asymmetric Dimer Enolase-2-phospho-D-glycerate/Enolase—Phosphoenolpyruvate at 2.0 Å Resolution," *Biochemistry*, **1997**, *36*, 12526–12534.
13. Poyner, R. R.; Cleland, W. W.; Reed, G. H., "Role of Metal Ions in Catalysis by Enolase: An Ordered Kinetic Mechanism for a Single Substrate Enzyme," *Biochemistry*, **2001**, *40*, 8009–8017.
14. Bollenbach, T. J.; Mesecar, A. D.; Nowak, T., "Role of Lysine 240 in the Mechanism of Yeast Pyruvate Kinase Catalysis," *Biochemistry*, **1999**, *38*, 9137–9145.
15. Jordan, F., "Current Mechanistic Understanding of Thiamin Diphosphate-Dependent Enzymatic Reactions," *Nat'l. Prod. Reports*, **2003**, *20*, 184–201.
16. Liu, S.; Gong, X.; Yan, X.; Peng, T.; Baker, J. C.; Li, L.; Robben, P. M.; Ravindran, S.; Andersson, L. A.; Cole, A. B.; Roche, T. E., "Reaction Mechanism for Mammalian Pyruvate Dehydrogenase Using Natural Lipoyl Domain Substrates," *Archiv. Biochem. Biophys.*, **2001**, *386*, 123–135.
17. Argyrou, A.; Blanchard, J. S.; Palfey, B. A., "The Lipoamide Dehydrogenase from *Mycobacterium tuberculosis* Permits the Direct Observation of Flavin Intermediates in Catalysis," *Biochemistry*, **2002**, *41*, 14580–14590.
18. Karpusas, M.; Branchaud, B.; Remington, S. J., "Proposed Mechanism for the Condensation Reaction of Citrate Synthase: 1.9 Å Structure of the Ternary Complex with Oxaloacetate and Carboxymethyl Coenzyme A," *Biochemistry*, **1990**, *29*, 2213–2219.
19. Lauble, H.; Kennedy, M. C.; Emptage, M. H.; Beinert, H.; Stout, C. D., "The Reaction of Fluorocitrate with Aconitase and the Crystal Structure of the Enzyme–Inhibitor Complex," *Proc. Natl. Acad. Sci., USA*, **1996**, *93*, 13699–13703.
20. Attwood, P. V., "The Structure and the Mechanism of Action of Pyruvate Carboxylase," *Internat. J. Biochem. Cell Bio.*, **1995**, *27*, 231–249.
21. Matte, A.; Tari, L. W.; Goldie, H.; Delbaere, L. T. J., "Structure and Mechanism of Phosphoenolpyruvate Carboxykinase," *J. Biol. Chem.*, **1997**, *272*, 8105–8108.

22. Choe, J.-Y.; Fromm, H. J.; Honzatko, R. B., "Crystal Structures of Fructose 1,6-Bisphosphatase: Mechanism of Catalysis and Allosteric Inhibition Revealed in Product Complexes," *Biochemistry*, **2000**, *39*, 8565–8574.
23. Choe, J.-Y.; Iancu, C. V.; Fromm, H. J.; Honzatko, R. B., "Metaphosphate in the Active Site of Fructose 1,6-Bisphosphatase," *J. Biol. Chem.*, **2003**, *278*, 16015–16020.
24. Cosgrove, M. S.; Naylor, C.; Paludan, S.; Adams, M. J.; Levy, H. R., "On the Mechanism of the Reaction Catalyzed by Glucose 6-Phosphate Dehydrogenase," *Biochemistry*, **1998**, *37*, 2759–2767.
25. Price, N. E.; Cook, P. F., "Kinetic and Chemical Mechanisms of the Sheep Liver 6-Phosphogluconate Dehydrogenase," *Archiv. Biochem. Biophys.*, **1996**, 336, 215–223.
26. Karsten, W. E.; Chooback, L.; Cook, P. F., "Glutamate 190 Is a General Acid Catalyst in the 6-Phosphogluconate-Dehydrogenase-Catalyzed Reaction," *Biochemistry*, **1998**, *37*, 15691–15697.
27. Zhang, R.-G.; Andersson, C. E.; Savchenko, A.; Skarina, T.; Evdokimova, E.; Beasley, S.; Arrowsmith, C. H.; Edwards, A. M.; Joachimiak, A.; Mowbray, S. L., "Structure of *Escherichia coli* Ribose-5-Phosphate Isomerase. A Ubiquitous Enzyme of the Pentose Phosphate Pathway and the Calvin Cycle," *Structure*, **2003**, *11*, 31–42.
28. Jelakovic, S.; Kopriva, S.; Suss, K.-H.; Schulz, G. E., "Structure and Catalytic Mechanism of the Cytosolic D-Ribulose-5-phosphate 3-Epimerase from Rice," *J. Molec. Biol.*, **2003**, *326*, 127–135.
29. Schneider, G.; Lindqvist, Y., "Crystallography and Mutagenesis of Transketolase: Mechanistic Implications for Enzymic Thiamin Catalysis," *Biochim. et Biophys. Acta*, **1998**, *1385*, 387–398.
30. Cleland, W. W.; Andrews, T. J.; Gutteridge, S.; Hartman, F. C.; Lorimer, G. H., "Mechanism of Rubisco: The Carbamate as General Base," *Chem. Reviews*, **1998**, *98*, 549–561.
31. Roy, H.; Andrews, T. J., "Rubisco: Assembly and Mechanism," *Advances in Photosynthesis*, **2000**, *9*, 53–83.

Problems

4.1 What coenzyme is typically associated with each of the following transformations?

(a) The phosphorylation of an alcohol to give a phosphate.

(b) The oxidative decarboxylation of an α-keto acid to give a thioester.

(c) The carboxylation of a thioester to give a β-keto thioester.

4.2 Show the mechanism of step 4 in the citric acid cycle, the conversion of α-ketoglutarate to succinyl CoA.

4.3 Show the mechanism of step 7 in the RPP cycle, the transketolase-catalyzed reaction of fructose 6-phosphate with glyceraldehyde 3-phosphate to give xylulose 5-phosphate and erythrose 4-phosphate.

4.4 Step 7 of the citric acid cycle is the conjugate nucleophilic addition of water to fumarate to yield (S)-malate. Does water add to the *re* face or the *si* face of fumarate?

4.5 Show the mechanism of step 10 in the RPP cycle, the transketolase-catalyzed reaction of sedoheptulose 7-phosphate with glyceraldehyde 3-phosphate to give xylulose 5-phosphate and ribose 5-phosphate.

4.6 Mannose, a component of dietary glycoproteins, is metabolized by an initial phosphorylation and isomerization to fructose 6-phosphate. Propose a mechanism for the isomerization.

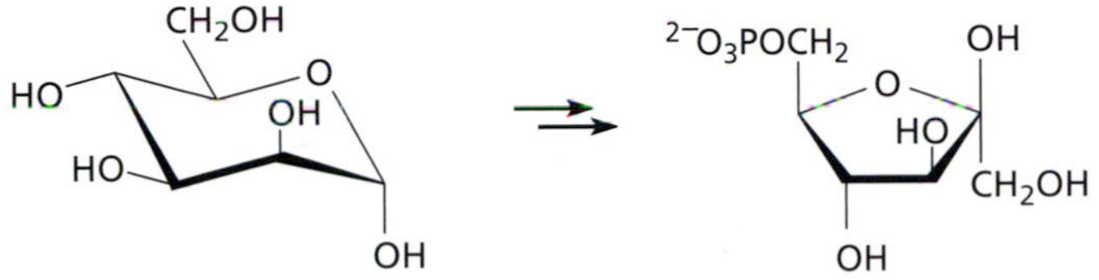

CH_2OH HO OH HO OH O → $^{2-}O_3POCH_2$ O OH HO CH_2OH OH

Mannose **Fructose 6-phosphate**

4.7 Plants, but not animals, are able to synthesize glucose from acetyl CoA by a pathway that begins with the *glyoxalate cycle*. One of the steps in the cycle is the conversion of isocitrate to glyoxalate plus succinate, a process catalyzed by isocitrate lyase. Propose a mechanism for the reaction.

H CO_2^- ^-O_2C CO_2^- H OH —Isocitrate lyase→ O ‖ ^-O_2C–C–H + ^-O_2C CO_2^-

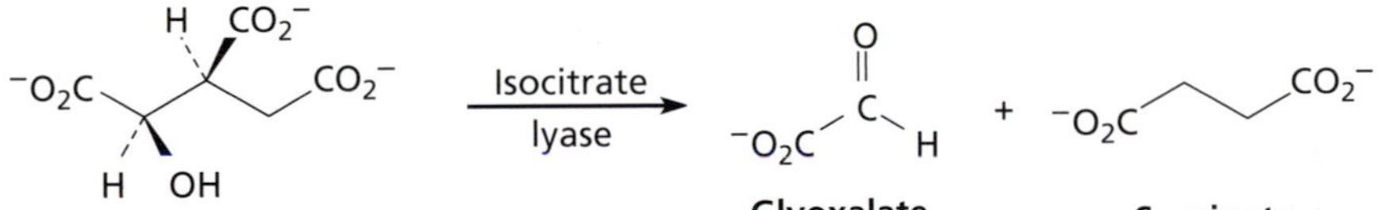

Isocitrate **Glyoxalate** **Succinate**

4.8 Galactose, a constituent of the disaccharide lactose found in dairy products, is metabolized by a pathway that includes the epimerization of UDP-galactose to UDP-glucose, where UDP = uridylyl diphosphate. The epimerase enzyme uses NAD^+ as cofactor. Propose a mechanism for the reaction.

HO CH_2OH O HO OH O–P(=O)(O⁻)–O–P(=O)(O⁻)–O—Uridine → CH_2OH O HO HO OH O–P(=O)(O⁻)–O–P(=O)(O⁻)–O—Uridine

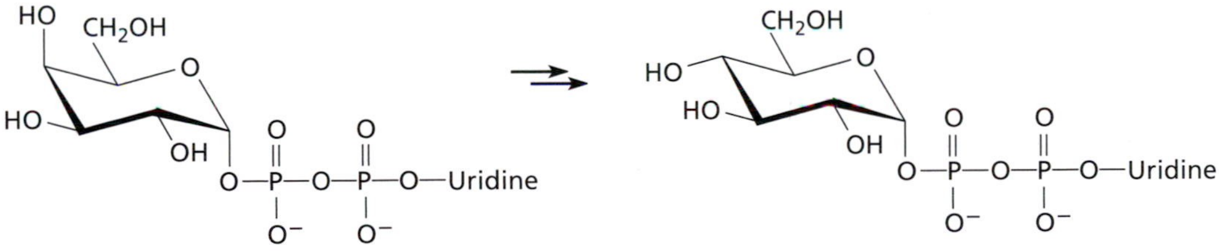

UDP-Galactose **UDP-Glucose**

4.9 Propose a mechanism for the conversion of 6-phosphogluconate to 2-keto-3-deoxy-6-phosphogluconate, a step in the Entner–Douderoff bacterial pathway for glucose catabolism.

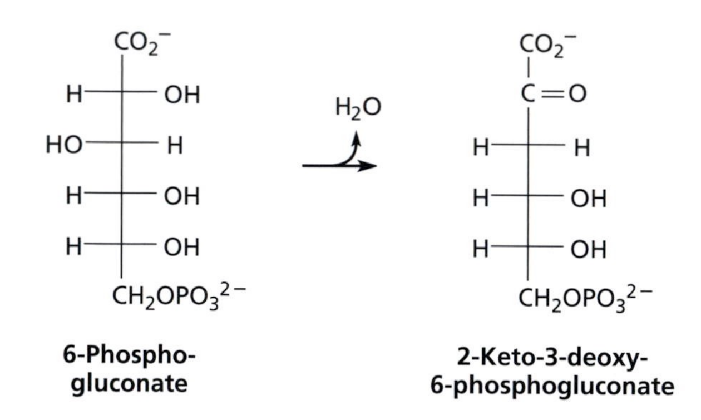

5 Amino Acid Metabolism

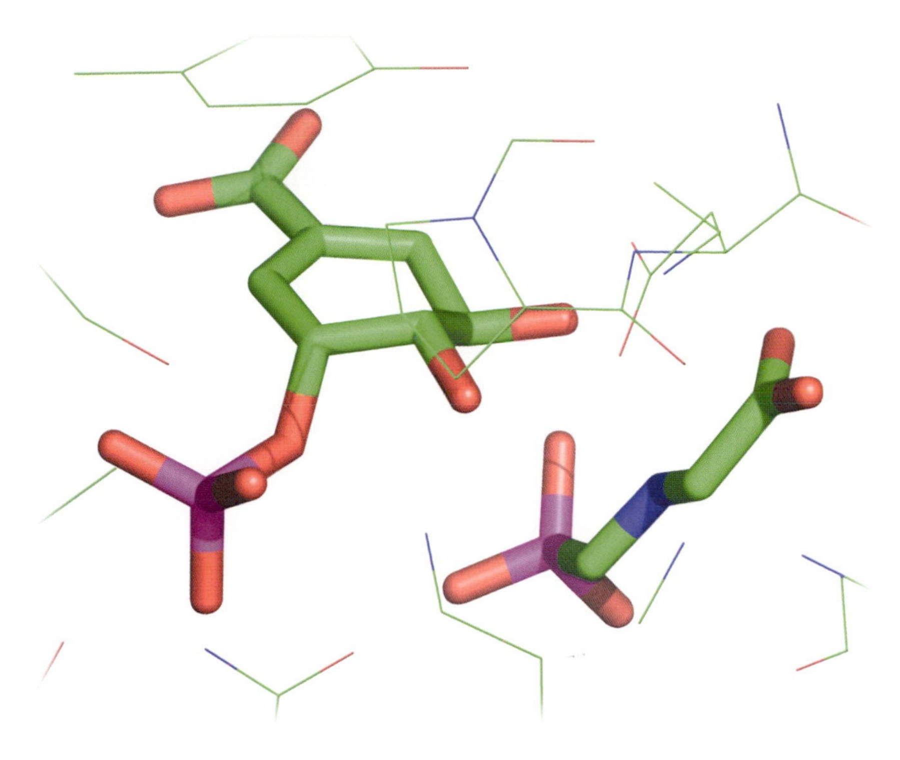

Humans derive most of their amino acids from digested proteins, while plants biosynthesize their own amino acids. Inhibition of amino acid biosynthesis is therefore an effective strategy for selectively killing plants with compounds that don't harm mammals. An example is the herbicide glyphosate, sold commercially as Roundup, which binds to 5-enolpyruvylshikimate-3-phosphate synthase and kills plants by inhibiting the biosynthesis of aromatic amino acids. The structure shows the active site of the enzyme bound to shikimate-3-phosphate and glyphosate.

The biochemistry of amino acids is more complex than that of triacylglycerols or carbohydrates because each of the 20 common α amino acids is biologically synthesized and degraded by its own unique pathway. Nevertheless, there are some common themes that run through amino acid metabolism, as well as many familiar substances seen previously in carbohydrate metabolism. We'll look first at the metabolic breakdown of amino acids, which occurs in three stages: removal of the α amino group as either ammonia or aspartate, conversion of ammonia and the aspartate nitrogen into urea, and conversion of the remaining amino acid carbon skeleton (often an α keto acid) into an intermediate that can enter the citric acid cycle.

$R-CH(\overset{+}{N}H_3)-CO_2^-$ ⇌ $R-CO-CO_2^-$ + NH_3 or $^-O_2CCH_2-CH(\overset{+}{N}H_3)-CO_2^-$

An α amino acid ⇌ **An α keto acid** + **Ammonia** or **Aspartate**

An α keto acid ⇌ Pyruvate, oxaloacetate, α-ketoglutarate, succinyl CoA, fumarate, acetoacetate, or acetyl CoA

Ammonia or **Aspartate** ⇌ $H_2N-CO-NH_2$ **Urea**

Following the discussion of amino acid breakdown, we'll cover the pathways by which the different amino acids are biosynthesized.

5.1 Deamination of Amino Acids

Transamination of Amino Acids

The first stage in the metabolic breakdown of most α amino acids is **deamination**, the removal of the α amino group and its replacement by a carbonyl. Deamination is usually accomplished by a **transamination** reaction, in which the amino acid —NH_2 group is transferred to the α carbon of α-ketoglutarate, forming a new α keto acid plus glutamate. The overall process occurs in two parts, is catalyzed by aminotransferases, and requires participation of pyridoxal phosphate (PLP), a derivative of pyridoxine (vitamin B_6), as coenzyme. Different aminotransferases differ in their specificity for amino acids, but the mechanism remains the same.

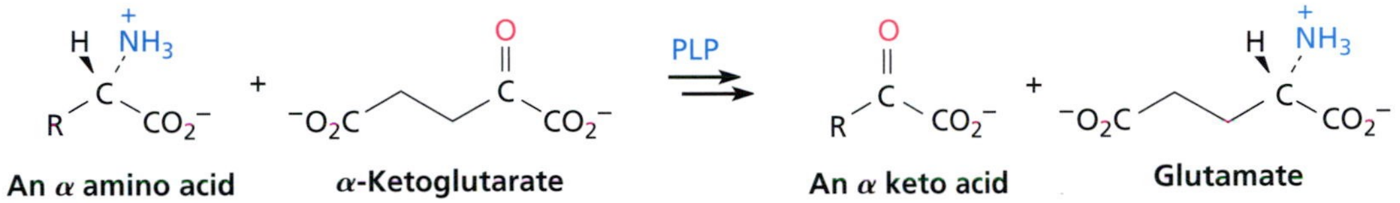

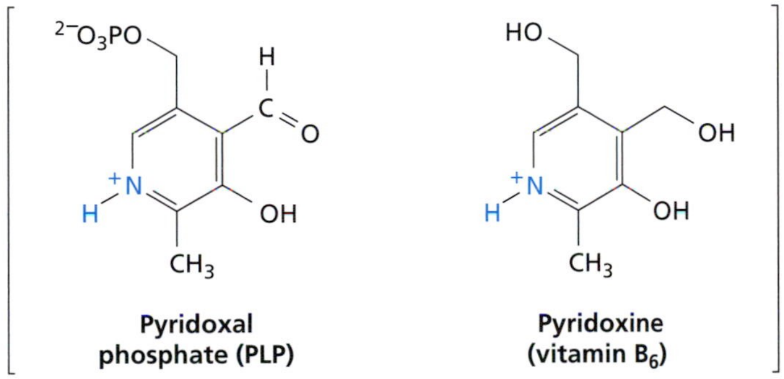

An α amino acid + **α-Ketoglutarate** $\xrightarrow{PLP}$ **An α keto acid** + **Glutamate**

Pyridoxal phosphate (PLP) **Pyridoxine (vitamin B_6)**

The mechanism of the first part of transamination is shown in Figure 5.1. The process begins with reaction between the α amino acid and pyridoxal phosphate, which is covalently bonded to the enzyme by an imine linkage between a lysine residue and the PLP aldehyde group. Deprotonation/reprotonation of the PLP–amino acid imine in steps 2 and 3 effects a tautomerization of the imine C═N double bond, and hydrolysis of the tautomerized imine in step 4 gives an α keto acid plus pyridoxamine phosphate (PMP).

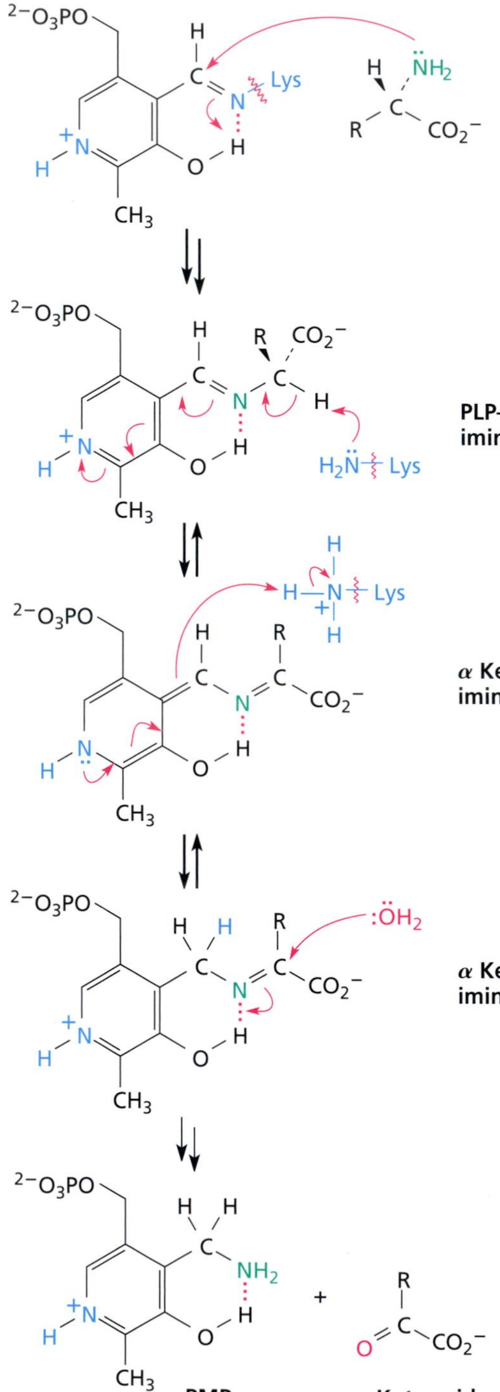

$^{2-}O_3PO$
H
Lys
H
$\ddot{N}H_2$
N
R
C
CO_2^-
H
O
CH3
Step 1 An amino acid reacts with the enzyme-bound PLP imine by nucleophilic addition of its —NH_2 group to the C═N bond of the imine, giving a PLP–amino acid imine and releasing the lysine amino group.
PLP–amino acid imine (Schiff base)
$H_2\ddot{N}$ Lys
Step 2 Base-induced deprotonation of the acidic α carbon of the amino acid gives an intermediate α keto acid imine...
α Keto acid imine
Step 3 ...that is reprotonated on the PLP carbon. The net result of this deprotonation/reprotonation sequence is tautomerization of the imine C═N bond.
$:\ddot{O}H_2$
α Keto acid imine tautomer
Step 4 Hydrolysis of the α keto acid imine by nucleophilic addition of water to the C═N bond gives the transamination products pyridoxamine phosphate (PMP) and α keto acid.
NH_2
PMP
α Keto acid

FIGURE 5.1 Mechanism of enzyme-catalyzed, PLP-dependent transamination of an α amino acid to give an α keto acid. Individual steps are explained in the text.

Step 1. Transimination The first step in trans*amination* is trans*imination*—the reaction of the PLP–enzyme imine with an α amino acid to give a PLP–amino acid imine plus released lysine amino group. The reaction occurs by nucleophilic addition of the amino acid —NH_2 group to the C═N double bond of the PLP imine, much as an amine adds to the C═O bond of a ketone or aldehyde (Section 1.5, Figure 1.10). The protonated diamine intermediate then undergoes a proton transfer and expels the lysine amino group to complete the step.

PLP–enzyme imine

Amino acid

Diamine intermediate

Diamine intermediate

PLP–amino acid imine

Steps 2–4. Tautomerization and hydrolysis Following formation of the PLP–amino acid imine, a tautomerization of the C═N double bond occurs. The lysine residue released during transimination deprotonates the acidic α position of the amino acid, with the protonated pyridine ring of PLP acting as the electron acceptor. Reprotonation then occurs on the carbon atom next to the ring, generating a product that is the imine of an α keto acid with pyridoxamine phosphate (PMP). Hydrolysis of this PMP–α keto acid imine by the usual mechanism (the exact reverse of imine formation) completes the first part of the deamination reaction.

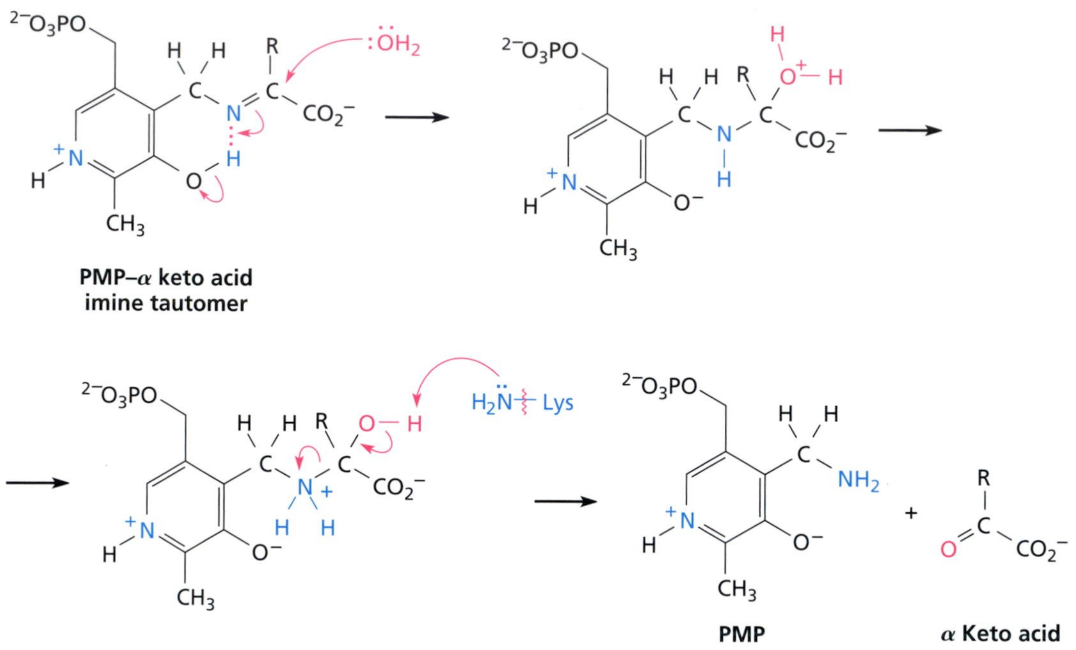

PMP–α keto acid imine tautomer

PMP

α Keto acid

With PLP plus the α amino acid now converted into PMP plus an α keto acid, PMP must be transformed back into PLP to complete the catalytic cycle. The conversion occurs by another transamination reaction, this one between PMP and an α keto acid, usually α-ketoglutarate or, less frequently, oxaloacetate or pyruvate. PLP plus glutamate (or aspartate) are the products, and the mechanism of the process is the exact reverse of that shown in Figure 5.1. That is, PMP and α-ketoglutarate give an imine; the PMP–ketoglutarate imine undergoes tautomerization of the C═N double bond to give a PLP–glutamate imine; and the PLP–glutamate imine reacts with a lysine residue on the enzyme in a transimination process to yield PLP–enzyme imine plus glutamate.

PMP + α-Ketoglutarate ⇌ PLP–enzyme imine + Glutamate

[or Oxaloacetate ⇌ Aspartate]

Oxidative Deamination of Glutamate

Following transferral of the —NH_2 group from an amino acid to α-ketoglutarate, the glutamate product is oxidatively deaminated by glutamate dehydrogenase[1] to give ammonia plus regenerated α-ketoglutarate. The reaction occurs by oxidation of the primary amine to an imine followed by hydrolysis, and the mechanism of the amine oxidation is analogous to that of alcohol oxidation (Section 1.9, Figure 1.15). Either NAD^+ or $NADP^+$ can function as the oxidizing coenzyme, depending on the organism, with a hydrogen transferred as hydride ion from the α carbon of glutamate.

NAD^+ + Glutamate → NADH + α-Iminoglutarate $\xrightarrow{H_2O}$ α-Ketoglutarate + NH_4^+

5.2 The Urea Cycle

The ammonia that results from amino acid catabolism is eliminated in one of three ways depending on the organism. Fish and other aquatic animals simply excrete the ammonia to their aqueous surroundings, but terrestrial organisms must first convert the ammonia into a nontoxic substance, either urea for mammals or uric acid for birds and reptiles.

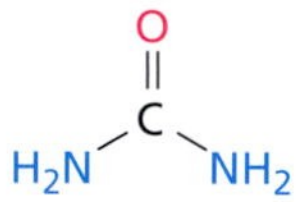

Urea 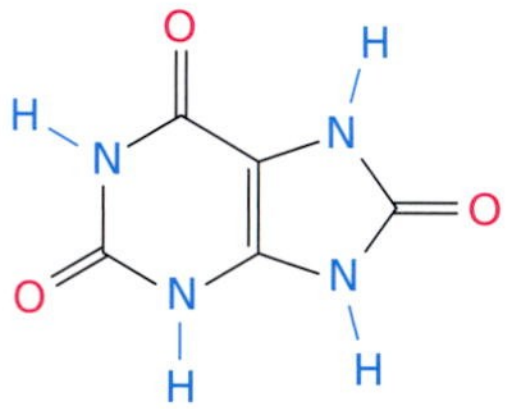 **Uric acid**

The conversion of ammonia into urea begins with its reaction with bicarbonate ion and ATP to give carbamoyl phosphate. The reaction is catalyzed by carbamoyl phosphate synthetase I and begins by activation of HCO_3^- with ATP, as occurs in the carboxylation of biotin (Section 3.4, Figure 3.13). Nucleophilic acyl substitution of the resultant carboxyphosphate with ammonia and subsequent phosphorylation of carbamate by a second equivalent of ATP gives the product.[2]

$$HCO_3^- + ATP \; (Mg^{2+}) \longrightarrow ADP + \text{Carboxyphosphate} \; (^{2-}O_3P\text{–O–C(=O)–OH}) \xrightarrow{:NH_3} {}^{2-}O_3P\text{–O–C(}^-\text{O)(}\overset{+}{N}H_3\text{)–OH} \xrightarrow{-P_i}$$

$$\longrightarrow \text{Carbamate} \; (H_2N\text{–C(=O)–O}^-) \xrightarrow{ATP \;\; ADP} \text{Carbamoyl phosphate} \; (H_2N\text{–C(=O)–OPO}_3^{2-})$$

Carbamoyl phosphate then enters the four-step **urea cycle**, whose overall result can be summarized as

$$NH_3 + HCO_3^- + 2\ ATP \longrightarrow \text{Carbamoyl phosphate} \; (H_2N\text{–C(=O)–OPO}_3^{2-})$$

$$\text{Carbamoyl phosphate} + \text{Aspartate} \xrightarrow{ATP \;\; AMP + PP_i, \; P_i} \text{Urea} \; (H_2N\text{–C(=O)–NH}_2) + \text{Fumarate} \; (^-O_2C\text{–CH=CH–CO}_2^-)$$

Note that only one of the two nitrogen atoms in urea comes from ammonia; the other nitrogen comes from aspartate, which is itself produced from glutamate by transamination with oxaloacetate. The reactions of the urea cycle are shown in Figure 5.2.

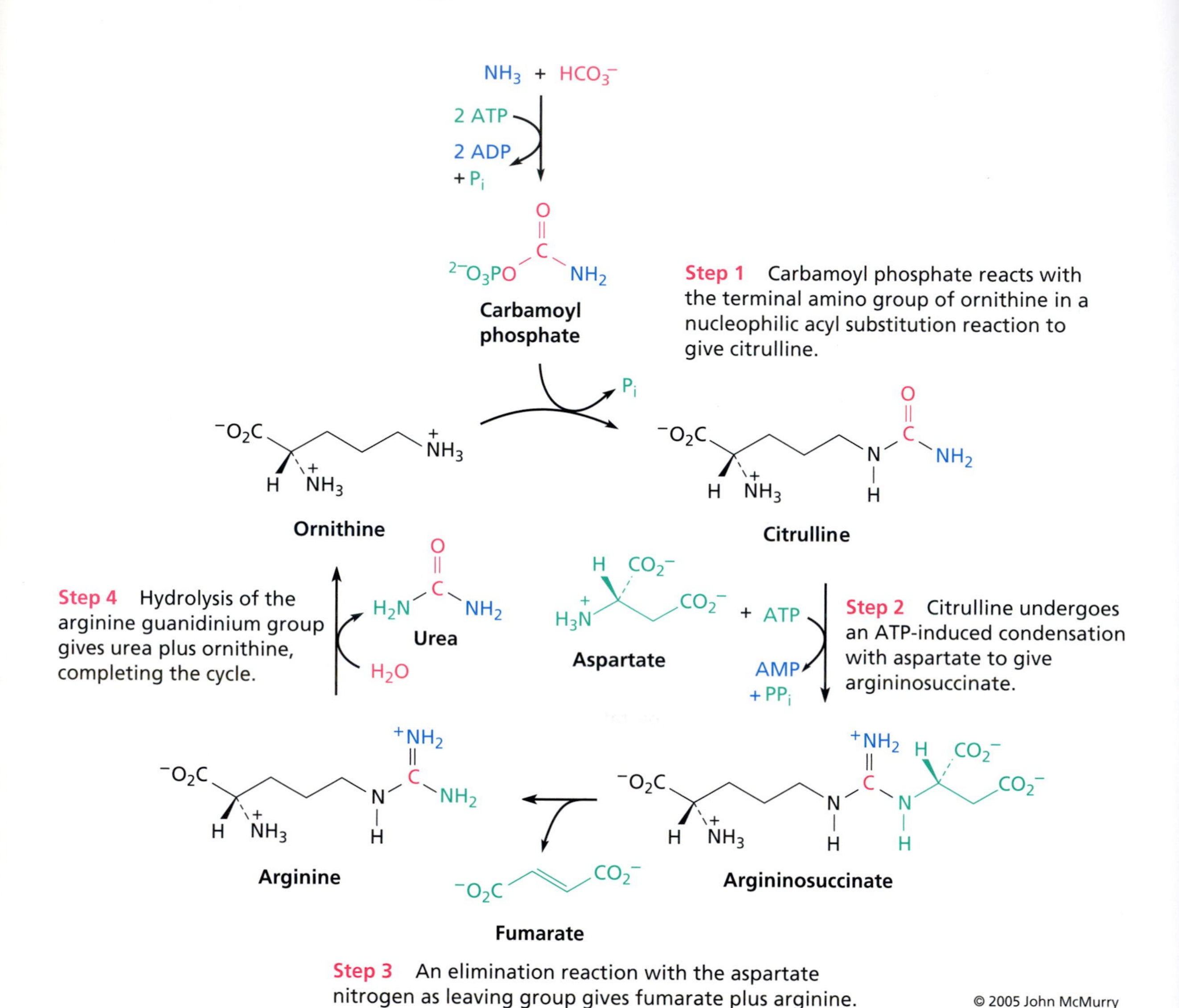

FIGURE 5.2 The urea cycle is a four-step series of reactions that converts ammonia into urea. Individual steps are explained in more detail in the text.

Steps 1–2. Argininosuccinate synthesis The cycle begins with a nucleophilic acyl substitution reaction of ornithine with carbamoyl phosphate to produce citrulline. The side-chain —NH_2 group of ornithine is the nucleophile, phosphate is the leaving group, and the reaction is catalyzed by ornithine transcarbamoylase.[3]

Citrulline then undergoes a condensation with aspartate in step 2 to yield argininosuccinate. The reaction is catalyzed by argininosuccinate synthetase[4] and occurs by the mechanism shown in Figure 5.3. The process is essentially a nucleophilic acyl substitution reaction in which the amide group of citrulline is first activated by reaction with ATP. Nucleophilic addition of aspartate to the C=N$^+$ double bond then gives a typical tetrahedral intermediate, which expels AMP as the leaving group.

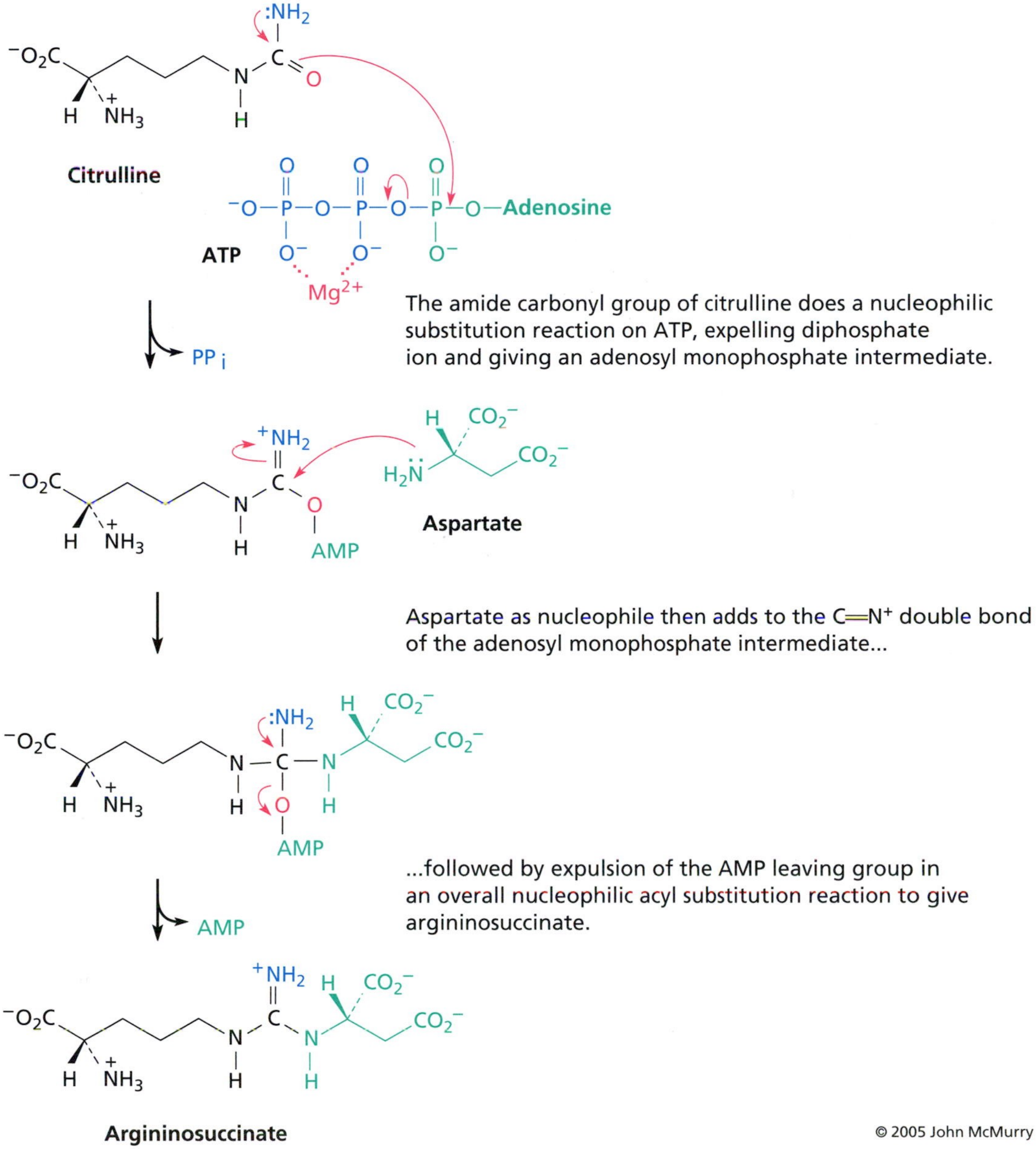

Citrulline

ATP

The amide carbonyl group of citrulline does a nucleophilic substitution reaction on ATP, expelling diphosphate ion and giving an adenosyl monophosphate intermediate.

Aspartate

Aspartate as nucleophile then adds to the C=N$^+$ double bond of the adenosyl monophosphate intermediate...

...followed by expulsion of the AMP leaving group in an overall nucleophilic acyl substitution reaction to give argininosuccinate.

Argininosuccinate

FIGURE 5.3 Mechanism of step 2 in the urea cycle, the reaction of citrulline with aspartate to give argininosuccinate.

Step 3. Fumarate elimination The third step in the urea cycle, the conversion of argininosuccinate to arginine plus fumarate, is an elimination reaction catalyzed by argininosuccinate lyase.[5, 6] The process occurs by an E1cB mechanism (Section 1.8) with loss of the *pro-R* hydrogen and with *anti* stereochemistry. A histidine residue on the enzyme acts as the base to carry out the deprotonation.

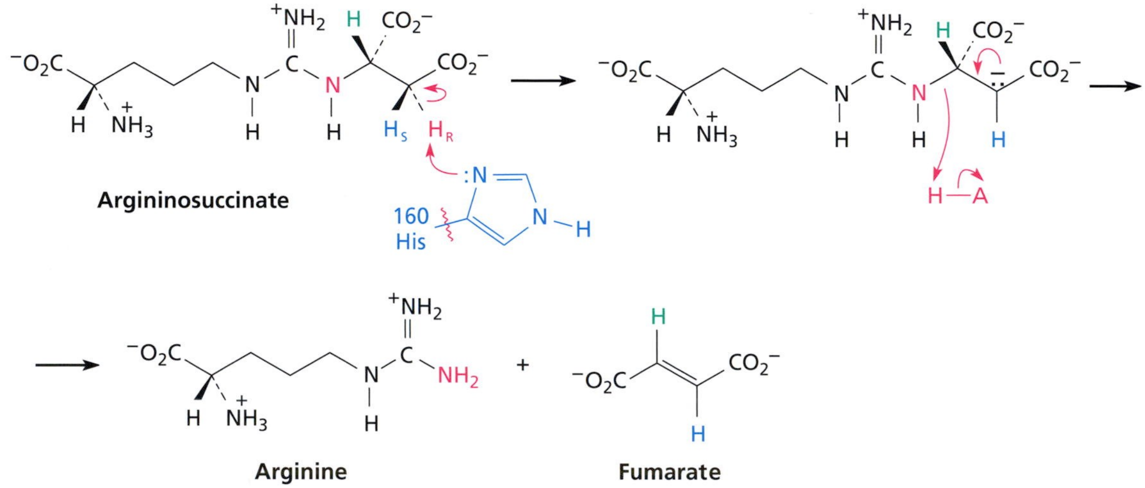

Step 4. Arginine hydrolysis The final step to close the urea cycle is the hydrolysis of arginine to give ornithine and urea. The reaction is catalyzed by the Mn^{2+}-containing enzyme arginase[7, 8] and occurs by addition of H_2O to the $C{=}N^+$ double bond, followed by elimination of ornithine from the tetrahedral intermediate.

5.3 Catabolism of Amino Acid Carbon Chains

With the nitrogen atom removed by transamination and the ammonia converted into urea, the final stage of amino acid catabolism is degradation of the carbon chain. As indicated in Figure 5.4, the carbon chains are commonly converted into one of seven intermediates that are further degraded in the citric acid cycle. Those amino acids that are directly degraded to pyruvate or to a citric acid cycle intermediate are called **glucogenic** because they can also enter the gluconeogenesis pathway (Section 4.5). Those that are converted to either acetoacetate or acetyl CoA are called **ketogenic** because they can also enter the fatty acid biosynthesis pathway (Section 3.4) or be converted into so-called *ketone bodies* (acetoacetate, β-hydroxybutyrate, and acetone). Note that several amino acids are both glucogenic and ketogenic.

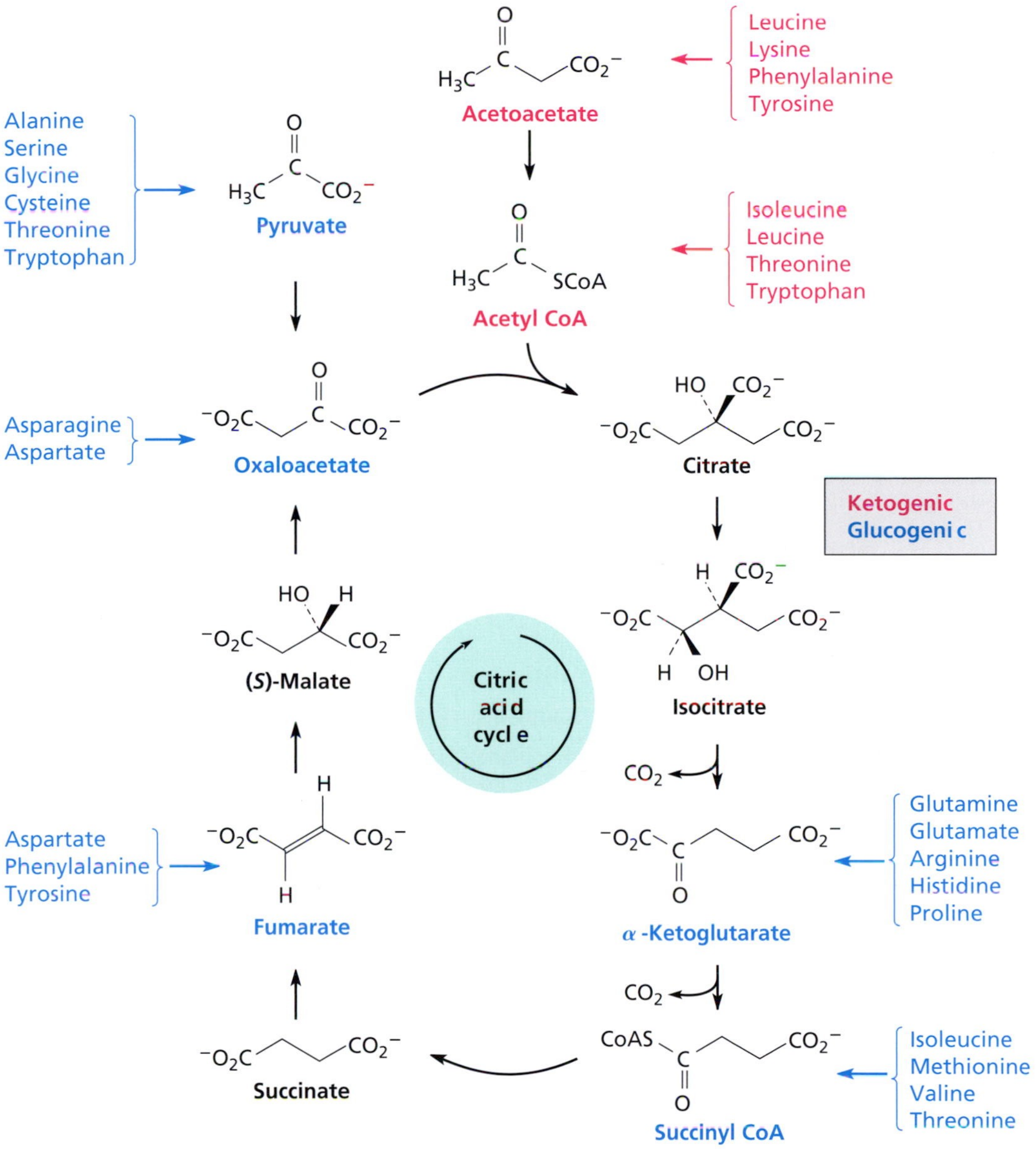

Leucine, Lysine, Phenylalanine, Tyrosine → Acetoacetate

Isoleucine, Leucine, Threonine, Tryptophan → Acetyl CoA

Alanine, Serine, Glycine, Cysteine, Threonine, Tryptophan → Pyruvate

Asparagine, Aspartate → Oxaloacetate

Citrate

Ketogenic
Glucogenic

(S)-Malate

Citric acid cycle

Isocitrate

Aspartate, Phenylalanine, Tyrosine → Fumarate

Glutamine, Glutamate, Arginine, Histidine, Proline → α-Ketoglutarate

Succinate

Isoleucine, Methionine, Valine, Threonine → Succinyl CoA

FIGURE 5.4 Carbon chains of the various amino acids are commonly converted into one of seven intermediates for further breakdown in the citric acid cycle. Ketogenic amino acids are shown in red; glucogenic amino acids in blue.

Alanine, Serine, Glycine, Cysteine, Threonine, and Tryptophan

Six amino acids are catabolized to pyruvate: alanine, serine, glycine, cysteine, threonine, and tryptophan. **Alanine** gives pyruvate directly by a transamination reaction with α-ketoglutarate:

Alanine + α-Ketoglutarate $\xrightarrow{PLP}$ Pyruvate + Glutamate

Serine is converted to pyruvate by two pathways. In one, serine is dehydrated by the PLP-dependent enzyme serine dehydratase to give an enamine, which is hydrolyzed to pyruvate:

Serine $\xrightarrow{-H_2O}$ Aminoacrylate (an enamine) $\xrightarrow{H_2O,\ -NH_3}$ Pyruvate

The reaction occurs by formation of a PLP–serine imine, followed by deprotonation of the acidic α carbon in the usual way (Figure 5.1). Elimination of the —OH leaving group, followed by cleavage of the imine, regenerates PLP and gives aminoacrylate, an enamine that tautomerizes to an imine and is then hydrolyzed to pyruvate (Figure 5.5).

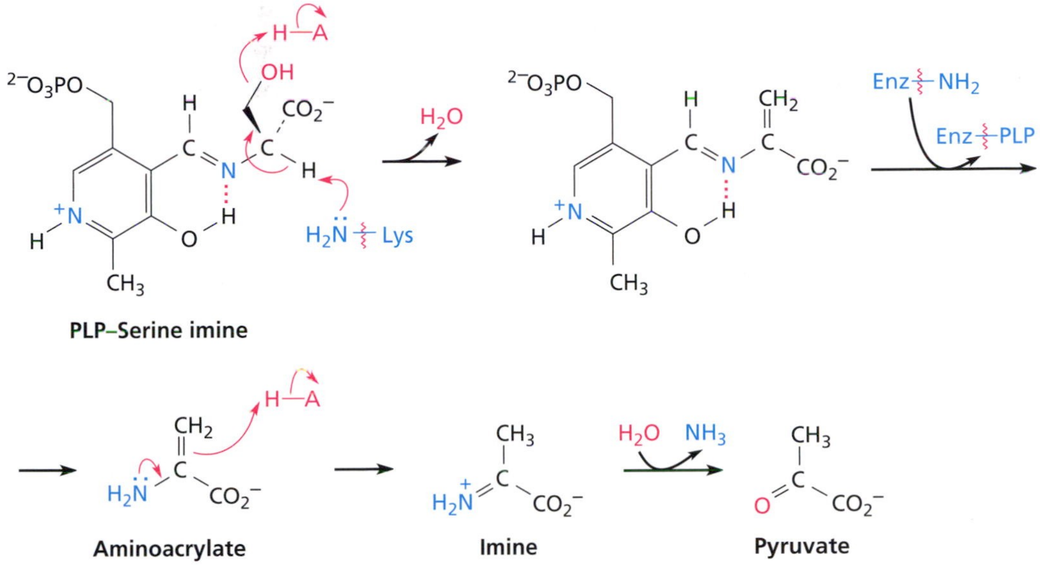

PLP–Serine imine

Aminoacrylate

Imine

Pyruvate

FIGURE 5.5 Mechanism of the PLP-dependent dehydration of serine.

In an alternative catabolic pathway, serine is converted to glycine by serine hydroxymethyltransferase,[9] another PLP-dependent enzyme that catalyzes the cleavage of CH_2O from the PLP–serine imine. A possible mechanism for this cleavage is a retro-aldol reaction to give formaldehyde, which then reacts with tetrahydrofolate (THF) and yields 5,10-methylenetetrahydrofolate (Figure 5.6). Alternatively, evidence has been presented[10] for a direct S_N2-like displacement of the protonated serine —OH group by a nitrogen atom of THF. (Tetrahydrofolate is the biologically active coenzyme derived from folic acid. As we'll see frequently in this and the next chapter, it is required by enzymes that catalyze the transfer of one-carbon units in the CH_3OH, CH_2O, or HCO_2H oxidation states.)

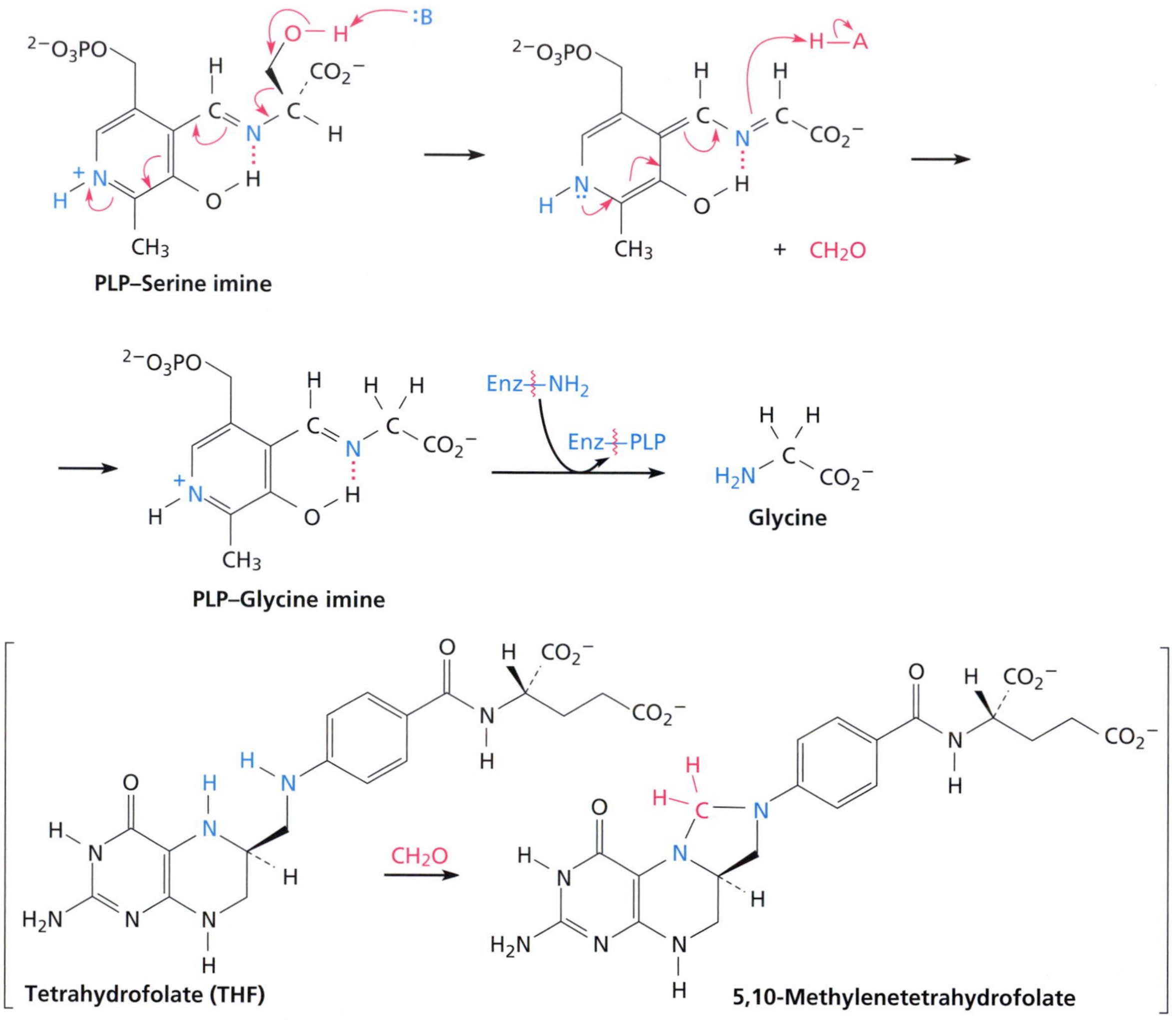

PLP–Serine imine

PLP–Glycine imine

Glycine

Tetrahydrofolate (THF)

5,10-Methylenetetrahydrofolate

FIGURE 5.6 Mechanism of serine catabolism to yield glycine.

Glycine can be catabolized by conversion to serine (the reverse of the serine catabolism pathway shown in Figure 5.6) and thence to pyruvate. Most commonly, however, it is degraded by the *glycine cleavage system*, a multienzyme complex similar to pyruvate dehydrogenase (Section 4.3, Figure 4.9). As shown in Figure 5.7, glycine catabolism occurs in several steps.

Steps 1–2. Imine Formation and decarboxylation Glycine reacts with enzyme-bound PLP to form a PLP–glycine imine, which undergoes decarboxylation by the typical retro-aldol mechanism.

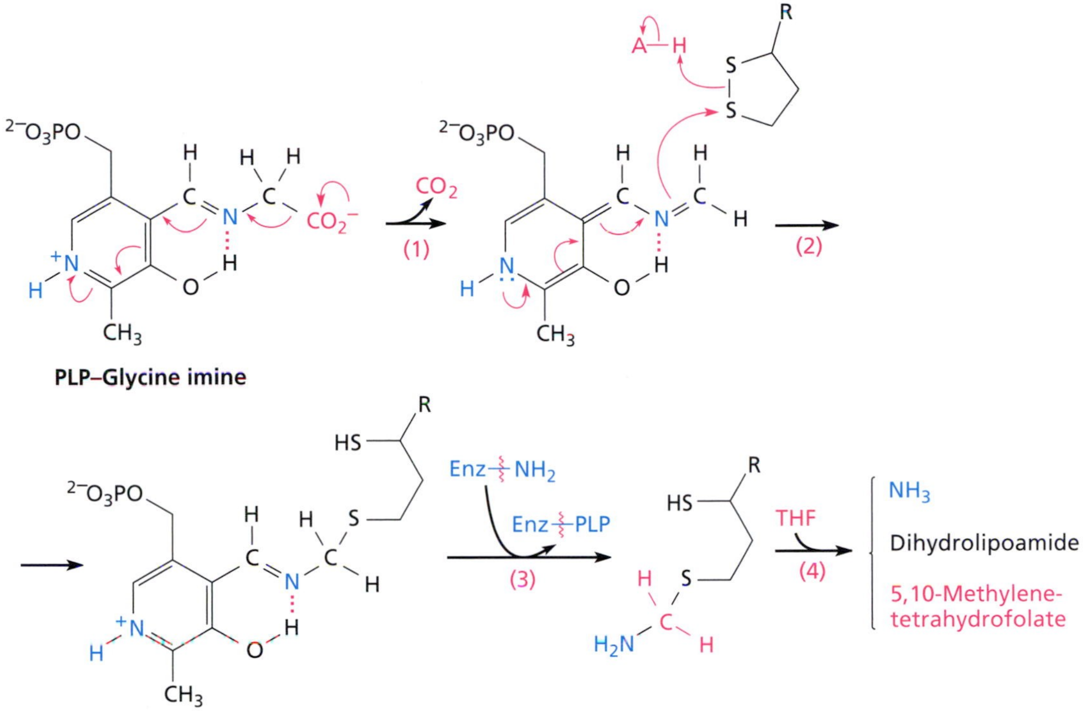

FIGURE 5.7 Mechanism of the glycine cleavage system, which results in the transfer of a CH_2 group to tetrahydrofolate.

Steps 3–4. Reaction with lipoamide, imine cleavage, and one-carbon transfer An S_N2-like reaction of the decarboxylated PLP–imine on a sulfur atom of lipoamide occurs, followed by imine cleavage and regeneration of enzyme-bound PLP. The methylene group is then transferred to tetrahydrofolate with concurrent loss of ammonia, release of dihydrolipoamide, and formation of 5,10-methylenetetrahydrofolate. Dihydrolipoamide is oxidized back to lipoamide by FAD, and the resulting $FADH_2$ is reoxidized by NAD^+, as described previously for the pyruvate dehydrogenase complex (Section 4.3).

Cysteine, like serine, is catabolized by several different pathways. The most common pathway in mammals begins with oxidation by cysteine dioxygenase to give cysteinesulfinate. (A **dioxygenase** incorporates both atoms from O_2 into the product as opposed to a **monooxygenase**, which incorporates only one of the O_2 atoms into the product.). Transamination of cysteinesulfinate with α-ketoglutarate and subsequent loss of SO_2 from the β-ketosulfinate then give pyruvate. The final step is analogous to the decarboxylation of a β keto acid (Section 3.4) and presumably occurs by a similar retro-aldol mechanism.

Cysteine $\xrightarrow{O_2}$ Cysteinesulfinate $\xrightarrow{\alpha\text{-Ketoglutarate} \rightarrow \text{Glutamate}}$ β-Ketosulfinate $\xrightarrow{H-A}$ Pyruvate + SO_2

Threonine has several catabolic pathways (Figure 5.8). Most commonly, it is oxidized by the NAD^+-containing enzyme threonine dehydrogenase to give 2-amino-3-ketobutyrate. A PLP-dependent retro-Claisen reaction with coenzyme A then gives acetyl CoA plus glycine, which is either converted to serine or further degraded by the glycine cleavage system. Alternatively, threonine is cleaved in a retro-aldol reaction to yield glycine plus acetaldehyde. The process is catalyzed by threonine aldolase and occurs by the same PLP-dependent pathway seen previously for serine hydroxymethyltransferase (Figure 5.6). The acetaldehyde is then oxidized to acetate and converted into acetyl CoA, so threonine is ketogenic as well as glucogenic.

In addition to the two routes just mentioned, threonine can also be catabolized by a third route, leading to succinyl CoA. Serine (threonine) dehydratase catalyzes dehydration of threonine to an enamine, which is hydrolyzed to give α-ketobutyrate. Oxidative decarboxylation of α-ketobutyrate yields propionyl CoA by a mechanism analogous to that we saw in Section 4.3, Figure 4.9, for the conversion of pyruvate to acetyl CoA, and carboxylation of propionyl CoA gives succinyl CoA.

Carboxylation of the β carbon of propionyl CoA to yield succinyl CoA is a four-step process that begins with biotin-dependent carboxylation of the α carbon. The reaction is catalyzed by propionyl-CoA carboxylase, yields (*S*)-methylmalonyl CoA as product, and occurs by a mechanism analogous to that of the acetyl CoA carboxylation we saw previously in Section 3.4, Figure 3.14. Interestingly, although the reaction produces the *S* enantiomer of methylmalonyl CoA, the *R* enantiomer is required for the subsequent rearrangement. Thus, an inversion of

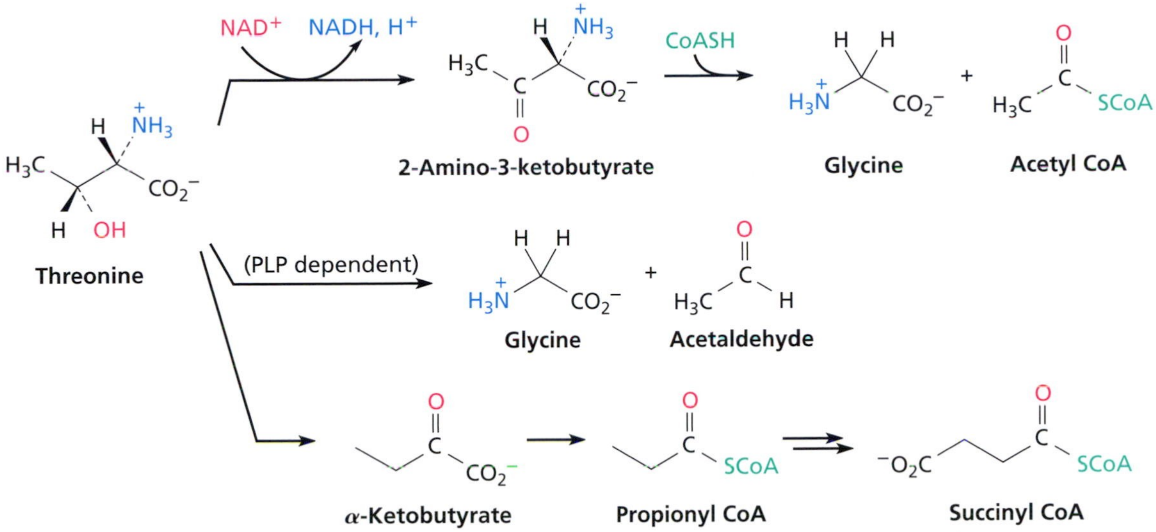

FIGURE 5.8 Three pathways for threonine catabolism.

configuration is needed, catalyzed by methylmalonyl-CoA racemase by deprotonation/reprotonation through an enolate-ion intermediate.

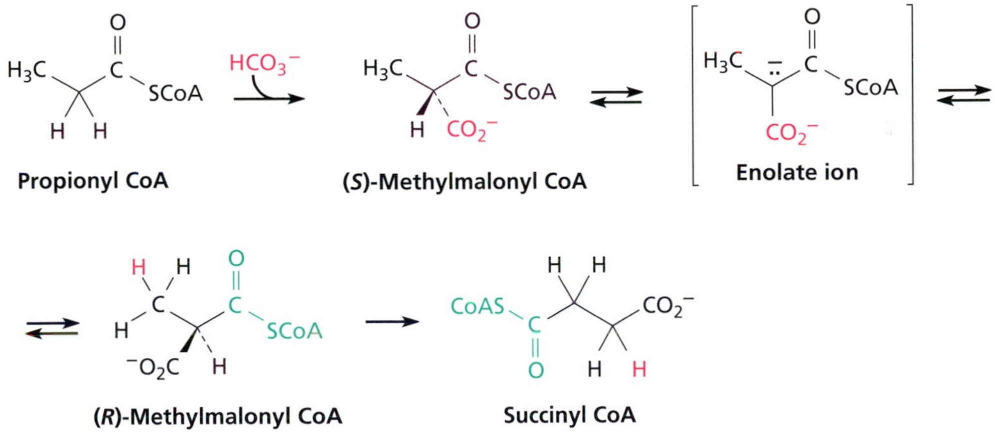

The final rearrangement of (R)-methylmalonyl CoA to succinyl CoA is catalyzed by methylmalonyl-CoA mutase,[11] an enzyme that uses 5′-deoxyadenosylcobalamin (coenzyme B_{12}) as cofactor. A radical rearrangement is involved in the transformation, and the likely mechanism is shown in Figure 5.9. Homolytic cleavage of the cobalt–carbon bond in coenzyme B_{12} first gives the deoxyadenosyl radical, which abstracts a hydrogen atom from the methyl group of methylmalonyl

CoA. The resultant methylmalonyl CoA radical next cyclizes to a cyclopropyloxy radical that ring-opens to an isomeric radical. Abstraction of a hydrogen from deoxyadenosine then gives succinyl CoA and regenerates the deoxyadenosyl radical, which recombines with cobalamin to complete the catalytic cycle.

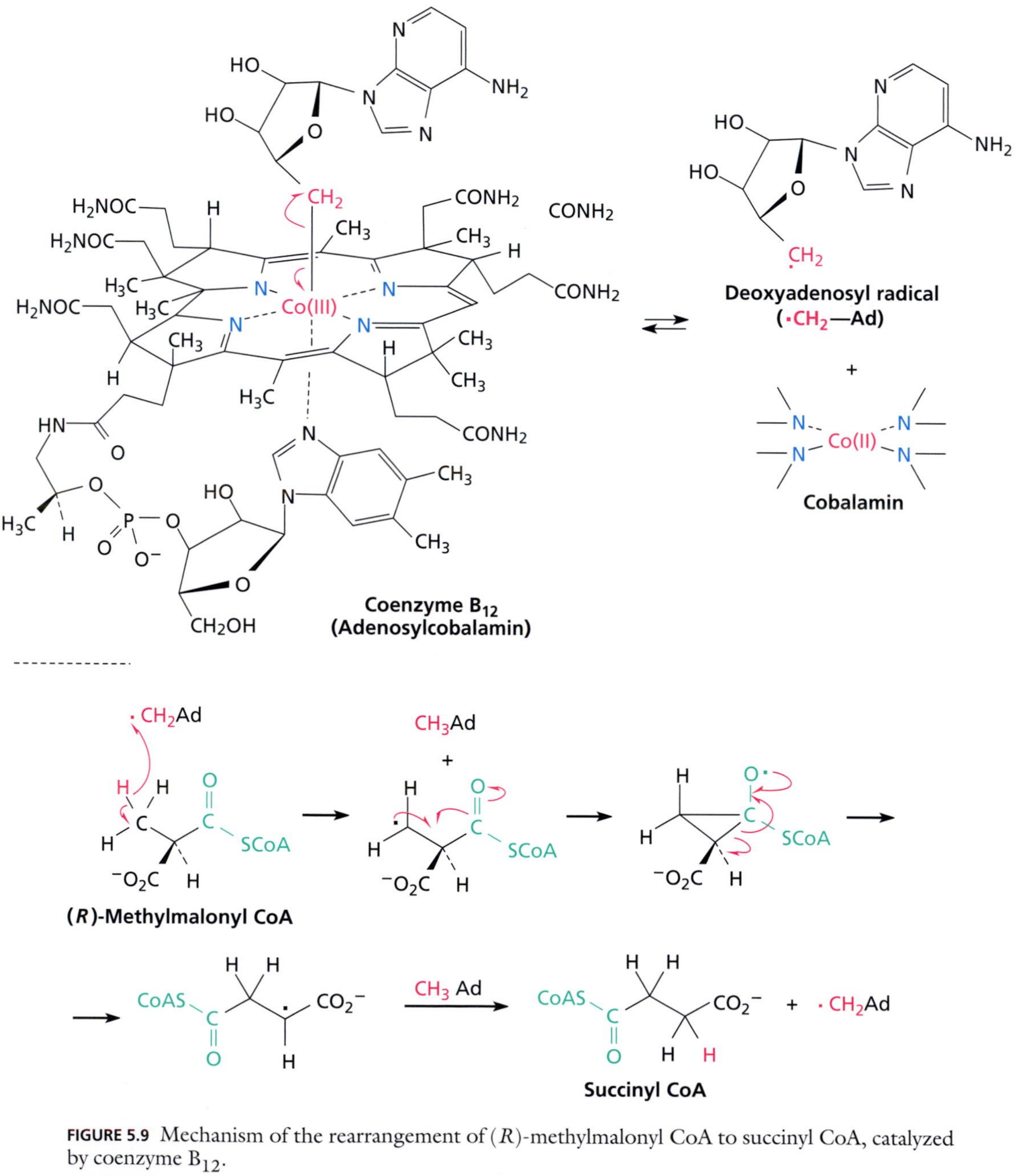

Coenzyme B_{12} (Adenosylcobalamin)

Deoxyadenosyl radical ($\cdot CH_2$—Ad)

Cobalamin

(*R*)-Methylmalonyl CoA

Succinyl CoA

FIGURE 5.9 Mechanism of the rearrangement of (*R*)-methylmalonyl CoA to succinyl CoA, catalyzed by coenzyme B_{12}.

In a general sense, coenzyme B_{12} catalyzes two kinds of reactions: exchange of a hydrogen with a group on the neighboring carbon, as occurs in the rearrangement of methylmalonyl CoA to succinyl CoA, and transfer of a methyl group between two molecules, as occurs in the biosynthesis of methionine that we'll see later in this chapter. We'll also look at the biosynthesis of coenzyme B_{12} in Section 7.5.

A Brief Note about Oxidation States in Iron Complexes

We'll see on numerous occasions in this and the remaining chapters that iron complexes are frequently involved in biological pathways, particularly in oxygenations and other redox reactions. To understand these processes, it's helpful to keep track of the electron transfers that occur during a reaction by looking at changes in the oxidation state of the iron. Because oxidation states are only a bookkeeping device and don't necessarily indicate anything about reaction mechanisms, a change in oxidation state is most easily determined by imagining a hypothetical step or series of steps that account for the observed result.

Assuming that the oxidation state of the iron in the reactant is known, the following guidelines can be used for determining a change in oxidation state:

1. Simple complexation or decomplexation of a ligand to a metal does not change the oxidation state. If, for instance, an Fe(II) complex having five ligands (L) and a vacant coordination site reacts with O_2 to give $FeL_5(O_2)$, one of the O_2 atoms uses two of its nonbonding electrons to complex as a ligand to iron and the oxidation state of the iron remains Fe(II). That is, for bookkeeping purposes, the electrons remain with the ligand. Note that O_2 is shown with two unpaired electrons to reflect the experimental observation that it is a diradical.

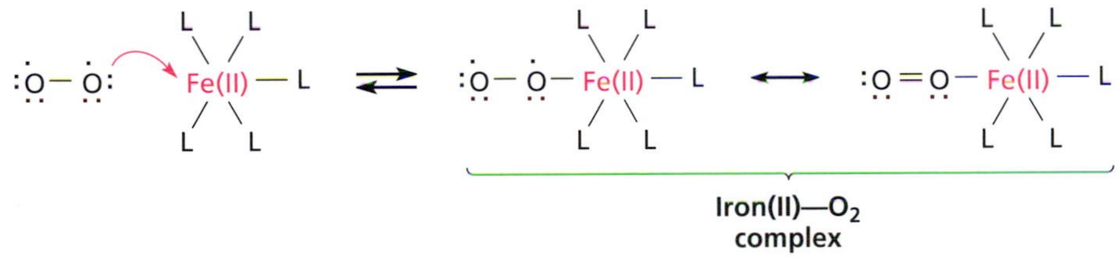

Iron(II)—O_2 complex

2. The oxidation state of the metal changes only if one or more electrons are removed directly from the iron atom or added directly onto the iron.
3. A metal complex can sometimes be drawn in different resonance forms, with the electrons in different places and with the iron in different oxidation states. If so, the most appropriate resonance form may depend on subsequent chemistry. Take the $FeL_5(O_2)$ complex, for instance. It's the same *complex*, but it can be shown in different resonance forms.

Iron(II)—O_2 complex ⟷ Iron(III) superoxide complex ⟷ Iron(IV) peroxide complex

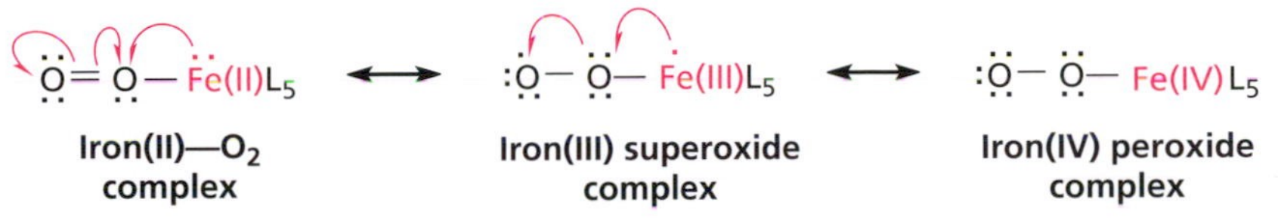

If experimental evidence shows that the $FeL_5(O_2)$ complex acts as an electrophile in its subsequent chemistry, then the Fe(II) resonance form is probably the best representation. This is, in fact, what is proposed for the first step of tryptophan catabolism (Figure 5.12).

Nu:

$$O{=}O{-}Fe(II)L_5 \longrightarrow Nu{-}O{-}O{-}Fe(II)L_5$$

Iron(II)—O_2 complex (electrophilic O)

If, on the other hand, the complex enters into a radical reaction, it might be better represented in the Fe(III) resonance form as the iron superoxide. This is thought to be what happens in step 5 of tryptophan catabolism (Figure 5.15).

R·

$$\cdot O{-}O{-}Fe(III)L_5 \longrightarrow R{-}O{-}O{-}Fe(III)L_5$$

Iron(III) superoxide (radical O)

Finally, if the complex acts as a nucleophile in its subsequent chemistry, it might best be drawn in the Fe(IV) resonance form as the iron peroxide. This is what happens in step 3 of phenylalanine catabolism (Figure 5.29).

E

$$:O{-}O{-}Fe(IV)L_5 \longrightarrow E{-}O{-}O{-}Fe(IV)L_5$$

Iron(IV) peroxide (nucleophilic O)

More complex reactions can be dealt with in the same way as those just described. Focus on the electrons in the reactant and the product, and imagine the changes necessary to bring about the observed results. We'll see later in this chapter, for instance, that phenylalanine undergoes hydroxylation of its aromatic ring to give tyrosine on reaction with an iron(IV)–oxo complex. We can imagine the process happening by the mechanism shown in Figure 5.10. Initial donation of two electrons from the aromatic ring to oxygen first forms a C—O bond and generates a carbocation. At the same time, a shift of two electrons from the Fe=O bond onto iron changes the Fe(IV) to Fe(II). Hydride migration and loss of the ketone ligand from iron then gives another Fe(II) complex, and tautomerization of the ketone gives tyrosine. The reaction may not necessarily happen this way, but breaking down the overall result into this series of simple steps is nevertheless a useful way of looking at things.

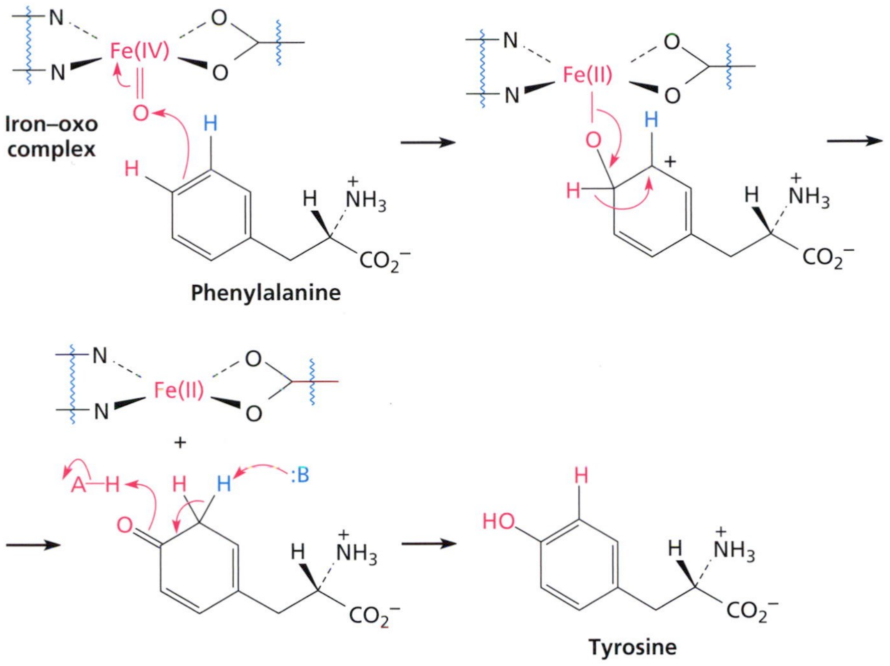

FIGURE 5.10 A possible mechanism for the hydroxylation of phenylalanine to give tyrosine.

Tryptophan has the most complex catabolic pathway of the 20 common amino acids because of its indole ring that must be degraded. Called the kynurenine pathway, the 14-step sequence of reactions shown in Figure 5.11 is required for complete degradation of tryptophan to two equivalents of acetyl CoA, three equivalents of CO_2, one formate, one ammonia, and one alanine, which is itself converted into pyruvate by transamination. Thus, tryptophan is both ketogenic and glucogenic.

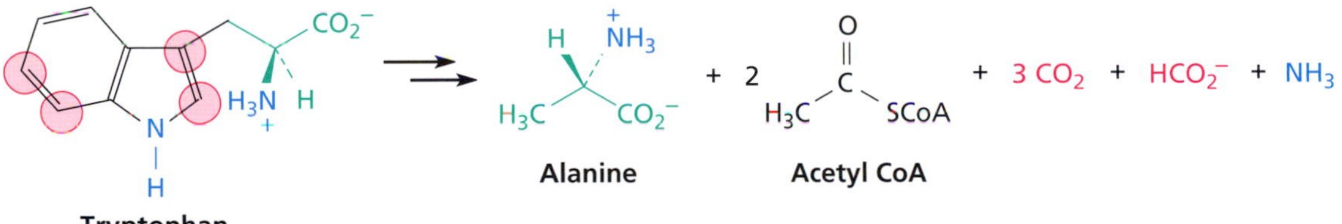

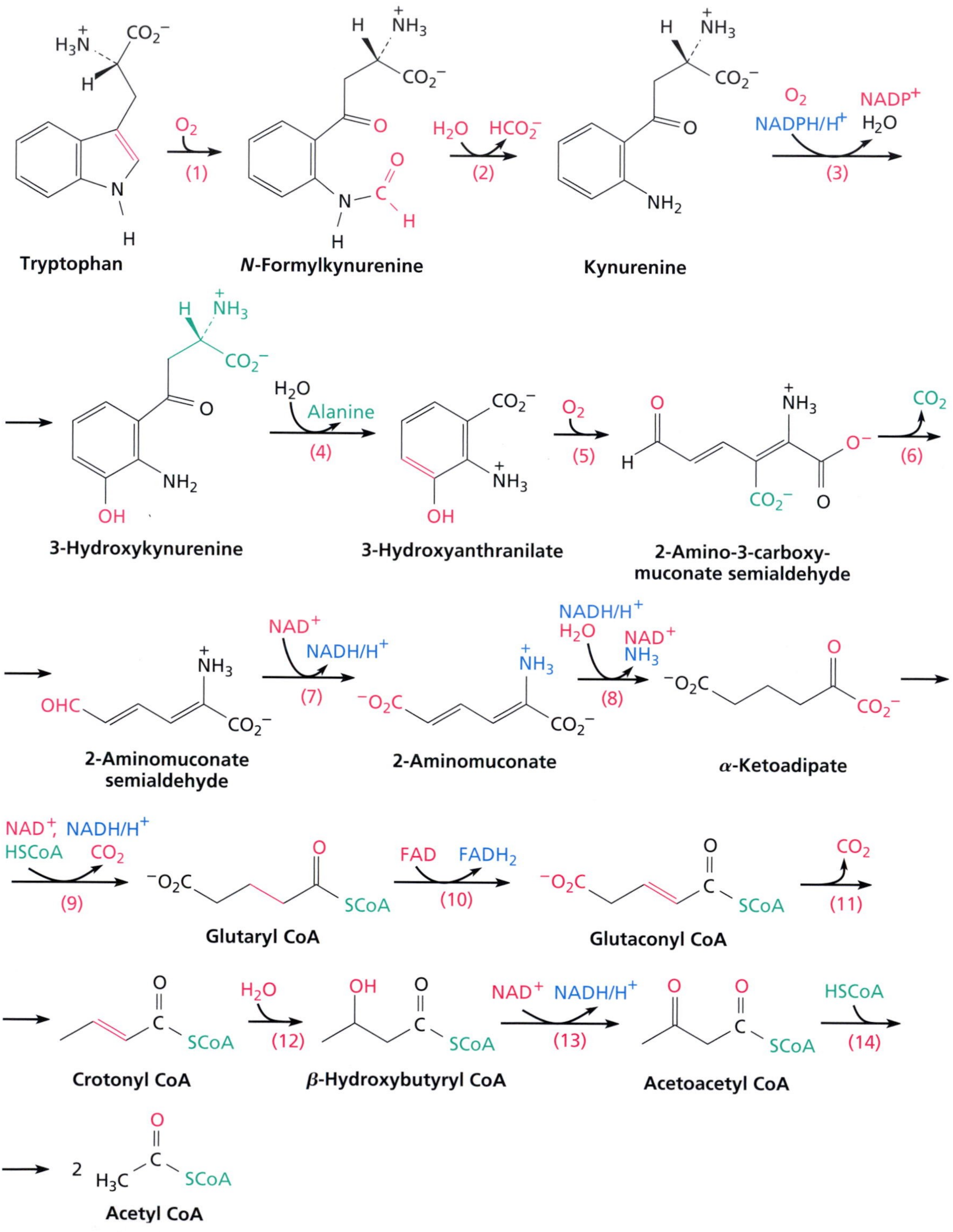

Tryptophan → (1) O_2 → ***N*-Formylkynurenine** → (2) H_2O, HCO_2^- → **Kynurenine** → (3) O_2, $NADPH/H^+$, $NADP^+$, H_2O → **3-Hydroxykynurenine** → (4) H_2O, Alanine → **3-Hydroxyanthranilate** → (5) O_2 → **2-Amino-3-carboxy-muconate semialdehyde** → (6) CO_2 → **2-Aminomuconate semialdehyde** → (7) NAD^+, $NADH/H^+$ → **2-Aminomuconate** → (8) $NADH/H^+$, H_2O, NAD^+, NH_3 → **α-Ketoadipate** → (9) NAD^+, $NADH/H^+$, HSCoA, CO_2 → **Glutaryl CoA** → (10) FAD, $FADH_2$ → **Glutaconyl CoA** → (11) CO_2 → **Crotonyl CoA** → (12) H_2O → **β-Hydroxybutyryl CoA** → (13) NAD^+, $NADH/H^+$ → **Acetoacetyl CoA** → (14) HSCoA → 2 **Acetyl CoA**

FIGURE 5.11 The 14-step kynurenine pathway for catabolism of tryptophan.

Steps 1–2. Ring cleavage and hydrolysis Tryptophan catabolism begins with oxidative cleavage of the indole ring to give *N*-formylkynurenine, a reaction catalyzed by the heme-containing enzyme tryptophan 2,3-dioxygenase.[12] Like coenzyme B_{12} (Figure 5.9), heme contains a metal ion—iron(II) rather than cobalt(III)—chelated by four nitrogen atoms arranged in a complex macrocyclic structure that we'll examine in Chapter 7 (Figure 7.12). Details of the tryptophan oxidative cleavage are not known, but a proposed mechanism is given in Figure 5.12. In this proposal, tryptophan undergoes electrophilic addition of a heme–O_2 complex at C3 (step i) to give an intermediate that undergoes migration of the C3 carbon atom to the neighboring oxygen, with concurrent loss of an iron oxide anion (step ii). Readdition of the iron oxide anion (step iii), followed by opening of the oxygen-containing ring (step iv), yields *N*-formylkynurenine, which is hydrolyzed by the Zn^{2+}-containing formylkynurenine formamidase[13] to give kynurenine plus formate (step v). This migration of the C3 carbon atom to oxygen in step ii is analogous to what occurs in the laboratory during a Baeyer–Villiger oxidation, in which a peroxyacid (RCO_3H) adds to a ketone carbonyl group and rearrangement occurs to yield a lactone.

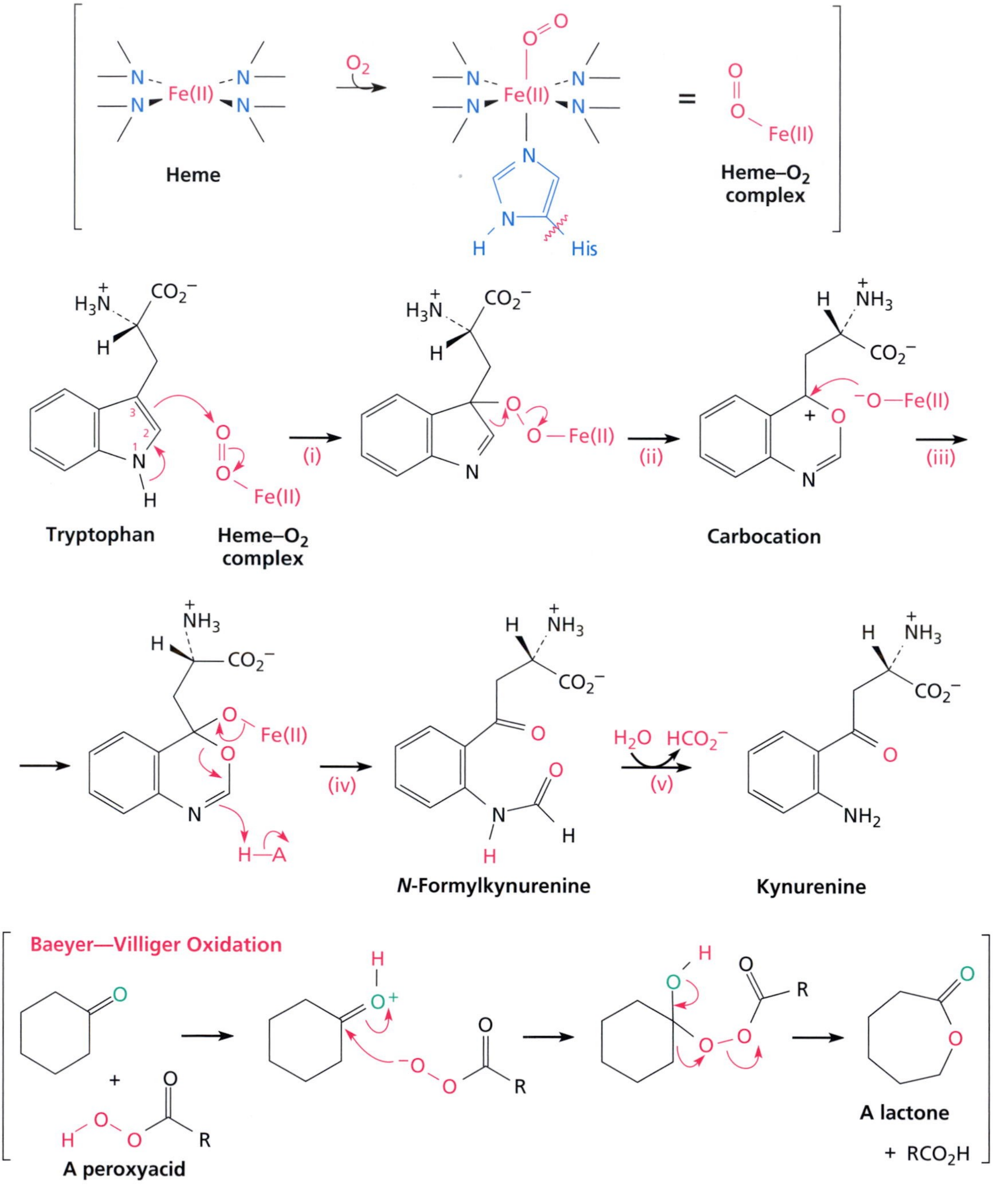

FIGURE 5.12 A proposed mechanism for the oxidative cleavage of tryptophan to *N*-formylkynurenine, the first step in tryptophan catabolism.

Step 3. Hydroxylation The hydroxylation of kynurenine to 3-hydroxykynurenine is catalyzed by kynurenine 3-monooxygenase[14] and requires both NADPH and a flavin cofactor. It's thought that the reaction occurs by the mechanism shown in Figure 5.13, which is similar to that of the better-studied oxidation of *p*-hydroxybenzoate to 3,4-dihydroxybenzoate.[15] FAD is first reduced by NADPH to give $FADH_2$, which reacts with O_2 to form a hydroperoxide. Electrophilic aromatic substitution of the hydroperoxide oxygen on the aromatic ring of kynurenine then gives 3-hydroxykynurenine.

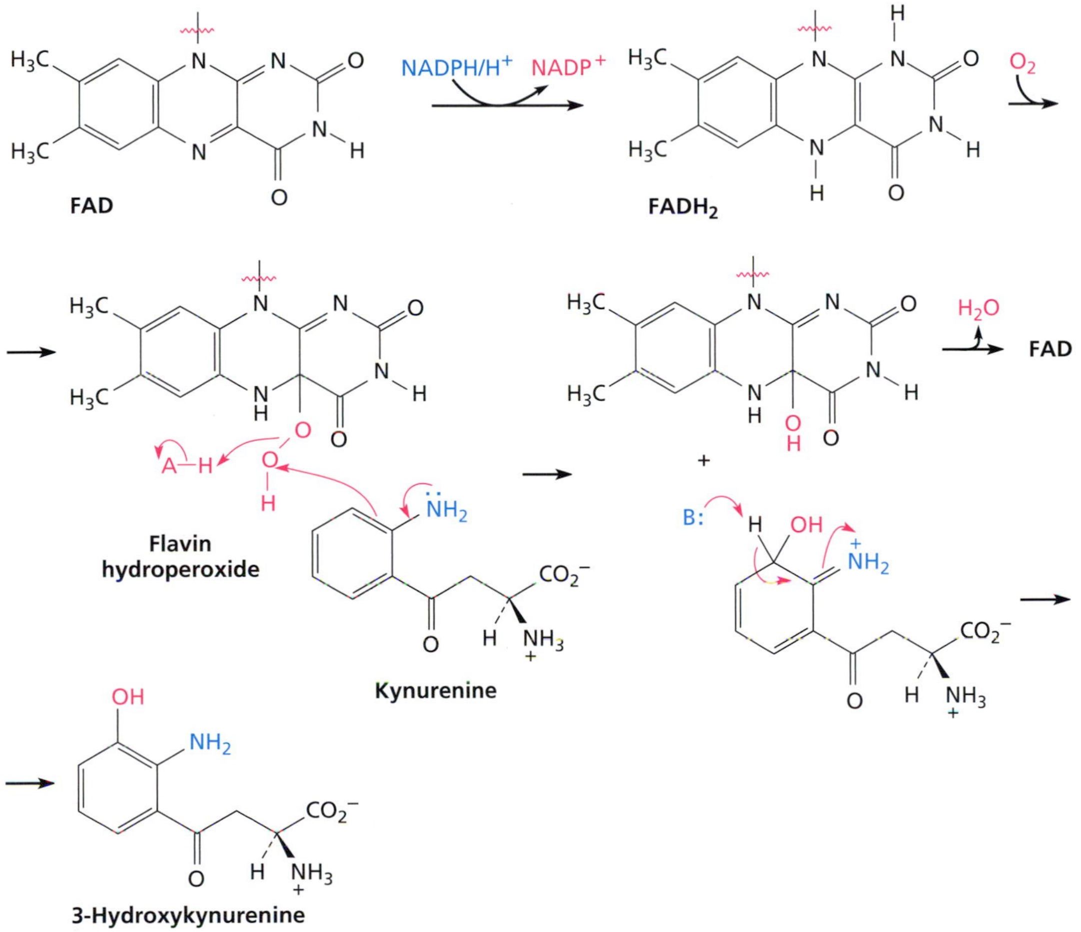

FIGURE 5.13 Proposed mechanism for the hydroxylation of kynurenine to give 3-hydroxykynurenine.

Step 4. Loss of alanine Conversion of 3-hydroxykynurenine to 3-hydroxyanthranilate plus alanine is a PLP-dependent process catalyzed by kynureninase.[16] The reaction shows yet another aspect of PLP chemistry—the ability to catalyze carbon–carbon bond cleavage α to a carbonyl group in a retro-Claisen process. As shown in Figure 5.14, 3-hydroxykynurenine forms an imine with PLP, which is deprotonated in the usual way to give an intermediate β keto imine. Addition of water to the ketone carbonyl group then effects a retro-Claisen cleavage to yield 3-hydroxyanthranilate, and reaction with the enzyme regenerates the enzyme–PLP imine with release of alanine.

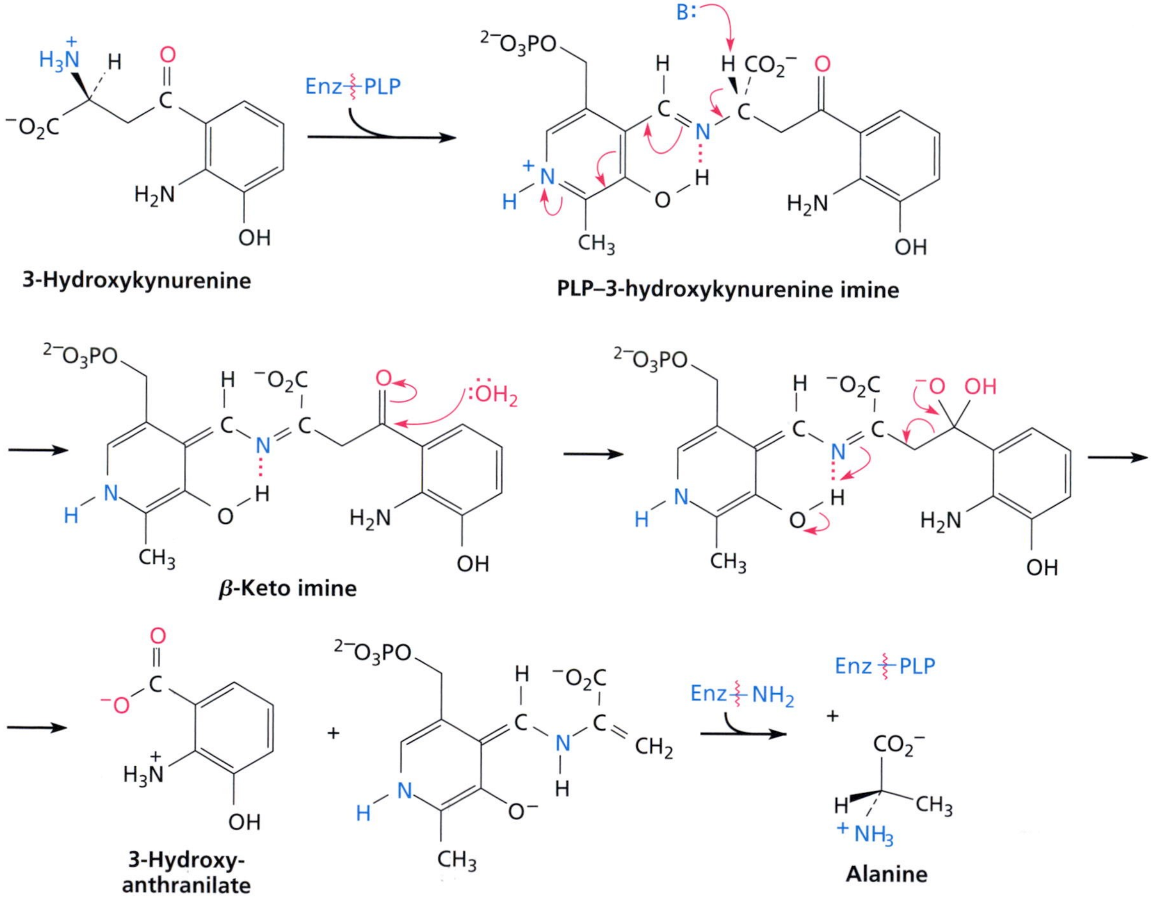

3-Hydroxykynurenine

PLP–3-hydroxykynurenine imine

***β*-Keto imine**

3-Hydroxyanthranilate

Alanine

FIGURE 5.14 Mechanism of the PLP-dependent conversion of 3-hydroxykynurenine to 3-hydroxyanthranilate plus alanine.

Step 5. Opening the aromatic ring Oxidative ring opening of 3-hydroxyanthranilate to give 2-amino-3-carboxymuconate semialdehyde is catalyzed by the nonheme-iron-containing 3-hydroxyanthranilate dioxygenase. The mechanism of the reaction is thought to be analogous to those of more well-studied extradiol catechol dioxygenases,[17] which catalyze the oxidative cleavage of a carbon–carbon bond adjacent to a phenolic hydroxyl group and yield an unsaturated aldehyde acid product. As shown in Figure 5.15, the reaction likely occurs through initial formation of an Fe(II) complex, which reacts with O_2 and forms a peroxide intermediate. Migration of a ring bond to oxygen then gives a lactone, which opens to the final product.

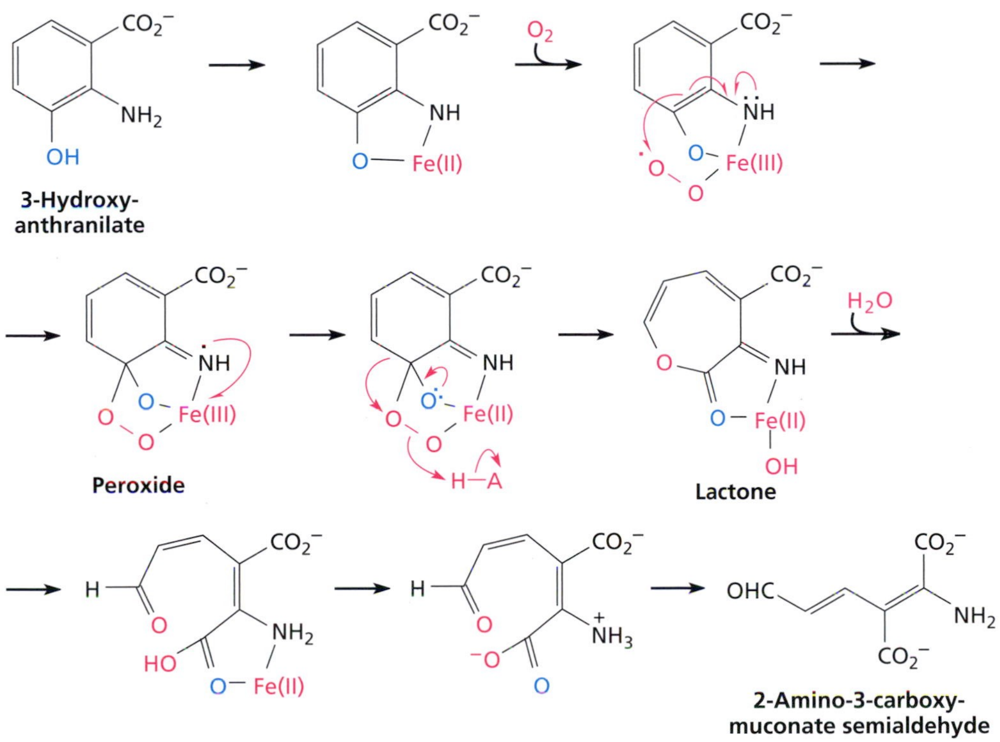

FIGURE 5.15 Proposed mechanism for the conversion of 3-hydroxyanthranilate to 2-amino-3-carboxymuconate semialdehyde.

Step 6. Decarboxylation Decarboxylation to yield 2-aminomuconate semialdehyde is catalyzed by 2-amino-3-carboxymuconate semialdehyde decarboxylase. The reaction is likely to occur by tautomerization to an iminium ion, followed by a retro-aldol reaction as typically found for a *β* keto acid.

2-Amino-3-carboxy-muconate semialdehyde ⇄ **Iminium ion tautomer** → (CO_2) **2-Aminomuconate semialdehyde**

Steps 7–8. Oxidation, hydrolysis, and reduction 2-Aminomuconate semialdehyde is oxidized in step 7 to the corresponding acid, 2-aminomuconate, by 2-aminomuconate semialdehyde dehydrogenase. The reaction requires NAD^+ as cofactor and is thought to occur by addition of water to the aldehyde to give an intermediate hydrate, which is oxidized in the usual way. 2-Aminomuconate then undergoes enamine hydrolysis to a ketone intermediate, followed by reduction of the conjugated double bond by NADH. Both steps are catalyzed by 2-aminomuconate oxidase.

2-Aminomuconate semialdehyde → (H_2O, NAD^+ → $NADH/H^+$) **2-Aminomuconate** → (H_2O → NH_3) → ($NADH/H^+$ → NAD^+) ***α*-Ketoadipate**

Steps 9–14. Conversion to acetoacetate The remaining steps 9–14 of tryptophan catabolism have all been seen previously in other pathways: In step 9, *α*-ketoadipate is oxidatively decarboxylated to glutaryl CoA by a multienzyme complex similar to pyruvate dehydrogenase (Section 4.3, Figure 4.9). In step 10, glutaryl CoA is dehydrogenated to give glutaconyl CoA, as in fatty acid oxidation (Section 3.3). In step 11, glutaconyl CoA decarboxylates to give crotonyl CoA by a typical retro-aldol pathway. In step 12, crotonyl CoA reacts with water by a conjugate addition pathway, as in fatty acid oxidation (Section 3.3). In step 13, β-hydroxybutyryl CoA is oxidized by NAD^+ to give acetoacetyl CoA by the usual mechanism. And in the final step 14, a retro-Claisen reaction yields the ultimate product, acetyl CoA.

A Brief Note about PLP Reactions

As a short breather at this point, it's interesting to note the remarkable variety in the kinds of reactions catalyzed by PLP-containing enzymes. Transaminations, eliminations, retro-aldol cleavages, decarboxylations, and retro-Claisen cleavages can all be accomplished by a variation of PLP chemistry, with the protonated pyridine ring acting as the electron acceptor and some part of an imine-bound amino acid acting as the electron donor. Figure 5.16 lists the examples we've seen.

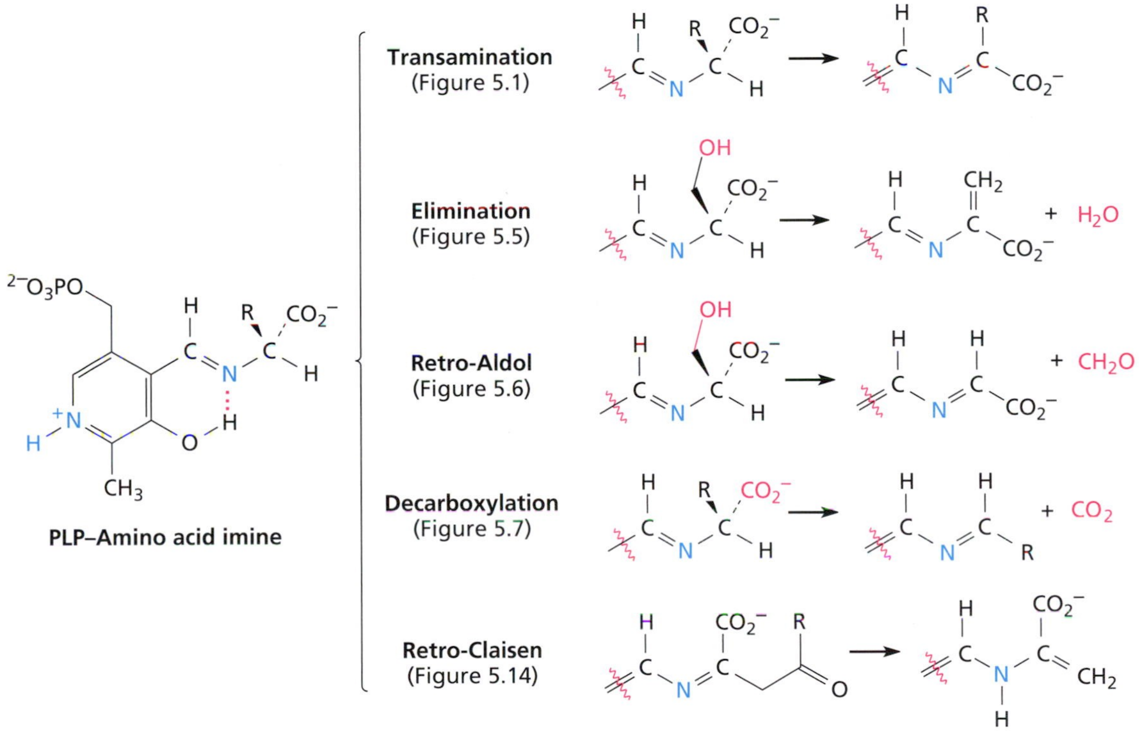

FIGURE 5.16 Some reactions catalyzed by PLP-containing enzymes.

Asparagine and Aspartate

Two amino acids, asparagine and aspartate, are catabolized to oxaloacetate. **Asparagine** is hydrolyzed by asparaginase to yield **aspartate**, which is converted to oxaloacetate by transamination. In addition, aspartate can undergo conversion into fumarate by loss of ammonia in some organisms. Catalyzed by aspartase,[18] the reaction is thought to occur by an E1cB mechanism through a stabilized enolate-ion intermediate.

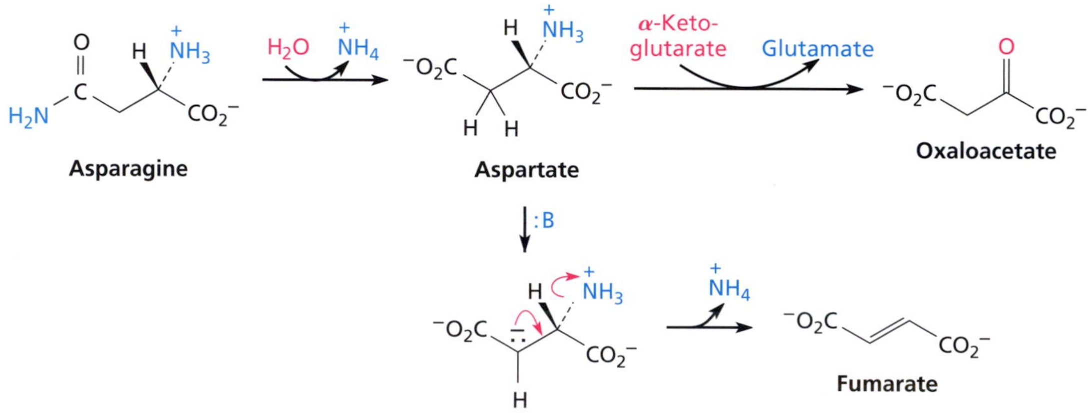

H_2O $\overset{+}{N}H_4$

α-Keto-glutarate Glutamate

Asparagine **Aspartate** **Oxaloacetate**

:B

$\overset{+}{N}H_4$

Fumarate

Glutamine, Glutamate, Arginine, Histidine, and Proline

Five amino acids are catabolized to α-ketoglutarate: glutamine, glutamate, arginine, histidine, and proline. **Glutamine** is hydrolyzed by glutaminase to yield **glutamate**, which is converted to α-ketoglutarate by oxidative deamination with glutamate dehydrogenase, as we saw previously in Section 5.1.

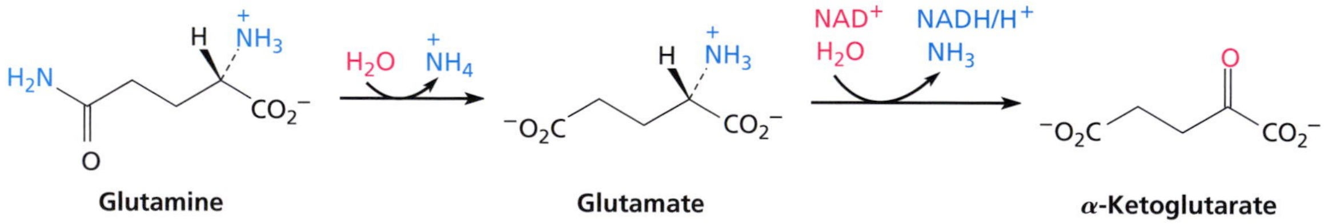

H_2O $\overset{+}{N}H_4$

NAD^+ H_2O $NADH/H^+$ NH_3

Glutamine **Glutamate** **α-Ketoglutarate**

Arginine is catabolized by initial hydrolysis to ornithine, as occurs in the urea cycle (Section 5.2). Transamination of the terminal —NH_2 group then gives the corresponding aldehyde glutamate 5-semialdehyde, which is oxidized to glutamate by NAD^+ (or $NADP^+$).

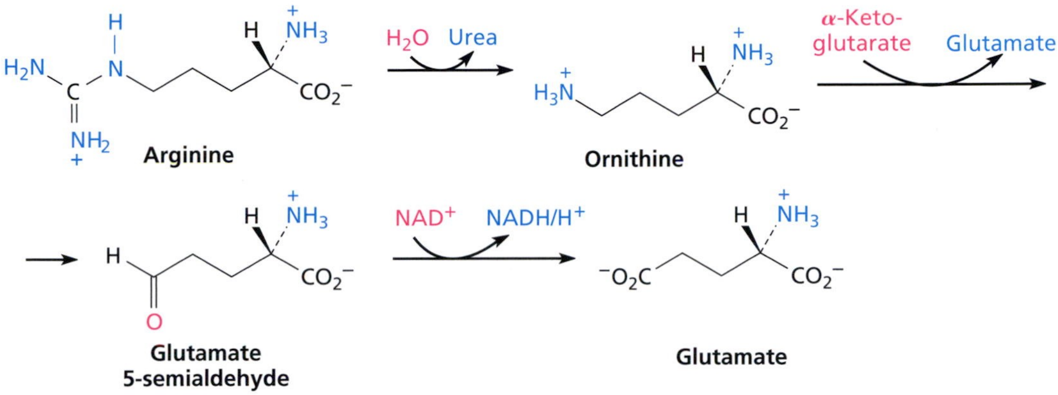

Histidine is converted to glutamate, and thence to α-ketoglutarate, by a four-step pathway that begins with elimination of the α amino group to give *trans*-urocanate. The reaction is catalyzed by histidine ammonia lyase[19] and involves an unusual 4-methylideneimidazol-5-one (MIO) cofactor, formed by cyclization of the segment —Ala(142)—Ser(143)—Gly(144)— in the proenzyme (Figure 5.17). Details of the amine elimination reaction are not yet clear, but one attractive suggestion is that histidine adds to the MIO in a conjugate nucleophilic addition (Michael reaction, Section 1.5). This addition acidifies the neighboring hydrogen and allows elimination of ammonia. A retro-Michael reaction then gives *trans*-urocanate plus regenerated MIO.

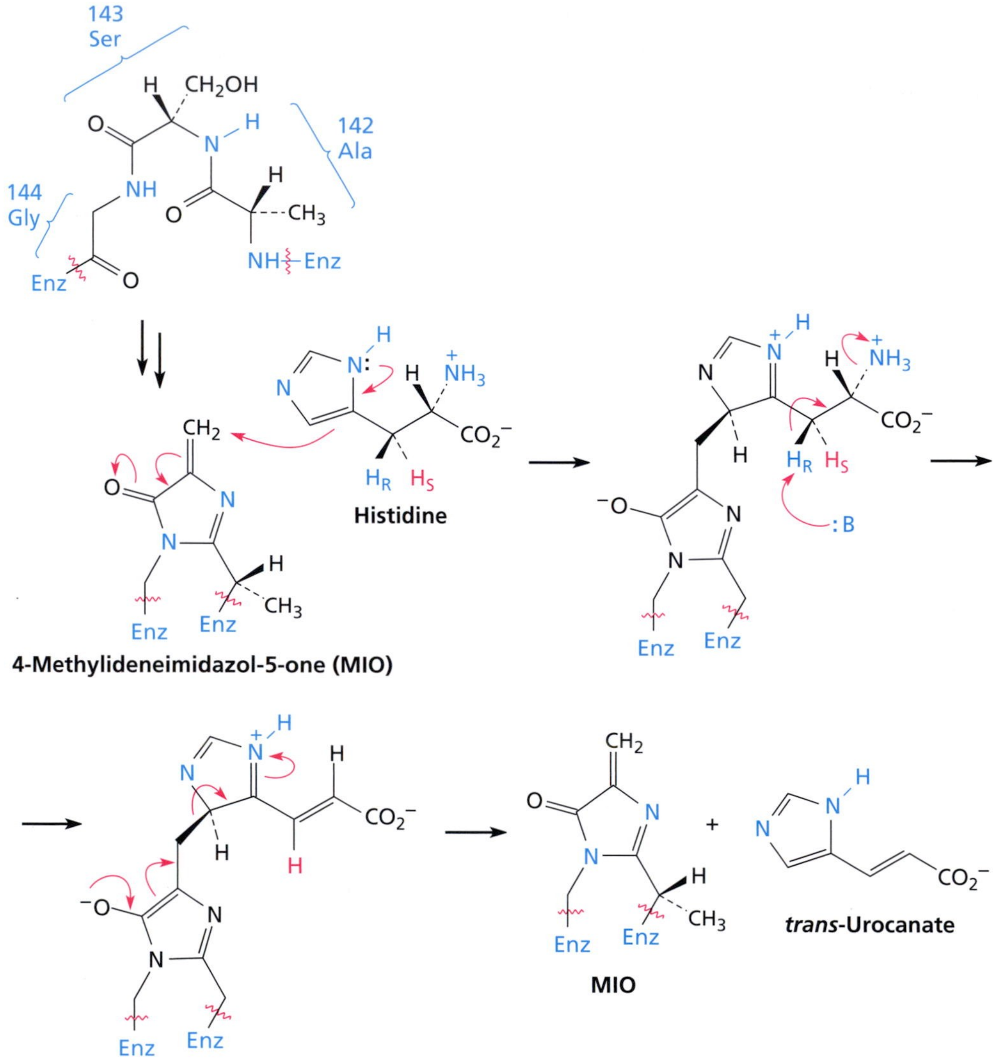

FIGURE 5.17 Formation of *trans*-urocanate, the first step in histidine catabolism, involves a 4-methylideneimidazol-5-one (MIO) cofactor, formed by cyclization of a segment of the peptide chain in the histidine ammonia lyase proenzyme.

trans-Urocanate is converted to glutamate in three additional steps: Conjugate addition of water followed by tautomerization yields an imidazolone; hydrolysis of the imidazolone opens the ring; and transfer of a carbon atom to tetrahydrofolate, as in serine catabolism, gives 5-formiminotetrahydrofolate, which is hydrolyzed to 5-formyl-THF plus ammonia. The initial hydration step is much more complex mechanistically than it might appear and involves a unique, nonredox reaction of NAD^+ acting as an electrophile.[20]

trans-Urocanate → (H_2O) → Imidazolone 5-propionate → (H_2O) → *N*-Formiminoglutamate → (THF) → Glutamate + 5-Formiminotetrahydrofolate

Proline is catabolized by oxidation to the imine 1-pyrroline-5-carboxylate, followed by hydrolysis of the imine to give glutamate 5-semialdehyde, the same intermediate formed in arginine catabolism.

Proline → (NAD^+ → $NADH/H^+$) → 1-Pyrroline-5-carboxylate → (H_2O) → Glutamate 5-semialdehyde

Valine, Isoleucine, and Leucine

Isoleucine, valine, and **leucine**, the three branched-chain amino acids, are catabolized by routes whose initial steps are identical. All undergo an initial transamination to give the corresponding α keto acid, an oxidative decarboxylation to give an acyl CoA, and a dehydrogenation to yield an unsaturated acyl CoA (Figure 5.18). The transaminations are catalyzed by branched-chain amino acid aminotransferase and take place through the usual mechanism involving PLP (Section 5.1, Figure 5.1). The oxidative decarboxylations are catalyzed by the α-ketoisovalerate dehydrogenase complex, with thiamin diphosphate (TPP) and lipoamide as coenzymes, analogous to the conversion of pyruvate to acetyl CoA (Section 4.3, Figure 4.9). The dehydrogenations are effected by acyl-CoA dehydrogenase with FAD as coenzyme, as occurs in fatty acid oxidation (Section 3.3).

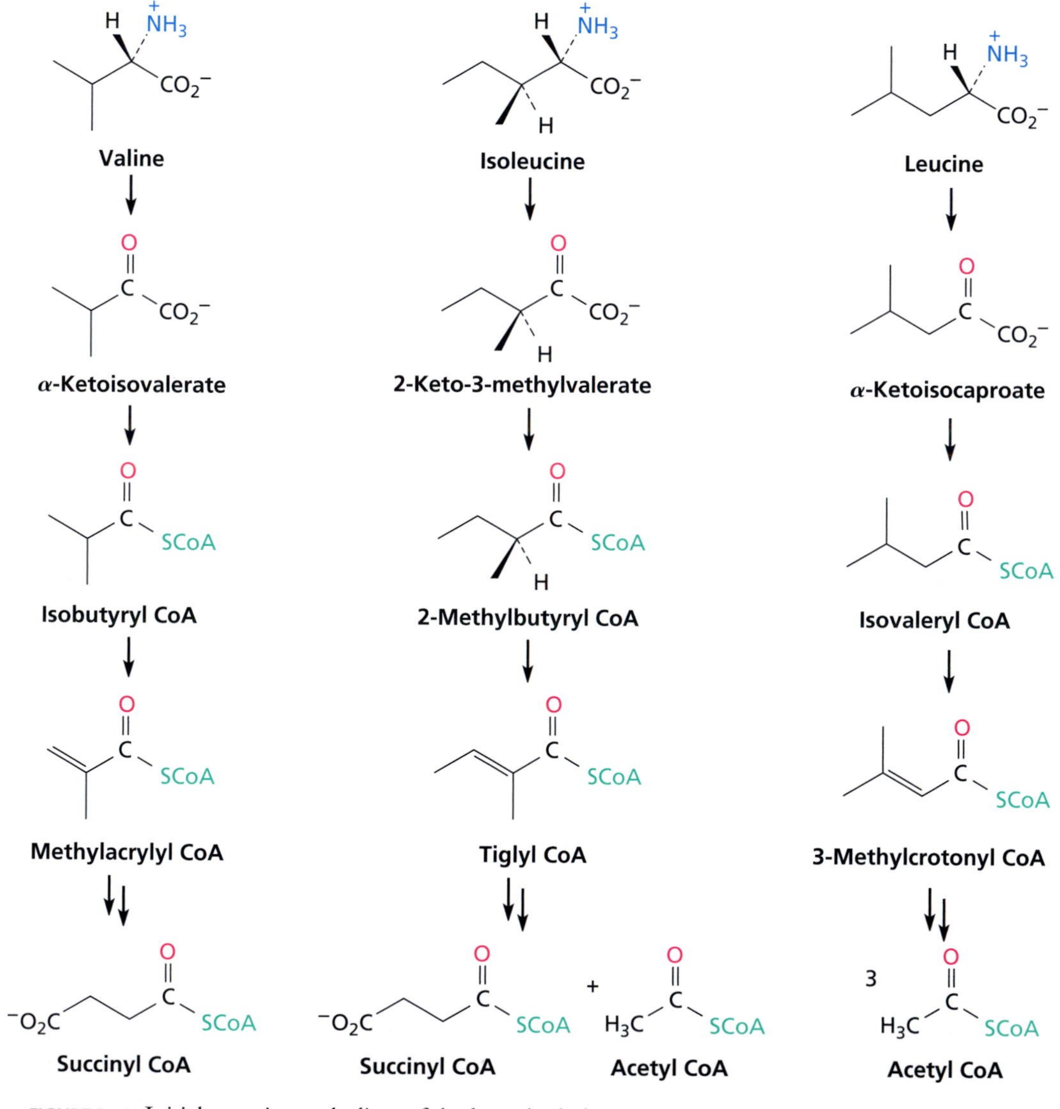

FIGURE 5.18 Initial steps in catabolism of the branched-chain amino acids valine, isoleucine, and leucine. Valine is glucogenic, isoleucine is both glucogenic and ketogenic, and leucine is ketogenic.

The methylacrylyl CoA resulting from valine catabolism is converted into succinyl CoA by a reaction sequence of seven steps, all of which we've seen previously (Figure 5.19). In the first step, conjugate addition of H_2O to methylacrylyl CoA gives β-hydroxyisobutyryl CoA by a mechanism analogous to that in the second step of fatty acid catabolism (Section 3.3). Hydrolysis of β-hydroxyisobutyryl CoA gives β-hydroxyisobutyrate by two sequential nucleophilic acyl substitution reactions: Initial reaction with a glutamate residue on the β-hydroxyisobutyryl CoA hydrolase gives an acyl enzyme complex that then reacts with water. Oxidation of the alcohol group by NAD^+ in the usual way (Section 1.9) next gives an aldehyde. Conversion of the aldehyde group to a thioester is thought to occur by addition of a thiol residue on the methylmalonate semialdehyde dehydrogenase to give a hemithioacetal, followed by oxidation with NAD^+ to an enzyme-bound thioester, as occurs in step 6 of glycolysis (Section 4.2, Figure 4.7). Nucleophilic acyl substitution by coenzyme A and decarboxylation by the usual retro-aldol mechanism complete the process, giving propionyl CoA that goes on to succinyl CoA as described previously for threonine catabolism.

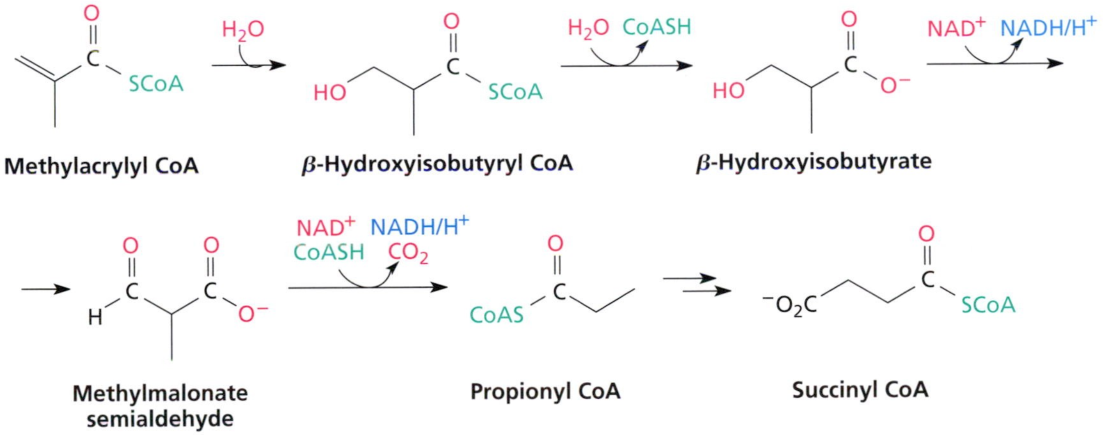

Methylacrylyl CoA → (H_2O) → β-Hydroxyisobutyryl CoA → (H_2O, CoASH) → β-Hydroxyisobutyrate → (NAD^+, $NADH/H^+$) → Methylmalonate semialdehyde → (NAD^+, $NADH/H^+$, CoASH, CO_2) → Propionyl CoA ⇉ Succinyl CoA

FIGURE 5.19 Mechanism of the conversion of methylacrylyl CoA to succinyl CoA during the catabolism of valine.

Tiglyl CoA resulting from isoleucine catabolism is converted into both succinyl CoA and acetyl CoA by a reaction sequence that we've seen previously in fatty acid metabolism. As shown in Figure 5.20, conjugate addition of water to the unsaturated acyl CoA and oxidation of the resultant alcohol give 2-methyl-3-ketobutyryl CoA. Retro-Claisen reaction by nucleophilic addition of coenzyme A

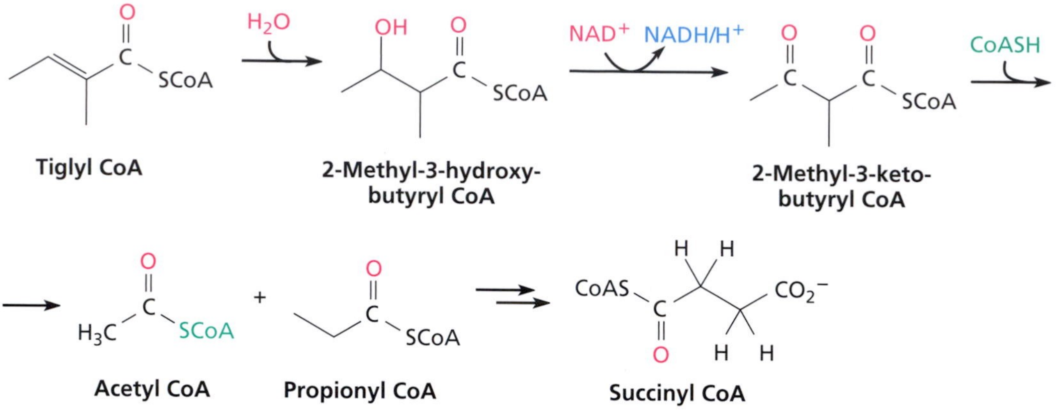

Tiglyl CoA → 2-Methyl-3-hydroxybutyryl CoA → 2-Methyl-3-ketobutyryl CoA → Acetyl CoA + Propionyl CoA → Succinyl CoA

FIGURE 5.20 Mechanism of the conversion of tiglyl CoA to succinyl CoA during the catabolism of isoleucine.

to the ketone carbonyl group of 2-methyl-3-ketobutyryl CoA then produces acetyl CoA and propionyl CoA, which is converted into succinyl CoA as in threonine catabolism.

The 3-methylcrotonyl CoA resulting from leucine catabolism is converted into acetyl CoA and acetoacetate by a three-step reaction sequence that begins with biotin-dependent carboxylation. Addition of water to the unsaturated acyl CoA in the usual way followed by a retro-aldol cleavage of 3-hydroxy-3-methylglutaryl CoA completes the process. The carboxylation step is slightly different from that seen in the carboxylation of acetyl CoA during fatty acid synthesis (Section 3.4, Figure 3.14) in that it occurs on an acidic γ carbon rather than the usual α carbon, but the mechanism is similar.

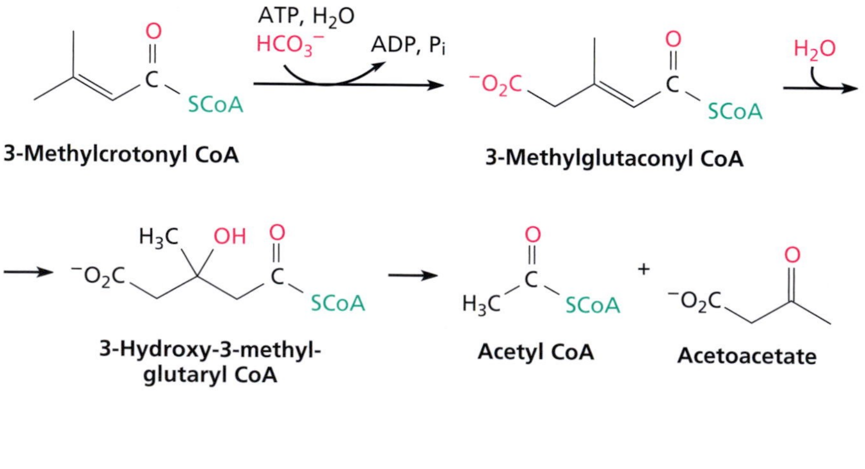

3-Methylcrotonyl CoA → 3-Methylglutaconyl CoA → 3-Hydroxy-3-methylglutaryl CoA → Acetyl CoA + Acetoacetate

Methionine

Methionine catabolism occurs by a complex pathway that leads first to *S*-adenosylmethionine, the primary biological methyl-group donor, and then to a mixture of cysteine and α-ketobutyrate (Figure 5.21). The α-ketobutyrate is then converted into succinyl CoA, as occurs in threonine catabolism.

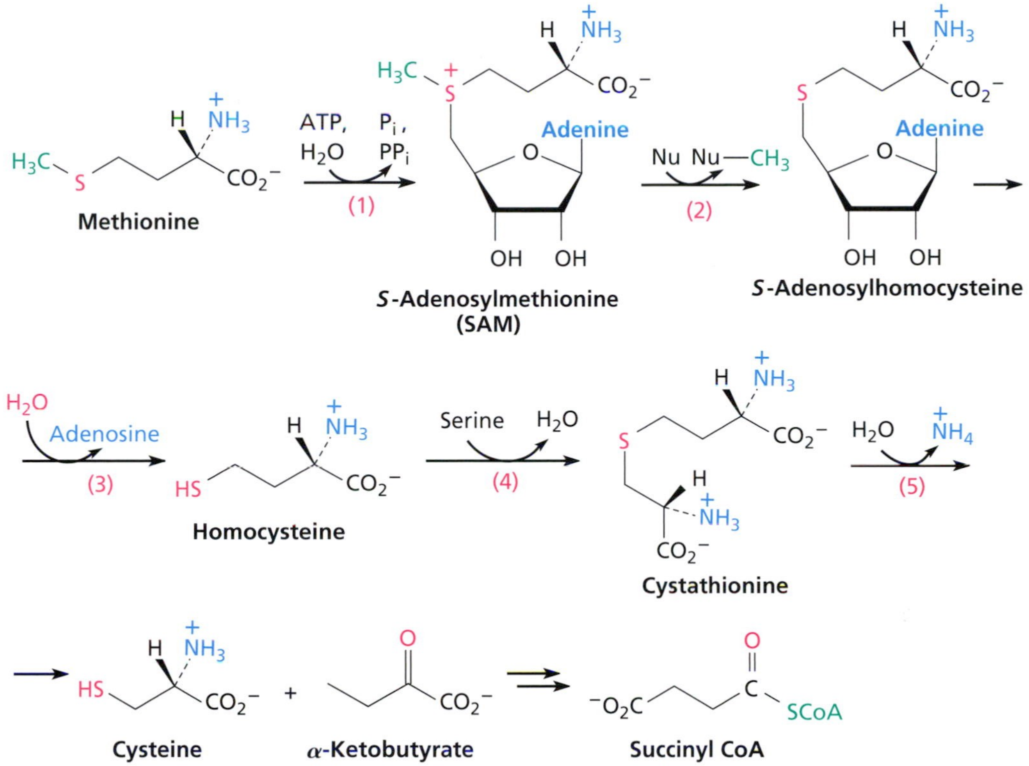

FIGURE 5.21 The catabolic pathway for methionine, which produces cysteine and succinyl CoA.

Step 1. SAM synthesis Methionine catabolism begins with reaction between methionine and ATP to give *S*-adenosylmethionine (SAM). The reaction is an S_N2 displacement of the triphosphate leaving group, followed by subsequent hydrolysis of triphosphate to monophosphate plus diphosphate.

Step 2. Methyl transfer *S*-Adenosylmethionine, the biological source of methyl groups in almost all biosynthetic pathways, next transfers its —CH_3 group to a nucleophile in a second S_N2 reaction, producing *S*-adenosylhomocysteine. (The prefix "homo-" generally means "having one more carbon than," so homocysteine has one more carbon than cysteine.)

Step 3. Hydrolysis Hydrolysis of *S*-adenosylhomocysteine gives homocysteine plus adenosine. Catalyzed by *S*-adenosylhomocysteine hydrolase,[21] the hydrolysis reaction is more complex than it might appear. Rather than being a straightforward S_N2 attack of water on C5 of *S*-adenosylhomocysteine with simultaneous expulsion of a protonated sulfur (thiol) leaving group, the reaction actually occurs in four steps (Figure 5.22). Oxidation of the C3—OH group by NAD^+ first gives a ketone, thereby acidifying the neighboring hydrogen at C4 and allowing a base-catalyzed elimination (E1cB mechanism) with homocysteine as the leaving group. Conjugate addition of water to the unsaturated ketone produced during elimination introduces an —OH group at C5, and reduction of the carbonyl group at C3 by NADH gives adenosine.

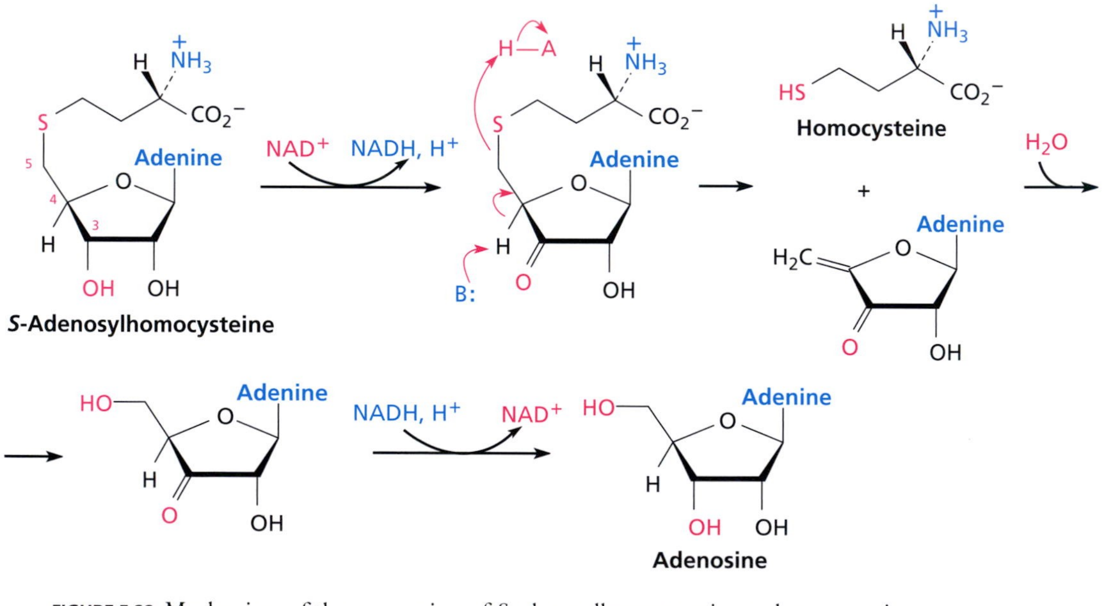

FIGURE 5.22 Mechanism of the conversion of *S*-adenosylhomocysteine to homocysteine.

Step 4. Addition of serine Homocysteine next reacts with serine to give cystathionine in a process catalyzed by the PLP-dependent enzyme cystathionine β-synthase.[22] Once again, the reaction requires several steps and is more complex than it might appear (Figure 5.23). The reaction begins with formation of a

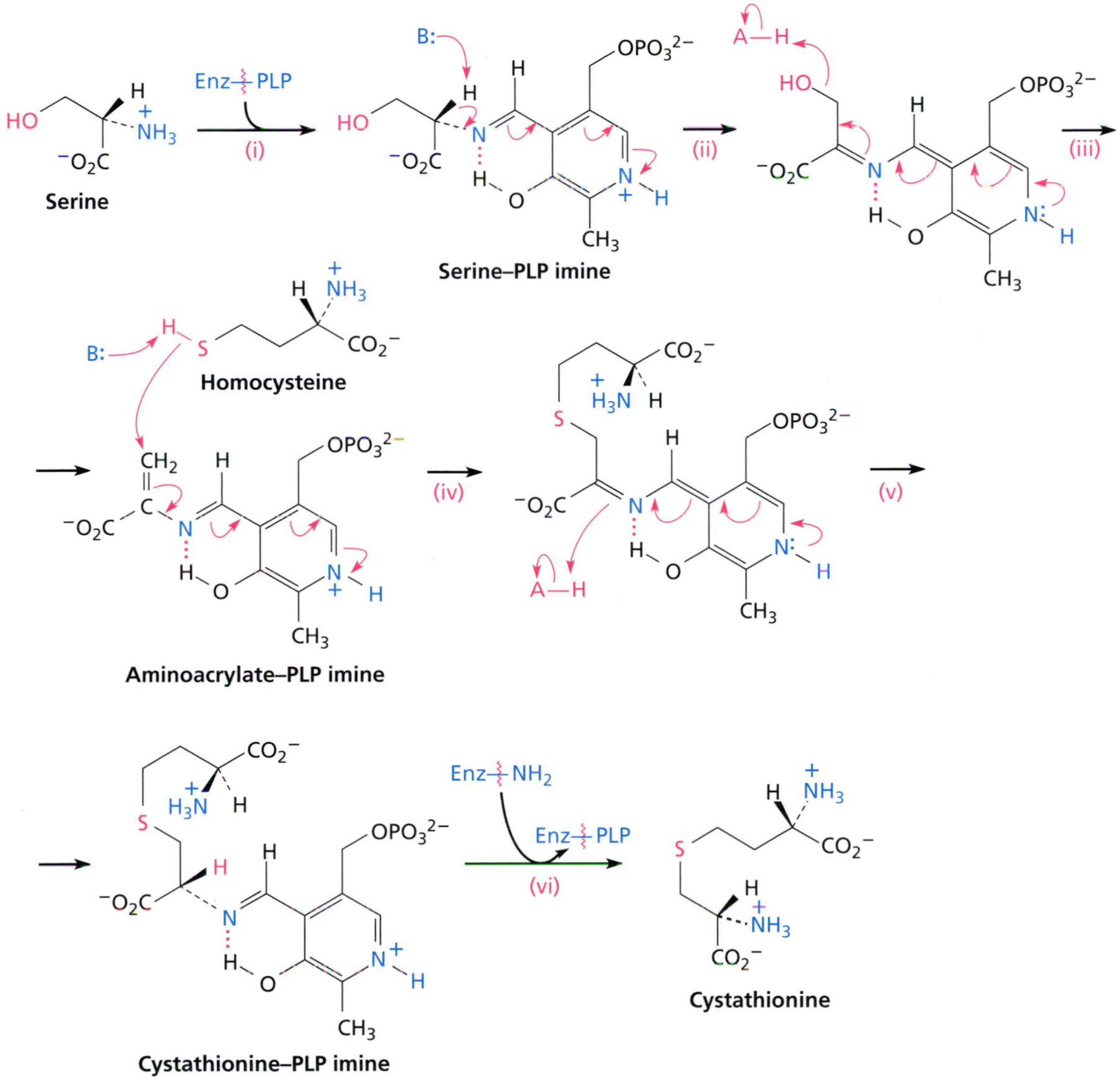

FIGURE 5.23 The mechanism of the conversion of homocysteine plus serine to cystathionine, catalyzed by the PLP-dependent enzyme cystathionine β-synthase.

serine–PLP imine (step i), which is deprotonated at the acidic α carbon in the usual way (step ii). Loss of water gives an aminoacrylate–PLP imine (step iii), and conjugate addition of homocysteine as nucleophile to the aminoacrylate double bond occurs (step iv). Protonation of the α carbon in the usual way then gives cystathionine–PLP imine (step v), which undergoes a transimination reaction with the synthase and releases cystathionine (step vi). One interesting note about the mechanism in Figure 5.23 is that, according to isotope-labeling studies, the addition of homocysteine in step iv occurs from the same face of the aminoacrylate double bond as the elimination of water in step iii, so the overall substitution takes place with retention of stereochemistry.

Step 5. Cleavage of cystathionine The final step in methionine catabolism is the conversion of cystathionine to cysteine plus α-ketobutyrate (Figure 5.24). This too is a multistep, PLP-dependent process and is catalyzed by cystathionine γ-lyase.[23] The reaction begins with formation of a cystathionine–PLP imine (step i), which is deprotonated at the acidic α carbon in the usual way to give an isomeric imine (step ii). Deprotonation of the imine gives an enamine (step iii), which expels cysteine as the leaving group and yields an unsaturated imine (step iv). Protonation on the terminal carbon then gives 2-aminocrotonate–PLP imine (step v), transfer of the PLP group to the enzyme yields 2-aminocrotonate (step vi), and hydrolysis of 2-aminocrotonate gives α-ketobutyrate (step vii).

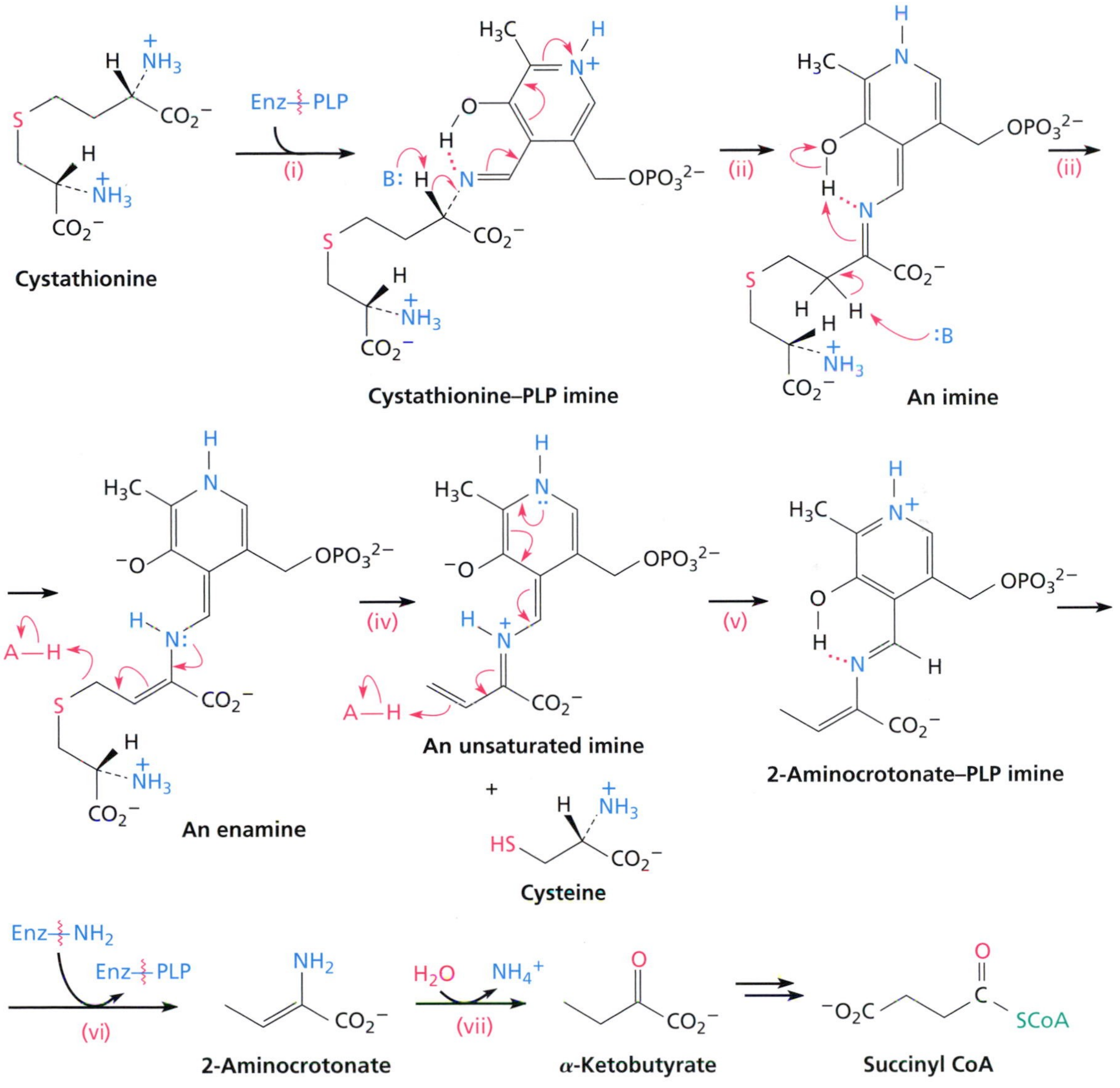

Enz–PLP (i)

Cystathionine

(ii)

Cystathionine–PLP imine

(iii)

An imine

An enamine

(iv)

An unsaturated imine

+

Cysteine

(v)

2-Aminocrotonate–PLP imine

Enz–NH_2 Enz–PLP (vi)

2-Aminocrotonate

H_2O NH_4^+ (vii)

α-Ketobutyrate

Succinyl CoA

FIGURE 5.24 Mechanism of the conversion of cystathionine to cysteine plus α-ketobutyrate.

Lysine

Lysine catabolism occurs through a 10-step pathway that begins with a reductive amination between lysine and α-ketoglutarate to give saccharopine and leads ultimately to acetyl CoA (Figure 5.25).

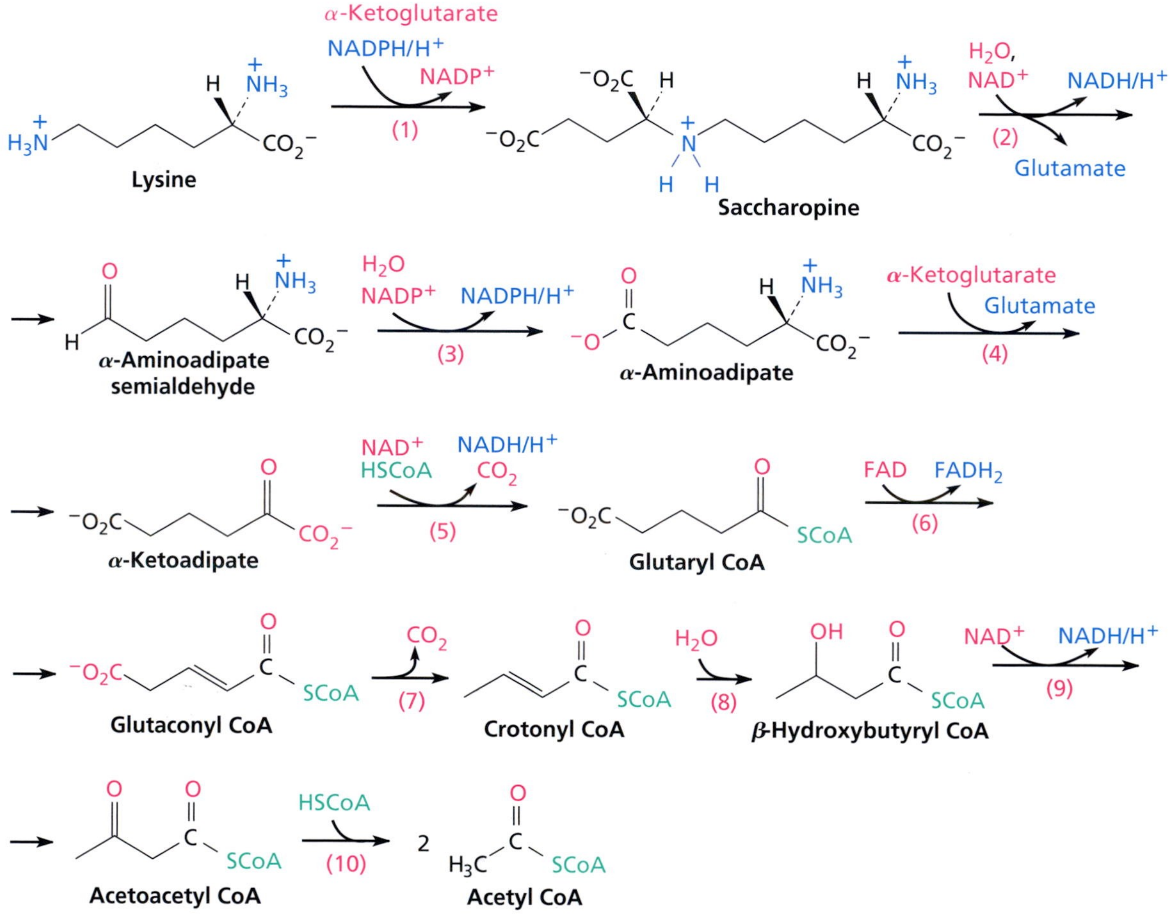

FIGURE 5.25 The 10-step catabolic pathway for conversion of lysine to acetoacetate.

Step 1. Reductive amination Catalyzed by saccharopine dehydrogenase, the reductive amination involves formation of an imine between the terminal $-NH_2$ group of lysine and the ketone carbonyl group of α-ketoglutarate, followed by NADPH reduction of the C=N double bond.

Step 2. Oxidative deamination Saccharopine is oxidatively deaminated in step 2 to give α-aminoadipate semialdehyde plus glutamate. The oxidative deamination is catalyzed by saccharopine reductase[24] and occurs by transfer of a hydride ion from saccharopine to NAD^+, giving an imine that is then hydrolyzed. The mechanism is analogous to that by which glutamate is oxidatively deaminated (Section 5.1).

NAD^+

Saccharopine

$NADH/H^+$

H_2O Glutamate

α-Aminoadipate semialdehyde

Step 3. Oxidation α-Aminoadipate semialdehyde is oxidized by $NADP^+$ to the corresponding carboxylic acid, α-aminoadipate. Details of the oxidation are not known, but the reaction presumably proceeds through an intermediate hydrate, formed by nucleophilic addition of water to the aldehyde carbonyl group.

Steps 4–10. Conversion to acetoacetate In step 4, α-aminoadipate undergoes PLP-dependent transamination by reaction with α-ketoglutarate, as described in Section 5.1, Figure 5.1. The remaining steps 5–10 of lysine catabolism are identical to steps 9–14 of tryptophan catabolism (Figure 5.11).

Phenylalanine and Tyrosine

Phenylalanine undergoes hydroxylation at its para position to give **tyrosine**, which is then catabolized by a six-step pathway to give fumarate and acetoacetate (Figure 5.26).

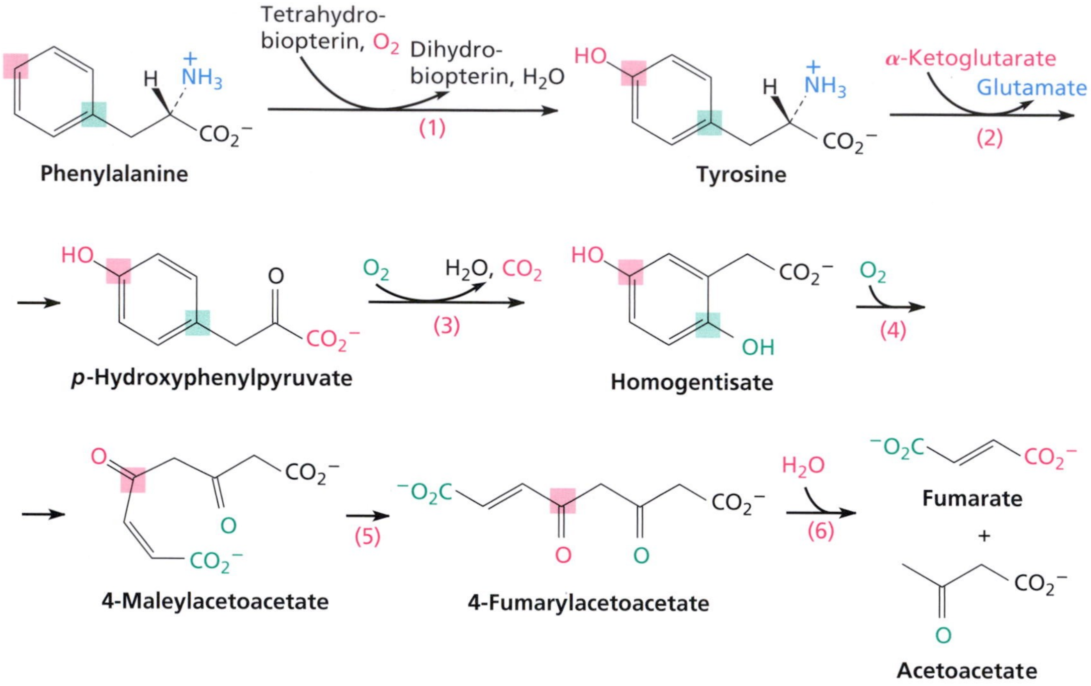

Tetrahydrobiopterin, O_2 — Dihydrobiopterin, H_2O (1)
Phenylalanine → Tyrosine
α-Ketoglutarate — Glutamate (2)
→ *p*-Hydroxyphenylpyruvate
O_2 — H_2O, CO_2 (3)
→ Homogentisate
O_2 (4)
→ 4-Maleylacetoacetate
(5)
→ 4-Fumarylacetoacetate
H_2O (6)
→ Fumarate + Acetoacetate

FIGURE 5.26 The pathway for catabolism of the aromatic amino acids phenylalanine and tyrosine.

Step 1. Hydroxylation Hydroxylation of phenylalanine is catalyzed by the iron-containing enzyme phenylalanine hydroxylase,[25] a monooxygenase that requires the cofactor 5,6,7,8-tetrahydrobiopterin. Tetrahydrobiopterin is structurally related to both tetrahydrofolate and flavin coenzymes (Table 2.5), although it occurs less commonly than either. Its function in phenylalanine hydroxylation is to participate in the activation of O_2 by the complexed Fe(II) atom in the enzyme.[26] This activation converts the iron(II) to an iron(IV)–oxo species that is thought to be the active hydroxylating agent. Dehydration of the hydroxylated biopterin by-product yields a quinoid structure, which tautomerizes to 7,8-dihydrobiopterin and is then reduced by NADH to regenerate tetrahydrobiopterin (Figure 5.27).

The mechanism of the hydroxylation step is not certain;[26, 27] both polar and radical pathways have been proposed. Among the experimental results to be accounted for is the observation that a deuterium originally bonded to the para position of phenylalanine ends up partially in the meta position of tyrosine—a so-called **NIH shift**, named after the U.S. National Institutes of Health where it

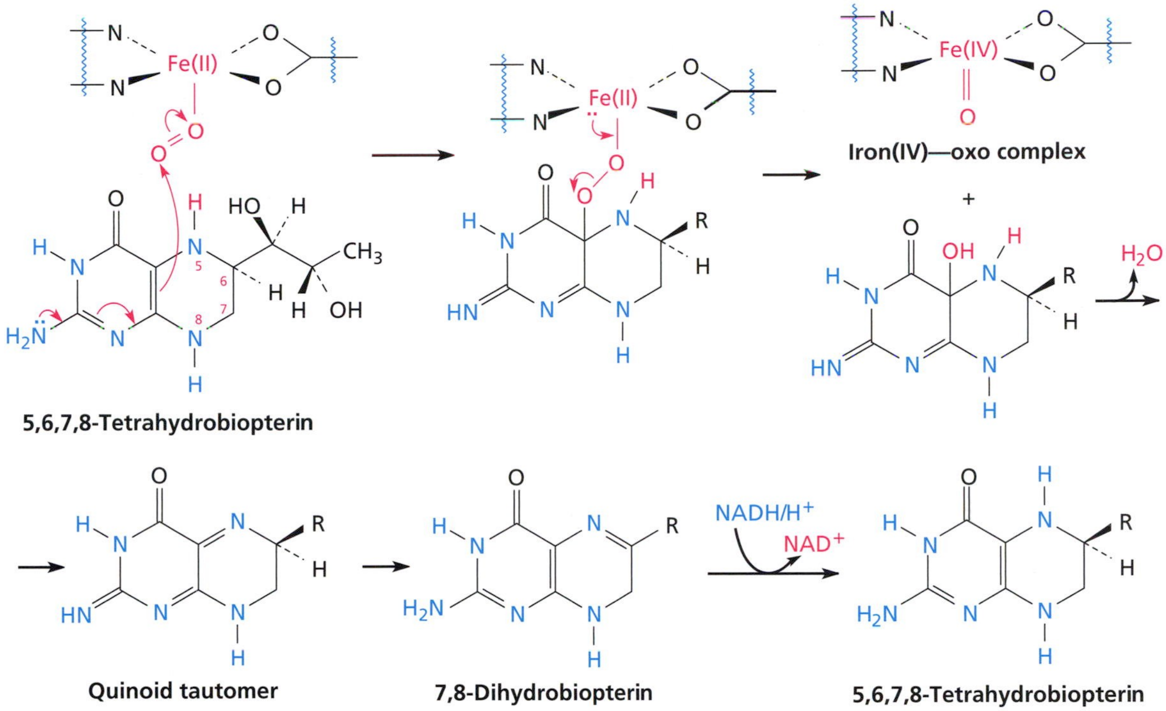

FIGURE 5.27 Mechanism of formation of the iron(IV)–oxo species involved in the hydroxylation of phenylalanine.

was first studied. One possibility, shown in Figure 5.28, begins with electrophilic addition to the ring to give a carbocation. Loss of iron and concurrent hydride migration of the hydrogen at the para position then gives a ketone, which can tautomerize by loss of either proton from the neighboring carbon. Alternatively, it has also been suggested that an epoxide intermediate might be formed initially, followed by acid-catalyzed epoxide opening and rearrangement.

Steps 2–3. Deamination, decarboxylation, and hydroxylation Tyrosine catabolism begins with PLP-dependent deamination by the usual mechanism (reaction with α-ketoglutarate) to yield *p*-hydroxyphenylpyruvate. Oxidative decarboxylation and ring hydroxylation of *p*-hydroxyphenylpyruvate then give homogentisate in a reaction catalyzed by the iron-containing enzyme *p*-hydroxyphenylpyruvate dioxygenase.[28]

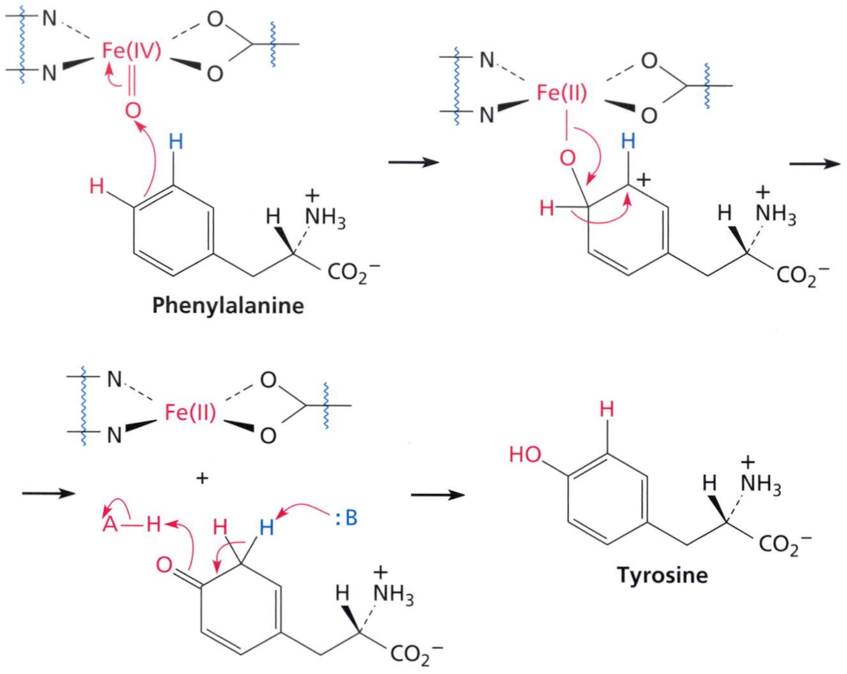

FIGURE 5.28 A possible mechanism of the hydroxylation of phenylalanine to give tyrosine. The reaction occurs with an NIH shift of the hydrogen (red) from the para carbon to the meta carbon.

p-Hydroxyphenylpyruvate dioxygenase is one of a family of nonheme-iron-dependent enzymes[29] called **α keto acid–dependent oxygenases**, which use molecular oxygen to couple the oxidative decarboxylation of an α keto acid with substrate hydroxylation at an unactivated C—H bond. Ascorbate is also required to keep the iron in its Fe(II) oxidation state and to protect the enzyme from oxidation.

A possible mechanism for the reaction, shown in Figure 5.29, is similar in many respects to that proposed for phenylalanine hydroxylation (Figure 5.28). Complexation of *p*-hydroxyphenylpyruvate and O_2 to the iron takes place, followed by intramolecular addition to the ketone carbonyl group (step i). Decarboxylation and formation of an iron(IV)–oxo species then occur (steps ii, iii), and the iron–oxo complex reacts with the aromatic ring in an electrophilic substitution reaction to give a carbocation (step iv). (Alternatively, an epoxide intermediate has been proposed.[30]) Migration of the two-carbon side chain in an NIH shift gives a ketone (step v), and loss of a proton from the neighboring carbon regenerates the aromatic ring (step vi).

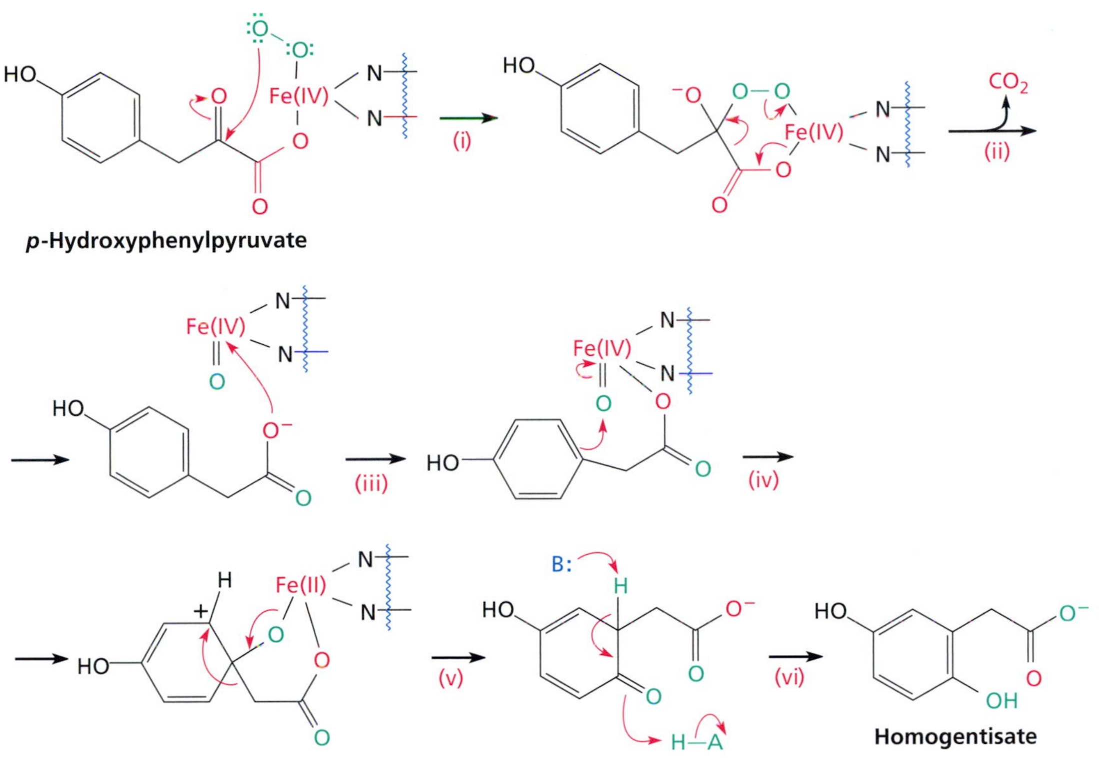

p-Hydroxyphenylpyruvate

Homogentisate

FIGURE 5.29 A mechanism for the conversion of *p*-hydroxyphenylpyruvate to homogentisate, catalyzed by a nonheme-iron-containing, α keto acid–dependent dioxygenase.

Step 4. Ring cleavage The aromatic ring of homogentisate is cleaved in step 4 by homogentisate dioxygenase,[31] which is thought to function by the mechanism shown in Figure 5.30. Coordination of homogentisate and O_2 to the active-site iron(II) yields a cyclic intermediate (steps i, ii), which undergoes nucleophilic attack of oxygen on the aromatic ring and cleavage of an Fe—O bond (step iii). Migration of a ring bond to oxygen gives a seven-membered unsaturated keto ester (step iv), and hydrolysis completes the process (step v). Note that one of the O_2 atoms becomes a ketone oxygen and one becomes a carboxyl oxygen. Note also that the migration of the ring bond to oxygen in step iv is analogous to what occurs in the oxidative cleavage of tryptophan to give *N*-formylkynurenine, the first step in tryptophan catabolism that we saw in Figure 5.12.

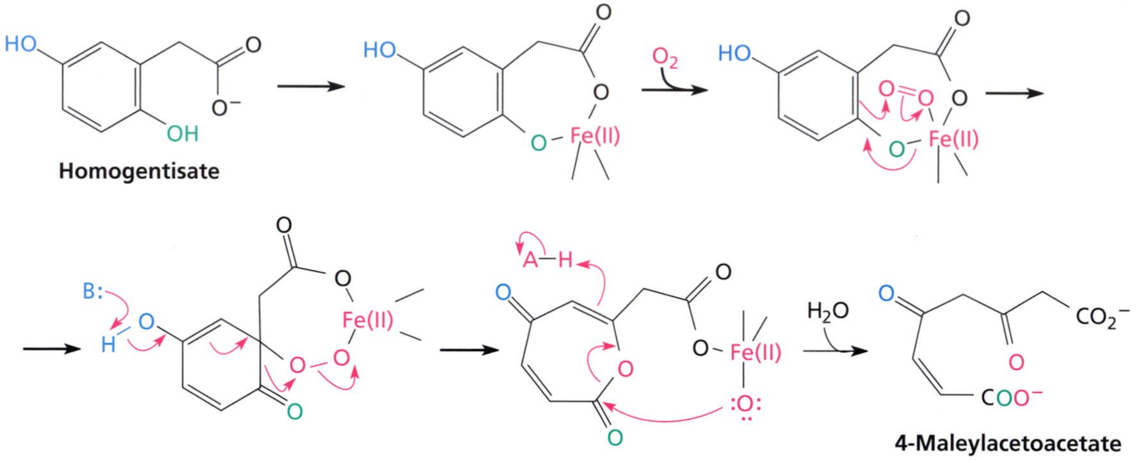

HO, OH, O⁻, Homogentisate, Fe(II), O_2, B:, H, H_2O, A—H, :O:, CO_2^-, COO⁻, 4-Maleylacetoacetate

FIGURE 5.30 Mechanism of the oxidation of homogentisate to 4-maleylacetoacetate, catalyzed by the iron-containing enzyme homogentisate dioxygenase.

Step 5. Isomerization Cis–trans isomerization of 4-maleylacetoacetate to 4-fumarylacetoacetate is catalyzed by maleylacetoacetate isomerase[32, 33] and uses the cofactor glutathione (GSH). The reaction occurs by conjugate addition of the nucleophilic thiol group of glutathione to the unsaturated ketone, followed by bond rotation around what is now a single bond, and expulsion of glutathione. Note that glutathione is a tripeptide (L-γ-glutamyl-L-cysteinylglycine) with an unusual γ amide link between glutamate and cysteine. It is involved in many metabolic processes and often has a protective role to prevent cell damage by various peroxides and reactive electrophiles.

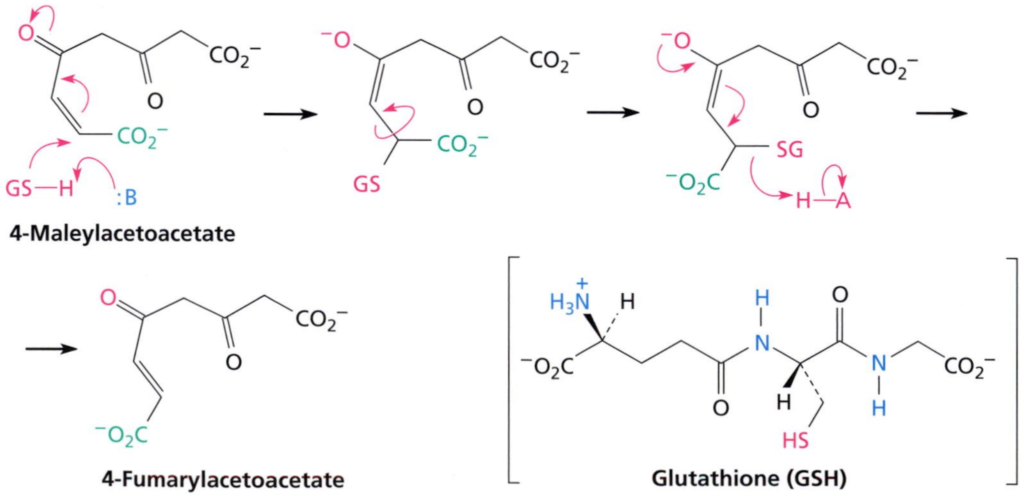

CO_2^-, GS—H, :B, 4-Maleylacetoacetate, ⁻O, GS, SG, ^-O_2C, H—A, 4-Fumarylacetoacetate, $H_3\overset{+}{N}$, H, N, HS, Glutathione (GSH)

Step 6. Retro-Claisen The final step in tyrosine catabolism is cleavage of 4-fumarylacetoacetate by a retro-Claisen reaction. Catalyzed by the magnesium-containing enzyme fumarylacetoacetate hydrolase,[34] the cleavage occurs by straightforward addition of water to the fumaryl carbonyl group to give a tetrahedral alkoxide intermediate that expels acetoacetate as the leaving group.

H
N
His
:N
H O H
^-O_2C
CO_2^-
O O
Mg^{2+}
4-Fumarylacetoacetate
H O
^-O_2C
CO_2^-
^-O O
Mg^{2+}
^-O_2C
CO_2^-
Fumarate
+
CO_2^-
O
Acetoacetate

5.4 Biosynthesis of Nonessential Amino Acids

Of the 20 amino acids in proteins, humans are able to synthesize only 11, called *nonessential amino acids*; the other 9, called *essential amino acids*, must be obtained in the diet. The division between essential and nonessential amino acids is not clearcut, however: Tyrosine, for instance, is sometimes considered nonessential because humans can produce it from phenylalanine, but phenylalanine itself is essential and must be obtained in the diet. Arginine can be synthesized by humans, but much of the arginine in proteins also comes from the diet. Figure 5.31 shows the common precursors of the 20 amino acids.

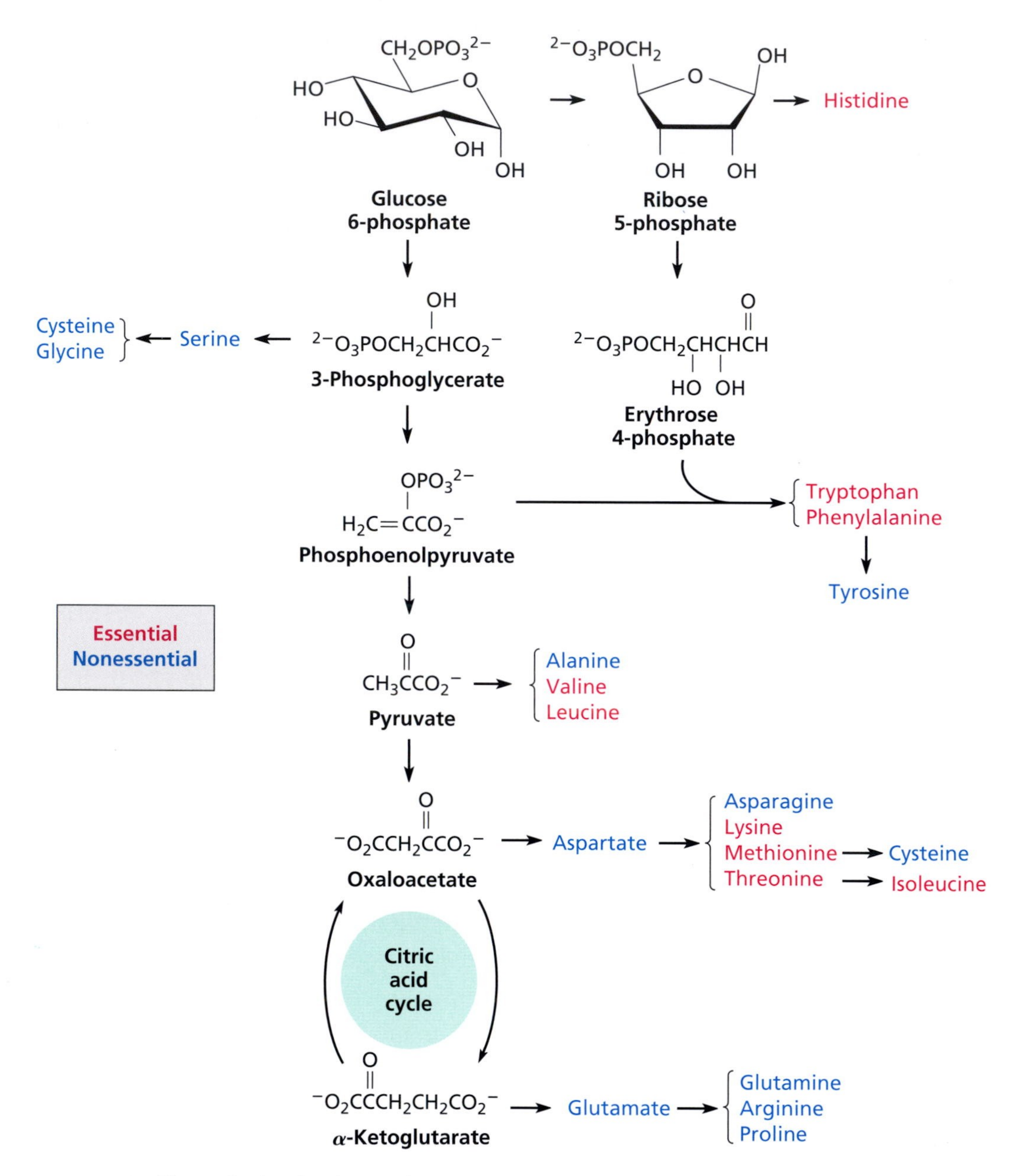

FIGURE 5.31 Biosynthesis of amino acids. Essential amino acids (red) are synthesized in plants and bacteria and must be obtained in the diet. Humans can synthesize only the nonessential amino acids (blue).

Alanine, Aspartate, Glutamate, Asparagine, Glutamine, Arginine, and Proline

Seven nonessential amino acids—alanine, aspartate, glutamate, asparagine, glutamine, arginine, and proline—are synthesized either from pyruvate or from the citric acid cycle intermediates oxaloacetate and α-ketoglutarate. **Alanine** results from PLP-dependent transamination of pyruvate, **aspartate** from oxaloacetate, and **glutamate** from α-ketoglutarate.

Pyruvate → Alanine (Amino acid → α Keto acid)

Oxaloacetate → Aspartate (Amino acid → α Keto acid)

α-Ketoglutarate → Glutamate (Amino acid → α Keto acid)

Asparagine and glutamine are synthesized by amide formation from aspartate and glutamate, respectively. **Asparagine** synthesis is catalyzed by asparagine synthetase,[35] which requires ATP as cofactor. The reaction proceeds through formation of a β-aspartyl adenosyl monophosphate, which then undergoes nucleophilic acyl substitution by ammonia. The ammonia is itself produced from glutamine by a nucleophilic acyl substitution reaction with a cysteine residue in the enzyme, followed by hydrolysis of the thioacyl enzyme intermediate to give glutamate. (You might recall from Section 3.1, Figure 3.3, that fatty acids are also activated by conversion into acyl adenosyl phosphates.)

Glutamine synthesis is catalyzed by glutamine synthetase,[36] and occurs by formation of an acyl phosphate followed by nucleophilic acyl substitution reaction with ammonia. The difference in activation strategies for the asparagine and glutamine pathways probably has no chemical significance but is simply the result of different evolutionary histories for the two enzymes.

Glutamine → Glutamate + NH_3

Aspartate — ATP, PP_i → Adenine — AMP → Asparagine

Glutamate — ATP, ADP → OPO_3^{2-} — P_i → Glutamine

In humans, **arginine** is synthesized from glutamate by the pathway shown in Figure 5.32. Reaction of glutamate with ATP gives an acyl phosphate, which is reduced by NADH in a nucleophilic acyl substitution reaction to yield glutamate 5-semialdehyde. PLP-mediated transamination of the aldehyde carbonyl group by reaction with glutamate then gives ornithine, which is converted to arginine in the urea cycle (Section 5.2, Figure 5.2), as previously discussed. **Proline** is also synthesized from glutamate 5-semialdehyde, by nonenzymatic formation of a cyclic imine followed by reduction of the C=N double bond with NADH (Figure 5.32). This pathway is the reverse of that for proline catabolism.

Serine, Cysteine, and Glycine

Three nonessential amino acids—serine, cysteine, and glycine—are formed from 3-phosphoglycerate, an intermediate in the glycolysis pathway (Section 4.2, Figure 4.3).

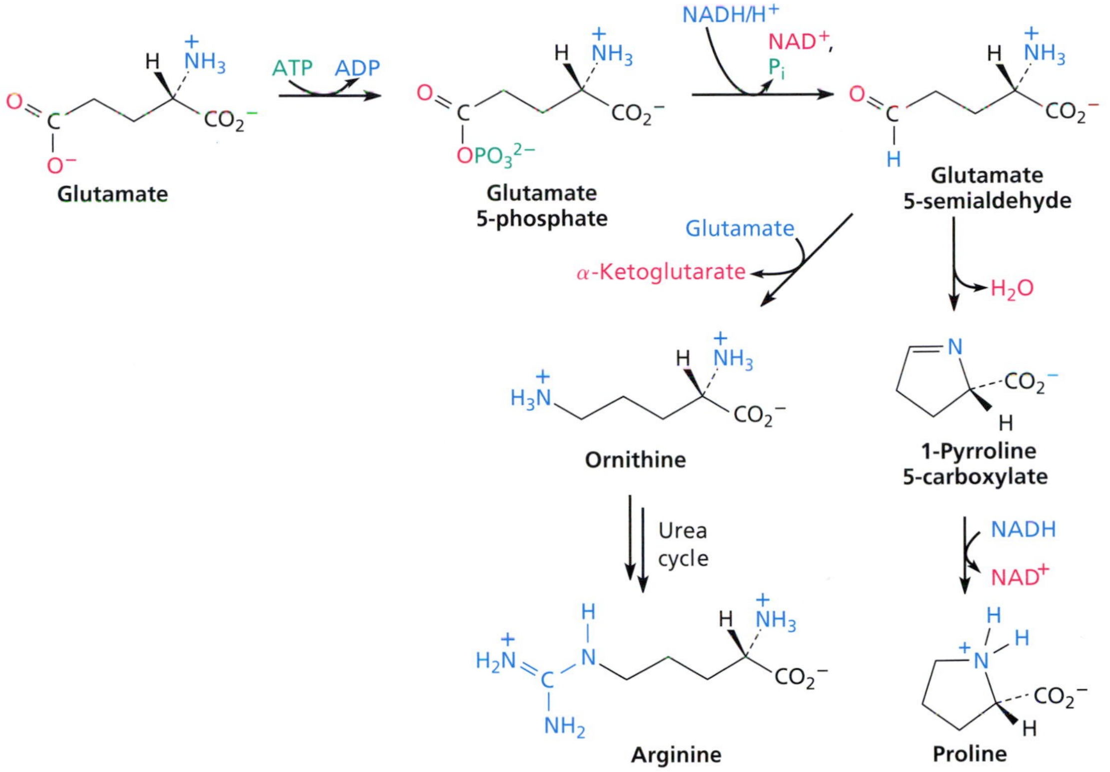

FIGURE 5.32 Biosynthesis of arginine and proline from glutamate.

Serine is formed by an initial oxidation of the —OH group in 3-phosphoglycerate to a ketone, followed by PLP-dependent transamination to give 3-phosphoserine. Hydrolysis of the phosphate group then yields serine. Catalyzed by phosphoserine phosphatase,[37] the hydrolysis occurs by nucleophilic attack of an aspartate residue in the enzyme on the 3-phosphoserine phosphorus to give an acyl phosphate intermediate, which then reacts with water.

NAD^+ $NADH/H^+$ Glutamate α-Ketoglutarate

3-Phosphoglycerate → **3-Phosphohydroxy-pyruvate** → **3-Phosphoserine**

H_2O P_i

3-Phosphoserine → **Serine**

Cysteine is synthesized from serine by the route discussed in the previous section in connection with methionine catabolism. That is, serine reacts with homocysteine to give cystathionine (Figure 5.23), which is then converted to cysteine (Figure 5.24). Note that the sulfur atom in cysteine comes originally from the essential amino acid methionine, so cysteine might also be considered essential. **Glycine** is produced from serine by transfer of CH_2O to tetrahydrofolate in a reaction catalyzed by serine hydroxymethyltransferase, as discussed and shown previously in Figure 5.6. In addition, glycine is produced by transamination of glyoxalate, $OHCCO_2^-$.

5.5 Biosynthesis of Essential Amino Acids

Lysine, Methionine, and Threonine

Essential amino acids are synthesized only in plants and microorganisms, not in humans or other higher organisms. Three essential amino acids—lysine, methionine, and threonine—are synthesized from the citric-acid-cycle intermediate oxaloacetate by way of aspartate. **Lysine** is synthesized from aspartate by an 11-step path (Figure 5.33) that uses many reactions we've seen previously.

Steps 1–3. Phosphorylation, reduction, and condensation Lysine biosynthesis begins with phosphorylation by ATP to give aspartyl β-phosphate. Reduction with NADPH gives aspartate semialdehyde, which undergoes an aldol condensation reaction with pyruvate.

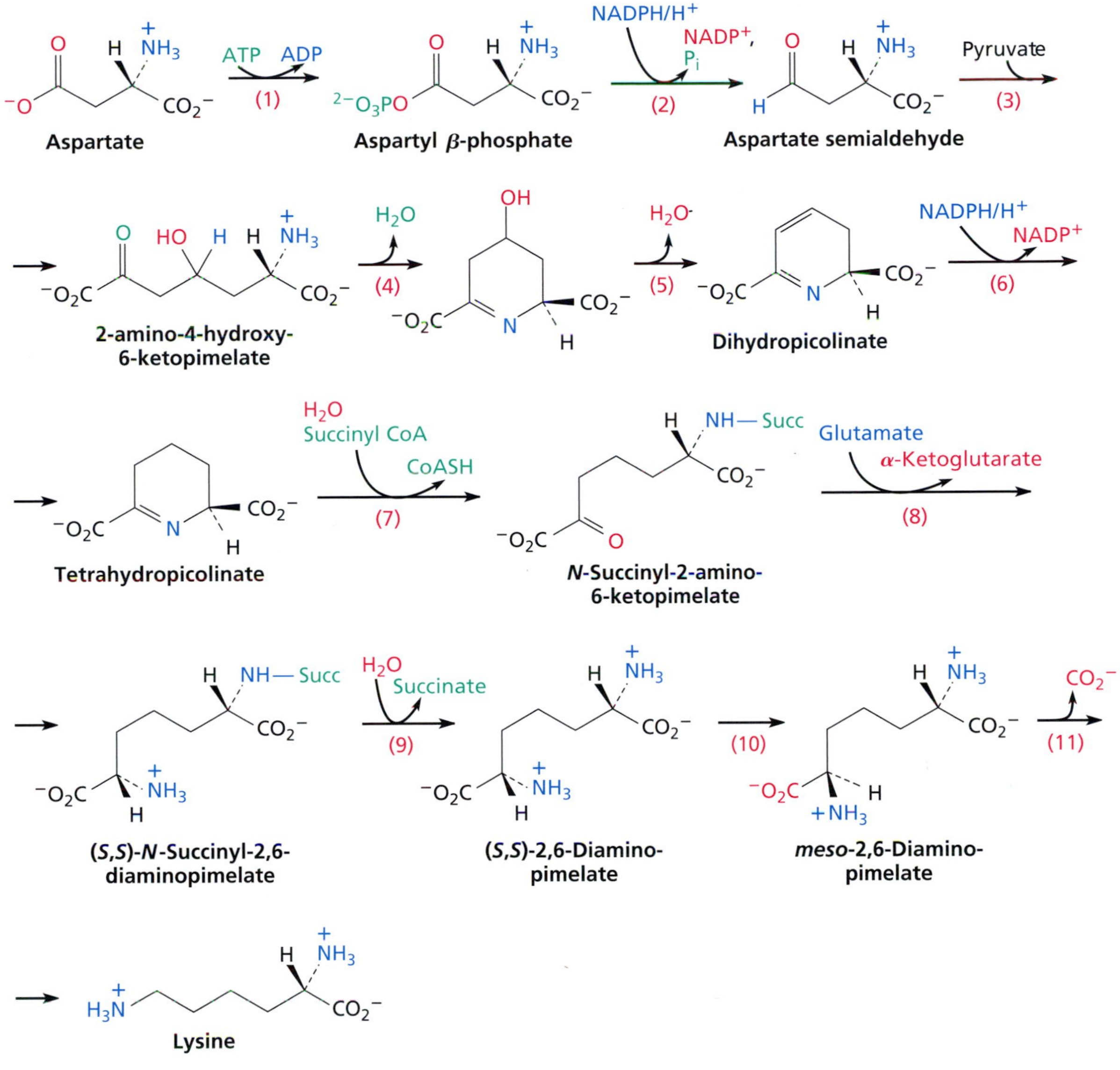

FIGURE 5.33 Pathway for the biosynthesis of lysine from aspartate.

Steps 4–6. Cyclization, dehydration, and reduction The product of cyclization in step 3, an amino β-hydroxy ketone (or its related Schiff base), cyclizes by internal imine formation, and then eliminates water by an E1cB reaction to give dihydropicolinate. Reduction of the C═C double bond of dihydropicolinate by reaction with NADPH in a conjugate addition reaction of hydride ion to the C═C—C═N$^+$ group yields tetrahydropicolinate.

Steps 7–9. Succinylation, hydrolysis, transamination, and hydrolysis Reaction of the amino group of tetrahydropicolinate with succinyl CoA, followed by hydrolysis, gives *N*-succinyl-2-amino-6-ketopimelate. Transamination of the keto group with glutamate then yields (*S*,*S*)-*N*-succinyl-2,6-diaminopimelate, which is hydrolyzed to the corresponding diamino diacid.

Step 10. Epimerization The (*S*,*S*)-*N*-succinyl-2,6-diaminopimelate produced in step 9 is next epimerized to the *meso*-2,6-diaminopimelate isomer. The epimerization is catalyzed by diaminopimelate epimerase,[38] a non-PLP enzyme that is thought to function by a straightforward deprotonation–reprotonation mechanism. (Many amino acid epimerases involve PLP imines.) A thiolate ion on one side of the molecule removes the acidic α proton, and immediate reprotonation by a cysteine residue from the opposite side accomplishes the epimerization.

(*S*, *S*)-2,6-Diaminopimelate → → ***meso*-2,6-Diaminopimelate**

Step 11. Decarboxylation The final decarboxylation of *meso*-2,6-diaminopimelate to give lysine occurs specifically at the *R* stereocenter and is a PLP-dependent process catalyzed by diaminopimelate decarboxylase.[39] Formation of the PLP imine takes place, followed by decarboxylation, with the protonated pyridinium ring acting as the electron acceptor. Protonation on the α carbon and removal of PLP complete the reaction (Figure 5.34). The process is similar to what occurs in the glycine cleavage system (Section 5.3, Figure 5.7).

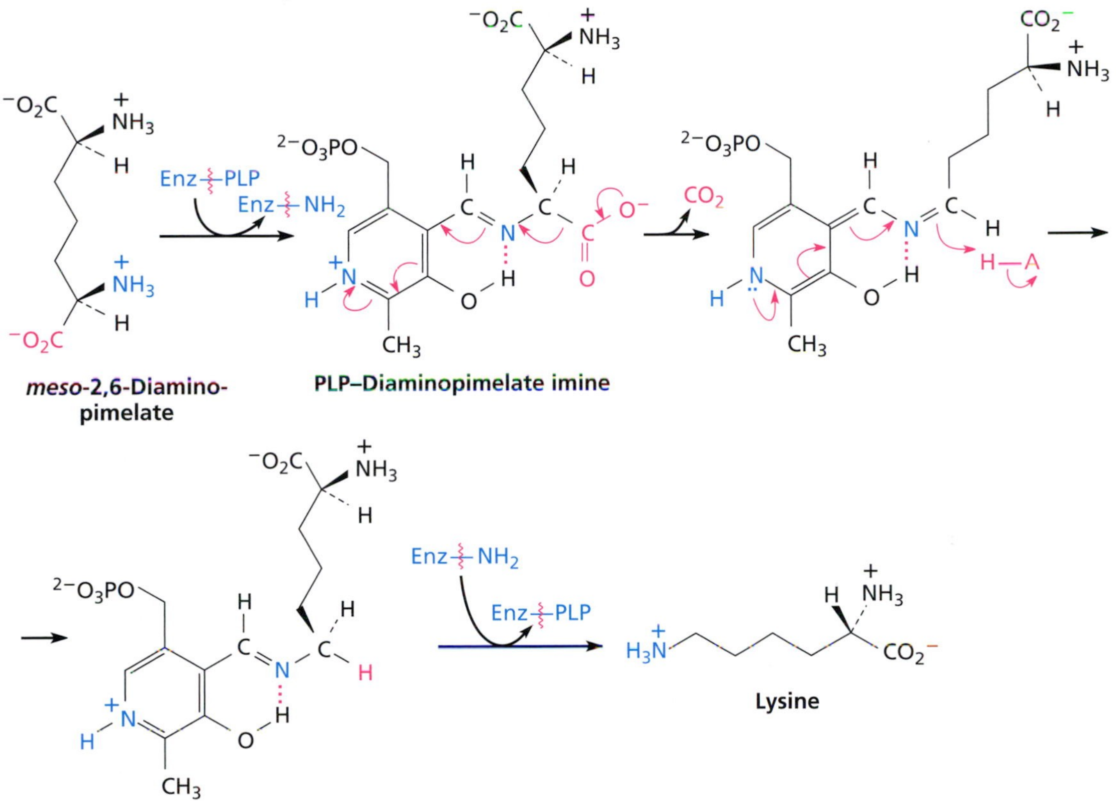

FIGURE 5.34 Mechanism of the PLP-dependent decarboxylation of *meso*-2,6-diaminopimelate to give lysine.

Methionine, like lysine, is synthesized from aspartate semialdehyde (Figure 5.35). Five steps are needed.

Steps 1–2. Reduction and succinylation Reduction of aspartate semialdehyde with NADPH gives homoserine, which is succinylated by reaction with succinyl CoA. The reaction is catalyzed by homoserine transsuccinylase[40] and occurs by initial reaction of succinyl CoA with a cysteine residue in the enzyme followed by transfer of the succinyl group to homoserine in a nucleophilic acyl substitution reaction.

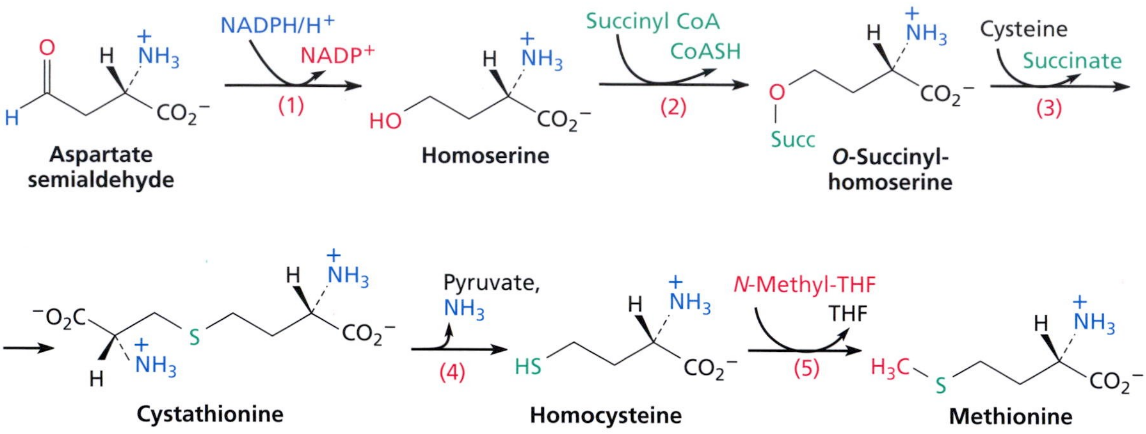

FIGURE 5.35 Pathway for the biosynthesis of methionine from aspartate.

Step 3. Addition of cysteine Reaction of *O*-succinylhomoserine with cysteine to give cystathionine occurs by an elimination–addition mechanism catalyzed by the PLP-dependent enzyme cystathionine γ-synthase[41] (Figure 5.36). Formation of a PLP–*O*-homoserine imine followed by deprotonation of the α position in the usual way gives an intermediate that is deprotonated on the β position to give an enamine. Elimination of succinate and conjugate addition of cysteine then produces another enamine, which is cleaved to give cystathionine.

Step 4. Cystathionine cleavage The cleavage of cystathionine to homocysteine plus pyruvate is catalyzed by yet another PLP-dependent enzyme, cystathionine β-lyase.[42] The reaction mechanism involves an elimination analogous to that occurring in serine catabolism (Section 5.3, Figure 5.5).

Step 5. Methylation Methylation of the homocysteine —SH group by reaction with 5-methyltetrahydrofolate is catalyzed by the coenzyme B_{12}–dependent enzyme methionine synthase.[43] As noted in Section 5.3 in connection with threonine catabolism, coenzyme B_{12} is involved in two kinds of reactions: exchange of a hydrogen with a group on the neighboring carbon and transfer of a methyl group between two molecules.

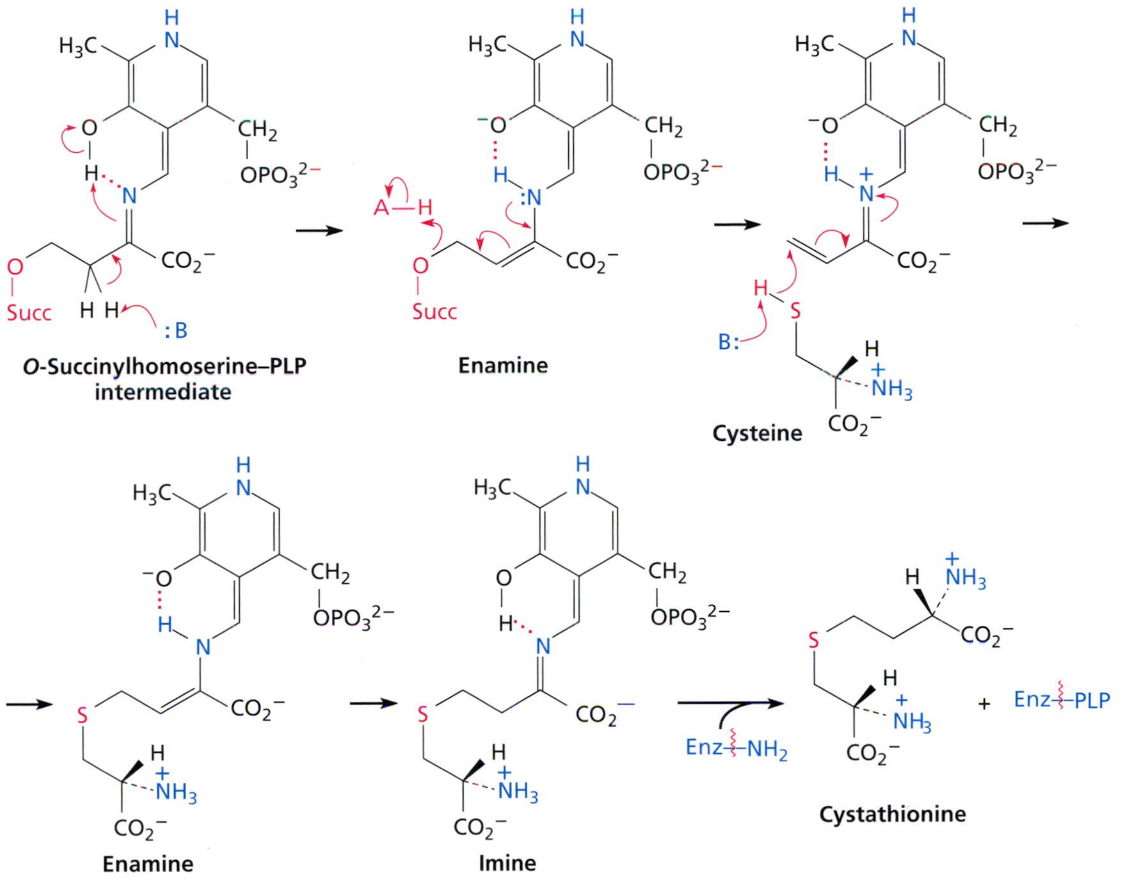

Figure 5.36 Mechanism of the reaction of O-succinylhomoserine with cysteine to give cystathionine.

The mechanism of the homocysteine methylation in many bacteria involves an initial transfer of a —CH_3 group from 5-methyltetrahydrofolate to the cobalt atom of cobalamin, an intermediate derived from coenzyme B_{12} by loss of adenosine. Methylcobalamin, the product of this first methylation, then transfers the methyl group to homocysteine. The overall reaction occurs with retention of configuration at the methyl carbon, which is suggestive but not definitive for two sequential S_N2 substitutions. In fact, the cobalt(I) atom in cobalamin *is* an excellent nucleophile and takes part in numerous nonbiological S_N2 reactions. The reaction is shown in Figure 5.37, along with the structure of methylcobalamin.

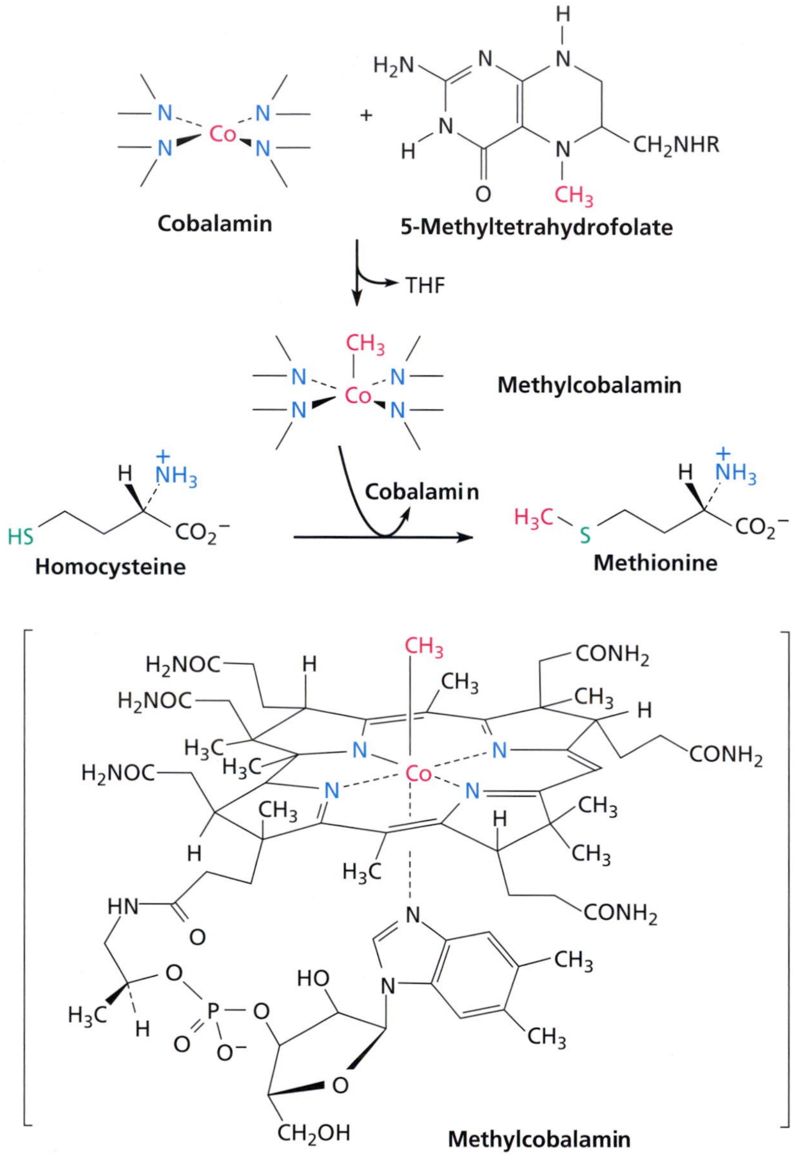

Cobalamin + 5-Methyltetrahydrofolate → THF; Methylcobalamin

Homocysteine → Methionine (Methylcobalamin → Cobalamin)

Methylcobalamin

FIGURE 5.37 The biosynthesis of methionine from homocysteine, and the structure of methylcobalamin.

Threonine is synthesized in two steps from homoserine: initial phosphorylation with ATP to give phosphohomoserine, followed by rearrangement.

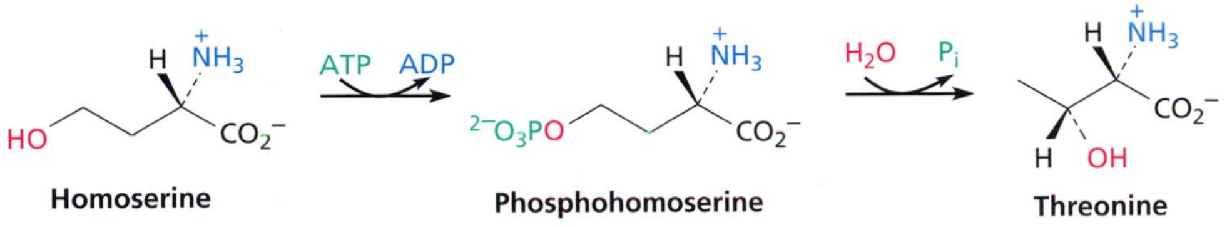

H $\overset{+}{N}H_3$
HO
CO_2^-
Homoserine
ATP ADP
H $\overset{+}{N}H_3$
$^{2-}O_3PO$
CO_2^-
Phosphohomoserine
H_2O P_i
H $\overset{+}{N}H_3$
CO_2^-
H OH
Threonine

Shown in Figure 5.38, the rearrangement of phosphohomoserine to threonine is catalyzed by the PLP-dependent enzyme threonine synthase[44] and begins with an elimination similar to that given in Figure 5.36 for cystathionine synthesis. Tautomerization of the elimination product, conjugate addition of water, and cleavage of the alcohol addition product complete the synthesis.

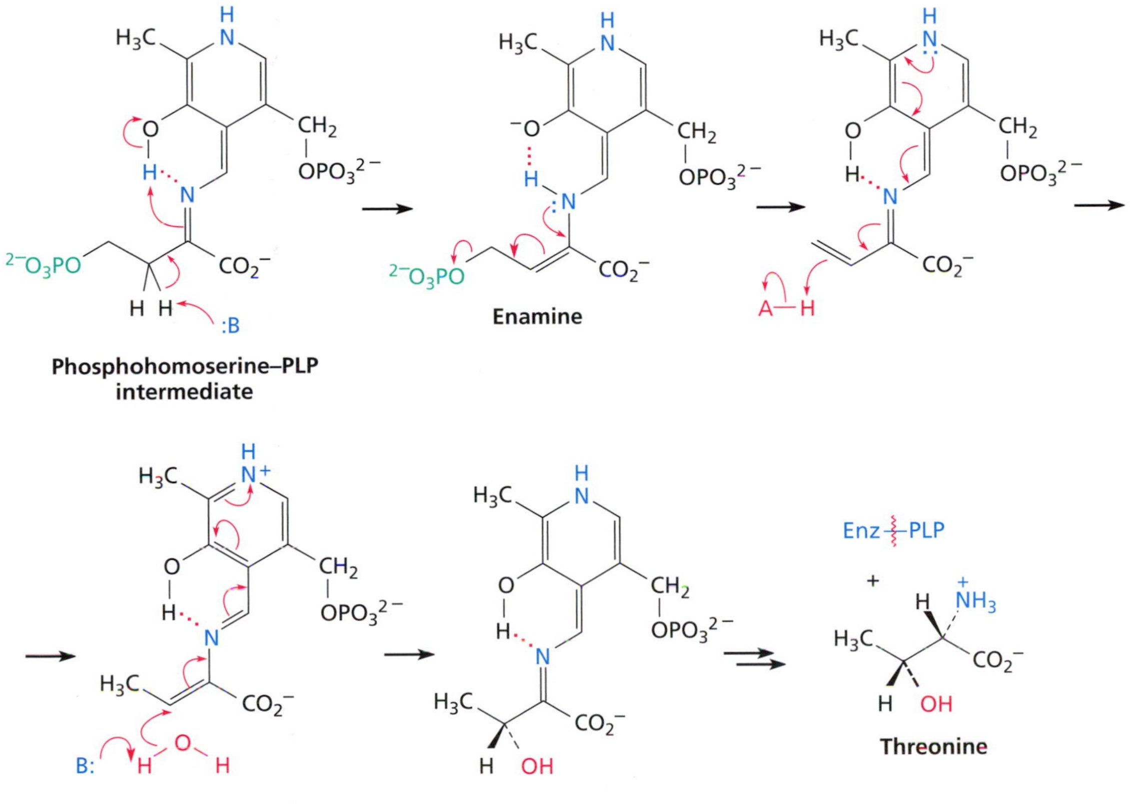

H
H_3C N
O
CH_2
H N
OPO_3^{2-}
$^{2-}O_3PO$
CO_2^-
H H :B
Phosphohomoserine–PLP intermediate

H
H_3C N
^-O
CH_2
H N
OPO_3^{2-}
$^{2-}O_3PO$
CO_2^-
Enamine

H
H_3C N
O
CH_2
H N
OPO_3^{2-}
CO_2^-
A—H

H
H_3C N^+
O
CH_2
H N
OPO_3^{2-}
H_3C
CO_2^-
B: H O H

H
H_3C N
O
CH_2
H N
OPO_3^{2-}
H_3C
CO_2^-
H OH

Enz–PLP
+
H $\overset{+}{N}H_3$
H_3C
CO_2^-
H OH
Threonine

FIGURE 5.38 Mechanism of threonine biosynthesis from phosphohomoserine.

Isoleucine, Valine, and Leucine

Three essential amino acids—isoleucine, valine, and leucine—are synthesized from pyruvate. As shown in Figure 5.39, **isoleucine** synthesis begins with addition of thiamin diphosphate to pyruvate and decarboxylation in the usual way (Section 4.3, Figure 4.9) to give hydroxyethylthiamin diphosphate (HETPP). Nucleophilic addition to α-ketobutyrate, itself produced by catabolism of threonine (Section 5.3), then gives 2-ethyl-2-hydroxy-3-ketobutyrate after expulsion of TPP. Rearrangement of 2-ethyl-2-hydroxy-3-ketobutyrate yields 2-keto-3-hydroxy-3-methylvalerate, which is reduced by NADH to 2,3-dihydroxy-3-methylvalerate. Dehydration of 2,3-dihydroxy-3-methylvalerate then gives an enol, tautomerization yields the ketone 2-keto-3-methylvalerate, and transamination with glutamate gives isoleucine.

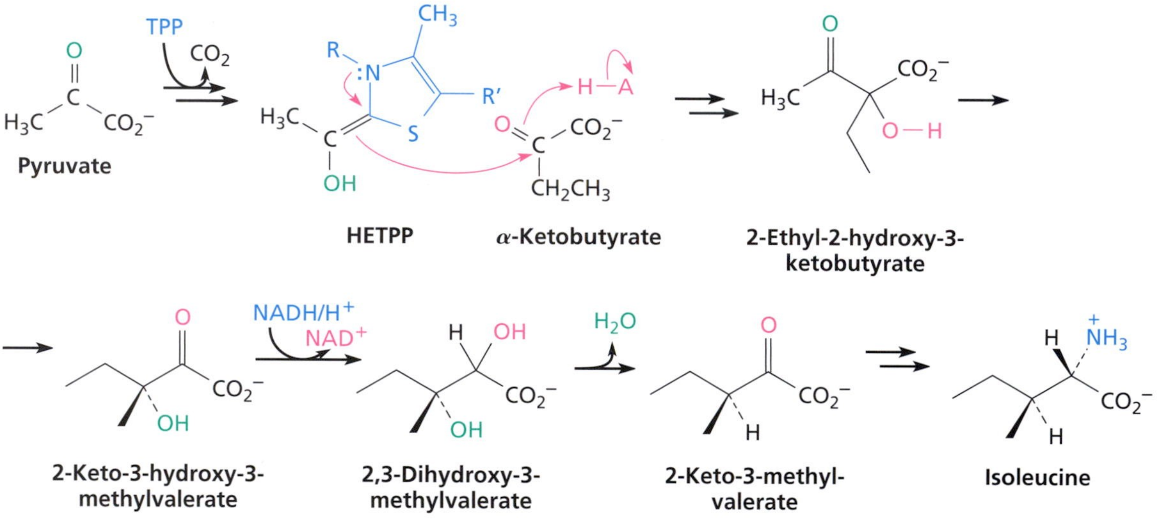

FIGURE 5.39 Pathway for isoleucine biosynthesis from pyruvate.

The rearrangement and subsequent reduction of 2-ethyl-2-hydroxy-3-ketobutyrate to 2,3-dihydroxy-3-methylvalerate in Figure 5.39 are catalyzed by acetohydroxy acid isomeroreductase.[45] The rearrangement occurs by an acyloin rearrangement mechanism (Section 3.5), in which an acid (Mg^{2+} in the enzyme) coordinates to the ketone carbonyl group, a base deprotonates the hydroxyl, and migration of the ethyl group from C2 to C3 occurs, giving an isomeric hydroxy ketone as product.

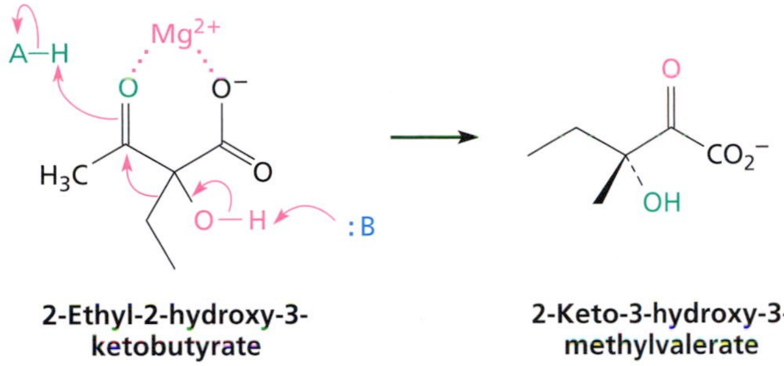

2-Ethyl-2-hydroxy-3-ketobutyrate

2-Keto-3-hydroxy-3-methylvalerate

Valine biosynthesis follows a six-step pathway analogous to that for isoleucine (Figure 5.40). Addition of TPP to pyruvate followed by decarboxylation gives HETPP, and nucleophilic addition to a second molecule of pyruvate gives 2-methyl-2-hydroxy-3-ketobutyrate. Acyloin rearrangement yields 2-keto-3-hydroxyisovalerate, which is reduced by NADH to 2,3-dihydroxyisovalerate. Dehydration then gives an enol that tautomerizes to α-ketoisovalerate, and PLP-dependent transamination with glutamate gives valine.

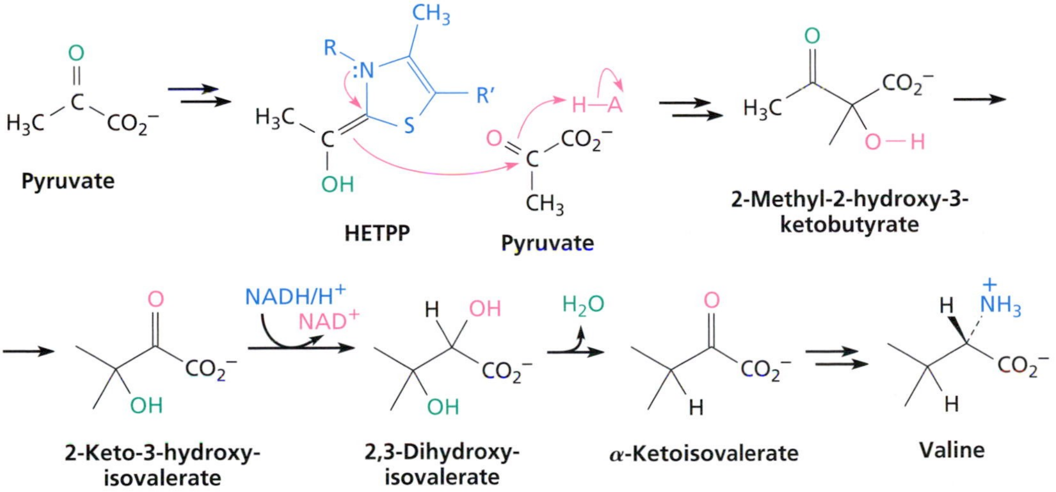

Pyruvate

HETPP

Pyruvate

2-Methyl-2-hydroxy-3-ketobutyrate

NADH/H^+

NAD^+

H_2O

2-Keto-3-hydroxy-isovalerate

2,3-Dihydroxy-isovalerate

α-Ketoisovalerate

Valine

FIGURE 5.40 Pathway for valine biosynthesis from pyruvate.

Leucine, like valine, is biosynthesized from pyruvate by a pathway that proceeds through α-ketoisovalerate as an intermediate (Figure 5.41). Aldol condensation of α-ketoisovalerate with acetyl CoA gives 1-isopropylmalate, which rearranges by dehydration and rehydration in the opposite sense to give 2-isopropylmalate. The process is analogous to what occurs in step 2 of the citric acid cycle (Section 4.4, Figure 4.13) in which citrate isomerizes to isocitrate. Oxidation of 2-isopropylmalate by NAD^+ and decarboxylation of the intermediate β keto acid, as in step 3 of the citric acid cycle, yields α-ketoisocaproate. Transamination then gives leucine.

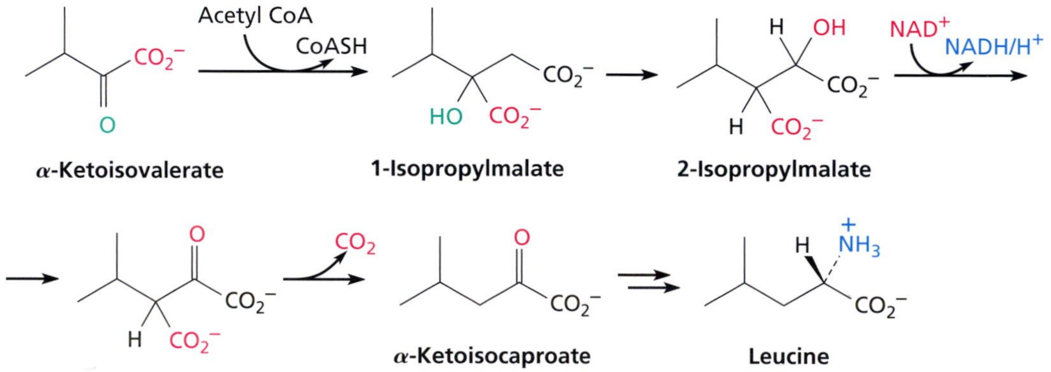

FIGURE 5.41 Pathway for leucine biosynthesis from pyruvate.

Tryptophan, Phenylalanine, and Tyrosine

The three aromatic amino acids—tryptophan, phenylalanine, and tyrosine—are biosynthesized from phosphoenolpyruvate (PEP) and erythrose 4-phosphate through a common intermediate called *chorismate*, whose biosynthesis is shown in Figure 5.42.

Steps 1–3. Formation of 5-dehydroquinate Chorismate biosynthesis begins with nucleophilic addition of phosphoenolpyruvate to erythrose 4-phosphate, followed by cyclization to a hemiacetal.[46] The product, 3-deoxyarabinoheptulosonate 7-phosphate (DAHP), then undergoes a remarkable series of transformations to give 5-dehydroquinate. Catalyzed by dehydroquinate synthase,[47] a

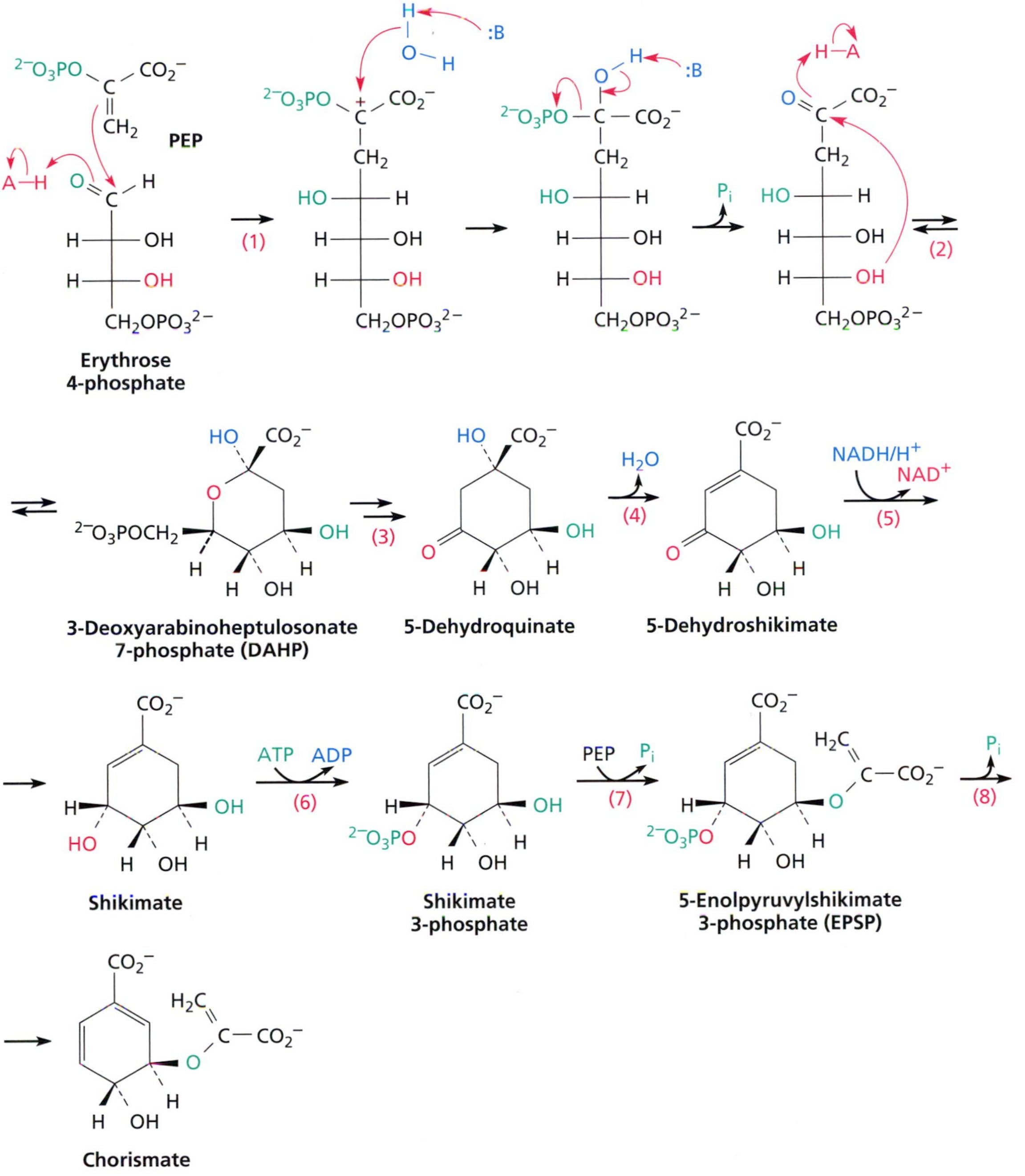

PEP

Erythrose 4-phosphate

(1)

P_i

(2)

(3)

3-Deoxyarabinoheptulosonate 7-phosphate (DAHP)

5-Dehydroquinate

H_2O

(4)

5-Dehydroshikimate

$NADH/H^+$ NAD^+

(5)

Shikimate

ATP ADP

(6)

Shikimate 3-phosphate

PEP P_i

(7)

5-Enolpyruvylshikimate 3-phosphate (EPSP)

P_i

(8)

Chorismate

FIGURE 5.42 Pathway for the biosynthesis of chorismate, an intermediate in the biosynthesis of aromatic amino acids.

five-step series of transformations occurs, as shown in Figure 5.43. Oxidation of the C5 hydroxyl by NAD^+ first yields a ketone (step i), which undergoes elimination of phosphate to give an α,β-unsaturated ketone (step ii). Reduction of the carbonyl group (step iii) and opening of the hemiacetal ring (step iv) then gives an enolate ion intermediate, which cyclizes by an internal aldol reaction to yield 5-dehydroquinate (step v).

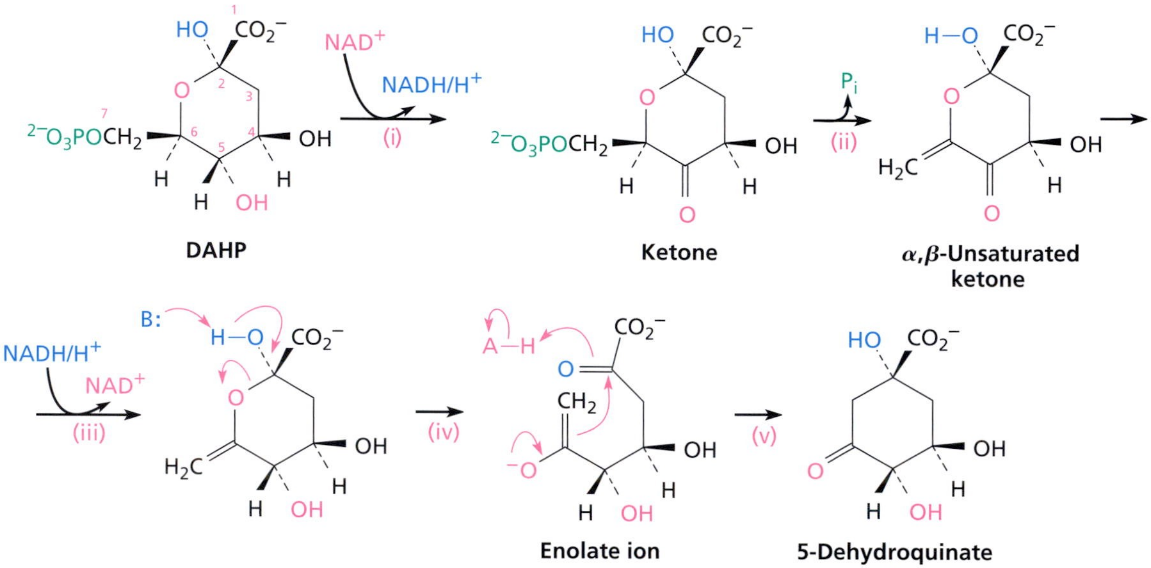

FIGURE 5.43 Mechanism of the conversion of 3-deoxyarabinoheptulosonate 7-phosphate to 5-dehydroquinate, catalyzed by dehydroquinate synthase.

Steps 4–6. Dehydration, reduction, and phosphorylation The next three steps of chorismate biosynthesis are relatively straightforward. Dehydration of 5-dehydroquinate occurs by an E1cB mechanism and yields 5-dehydroshikimate, which is reduced by NADH to the alcohol shikimate. Phosphorylation by ATP then gives shikimate 3-phosphate.

Steps 7–8. Formation of chorismate The reaction of shikimate 3-phosphate with phosphoenol pyruvate (PEP) to give 5-enolpyruvylshikimate 3-phosphate (EPSP) is catalyzed by 5-enolpyruvylshikimate 3-phosphate synthase (EPSPS)[48] and occurs by a two-step, addition–elimination sequence (Figure 5.44). Protonation of the phosphoenolpyruvate double bond on the *si* face by Glu-341 and concurrent nucleophilic addition of shikimate 3-phosphate to the *re* face first gives a tetrahedral intermediate with *S* stereochemistry at the chiral center. A *syn* elimination of phosphate then yields 5-enolpyruvylshikimate 3-phosphate. Note that the same hydrogen added to PEP in the first step is eliminated in the second step, so the double bond retains its stereochemistry. That is, the hydrogen cis to the carboxyl group in PEP remains cis to the carboxyl group in the EPSP product.

The last step in chorismate biosynthesis, loss of phosphate from 5-enolpyruvylshikimate 3-phosphate, is catalyzed by chorismate synthase[49] and occurs by a 1,4-*anti* elimination pathway. The mechanism of the reaction is poorly understood, but one possibility is the E1 pathway shown in Figure 5.44. Surprisingly, however, the reaction requires FMN as cofactor, implying the possibility that radical intermediates may be involved.

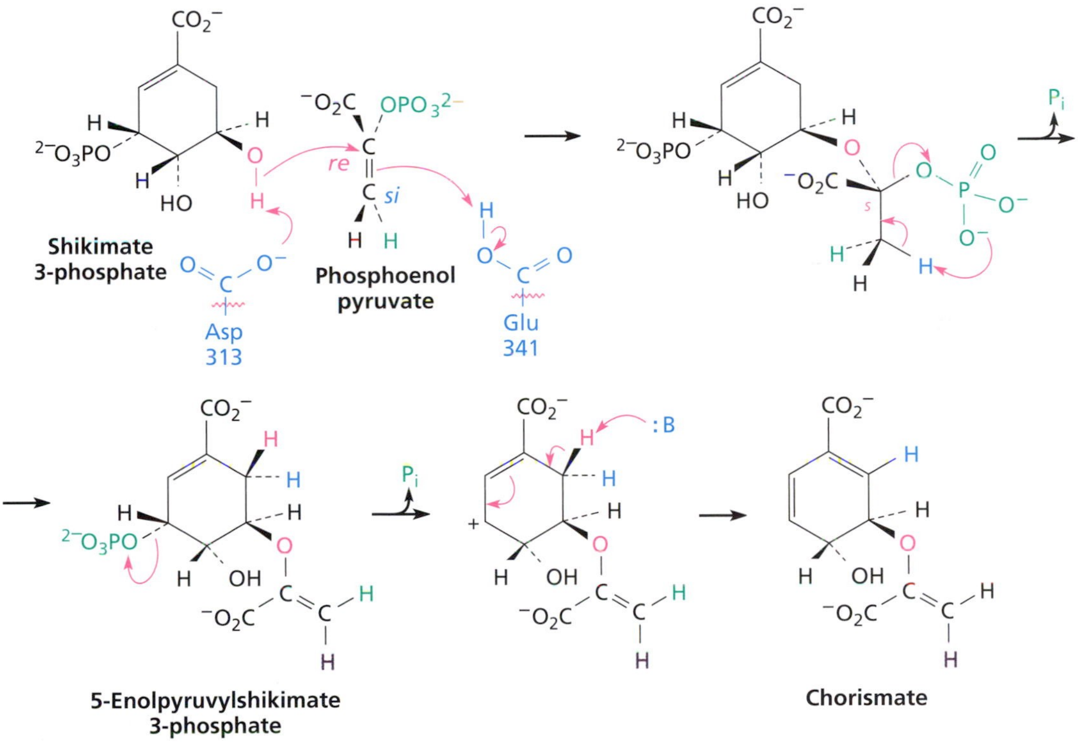

Shikimate 3-phosphate

Phosphoenol pyruvate

Asp 313

Glu 341

5-Enolpyruvylshikimate 3-phosphate

Chorismate

FIGURE 5.44 Mechanism of the conversion of shikimate 3-phosphate to chorismate.

Tryptophan is biosynthesized from chorismate by the series of reactions shown in Figure 5.45.

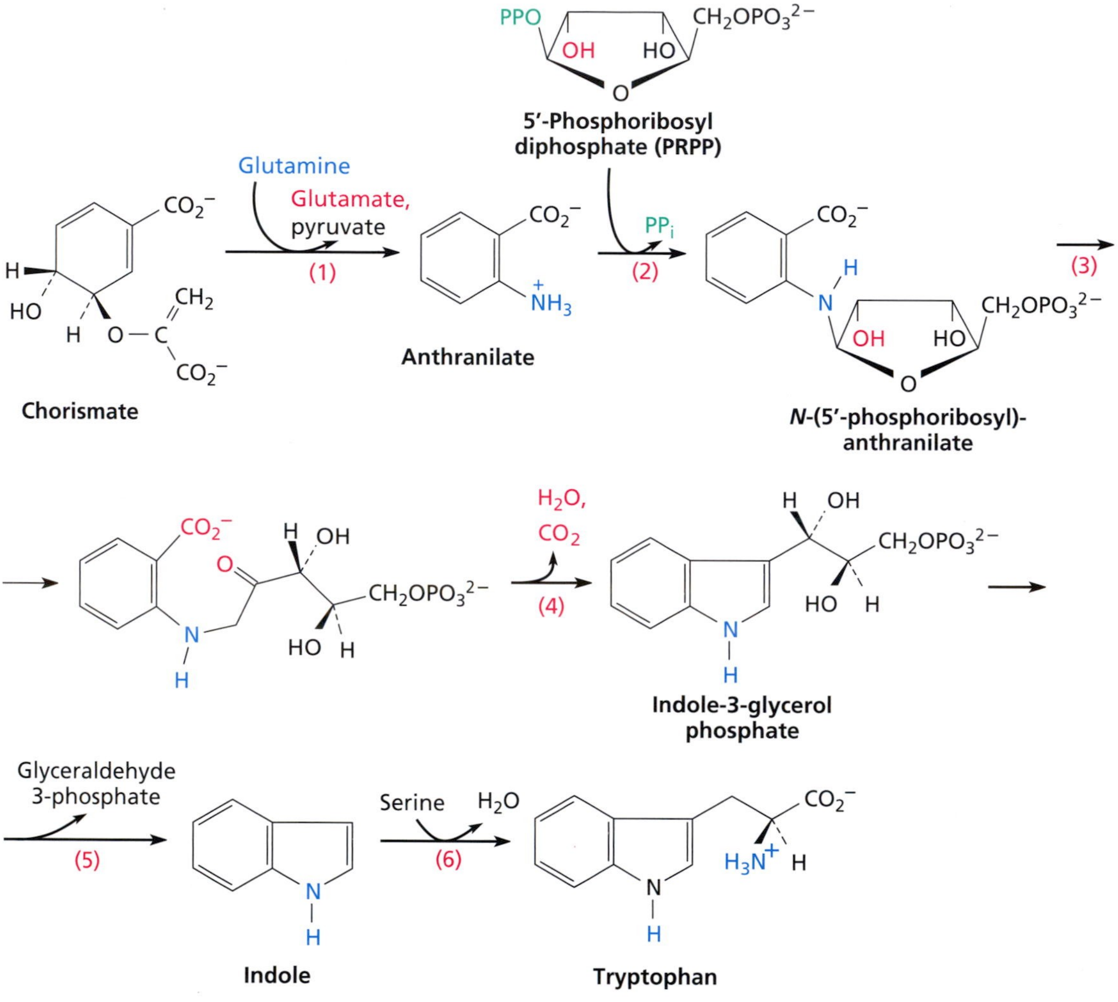

FIGURE 5.45 Pathway for the biosynthesis of tryptophan from chorismate.

Step 1. Formation of anthranilate The conversion of chorismate to anthranilate is catalyzed by anthranilate synthase[50] and appears to involve a nucleophilic allylic substitution of OH by NH_3, itself generated in the enzyme by hydrolysis of glutamine. Loss of pyruvate completes the process. Both steps require the presence of Mg^{2+} ion.

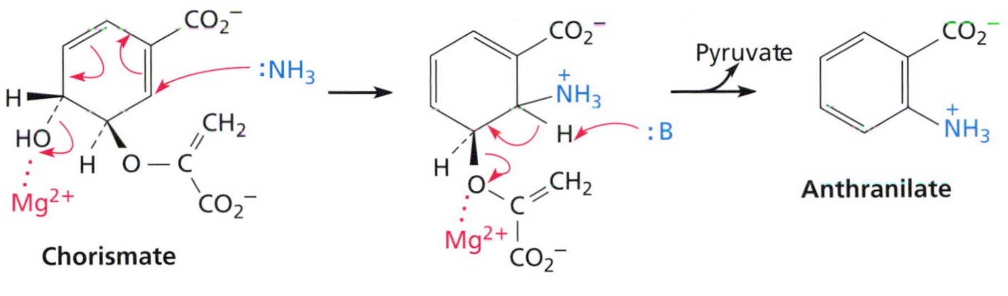

Chorismate

Anthranilate

Steps 2–3. Ribosylation and isomerization Ribosylation of anthranilate by reaction with 5-phosphoribosyl diphosphate (PRPP) gives *N*-(5′-phosphoribosyl)anthranilate, which undergoes isomerization to a ring-opened ketone. The isomerization is catalyzed by phosphoribosyl anthranilate isomerase[51] and occurs by opening of the ribose ring to give an iminium ion, loss of a proton to give an enol, and tautomerization of the enol to a ketone.

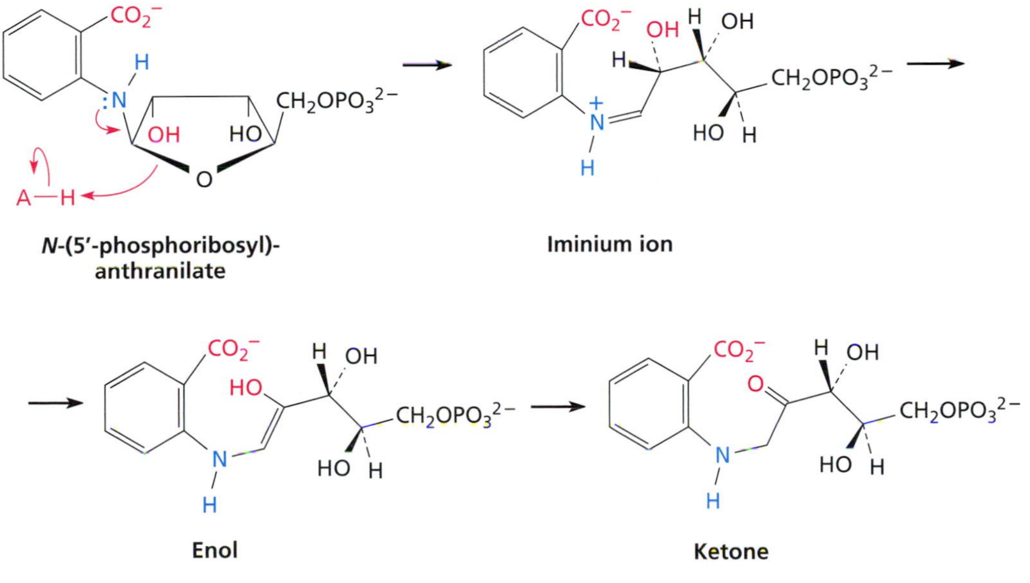

N-(5′-phosphoribosyl)-anthranilate

Iminium ion

Enol

Ketone

Step 4. Cyclization and decarboxylation The ketone resulting from isomerization of phosphoribosyl anthranilate is next cyclized and decarboxylated to give indole-3-glycerol phosphate. Catalyzed by indole-3-glycerol phosphate synthase,[52] the reaction occurs by nucleophilic attack of the aromatic ring on the ketone carbonyl group, decarboxylation of the iminium ion intermediate, and dehydration to give the aromatic indole ring.

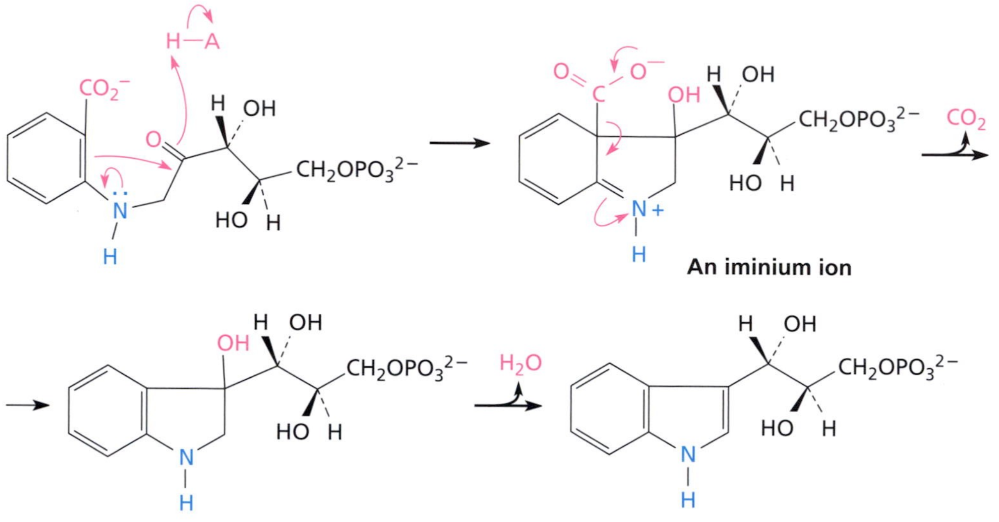

An iminium ion

Indole-3-glycerol phosphate

Steps 5–6. Elimination of glyceraldehyde 3-phosphate and addition of serine

Steps 5 and 6 of tryptophan biosynthesis are both catalyzed by tryptophan synthase.[53–55] Indole-3-glycerol phosphate is first protonated to give an iminium ion, which then undergoes a retro-aldol cleavage to give indole plus glyceraldehyde 3-phosphate. Conjugate addition of indole (a cyclic enamine) to the double bond of aminoacrylate–PLP imine produced from serine then yields tryptophan (Figure 5.46). The formation of the aminoacrylate–PLP imine from serine and subsequent conjugate nucleophilic addition reaction is analogous to what happens in methionine metabolism when homocysteine reacts with serine to give cystathionine (Section 5.3, Figure 5.23).

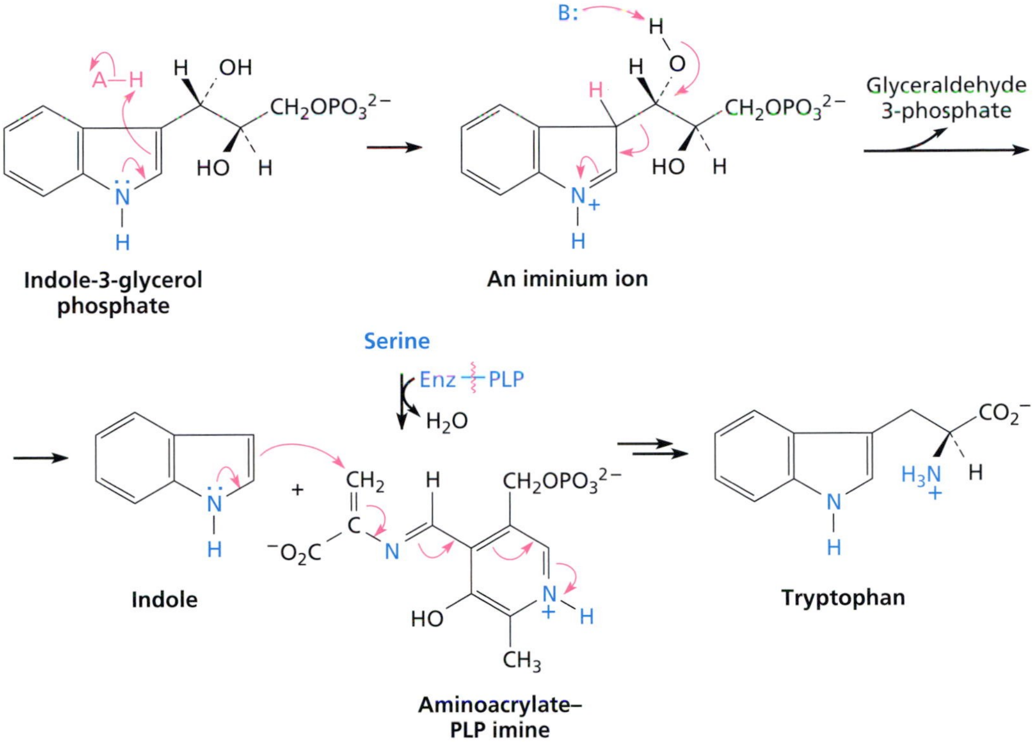

FIGURE 5.46 Mechanism of the conversion of indole-3-glycerol phosphate to tryptophan.

Phenylalanine and **tyrosine**, like tryptophan, are biosynthesized from chorismate (Figure 5.47). The key step is a Claisen rearrangement of chorismate to prephenate, catalyzed by chorismate mutase.[56] Although unusual in biochemical pathways, a Claisen rearrangement is a well-known and much-studied laboratory process that results in the conversion of an allylic vinyl ether to an unsaturated ketone. The reaction occurs in a single step and involves a redistribution of bonding electrons through a cyclic transition state. Prephenate then loses CO_2 and H_2O in the usual way to give phenylpyruvate, which undergoes transamination with glutamate to yield phenylalanine.

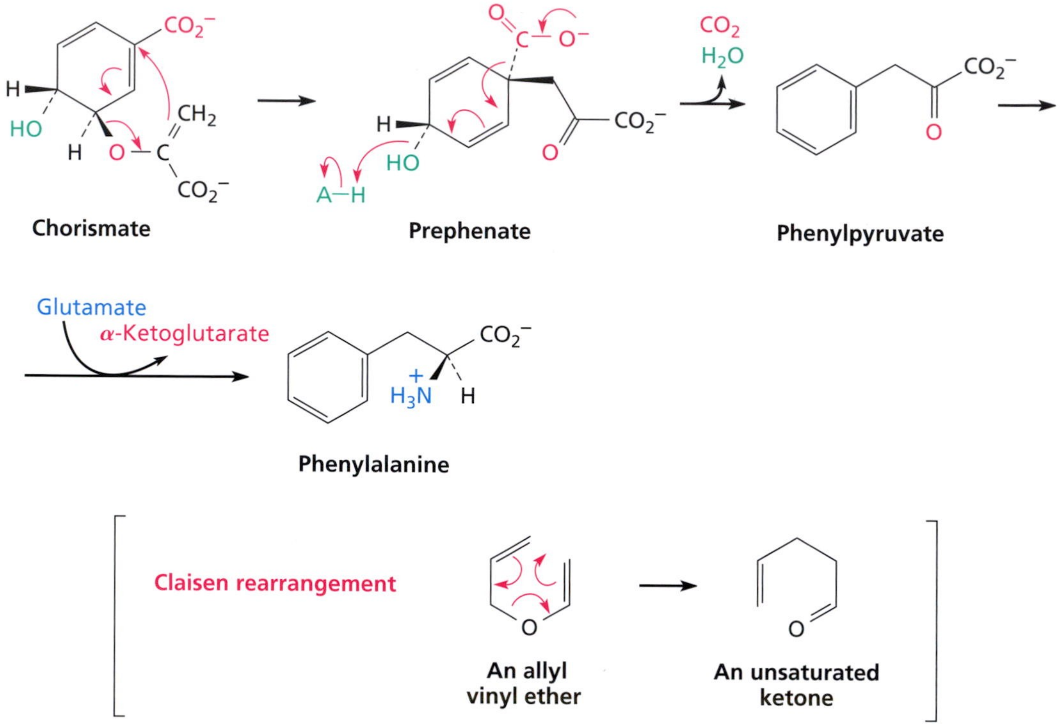

FIGURE 5.47 Pathway for the biosynthesis of phenylalanine from chorismate.

In mammals, tyrosine results from hydroxylation of phenylalanine, as we saw previously in Section 5.3 (Figure 5.28). In bacteria, however, tyrosine is produced directly from prephenate by the pathway shown in Figure 5.48. Oxidation of prephenate by NAD^+ first gives a cyclohexadienone intermediate, which loses CO_2 by the usual retro-aldol mechanism. The 4-hydroxyphenylpyruvate that results then undergoes transamination with glutamate. In plants, a similar pathway is followed, but the order of steps is reversed. That is, prephenate undergoes an initial transamination, which is followed by oxidation and decarboxylation.

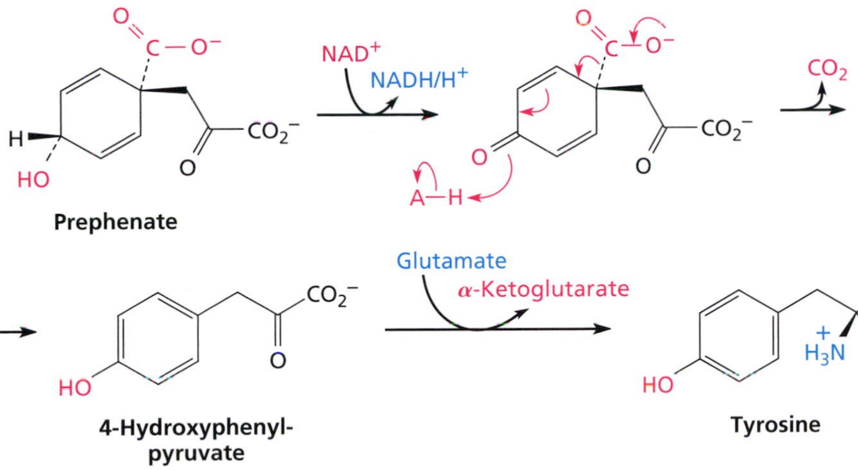

FIGURE 5.48 Pathway for the conversion of prephenate to tyrosine in bacteria.

Histidine

Histidine biosynthesis, shown in Figure 5.49, begins with nucleophilic substitution reaction of a nitrogen atom on ATP with phosphoribosyl diphosphate (step 1). The reaction proceeds by loss of diphosphate ion to give an oxonium-ion intermediate, followed by addition of ATP, as typically occurs in glycoside hydrolysis (Section 4.1, Figure 4.2). Hydrolysis of the triphosphate linkage then gives the corresponding monophosphate (step 2). Further hydrolysis opens the six-membered ring of adenine (step 3), and isomerization to a ketone occurs with opening of the ribose ring (step 4), as in step 2 of tryptophan biosynthesis (Figure 5.45). Reaction with ammonia, produced by hydrolysis of glutamine, splits off the original five-membered ring of adenine and yields imidazole glycerol phosphate (step 5), which is dehydrated and tautomerized to the ketone imidazole acetol phosphate (step 6). Transamination with glutamate (step 7), hydrolysis of the terminal phosphate group (step 8), and oxidation of the primary alcohol (step 9) complete the histidine biosynthesis.

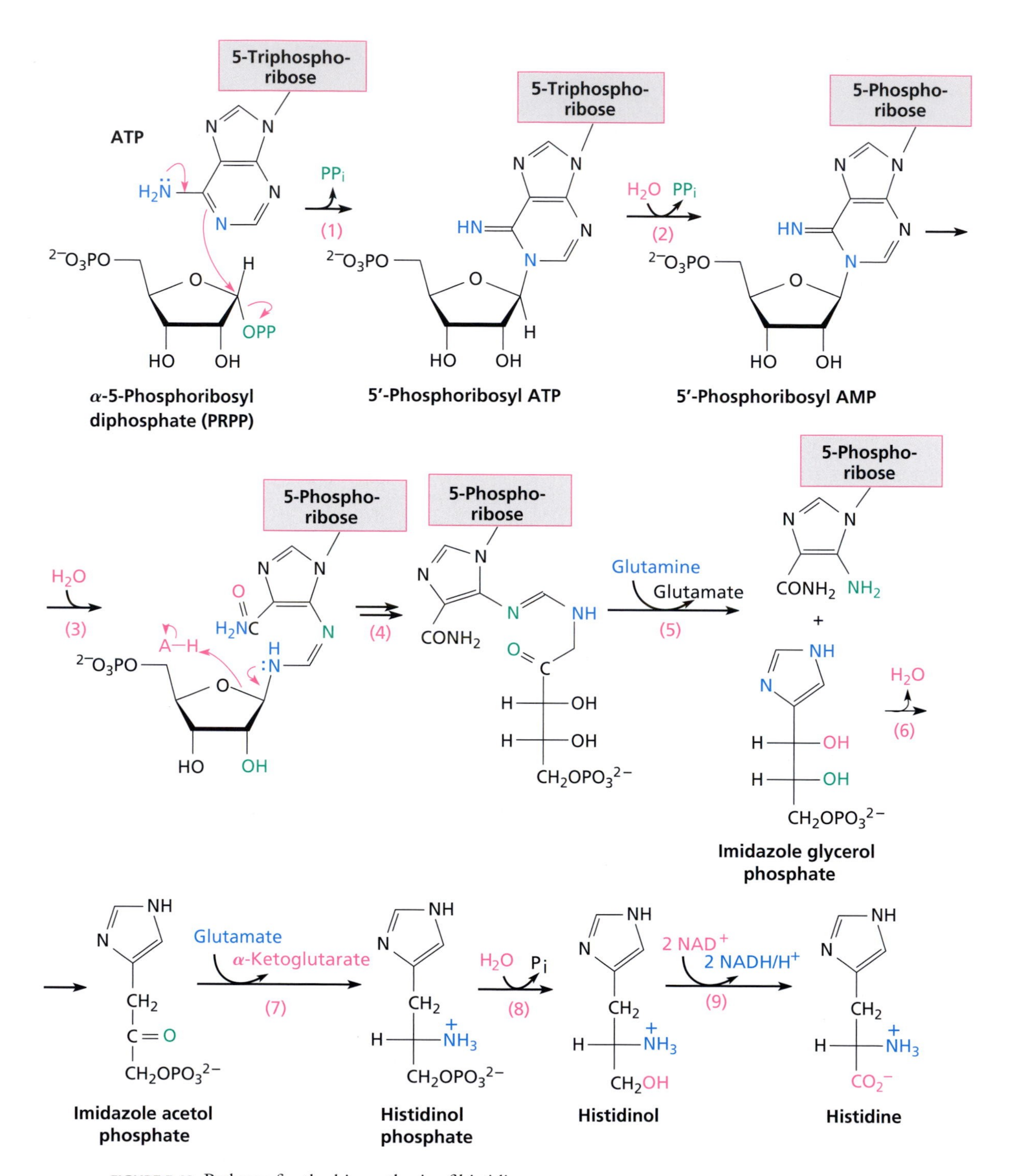

FIGURE 5.49 Pathway for the biosynthesis of histidine.

References

1. Perez-Pomares, F.; Ferrer, J.; Camacho, M.; Pire, C.; Llorca, F.; Bonete, M. J., "Amino Acid Residues Involved in the Catalytic Mechanism of NAD-Dependent Glutamate Dehydrogenase from *Halobacterium salinarum*," *Biochim. Biophys. Acta*, **1999**, *1426*, 513–525.
2. Holden, H. M.; Thoden, J. B.; Raushel, F. M., "Carbamoyl Phosphate Synthetase. An Amazing Biochemical Odyssey from Substrate to Product," *Cell. Mol. Life Sci.*, **1999**, *56*, 507–522.
3. Shi, D.; Morizono, H.; Ha, Y.; Aoyagi, M.; Tuchman, M.; Allewell, N. M., "1.85-Å Resolution Crystal Structure of Human Ornithine Transcarbamoylase Complexed with *N*-Phosphonacetyl-L-Ornithine. Catalytic Mechanism and Correlation with Inherited Deficiency," *J. Biol. Chem.*, **1998**, *273*, 34247–34254.
4. Lemke, C. T.; Howell, P. L., "The 1.6-Å Crystal Structure of *E. coli* Argininosuccinate Synthetase Suggests a Conformational Change during Catalysis," *Structure*, **2001**, *9*, 1153–1164.
5. Wu, C.-Y.; Lee, H.-J.; Wu, S.-H.; Chen, S.-T.; Chiou, S.-H.g; Chang, G.-G., "Chemical Mechanism of the Endogenous Argininosuccinate Lyase Activity of Duck Lens δ2-Crystallin," *Biochem. J.*, **1998**, *333*, 327–334.
6. Sampaleanu, L. M.; Yu, B.; Howell, P. L., "Mutational Analysis of Duck δ2 Crystallin and the Structure of an Inactive Mutant with Bound Substrate Provide Insight into the Enzymatic Mechanism of Argininosuccinate Lyase," *J. Biol. Chem.*, **2002**, *277*, 4166–4175.
7. Cox, J. D.; Cama, E.; Colleluori, D. M.; Pethe, S.; Boucher, J.-L.; Mansuy, D.; Ash, D. E.; Christianson, D. W., "Mechanistic and Metabolic Inferences from the Binding of Substrate Analogues and Products to Arginase," *Biochemistry*, **2001**, *40*, 2689–2701.
8. Bewley, M. C.; Jeffrey, P. D.; Patchett, M. L.; Kanyo, Z. F.; Baker, E. N., "Crystal Structures of *Bacillus caldovelox* Arginase in Complex with Substrate and Inhibitors Reveal New Insights into Activation, Inhibition and Catalysis in the Arginase Superfamily," *Structure*, **1999**, *7*, 435–448.
9. Scarsdale, J. N.; Kazanina, G.; Radaev, S.; Schirch, V.; Wright, H. T., "Crystal Structure of Rabbit Cytosolic Serine Hydroxymethyltransferase at 2.8-Å Resolution: Mechanistic Implications," *Biochemistry*, **1999**, *38*, 8347–8358.
10. Trivedi, V.; Gupta, A.; Jala, V. R.; Saravanan, P.; Rao, G. S. J.; Rao, N. A.; Savithri, H. S.; Subramanya, H. S., "Crystal Structure of Binary and Ternary Complexes of Serine Hydroxymethyltransferase from *Bacillus stearothermophilus*: Insights into the Catalytic Mechanism," J. *Biol. Chem.*, **2002**, *277*, 17161–17169.
11. Marsh, E. N. G.; Drennan, C. L., "Adenosylcobalamin-Dependent Isomerases: New Insights into Structure and Mechanism," *Curr. Opin. Chem. Biol.*, **2001**, *5*, 499–505.
12. Terentis, A. C.; Thomas, S. R.; Takikawa, O.; Littlejohn, T. K.; Truscott, R. J. W.; Armstrong, R. S.; Yeh, S.-R.; Stocker, R., "The Heme Environment of Recombinant

Human Indoleamine 2,3-Dioxygenase: Structural Properties and Substrate–Ligand Interactions," *J. Biol. Chem.*, **2002**, *277*, 15788–15794.

13. Pabarcus, M.; Casida, J., "Kynurenine Formamidase: Determination of Primary Structure and Modeling-Based Prediction of Tertiary Structure and Catalytic Triad," *Biochem. Biophys. Acta*, **2002**, *1596*, 201–211.
14. Breton, J.; Avanzi, N.; Magagnin, S.; Covini, N.; Magistrelli, G.; Cozzi, L.; Isacchi, A., "Functional Characterization and Mechanism of Action of Recombinant Human Kynurenine 3-Hydroxylase," *Eur. J. Biochem*, **2000**, *267*, 1092–1099.
15. Entsch B.; van Berkel W. J., "Structure and Mechanism of *para*-Hydroxybenzoate Hydroxylase," *FASEB J.*, **1995**, *9*, 476–483.
16. Phillips, R. S.; Sundararaju, B.; Koushik, S. V., "The Catalytic Mechanism of Kynureninase from *Pseudomonas fluorescens*: Evidence for Transient Quinonoid and Ketimine Intermediates from Rapid-Scanning Stopped-Flow Spectrophotometry," *Biochemistry*, **1998**, *37*, 8783–8789.
17. Bugg, T. D. H.; Lin, G., "Solving the Riddle of the Intradiol and Extradiol Catechol Dioxygenases: How Do Enzymes Control Hydroperoxide Rearrangements?" *Chem. Commun.*, **2001**, 941–952.
18. Yoon, M.-Y.; Thayer-Cook, K. A.; Berdis, A. J.; Karsten, W. E.; Schnackerz, K. D.; Cook, P. F., "Acid–Base Chemical Mechanism of Aspartase from *Hafnia alvei*," *Archiv. Biochem. Biophys.*, **1995**, *320*, 115–122.
19. Baedeker, M.; Schulz, G. E., "Structures of two Histidine Ammonia-Lyase Modifications and Implications for the Catalytic Mechanism," *European J. Biochemistry*, **2002**, *269*, 1790–1797.
20. Retey, J., "The Urocanase Story: A Novel Role of NAD^+ as Electrophile," *Arch. Biochem. and Biophys.*, **1994**, *314*, 1–16.
21. Takata, Y.; Yamada, T.; Huang, Y.; Komoto, J.; Gomi, T.; Ogawa, H.; Fujioka, M.; Takusagawa, F., "Catalytic Mechanism of *S*-Adenosylhomocysteine Hydrolase; Site-Directed Mutagenesis of Asp-130, Lys-185, Asp-189, and Asn-190," *J. Biol. Chem.*, **2002**, *277*, 22670–22676.
22. Jhee, K.-H.; McPhie, P.; Miles, E. W., "Yeast Cystathionine β-Synthase Is a Pyridoxal Phosphate Enzyme but, Unlike the Human Enzyme, Is Not a Heme Protein," *J. Biol. Chem.*, **2000**, *275*, 11541–11544.
23. Messerschmidt, A.; Worbs, M.; Steegborn, C.; Wahl, M. C.; Huber, R.; Laber, B.; Clausen, T., "Determinants of Enzymatic Specificity in the Cys-Met-Metabolism PLP-Dependent Enzymes Family: Crystal Structure of Cystathionine γ-Lyase from Yeast and Intrafamiliar Structure Comparison," *Biol. Chem.*, **2003**, *384*, 373–386.
24. Johansson, E.; Steffens, J. J.; Lindqvist, Y.; Schneider, G., "Crystal Structure of Saccharopine Reductase from *Magnaporthe grisea*, an Enzyme of the α-Aminoadipate Pathway of Lysine Biosynthesis," *Structure*, **2000**, *8*, 1037–1047.
25. Andreas-Anderson, O.; Flatmark, T.; Hough, E., "Crystal Structure of the Ternary Complex of the Catalytic Domain of Human Phenylalanine Hydroxylase with Tetrahydrobiopterin and 3-(2-Thienyl)-L-Alanine, and Its Implications for the Mechanism of Catalysis and Substrate Activation," *J. Mol. Biol.*, **2002**, *320*, 1095–1108.

26. Fitzpatrick, P. F., "Mechanism of Aromatic Amino Acid Hydroxylation," *Biochemistry*, **2000**, *42*, 14083–14091.
27. de Visser, S. P.; Shaik, S., "A Proton-Shuttle Mechanism Mediated by the Porphyrin in Benzene Hydroxylation by Cytochrome P450 Enzymes," *J. Amer. Chem. Soc.*, **2003**, *125*, 7413–7424.
28. Crouch, N. P.; Adlington, R. M.; Baldwin, J. E.; Lee, M.-H.; Mackinnon, C. H., "A Mechanistic Rationalization for the Substrate Specificity of Recombinant Mammalian 4-Hydroxyphenylpyruvate Dioxygenase (4-HPPD)," *Tetrahedron*, **1997**, *53*, 6993–7010.
29. Solomon, E. I.; Brunold, T. C.; Davis, M. I.; Kemsley, J. N.; Lee, S.-K.; Lehnert, N.; Neese, F.; Skulan, A. J.; Yang, Y.-S.; Zhou, J., "Geometric and Electronic Structure/Function Correlations in Non-Heme Iron Enzymes," *Chem. Rev.*, **2000**, *100*, 235–349.
30. Gunsior, M.; Ravel, J.; Challis, G. L.; Townsend, C. A., "Engineering *p*-Hydroxyphenylpyruvate Dioxygenase to a *p*-Hydroxymandelate Synthase and Evidence for the Proposed Benzene Oxide Intermediate in Homogentisate Formation," *Biochemistry*, **2004**, *43*, 663–674.
31. Titus, G. P.; Mueller, H. A.; Burgner, F.; De Cordoba, S. R.; Penalva, M. A.; Timm, D. E., "Crystal Structure of Human Homogentisate Dioxygenase," *Nature Struct. Biol.*, **2000**, *7*, 542–546.
32. Polekhina, G.; Board, P. G.; Blackburn, A. C.; Parker, M. W., "Crystal Structure of Maleylacetoacetate Isomerase/Glutathione Transferase Zeta Reveals the Molecular Basis for Its Remarkable Catalytic Promiscuity," *Biochemistry*, **2001**, *40*, 1567–1576.
33. Board, P. G.; Taylor, M. C.; Coggan, M.; Parker, M. W.; Lantum, H. B.; Anders, M. W., "Clarification of the Role of Key Active Site Residues of Glutathione Transferase Zeta/Maleylacetoacetate Isomerase by a New Spectrophotometric Technique," *Biochem. J.*, **2003**, *374*, 731–737.
34. Bateman, R. L.; Bhanumoorthy, P.; Witte, J. F.; McClard, R. W.; Grompe, M.; Timm, D. E., "Mechanistic Inferences from the Crystal Structure of Fumarylacetoacetate Hydrolase with a Bound Phosphorus-Based Inhibitor," *J. Biol. Chem.*, **2001**, *276*, 15284–15291.
35. Larsen, T. M.; Boehlein, S. K.; Schuster, S. M.; Richards, N. G. J.; Thoden, J. B.; Holden, H. M.; Rayment, I., "Three-Dimensional Structure of *Escherichia coli* Asparagine Synthetase B: A Short Journey from Substrate to Product," *Biochemistry*, **1999**, *38*, 16146–16157.
36. Liaw, S. H.; Eisenberg, D., "Structural Model for the Reaction Mechanism of Glutamine Synthetase, Based on Five Crystal Structures of Enzyme–Substrate Complexes," *Biochemistry*, **1994**, *33*, 675–681.
37. Wang, W.; Cho, Ho S.; Kim, R.; Jancarik, J.; Yokota, H.; Nguyen, H. H.; Grigoriev, I. V.; Wemmer, D. E.; Kim, S.-H., "Structural Characterization of the Reaction Pathway in Phosphoserine Phosphatase: Crystallographic 'Snapshots' of Intermediate States," *J. Mol. Biol.*, **2002**, *319*, 421–431.

38. Koo, C. W.; Blanchard, J. S., "Chemical Mechanism of *Haemophilus influenzae* Diaminopimelate Epimerase," *Biochemistry*, **1999**, *38*, 4416–4422.
39. Gokulan, K.; Rupp, B.; Pavelka, M. S.; Jacobs, W. R.; Sacchettini, J. C., "Crystal Structure of *Mycobacterium tuberculosis* Diaminopimelate Decarboxylase, an Essential Enzyme in Bacterial Lysine Biosynthesis," *J. Biol. Chem.*, **2003**, *278*, 18588–18596.
40. Born, T. L.; Blanchard, J. S., "Enzyme-Catalyzed Acylation of Homoserine: Mechanistic Characterization of the *Escherichia coli* metA-Encoded Homoserine Transsuccinylase," *Biochemistry*, **1999**, *38*, 14416–14423.
41. Clausen, T.; Huber, R.; Prade, L.; Wahl, M. C.; Messerschmidt, A., "Crystal Structure of *Escherichia coli* Cystathionine γ-Synthase at 1.5 Å Resolution," *EMBO J.*, **1998**, *17*, 6827–6838.
42. Clausen, T.; Huber, R.; Laber, B.; Pohlenz, H.-D.; Messerschmidt, A., "Crystal Structure of the Pyridoxal 5'-Phosphate-Dependent Cystathionine β-Lyase from *Escherichia coli* at 1.83 Å," *J. Mol. Biol.*, **1996**, *262*, 202–224.
43. Matthews, R. G., "Cobalamin-Dependent Methyltransferases," *Acc. Chem. Res.*, **2001**, *34*, 681–689.
44. Garrido-Franco, M.; Ehlert, S.; Messerschmidt, A.; Marinkovic, S.; Huber, R.; Laber, B.; Bourenkov, G. P.; Clausen, T., "Structure and Function of Threonine Synthase from Yeast," *J. Biol. Chem.*, **2002**, *277*, 12396–12405.
45. Dumas, R.; Biou, V.; Halgand, F.; Douce, R.; Duggleby, R. G., "Enzymology, Structure, and Dynamics of Acetohydroxy Acid Isomeroreductase," *Acc. Chem. Res.*, **2001**, *34*, 399–408.
46. Wang, J.; Duewel, H. S.; Woodard, R. W.; Gatti, D. L., "Structures of *Aquifex aeolicus* KDO8P Synthase in Complex with R5P and PEP, and with a Bisubstrate Inhibitor: Role of Active Site Water in Catalysis," *Biochemistry*, **2001**, *40*, 15676–15683.
47. Carpenter, E. P.; Hawkins, A. R.; Frost, J. W.; Brown, K. A., "Structure of Dehydroquinate Synthase Reveals an Active Site Capable of Multistep Catalysis," *Nature*, **1998**, *394*, 299–302.
48. An, M.; Maitra, U.; Neidlein, U.; Bartlett, P. A., "5-Enolpyruvylshikimate 3-Phosphate Synthase: Chemical Synthesis of the Tetrahedral Intermediate and Assignment of the Stereochemical Course of the Enzymatic Reaction," *J. Amer. Chem. Soc.*, **2003**, *125*, 12759–12767.
49. Maclean, J.; Ali, S., "The Structure of Chorismate Synthase Reveals a Novel Flavin Binding Site Fundamental to a Unique Chemical Reaction," *Structure*, **2003**, *11*, 1499–1511.
50. Knochel, T.; Ivens, A.; Hester, G.; Gonzalez, A.; Bauerle, R.; Wilmanns, M.; Kirschner, K.; Jansonius, J. N., "The Crystal Structure of Anthranilate Synthase from *Sulfolobus solfataricus*: Functional Implications," *Proc. Nat'l Acad. Sci. U.S.*, **1999**, *96*, 9479–9484
51. Hommel U.; Eberhard M.; Kirschner, K., "Phosphoribosyl Anthranilate Isomerase Catalyzes a Reversible Amadori Reaction," *Biochemistry*, **1995**, *34*, 5429–5439.

52. Hennig, M.; Darimont, B. D.; Jansonius, J. N.; Kirschner, K., "The Catalytic Mechanism of Indole-3-glycerol Phosphate Synthase: Crystal Structures of Complexes of the Enzyme from *Sulfolobus solfataricus* with Substrate Analogue, Substrate, and Product," *J. Mol. Biol.*, **2002**, *319*, 757–766.
53. Sachpatzidis, A.; Dealwis, C.; Lubetsky, J. B.; Liang, P.-H.; Anderson, K. S.; Lolis, E., "Crystallographic Studies of Phosphonate-Based α-Reaction Transition-State Analogues Complexed to Tryptophan Synthase," *Biochemistry*, **1999**, *38*, 12665–12674.
54. Woehl, E.; Dunn, M. F., "Mechanisms of Monovalent Cation Action in Enzyme Catalysis: The First Stage of the Tryptophan Synthase β-Reaction," *Biochemistry*, **1999**, *38*, 7118–7130.
55. Woehl, E.; Dunn, M. F., "Mechanisms of Monovalent Cation Action in Enzyme Catalysis: The Tryptophan Synthase α-, β-, and $\alpha\beta$-Reactions," *Biochemistry*, **1999**, *38*, 7131–7141.
56. Ganem, B., "Catalysis by Chorismate Mutases," *Comp. Nat. Prod. Chem.*, **1999**, *5*, 343–370.

Problems

5.1 Write the mechanism of the reaction of pyridoxamine phosphate (PMP) with α-ketoglutarate to give pyridoxal phosphate (PLP) and glutamate.

5.2 In the oxidative deamination of glutamate to α-ketoglutarate shown in Section 5.1, is the hydride ion transferred to NAD^+ with *pro-R* or *pro-S* stereochemistry?

5.3 Write the mechanism of the PLP-dependent transamination reaction of glutamate and oxaloacetate to yield α-ketoglutarate plus aspartate.

5.4 Propose a mechanism for the acid-catalyzed formation of 5,10-methylenetetrahydrofolate from tetrahydrofolate plus formaldehyde, a likely reaction in serine catabolism (Figure 5.6).

5.5 Show the mechanism of the conversion of 2-amino-3-ketobutyrate to glycine plus acetyl CoA, a step in threonine catabolism.

$$H_3C{-}C(=O){-}CH(\overset{+}{N}H_3){-}CO_2^- \xrightarrow{\text{CoASH}} \overset{+}{H_3N}{-}CH_2{-}CO_2^- + H_3C{-}C(=O){-}SCoA$$

2-Amino-3-ketobutyrate **Glycine** **Acetyl CoA**

5.6 Propose a mechanism for the synthesis of 4-methylideneimidazol-5-one by cyclization of a segment of the peptide chain in the histidine ammonia lyase enzyme.

4-Methylidene-imidazol-5-one

5.7 Propose a mechanism for the biotin-dependent carboxylation of tiglyl CoA to yield 3-methylglutaconyl CoA, a step in leucine catabolism.

ATP, H_2O, CO_2 → ADP, P_i

Tiglyl CoA → **3-Methylglutaconyl CoA**

5.8 Show the mechanism of the reductive amination between lysine and α-ketoglutarate to give saccharopine.

α-Ketoglutarate, NADPH/H^+ → $NADP^+$

Lysine → **Saccharopine**

5.9 Show the mechanism of the oxidative deamination of saccharopine.

H_2O, NAD^+ → NADH/H^+, Glutamate

Saccharopine → **α-Aminoadipate semialdehyde**

5.10 Propose a mechanism for tautomerization of the quinoid structure produced in phenylalanine hydroxylation to give 7,8-dihydrobiopterin.

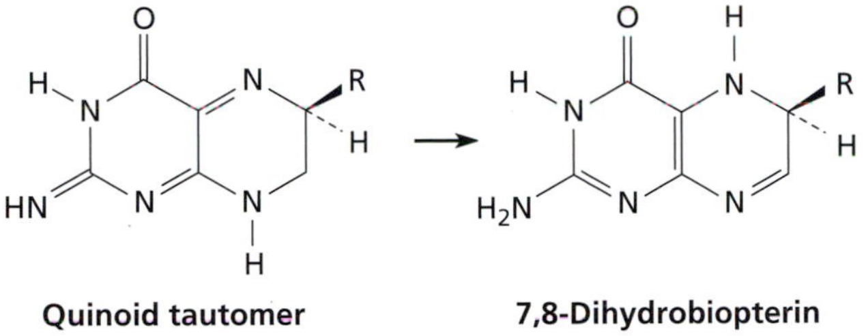

Quinoid tautomer → **7,8-Dihydrobiopterin**

5.11 Propose a mechanism for the PLP-dependent conversion of cystathionine to homocysteine, a step in methionine biosynthesis.

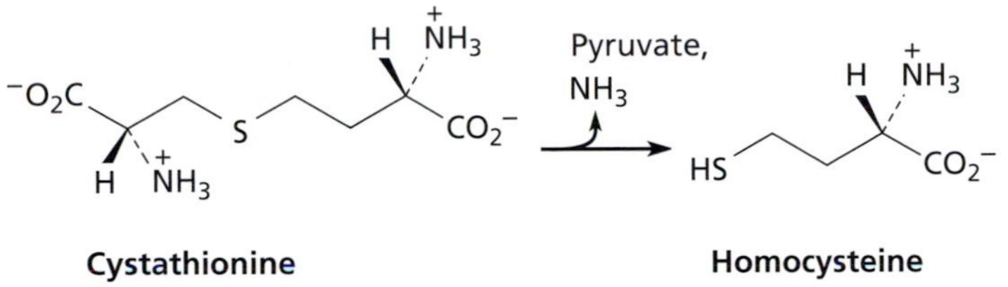

Cystathionine → (Pyruvate, NH_3) **Homocysteine**

5.12 Propose a mechanism for step 4 in histidine biosynthesis (Figure 5.49), the opening of the ribose ring to yield a ketone.

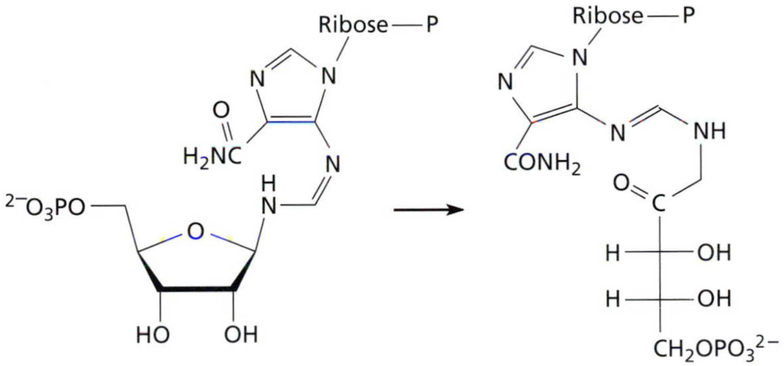

5.13 Propose a mechanism for step 5 in histidine biosynthesis (Figure 5.49), the synthesis of imidazole glycerol phosphate.

Ribose—P
Glutamine (NH_3)
Glutamate
Imidazole glycerol phosphate

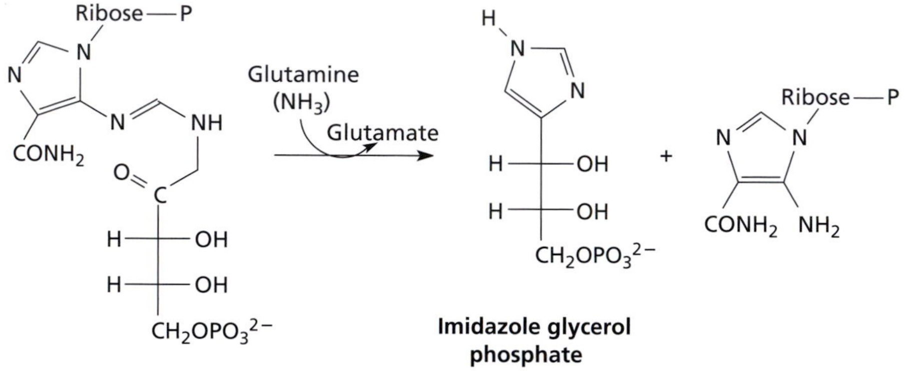

5.14 Polyamines such as putrescine (1,4-butanediamine) are biosynthesized from *S*-adenosylmethionine by a pathway that begins with the following step. What cofactor is likely to be needed for this reaction? Propose a mechanism.

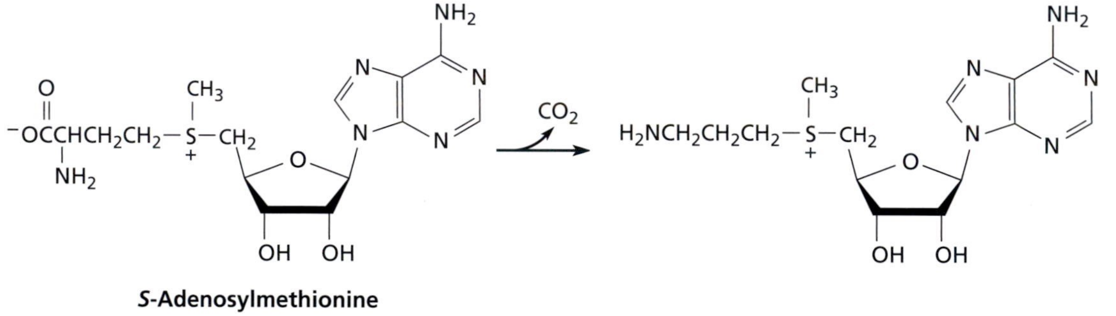

CO_2
***S*-Adenosylmethionine**

6 Nucleotide Metabolism

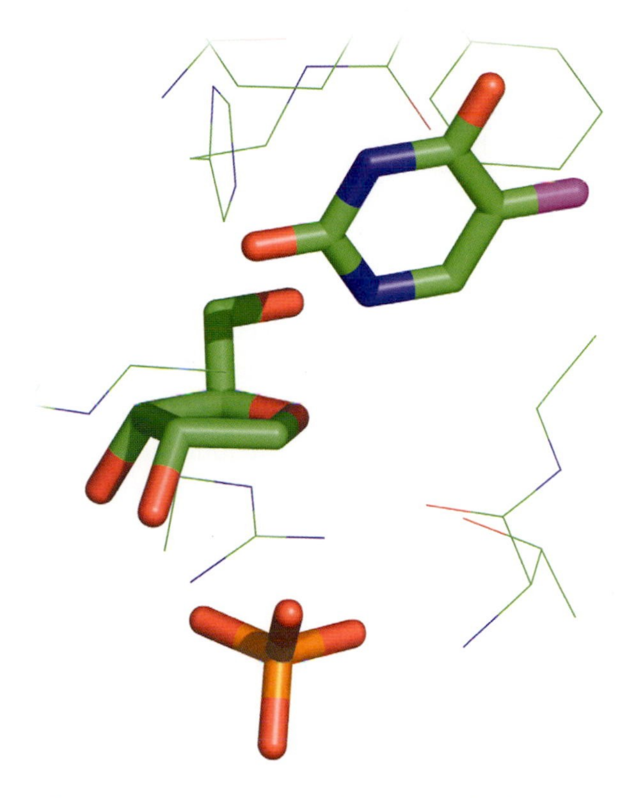

Extensive salvage pathways exist to conserve nucleotides, the building blocks for DNA and RNA. The structure shows the ribosyl oxonium ion trapped at the active site of uridine phosphorylase, using fluorouridine and sulfate as alternative substrates.

Nucleotides, the last major group of biomolecules that we'll consider, form the building blocks of nucleic acids just as amino acids form the building blocks of proteins. In addition, nucleoside triphosphates are involved as phosphorylating agents in many biochemical pathways and nucleotides are constituents of several important coenzymes, including NAD^+, FAD, and coenzyme A.

Recall from Section 2.5 and Figure 2.12 that there are four ribonucleotides and four deoxyribonucleotides, each consisting of a cyclic amine (a "base") bonded to a five-carbon sugar, with the sugar in turn bonded to a phosphate group. In ribonucleotides, two of the amine bases (cytosine and uracil) have a modified pyrimidine ring, and two (adenine and guanine) have a modified purine ring. In deoxyribonucleotides, cytosine, adenine, and guanine are still present but thymine replaces uracil.

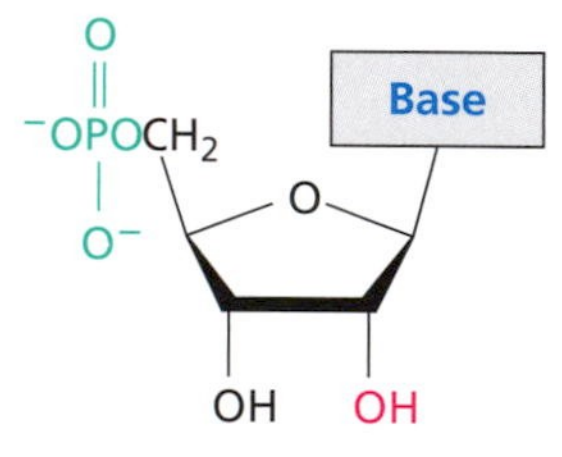

Ribonucleotides

Cytidine monophosphate (CMP)
Uridine monophosphate (UMP)
Adenosine monophosphate (AMP)
Guanosine monophosphate (GMP)

Deoxyribonucleotides

Deoxycytidine monophosphate (dCMP)
Thymidine monophosphate (dTMP or TMP)
Deoxyadenosine monophosphate (dAMP
Deoxyguanosine monophosphate (dGMP)

Pyrimidine

Purine

As in previous chapters, we'll look first at the metabolic pathways by which nucleotides are degraded and then cover their biosynthesis.

6.1 Nucleotide Catabolism

Dietary nucleic acids pass through the stomach to the intestines, where they are hydrolyzed to their constituent nucleotides by a variety of different nucleases. Further breakdown by various nucleotidases and phosphatases gives nucleosides, and a third hydrolysis by nucleosidases and nucleoside phosphorylases gives the constituent bases. A fraction of these bases are transported to tissues where they are reused for nucleic acid synthesis, but the rest are catabolized to produce intermediates of other metabolic processes.

Pyrimidines: Cytidine, Uridine, and Thymidine

The catabolism of **cytidine** begins with its hydrolytic deamination to give uridine. The reaction is catalyzed by cytidine deaminase[1] and occurs by nucleophilic addition of water to the C=N double bond, followed by expulsion of ammonia.

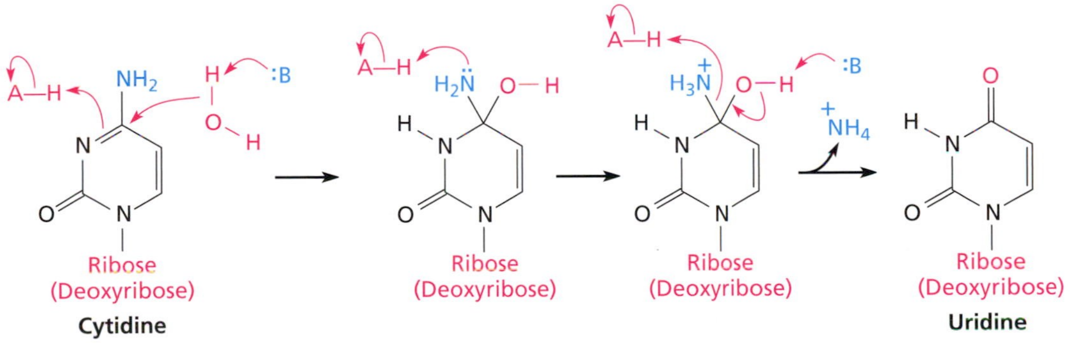

Uridine is cleaved by phosphorolysis to give uracil plus ribose 1-phosphate, and the uracil is then catabolized as the free base. Uridine catabolism occurs in six steps, as shown in Figure 6.1.

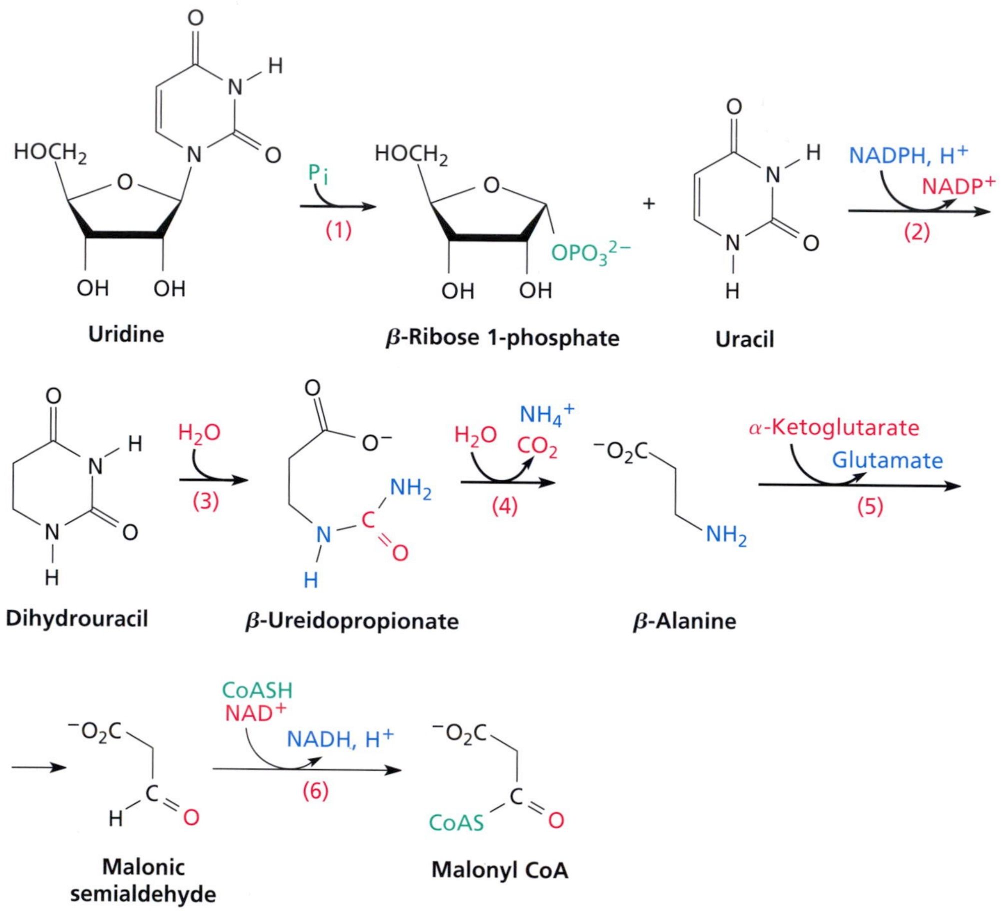

FIGURE 6.1 Pathway for the catabolism of uracil.

Steps 1–2. Phosphorolysis and reduction The phosphorolysis of uridine to give β-ribose 1-phosphate plus uracil is catalyzed by uridine phosphorylase and occurs by an S_N1-like replacement of uracil by phosphate ion through an oxonium-ion intermediate, analogous to the reaction of inverting glycosidases shown in Figure 4.2. Reduction of the C=C double bond in uracil then gives dihydrouracil in a reaction catalyzed by dihydropyrimidine dehydrogenase.[2] This reduction is substantially more complex than the result suggests. Rather than react directly with the substrate, NADPH first reduces a nonactive-site FAD by hydride transfer, and $FADH_2$ then reduces an active-site FMN (flavin mononucleotide) by a long-range electron transfer mediated by two iron–sulfur clusters. Reduced FMN transfers a hydride ion to the *si* face of the unsaturated carbonyl group in uracil, and protonation of the intermediate anion by Cys-671 also occurs on the *si* face giving dihydrouracil (Figure 6.2).

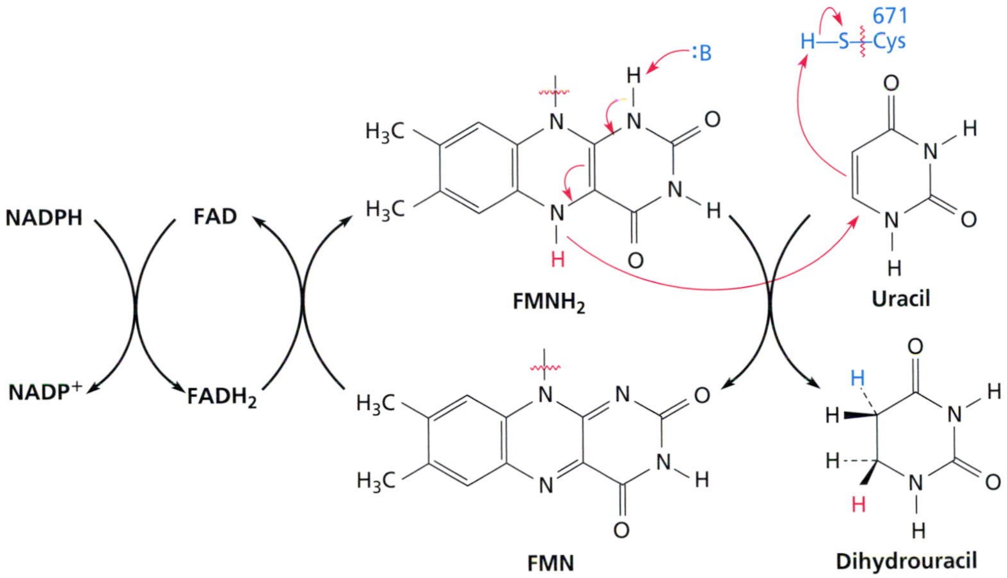

FIGURE 6.2 Mechanism of the reduction of uracil, catalyzed by dihydropyrimidine dehydrogenase.

An X-ray crystal structure of the active site in the enzyme–substrate complex of dihydropyrimidine dehydrogenase shows the flavin ring, the adjacent uracil substrate, and Cys-671 positioned for protonation (Figure 6.3).

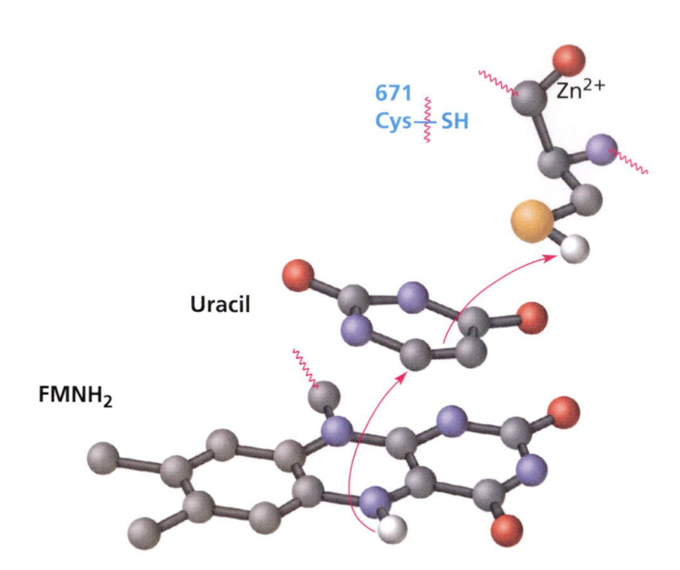

FIGURE 6.3 An X-ray crystal structure of the active site in the enzyme–substrate complex of dihydropyrimidine dehydrogenase. Reduced flavin transfers a hydride ion to uracil, and protonation by Cys-671 gives dihydrouracil.

Steps 3–4. Hydrolysis and decarboxylation Dihydrouracil undergoes hydrolysis by a nucleophilic acyl substitution mechanism to give the open-chain β-ureidopropionate. Further hydrolysis of the urea group yields ammonia and a carbamic acid ($R—NH—CO_2^-$), which decarboxylates to give β-alanine.

Steps 5–6. Transamination and oxidation PLP-dependent transamination of β-alanine by reaction with α-ketoglutarate in the usual way (Section 5.1) is followed by oxidation of the resultant malonic semialdehyde to yield malonyl CoA. Malonyl CoA then either decarboxylates to acetyl CoA or enters the pathway for fatty acid synthesis (Section 3.4, Figure 3.12). As in valine catabolism (Section 5.3), this final oxidation is thought to occur by addition to the aldehyde of a thiol residue on the dehydrogenase enzyme to give a hemithioacetal, followed by oxidation with NAD^+ and nucleophilic acyl substitution by coenzyme A.

Thymidine is cleaved to thymine and then degraded by a pathway analogous in all respects to that of uracil. The final product is methylmalonyl CoA, the same substance produced by threonine catabolism and ultimately converted to succinyl CoA (Section 5.3).

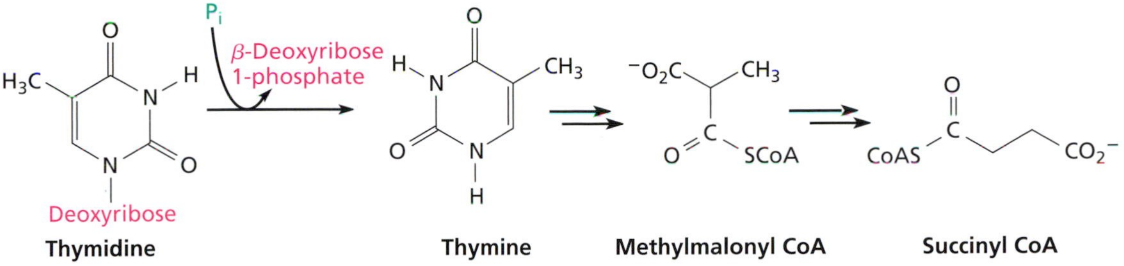

Thymidine → (P_i; β-Deoxyribose 1-phosphate) → Thymine → → Methylmalonyl CoA → → Succinyl CoA

Purines: Adenosine and Guanosine

In mammals, **adenosine** (or deoxyadenosine) is not cleaved directly to adenine. Instead, it is deaminated to give inosine by the same mechanism as that in cytidine catabolism, and inosine is cleaved by purine nucleoside phosphorylase to hypoxanthine. Hypoxanthine is then oxidized to yield xanthine, and a further oxidation produces uric acid, which is excreted in the urine (Figure 6.4).

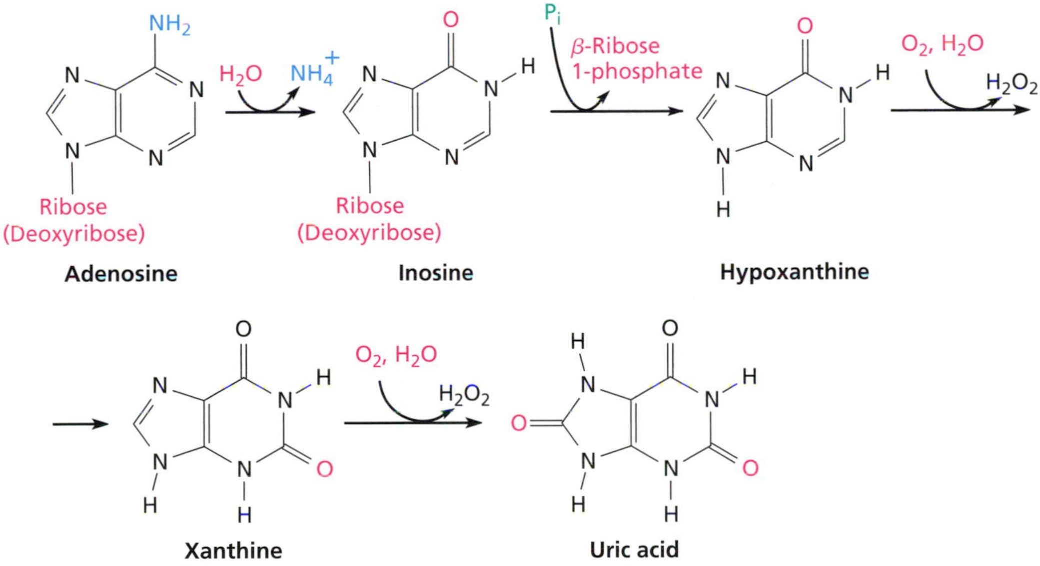

Adenosine → (H_2O, NH_4^+) → Inosine → (P_i; β-Ribose 1-phosphate) → Hypoxanthine → (O_2, H_2O; H_2O_2) → Xanthine → (O_2, H_2O; H_2O_2) → Uric acid

FIGURE 6.4 Pathway for the catabolism of adenosine to uric acid.

Hypoxanthine and xanthine oxidations are both catalyzed by xanthine oxidase,[3, 4] a complex enzyme that contains FAD, two iron–sulfur clusters, and an oxo–molybdenum(VI) cofactor. Numerous mechanisms have been proposed, but current evidence[3] suggests the process shown in Figure 6.5. In this mechanism, a glutamate residue in the enzyme deprotonates the Mo—OH group, and the resulting anion does a nucleophilic addition to a C═N double bond in hypoxanthine. The nitrogen anion then expels hydride ion, which adds to an S═Mo bond, thereby reducing the molybdenum center from Mo(VI) to Mo(IV). Subsequent hydrolysis of the O—Mo bond gives an enol that tautomerizes to xanthine, and the reduced molybdenum is reoxidized by O_2 in a complex redox pathway. Note that the expulsion of hydride ion by electrons on the neighboring nitrogen atom is analogous to what occurs during NADH reductions (Section 1.9).

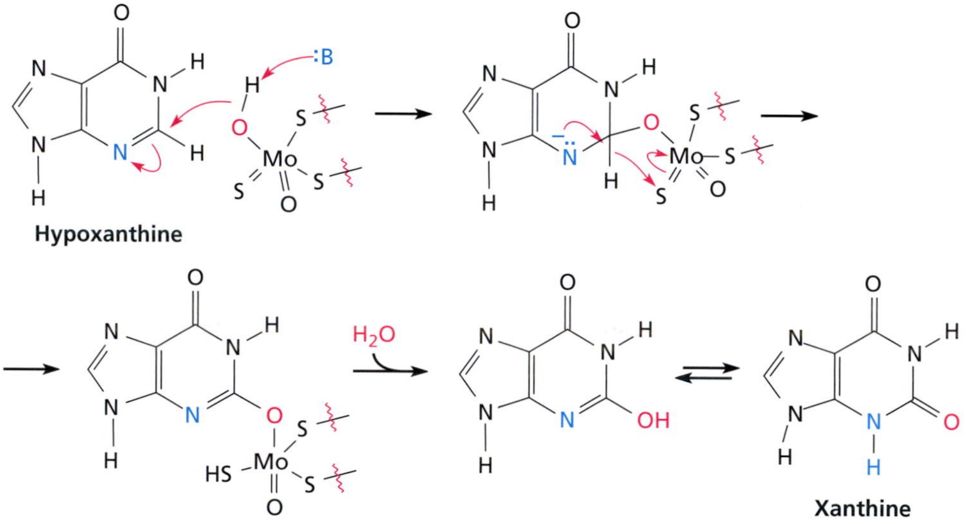

Hypoxanthine

Xanthine

FIGURE 6.5 Proposed mechanism for the oxidation of hypoxanthine to xanthine. A similar process occurs in the subsequent oxidation of xanthine to uric acid.

Guanosine (or deoxyguanosine) is cleaved by purine nucleoside phosphorylase to guanine, which is then hydrolytically deaminated to yield xanthine by the same mechanism as in cytidine deamination.

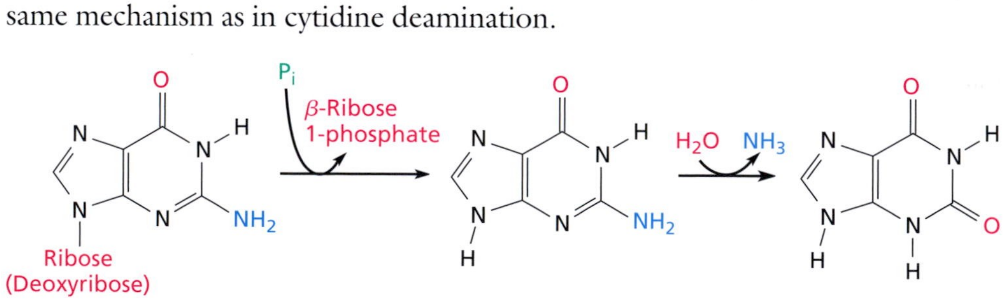

Ribose (Deoxyribose)

P_i → β-Ribose 1-phosphate

H_2O → NH_3

Guanosine

Guanine

Xanthine

6.2 Biosynthesis of Pyrimidine Ribonucleotides

Uridine Monophosphate

Uridine monophosphate (UMP) is biosynthesized in a six-step pathway from aspartate, bicarbonate, and ammonia, which itself comes from the amide nitrogen of glutamine (Figure 6.6).

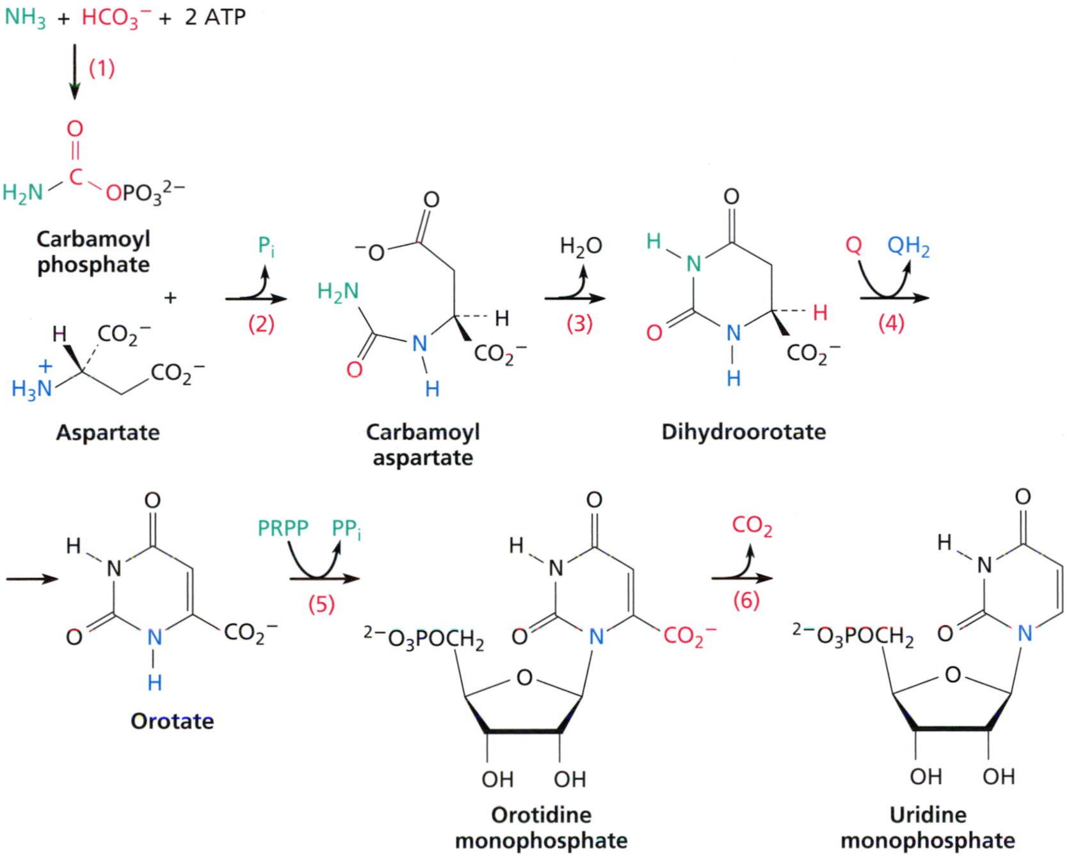

FIGURE 6.6 Pathway for the biosynthesis of uridine monophosphate (UMP) from carbamoyl phosphate and aspartate.

Step 1. Carbamoyl phosphate synthesis UMP biosynthesis begins with formation of carbamoyl phosphate, catalyzed by carbamoyl phosphate synthetase II.[5] The reaction is identical to that occurring in the urea cycle (Section 5.2) except that the ammonia used for pyrimidine synthesis comes from hydrolysis of glutamine within the synthetase enzyme rather than from free ammonia as in the urea cycle.

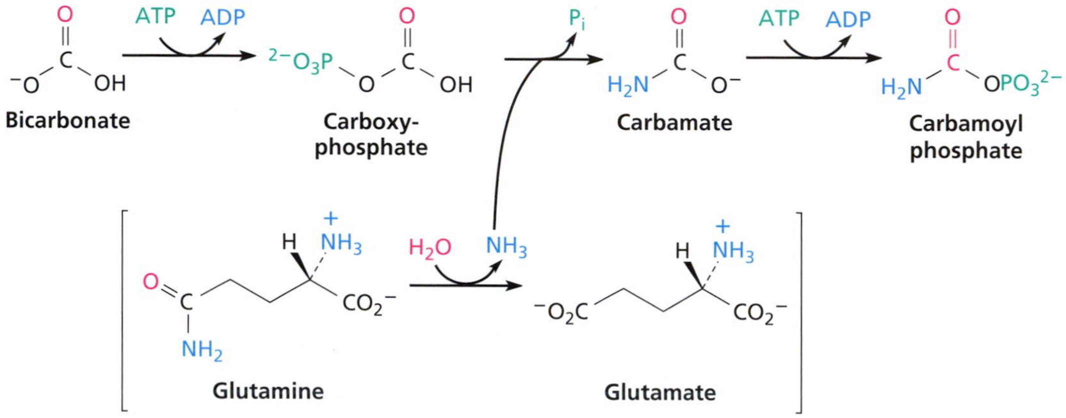

Steps 2–3. Reaction with aspartate and cyclization Carbamoyl phosphate reacts with aspartate in a nucleophilic acyl substitution reaction with phosphate as the leaving group to give carbamoyl aspartate. Cyclization then forms dihydroorotate. The cyclization is catalyzed by dihydroorotase[6] and is mechanistically interesting because it accomplishes the formation of an amide bond between a poor nucleophile (a urea-like nitrogen) and a poor electrophile (a carboxylate). What evidently happens is that the carboxylate is activated by coordination to two Lewis-acidic Zn^{2+} ions, and both reacting centers are surrounded by various charged groups within the enzyme that electrostatically stabilize the reaction intermediates. Deprotonation of the urea $—NH_2$ by an aspartate residue and concurrent addition to the carboxylate carbonyl group in a nucleophilic acyl substitution reaction gives the product. Figure 6.7 shows both the mechanism and an X-ray crystal structure of the substrate bound in the active site.

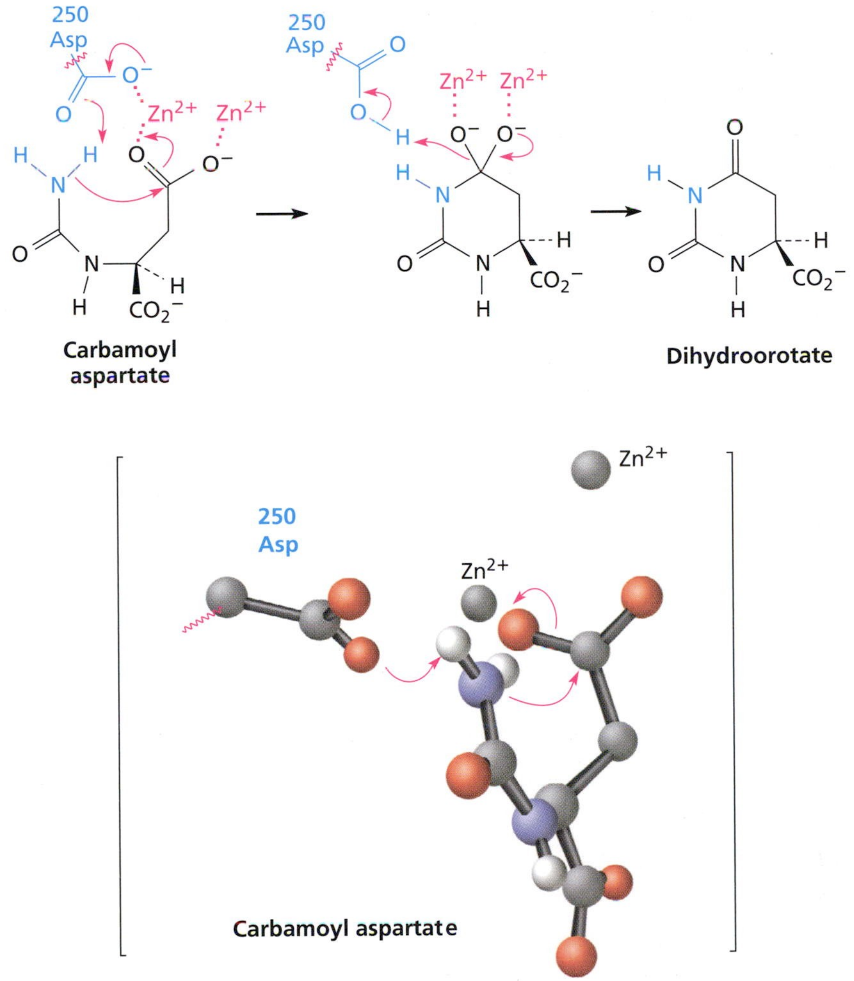

FIGURE 6.7 Mechanism of the cyclization of carbamoyl aspartate to dihydroorotate, along with an X-ray crystal structure of the substrate bound in the active site.

Step 4. Dehydrogenation Introduction of a double bond into dihydroorotate to give orotate is catalyzed by dihydroorotate dehydrogenase[7, 8] a flavin-dependent enzyme that, in humans, uses coenzyme Q, also called ubiquinone, as the ultimate electron acceptor. The reaction occurs by base abstraction of the *pro-S* hydrogen at C5 and donation of hydride ion from C6 to FMN. The $FMNH_2$ is then reoxidized by coenzyme Q. As shown in Figure 6.8, CoQ is a benzoquinone with a long hydrocarbon tail that allows it to dissolve readily in lipid membranes. Its function is to act as a redox agent in the transport of electrons between enzymes embedded in the inner mitochondrial membrane.

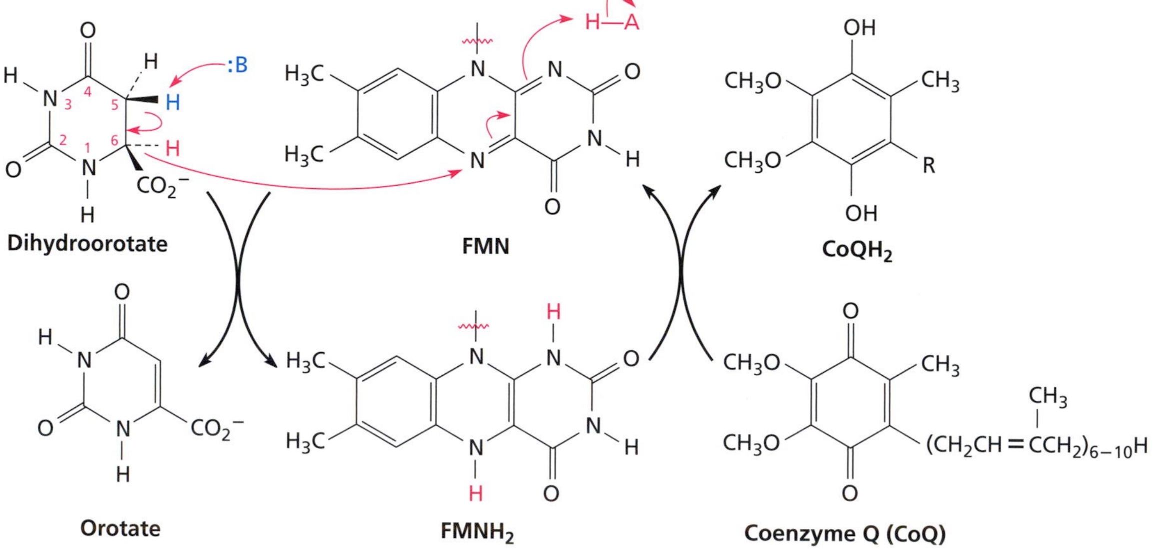

FIGURE 6.8 Mechanism of the dehydrogenation of dihydroorotate to orotate.

Step 5. Ribonucleotide formation Orotate reacts with 5-phosphoribosyl α-diphosphate (PRPP) to give the ribonucleotide orotidine monophosphate (OMP). This ribonucleotide formation takes place by a nucleophilic substitution reaction, catalyzed by orotate phosphoribosyltransferase.[9] Although the reaction occurs with an inversion of stereochemistry, the likely mechanism involves spontaneous, S_N1-like loss of diphosphate ion to give an oxonium-ion intermediate, much like what occurs in the hydrolysis of a polysaccharide catalyzed by an inverting glycosidase (Section 4.1, Figure 4.2). The 5-phosphoribosyl α-diphosphate precursor is formed from α-D-ribose 5-phosphate by reaction with ATP in the presence of PRPP synthetase (Figure 6.9).

Step 6. Decarboxylation The final step in UMP biosynthesis is the decarboxylation of OMP, catalyzed by orotidine monophosphate decarboxylase.[10] This enzyme contains no cofactors and holds the distinction of having the greatest

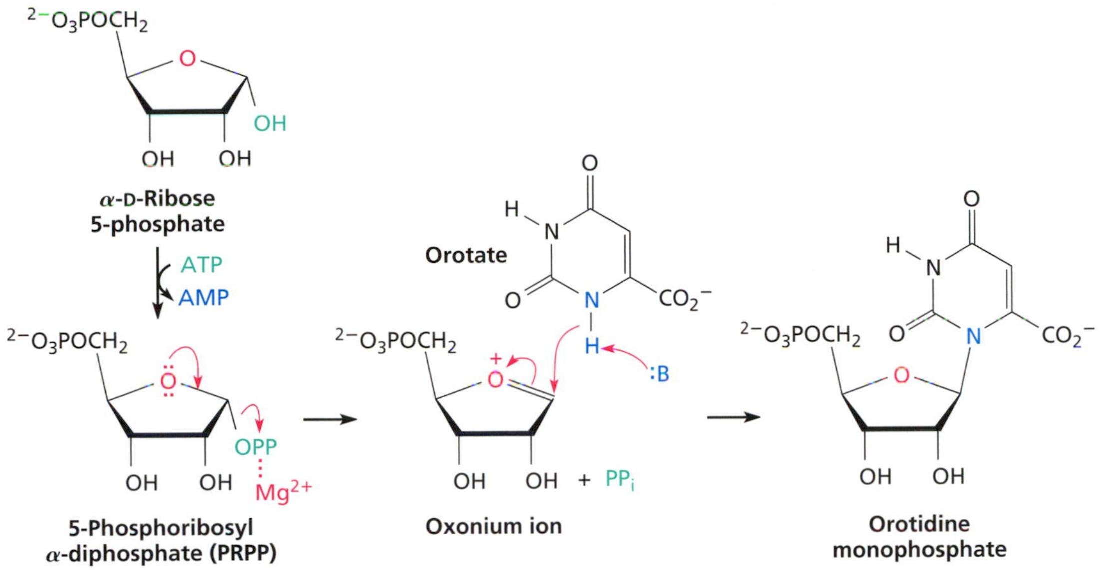

FIGURE 6.9 Mechanism of the formation of orotidine monophosphate.

experimentally determined rate acceleration known for any enzyme, a factor of 2×10^{23} M^{-1} for the catalyzed versus uncatalyzed reaction! Mechanistically, the decarboxylation is unusual because the substrate is not a β keto acid and has no obvious electron sink nearby to accept electrons as CO_2 leaves. It's thought instead that the decarboxylation occurs in a single step, driven by electrostatic interactions between the substrate and charged residues in the active site. An aspartate residue held near the carboxylate destabilizes the ground state, while a protonated lysine stabilizes the transition state and provides a proton as CO_2 departs.

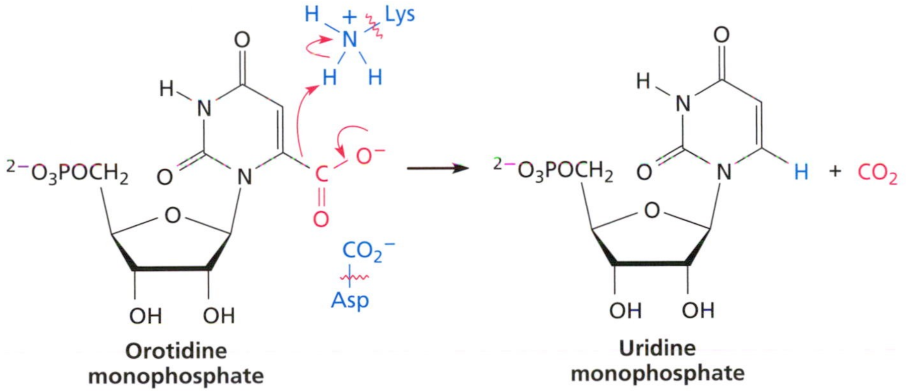

Cytidine Triphosphate

Following its synthesis from orotate, uridine monophosphate is converted into the corresponding triphosphate (UTP) by two sequential reactions with ATP. Uridine triphosphate is then converted into **cytidine triphosphate** by a reaction that is essentially the reverse of the cytidine → uridine conversion seen in cytidine catabolism (Section 6.1). The primary difference between the two processes is that the cytidine → uridine conversion requires no ATP while the uridine → cytidine conversion is coupled to ATP hydrolysis for energetic reasons. Catalyzed by CTP synthase,[11] glutamine is first hydrolyzed to glutamate plus ammonia at one site in the enzyme, a process similar to what occurs in carbamoyl phosphate synthesis (Section 6.2). The ammonia then moves through a channel in the enzyme to the next reaction site.

In the second site, uridine triphosphate is phosphorylated on the pyrimidine oxygen by ATP, and the resultant imino phosphate undergoes nucleophilic acyl substitution by addition of NH_3 to the C=N double bond followed by elimination of P_i.

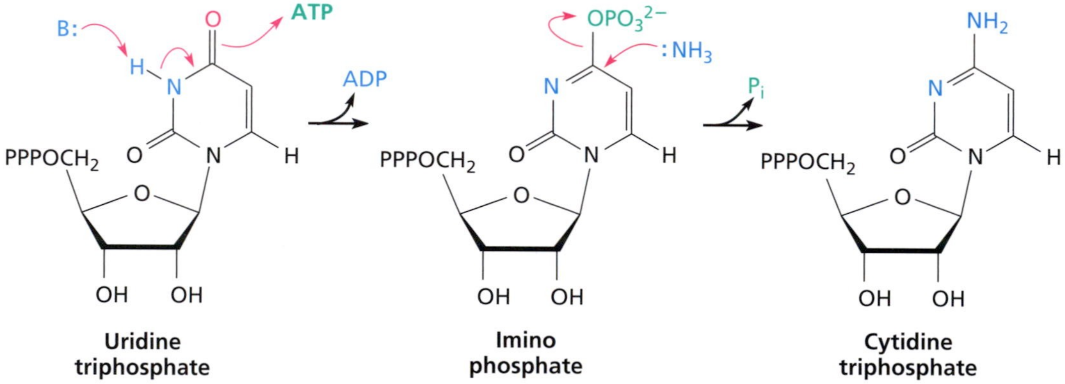

Uridine triphosphate → Imino phosphate → Cytidine triphosphate

6.3 Biosynthesis of Purine Ribonucleotides

As we've just seen, pyrimidine nucleotides are synthesized by an initial multistep formation of the pyrimidine base, followed by attachment of phosphoribose to the base. Purine nucleotides, in contrast, are formed by the initial attachment of an —NH_2 group to the phosphoribose, followed by multistep formation of the purine base. Inosine monophosphate (IMP) is the first fully formed purine ribonucleotide, with adenosine monophosphate (AMP) and guanosine monophosphate (GMP) then derived from it.

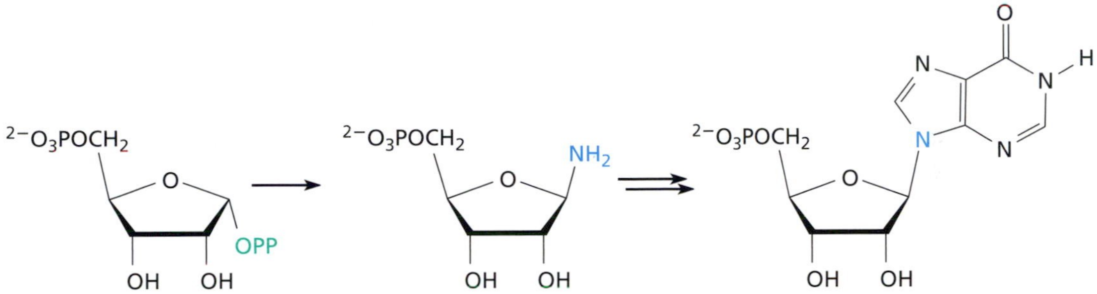

5-Phosphoribosyl α-diphosphate (PRPP)   β-5-Phosphoribosylamine   Inosine monophosphate (IMP)

Inosine Monophosphate

The biosynthetic pathway for inosine monophosphate is shown in Figure 6.10. The pathway has 11 steps starting from 5-phosphoribosyl α-diphosphate, which is itself prepared by reaction of α-D-ribose 5-phosphate with ATP, as noted in the previous section.

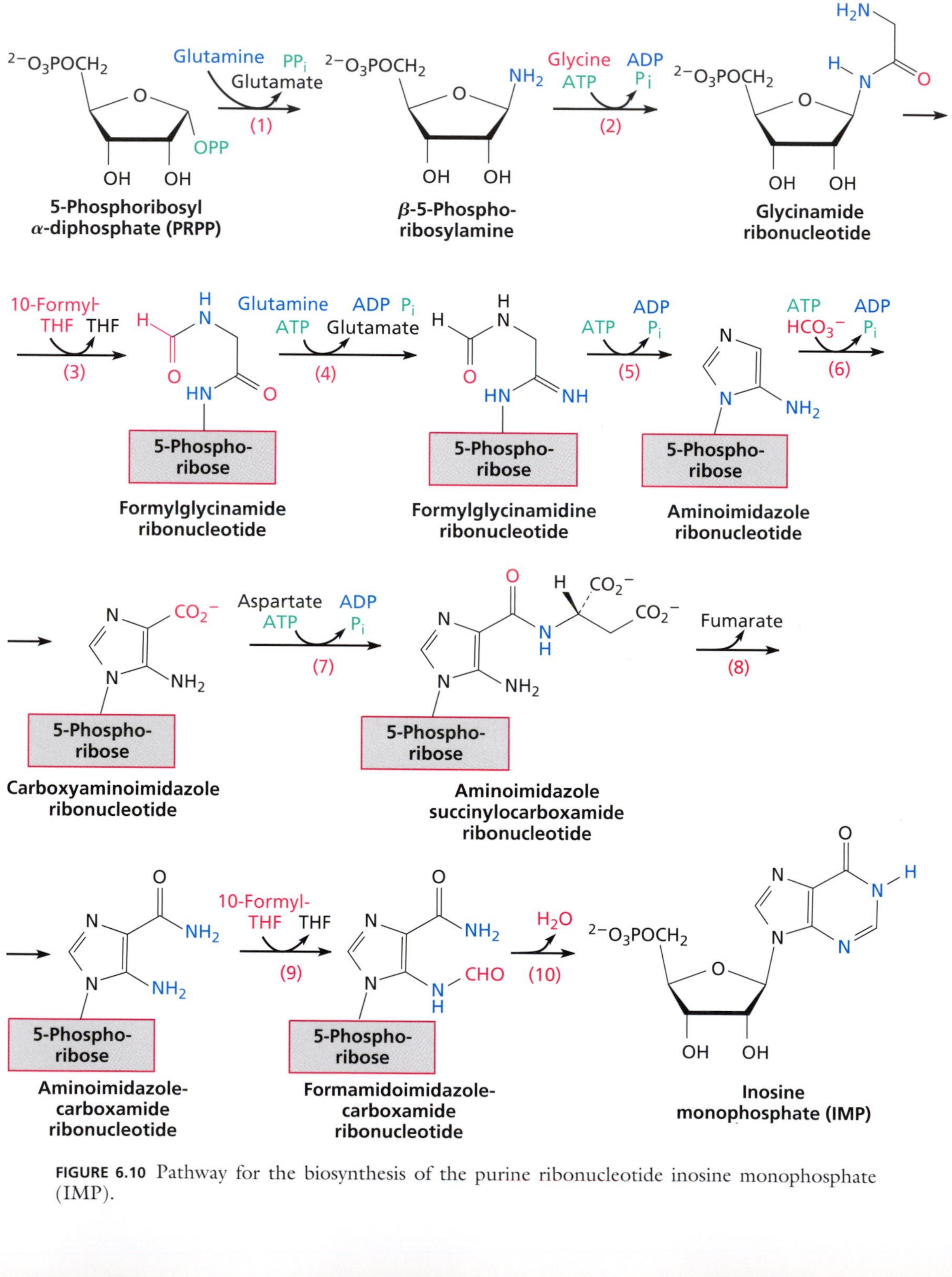

$^{2-}O_3POCH_2$ OPP OH OH

5-Phosphoribosyl α-diphosphate (PRPP)

Glutamine PP_i Glutamate (1)

$^{2-}O_3POCH_2$ NH_2 OH OH

β-5-Phospho-ribosylamine

Glycine ADP ATP P_i (2)

H_2N H N O $^{2-}O_3POCH_2$ OH OH

Glycinamide ribonucleotide

10-Formyl-THF THF (3)

H N H O HN O 5-Phospho-ribose

Formylglycinamide ribonucleotide

Glutamine ADP P_i ATP Glutamate (4)

H N H O HN NH 5-Phospho-ribose

Formylglycinamidine ribonucleotide

ADP ATP P_i (5)

N N NH_2 5-Phospho-ribose

Aminoimidazole ribonucleotide

ATP ADP HCO_3^- P_i (6)

N CO_2^- N NH_2 5-Phospho-ribose

Carboxyaminoimidazole ribonucleotide

Aspartate ADP ATP P_i (7)

O H CO_2^- N N H CO_2^- N NH_2 5-Phospho-ribose

Aminoimidazole succinylocarboxamide ribonucleotide

Fumarate (8)

O N NH_2 N NH_2 5-Phospho-ribose

Aminoimidazole-carboxamide ribonucleotide

10-Formyl-THF THF (9)

O N NH_2 N N H CHO 5-Phospho-ribose

Formamidoimidazole-carboxamide ribonucleotide

H_2O (10)

O N N H N N $^{2-}O_3POCH_2$ O OH OH

Inosine monophosphate (IMP)

FIGURE 6.10 Pathway for the biosynthesis of the purine ribonucleotide inosine monophosphate (IMP).

Step 1. Amine formation 5-Phosphoribosyl α-diphosphate is converted to β-5-phosphoribosylamine by glutamine PRPP amidotransferase.[12] As in carbamoyl phosphate synthesis (Section 6.2), glutamine is first hydrolyzed to ammonia at one site in the enzyme, and the ammonia then moves through a channel to a second site where it reacts with 5-phosphoribosyl α-diphosphate. Reaction with the ribosyl diphosphate takes place through an oxonium-ion intermediate and occurs with a net inversion of configuration, as in the synthesis of orotidine monophosphate (Figure 6.9).

Step 2. Glycinamide formation Glycine and β-5-phosphoribosylamine react to form the amide glycinamide ribonucleotide (GAR) in a reaction catalyzed by GAR synthetase.[13] The reaction occurs by initial formation of a glycyl phosphate, followed by nucleophilic acyl substitution with the ribosylamine. This mechanism is similar to that seen in the biosynthesis of glutamine from glutamate (Section 5.4).

Step 3. Formylation Formylation of the amino group in glycinamide ribonucleotide is catalyzed by GAR transformylase[14] and occurs by transfer of a formyl group from 10-formyltetrahydrofolate in a nucleophilic acyl substitution reaction.

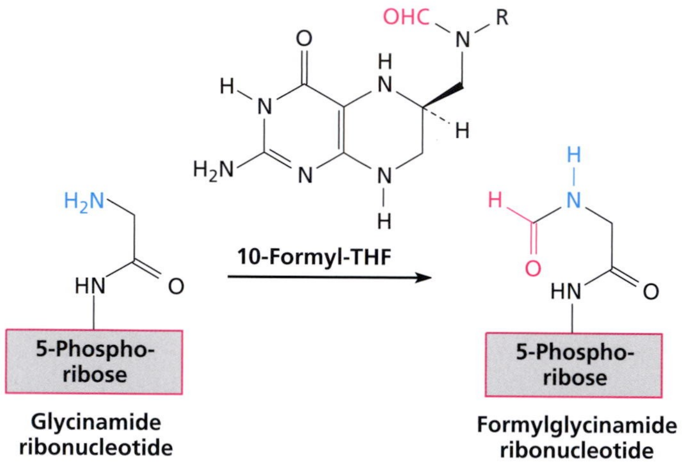

10-Formyl-THF

5-Phospho-ribose

Glycinamide ribonucleotide

5-Phospho-ribose

Formylglycinamide ribonucleotide

Step 4. Glycinamidine formation Formylglycinamide ribonucleotide is converted to formylglycinamidine ribonucleotide by reaction with ATP and ammonia (an amidine has the structure $R_2N—C═NH$). The reaction is catalyzed by formylglycinamidine (FGAM) synthetase and takes place by the mechanism shown in Figure 6.11. The process is very similar to what occurs in the conversion of uridine triphosphate to cytidine triphosphate (Section 6.2).

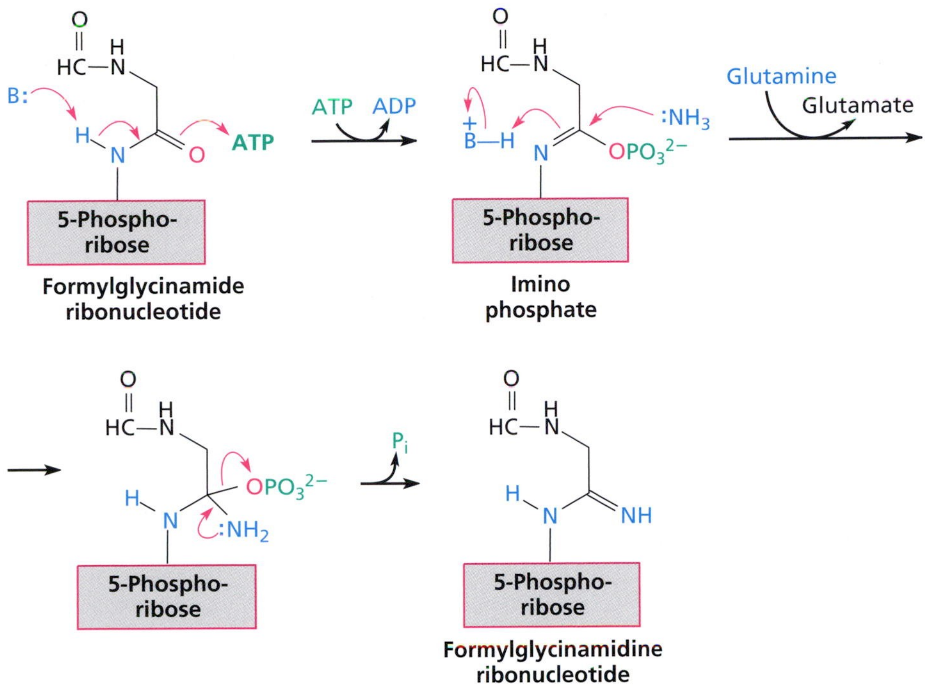

FIGURE 6.11 Mechanism of formylglycinamidine formation in step 4 of inosine biosynthesis.

Step 5. Imidazole formation Closure of the imidazole ring in step 5 is catalyzed by aminoimidazole (AIR) synthetase,[15] an ATP-dependent enzyme whose mechanism is analogous to that of FGAM synthetase in step 4 (Figure 6.11). Following ring closure, a tautomerization of the imine to an enamine occurs.

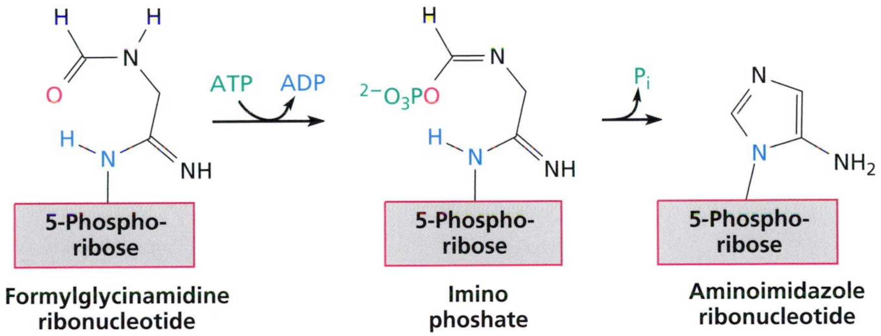

Step 6. Carboxylation Aminoimidazole ribonucleotide undergoes carboxylation by reaction with HCO_3^- in a process catalyzed by AIR carboxylase.[16, 17] Unlike most other carboxylations, however, the reaction does not require biotin (Section 3.4, Figure 3.14). Instead, the reaction takes place by nucleophilic addition of the amino group to carboxyphosphate (Section 3.4, Figure 3.13) to give an *N*-carboxyaminoimidazole ring, followed by loss of CO_2 and immediate readdition (Figure 6.12). The amino group of AIR thus serves the same purpose as biotin.

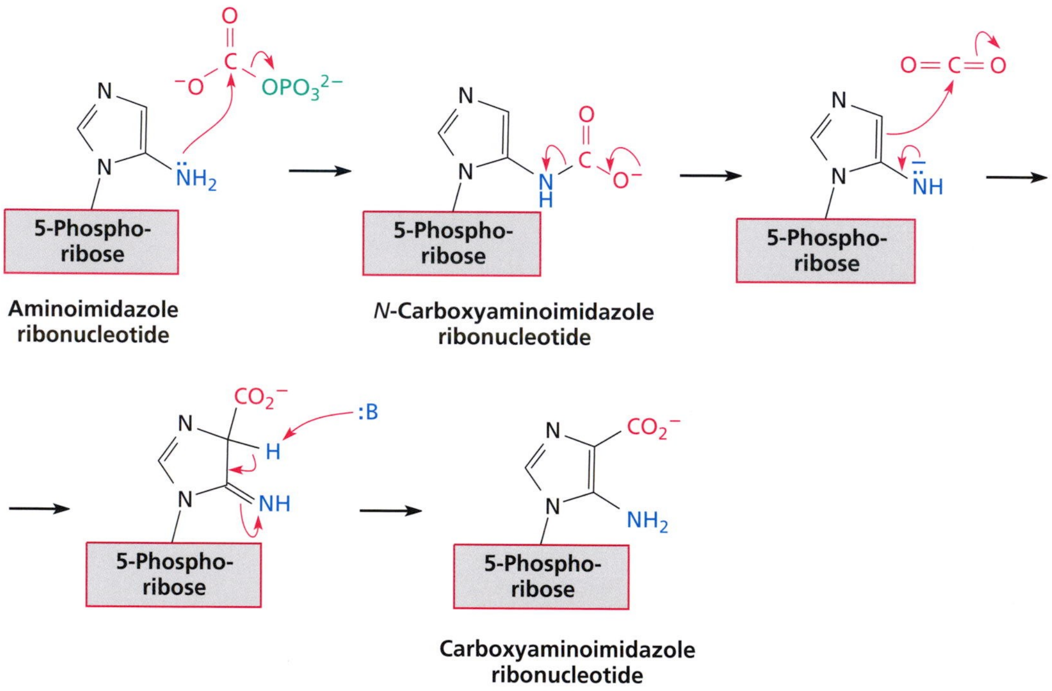

Aminoimidazole ribonucleotide

***N*-Carboxyaminoimidazole ribonucleotide**

Carboxyaminoimidazole ribonucleotide

FIGURE 6.12 Mechanism of the carboxylation of aminoimidazole ribonucleotide.

Steps 7–8. Succinylocarboxamide formation and fumarate elimination Carboxyaminoimidazole ribonucleotide reacts with aspartate in step 7 to form an amide. The reaction is catalyzed by aminoimidazole succinylocarboxamide ribonucleotide (SAICAR) synthetase,[18] requires ATP as cofactor, and is mechanistically analogous to glycinamide formation in step 2. SAICAR then undergoes elimination of fumarate in an E1cB reaction to give aminoimidazole carboxamide ribonucleotide (AICAR), a process catalyzed by adenylosuccinate lyase.[19] Note the similarity of steps 7 and 8 in inosine synthesis to steps 2 and 3 in the urea cycle (Section 5.2), in which citrulline is converted to arginine.

Step 9. Formylation The final atom needed for purine synthesis is added in step 9 by formylation of AICAR. The reaction is catalyzed by AICAR transformylase[20] and takes place by transfer of a formyl group from 10-formyltetrahydrofolate by a pathway analogous to that in step 3.

Step 10. Cyclization to form IMP The route for inosine monophosphate biosynthesis concludes with the cyclization of formamidoimidazole carboxamide ribonucleotide (FAICAR) in a reaction catalyzed by IMP cyclohydrolase.[21] Unlike the cyclization reaction in step 5, which forms the imidazole ring, this final cyclization occurs directly and does not require ATP.

Adenosine Monophosphate and Guanosine Monophosphate

Adenosine monophosphate and guanosine monophosphate are both derived from inosine monophosphate by straightforward transformations of the sort we've already encountered (Figure 6.13).

Adenosine monophosphate is synthesized in a two-step sequence from IMP: initial reaction with aspartate to yield adenylosuccinate, followed by loss of fumarate. The first step is catalyzed by adenylosuccinate synthetase,[22] requires

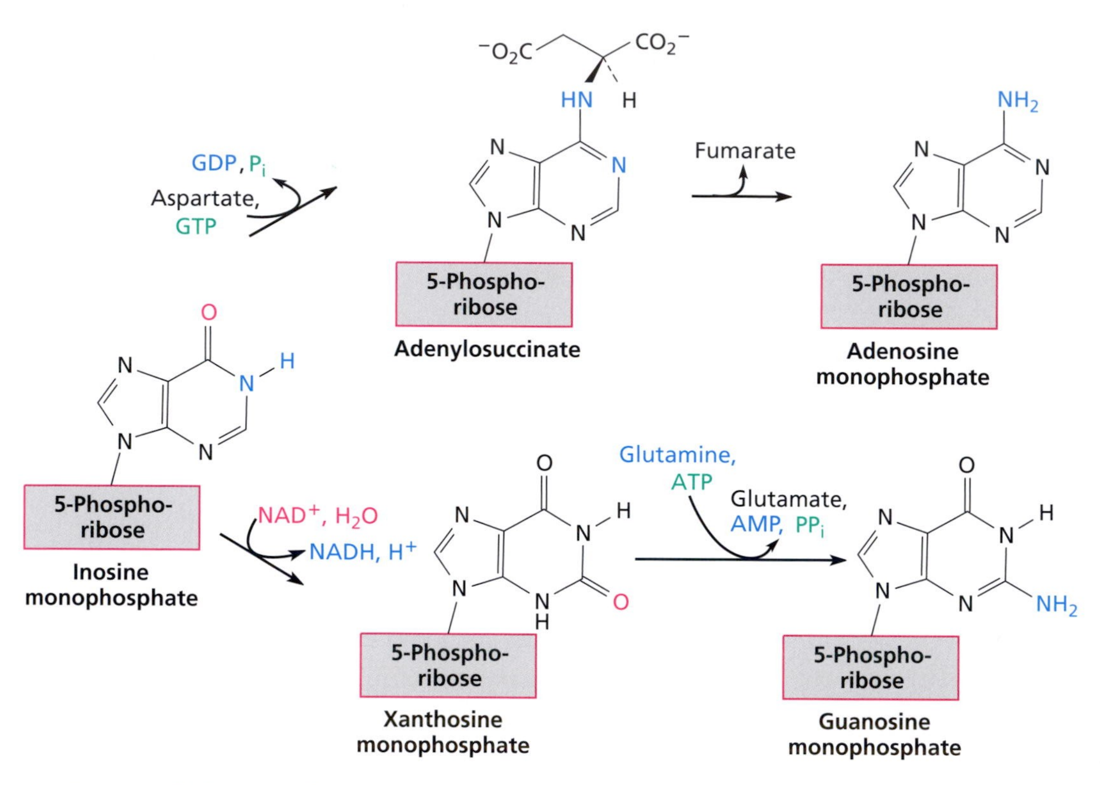

FIGURE 6.13 Pathway for the conversion of inosine monophosphate to adenosine monophosphate and guanosine monophosphate.

GTP as coenzyme, and is mechanistically analogous to the conversion of UTP to CTP discusssed at the end of Section 6.2. The second step is mechanistically similar to the eighth step in IMP synthesis and is catalyzed by the same adenylosuccinate lyase enzyme.[20]

Guanosine monophosphate is also synthesized from IMP in two steps: oxidation and hydrolysis to form xanthosine monophosphate (XMP), and amination. The oxidation is catalyzed by IMP dehydrogenase[23] and uses NAD^+ as coenzyme. As shown in Figure 6.14, a thiol group on the enzyme first reacts with the purine ring in a conjugate addition reaction, and the tetrahedral intermediate transfers hydride ion to NAD^+. Addition of water then replaces the thiol by —OH, and tautomerization of the product gives XMP. Conversion of XMP to GMP in the second step is catalyzed by GMP synthetase[24] and is mechanistically analogous to step 2 in the urea cycle, in which citrulline reacts with the amino group of aspartate to give argininosuccinate through an acyl adenosyl phosphate intermediate (Section 5.2, Figure 5.3).

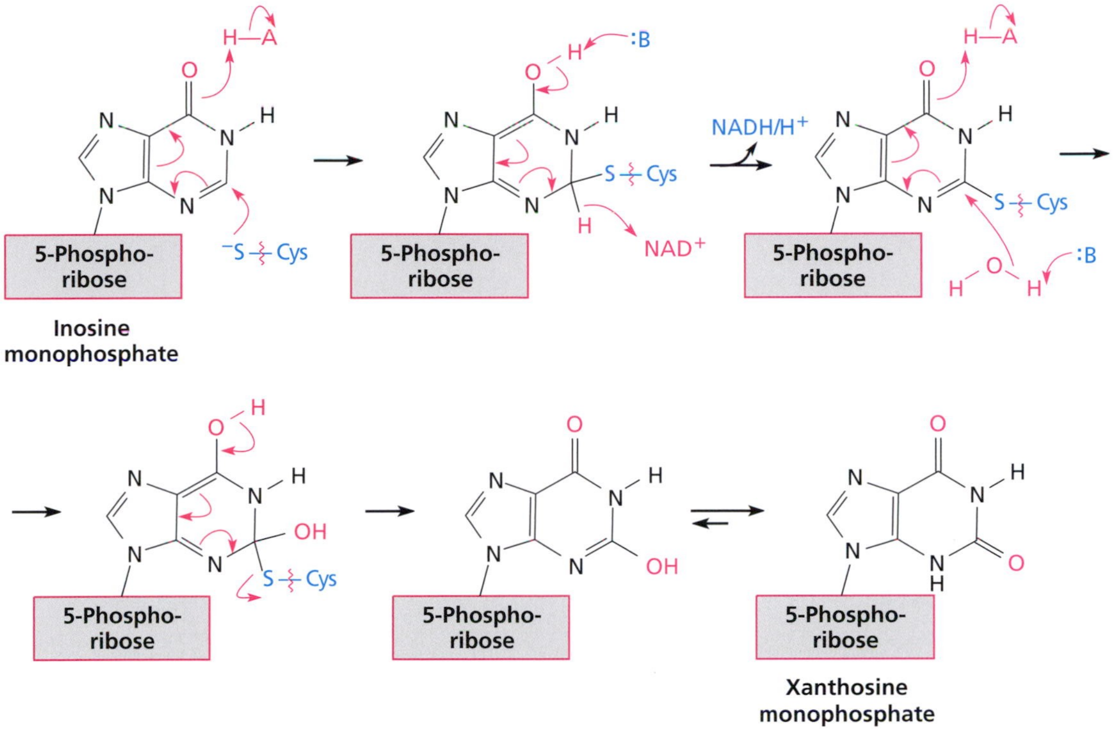

FIGURE 6.14 Mechanism of the oxidation of inosine monophosphate to xanthosine monophosphate.

6.4 Biosynthesis of Deoxyribonucleotides

Deoxyadenosine, Deoxyguanosine, Deoxycytidine, and Deoxyuridine Diphosphates

The deoxyribonucleoside diphosphates dADP, dGDP, dCDP, and dUDP arise biosynthetically through deoxygenation of the corresponding ribonucleoside diphosphates catalyzed by ribonucleotide reductase.[25, 26] Three classes of ribonucleotide reductases are known and are used by different organisms. All are metalloenzymes that use an active-site thiyl radical, but they differ in their mechanisms of radical generation and the nature of the metal species at their active sites. The mechanism of the deoxygenation reaction catalyzed by the non-heme Fe(III) enzyme in eukaryotes is shown in Figure 6.15.

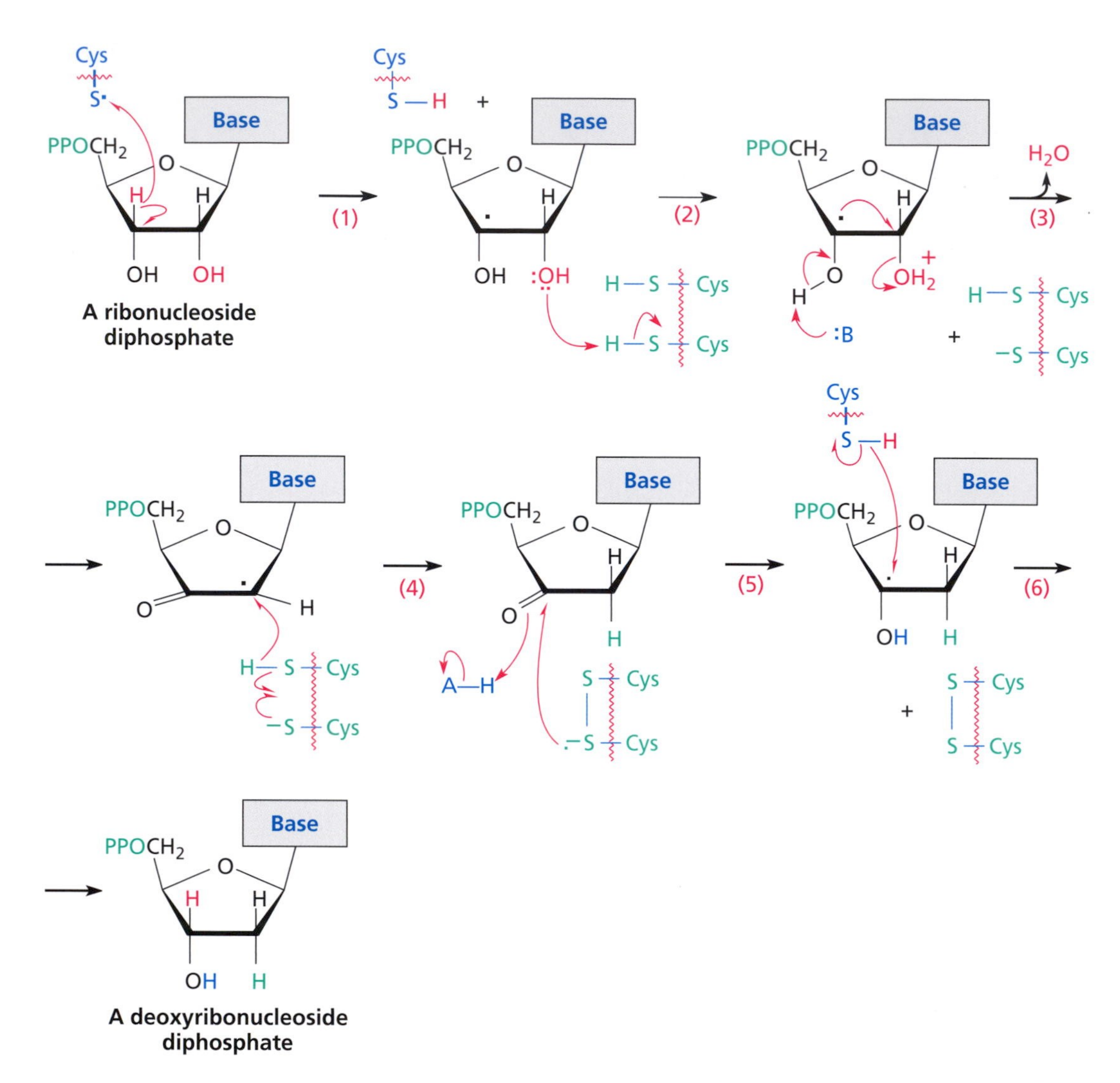

FIGURE 6.15 Mechanism of reduction of ribonucleoside diphosphates to deoxyribonucleoside diphosphates in a reaction catalyzed by ribonucleotide reductase.

Step 1. Hydrogen abstraction Ribonucleotide reduction begins with abstraction of the hydrogen atom at C3 by a thiyl radical center on a cysteine residue in the enzyme.

Steps 2–3. Dehydration The radical is protonated on the —OH group at C2 by one of a pair of cysteine residues, and the protonated alcohol loses water in an S_N1-like reaction. At the same time, the —OH group at C3 is deprotonated by a glutamate residue acting as a base, giving a neutral α keto radical as product.

Step 4. Hydrogen addition The α keto radical is reduced by addition of a hydrogen atom to the radical center at C2, yielding a neutral ketone. The second of the pair of cysteine residues is the hydrogen-atom donor, and the reaction occurs with formation of a sulfur anion–radical containing a disulfide bond between the two cysteines.

Step 5. Electron transfer The sulfur anion–radical transfers an electron to the nearby carbonyl group at C3, and the carbonyl oxygen is protonated to give a hydroxy radical.

Step 6. Hydrogen addition The hydrogen atom abstracted from C3 by a cysteine residue in the first step is re-added from the same side of the ribose ring in the final step to give the deoxyribonucleotide product. Following this final step, the disulfide produced in step 5 is reduced back to a thiol pair to regenerate active enzyme.

Thymidine Monophosphate

Thymidine monophosphate (dTMP or just TMP), the only DNA monomer without a direct RNA counterpart, is biosynthesized from deoxyuridine monophosphate (dUMP) in a complex process catalyzed by thymidylate synthase.[27, 28] 5,10-Methylenetetrahydrofolate (Figure 5.6) is the methyl donor, and the mechanism of the reaction is shown in Figure 6.16. An X-ray crystal structure of the enzyme–substrate active site is shown in Figure 6.17.

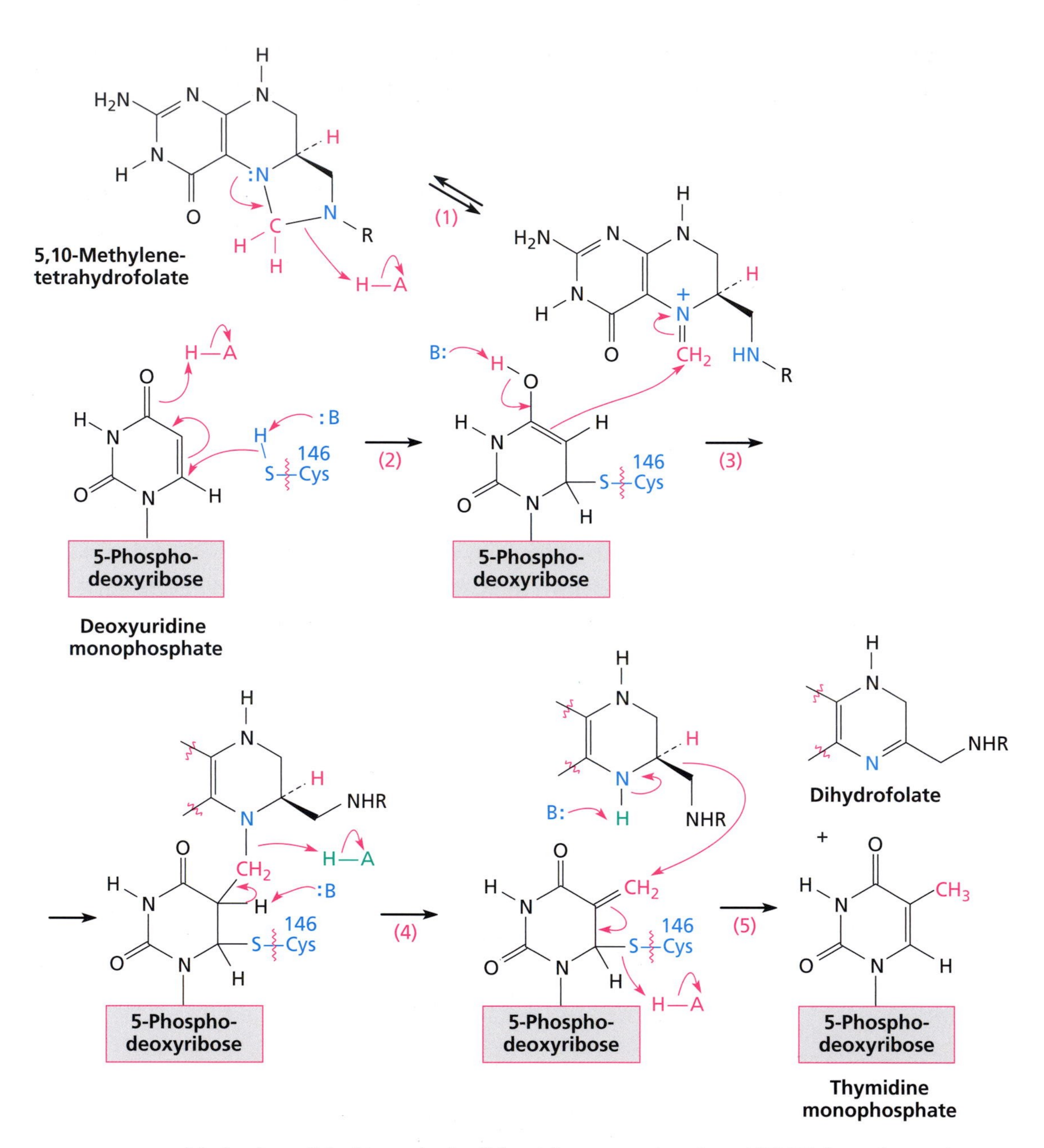

FIGURE 6.16 Mechanism of the biosynthesis of thymidine monophosphate (dTMP) from deoxyuridine monophosphate (dUMP), catalyzed by thymidylate synthase.

Step 1. Preliminary equilibrium 5,10-Methylenetetrahydrofolate undergoes reversible opening of the five-membered ring in a preliminary equilibrium to give an iminium ion.

Steps 2–3. Addition of cysteine and reaction with methylene-THF A cysteine residue (Cys-146) in the enzyme adds to the double bond in the uracil ring of dUMP in a conjugate addition reaction.The enol product then adds to the iminium ion of 5,10-methylene-THF.

Step 4. Elimination of THF Base-catalyzed elimination of tetrahydrofolate from the adduct generates a new unsaturated carbonyl group on the uracil ring.

Step 5. Reduction Tetrahydrofolate transfers a hydride ion to the uracil ring in a conjugate addition reaction, and the resulting enolate ion expels the cysteine thiolate ion, forming dTMP plus dihydrofolate (DHF). The dihydrofolate is converted back to 5,10-methylenetetrahydrofolate by a two-step pathway that involves initial reduction of DHF to THF by NADPH, followed by transfer of the serine —CH_2OH group (Figure 5.6).

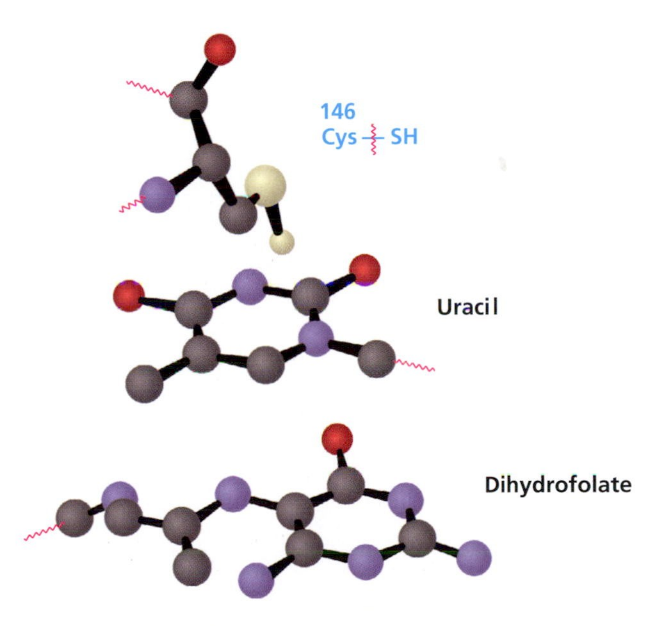

FIGURE 6.17 An X-ray crystal structure of the active site in the enzyme–substrate complex of thymidylate synthase. Cys-146 adds to the double bond of the uracil ring, and a methylene group is transferred from methylene-THF.

References

1. Betts, L.; Xiang, S.; Short, S. A.; Wolfenden, R.; Carter, C. W., Jr., "Cytidine Deaminase. The 2.3 Å Crystal Structure of an Enzyme: Transition-State Analog Complex," *J. Mol. Biol.*, **1994**, *235*, 635–56.
2. Dobritzsch, D.; Ricagno, S.; Schneider, G.; Schnackerz, K. D.; Lindqvist, Y., "Crystal Structure of the Productive Ternary Complex of Dihydropyrimidine Dehydrogenase with NADPH and 5-Iodouracil," *J. Biol. Chem.*, **2002**, *277*, 13155–13166.
3. Stockert, A. L.; Shinde, S. S.; Anderson, R. F.; Hille, R., "The Reaction Mechanism of Xanthine Oxidase: Evidence for Two-Electron Chemistry Rather Than Sequential One-Electron Steps," *J. Amer. Chem. Soc.*, **2002**, *124*, 14554–14555.
4. Hille, R., "The Mononuclear Molybdenum Enzymes," *Chem. Rev.*, **1996**, *96*, 2757–2816.
5. Holden, H. M.; Thoden, J. B.; Raushel, F. M., "Carbamoyl Phosphate Synthetase. An Amazing Biochemical Odyssey from Substrate to Product," *Cell. Mol. Life Sci.*, **1999**, *56*, 507–522.
6. Thoden, J. B.; Phillips, G. N., Jr.; Neal, T. M.; Raushel, F. M.; Holden, H. M., "Molecular Structure of Dihydroorotase: A Paradigm for Catalysis through the Use of a Binuclear Metal Center," *Biochemistry*, **2001**, *40*, 6989–6997.
7. Jiang, W.; Locke, G.; Harpel, M. R.; Copeland, R. A.; Marcinkeviciene, J., "Role of Lys100 in Human Dihydroorotate Dehydrogenase: Mutagenesis Studies and Chemical Rescue by External Amines," *Biochemistry*, **2000**, *39*, 7990–7997.
8. Palfey, B. A.; Bjoernberg, O.; Jensen, K. F., "Insight into the Chemistry of Flavin Reduction and Oxidation in *Escherichia coli* Dihydroorotate Dehydrogenase Obtained by Rapid Reaction Studies," *Biochemistry*, **2001**, *40*, 4381–4390.
9. Tao, W.; Grubmeyer, C.; Blanchard, J. S., "Transition-State Structure of *Salmonella typhimurium* Orotate Phosphoribosyltransferase," *Biochemistry*, **1996**, *35*, 14–21.
10. Begley, T. P.; Appleby, T. C.; Ealick, S. E., "The Structural Basis for the Remarkable Catalytic Proficiency of Orotidine 5′-Monophosphate Decarboxylase," *Curr. Opin. Struct. Biol.*, **2000**, *10*, 711–718.
11. Iyengar, A.; Bearne, S. L., "Aspartate-107 and Leucine-109 Facilitate Efficient Coupling of Glutamine Hydrolysis to CTP Synthesis by *Escherichia coli* CTP Synthase," *Biochem. J.*, **2003**, *369*, 497–507.
12. Smith, J. L., "Glutamine PRPP Amidotransferase: Snapshots of an Enzyme in Action," *Curr. Opin. Struct. Biol.*, **1998**, *8*, 686–694.
13. Wang, W.; Kappock, T. J.; Stubbe, J.; Ealick, S. E., "X-ray Crystal Structure of Glycinamide Ribonucleotide Synthetase from *Escherichia coli*," *Biochemistry*, **1998**, *37*, 15647–15662.
14. Zhang, Y.; Desharnais, J.; Greasley, S. E.; Beardsley, G. P.; Boger, D. L.; Wilson, I. A., "Crystal Structures of Human GAR Tfase at Low and High pH and with Substrate β-GAR," *Biochemistry*, **2002**, *41*, 14206–14215.

15. Li, C.; Kappock, T. J.; Stubbe, J.; Weaver, T. M.; Ealick, S. E., "X-ray Crystal Structure of Aminoimidazole Ribonucleotide Synthetase (PurM), from the *Escherichia coli* Purine Biosynthetic Pathway at 2.5 Å Resolution," *Structure*, **1999**, *7*, 1155–1166.
16. Firestine, S. M.; Poon, S.-W.; Mueller, E. J.; Stubbe, J.; Davisson, V. J., "Reactions Catalyzed by 5-Aminoimidazole Ribonucleotide Carboxylases from *Escherichia coli* and *Gallus gallus*: A Case for Divergent Catalytic Mechanisms?" *Biochemistry*, **1994**, *33*, 11927–11934.
17. Thoden, J. B.; Kappock, T. J.; Stubbe, J.; Holden, H. M., "Three-Dimensional Structure of N5-Carboxyaminoimidazole Ribonucleotide Synthetase: A Member of the ATP Grasp Protein Superfamily," *Biochemistry*, **1999**, *38*, 15480–15492.
18. Urusova, D. V.; Antonyuk, S. V.; Grebenko, A. I.; Lamzin, V. S.; Melik-Adamyan, V. R., "X-ray Diffraction Study of the Complex of the Enzyme SAICAR Synthase with Substrate Analogues," *Cryst. Rpts.*, **2003**, *48*, 763–767.
19. Toth, E. A.; Yeates, T. O., "The Structure of Adenylosuccinate Lyase, an Enzyme with Dual Activity in the de novo Purine Biosynthetic Pathway," *Structure*, **2000**, *8*, 16–174.
20. Wolan, D. W.; Greasley, S. E.; Wall, M. J.; Benkovic, S. J.; Wilson, I. A., "Structure of Avian AICAR Transformylase with a Multisubstrate Adduct Inhibitor β-DADF Identifies the Folate Binding Site," *Biochemistry*, **2003**, *42*, 10904–10914.
21. Vergis, J. M.; Bulock, K. G.; Fleming, K. G.; Beardsley, G. P., "Human 5-Aminoimidazole-4-carboxamide Ribonucleotide Transformylase/Inosine 5′-Monophosphate Cyclohydrolase. A Bifunctional Protein Requiring Dimerization for Transformylase Activity but Not for Cyclohydrolase Activity," *J. Biol. Chem.*, **2001**, *276*, 7727–7733.
22. Poland, B. W.; Fromm, H. J.; Honzatko, R. B., "Crystal Structures of Adenylosuccinate Synthetase from *Escherichia coli* Complexes with GDP, IMP, Hadacidin, NO_3^-, and Mg^{2+}," *J. Mol. Biol.*, **1996**, *264*, 1013–1027.
23. Kerr, K. M.; Digits, J. A.; Kuperwasser, N.; Hedstrom, L., "Asp338 Controls Hydride Transfer in *Escherichia coli* IMP Dehydrogenase," *Biochemistry*, **2000**, *39*, 9804–9810.
24. Tesmer, J. J. G.; Klem, T. J.; Deras, M. L.; Davisson, V. J.; Smith, J. L., "The Crystal Structure of GMP Synthetase Reveals a Novel Catalytic Triad and Is a Structural Paradigm for Two Enzyme Families," *Nature Struct. Biol.*, **1996**, *3*, 74–86.
25. Licht, S.; Stubbe, J., "Mechanistic Investigations of Ribonucleotide Reductases," Editor: Poulter, C. D., *Comprehensive Natural Products Chemistry*, **1999**, *5*, 163–203, Elsevier Science B.V., Amsterdam, Netherlands.
26. Jordan, A.; Reichard, P., "Ribonucleotide Reductases," *Ann. Rev. Biochem.*, **1998**, *67*, 71–98.
27. Finer-Moore, J. S.; Santi, D. V.; Stroud, R. M., "Lessons and Conclusions from Dissecting the Mechanism of a Bisubstrate Enzyme: Thymidylate Synthase Mutagenesis, Function, and Structure," *Biochemistry*, **2003**, *42*, 248–256.
28. Hyatt, D. C.; Maley, F.; Montfort, W. R., "Use of Strain in a Stereospecific Catalytic Mechanism: Crystal Structures of *Escherichia coli* Thymidylate Synthase Bound to FdUMP and Methylenetetrahydrofolate," *Biochemistry*, **1997**, *36*, 4585–4594.

Problems

6.1 Write a mechanism for hydrolysis of β-ureidopropionate to give β-alanine, a step in uracil catabolism (Figure 6.1).

H_2O → NH_4^+, CO_2

***β*-Ureidopropionate** **β-Alanine**

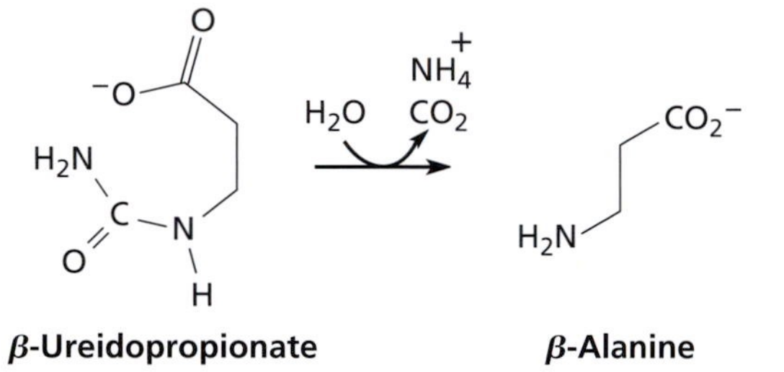

6.2 Write a mechanism for the oxidation of malonic semialdehyde to give malonyl CoA, the final step in uracil catabolism (Figure 6.1).

CoASH, NAD^+ → NADH, H^+

Malonic semialdehyde **Malonyl CoA**

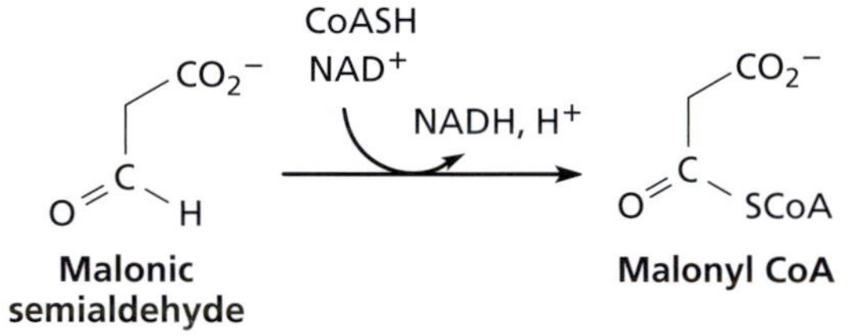

6.3 Show the steps and identify the intermediates in the catabolism of thymine to give methylmalonyl CoA.

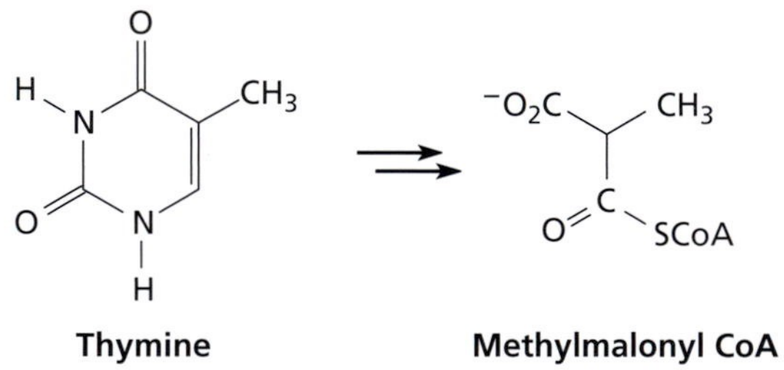

Thymine **Methylmalonyl CoA**

6.4 Write a mechanism for the formylation of glycinamide ribonucleotide, the third step in inosine biosynthesis (Figure 6.10).

10-Formyl-THF

5-Phospho-ribose

Glycinamide ribonucleotide

Formylglycinamide ribonucleotide

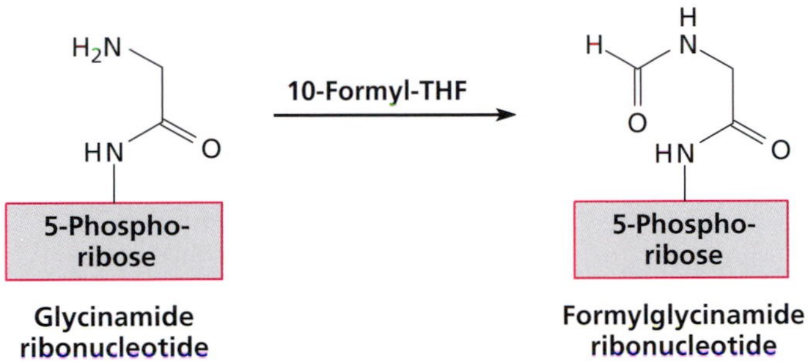

6.5 Write a mechanism for the formation of aminoimidazole ribonucleotide from formylglycinamidine ribonucleotide, the fifth step in inosine biosynthesis (Figure 6.10).

ATP ADP P_i

Formylglycinamidine ribonucleotide

Aminoimidazole ribonucleotide

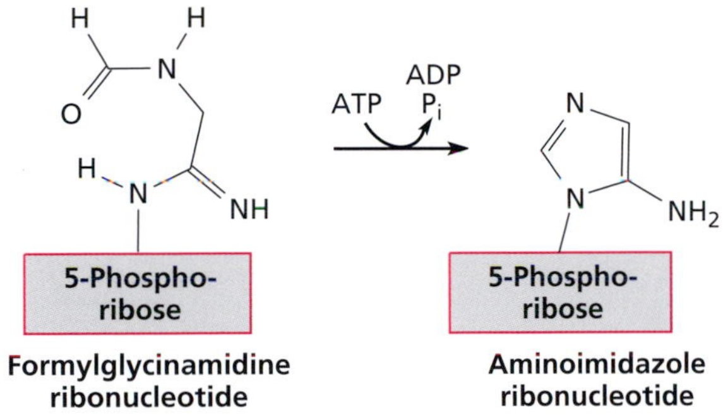

6.6 Write a mechanism for the formation of adenylosuccinate from inosine monophosphate.

Aspartate, GTP GDP, P_i

Inosine monophosphate

Adenylosuccinate

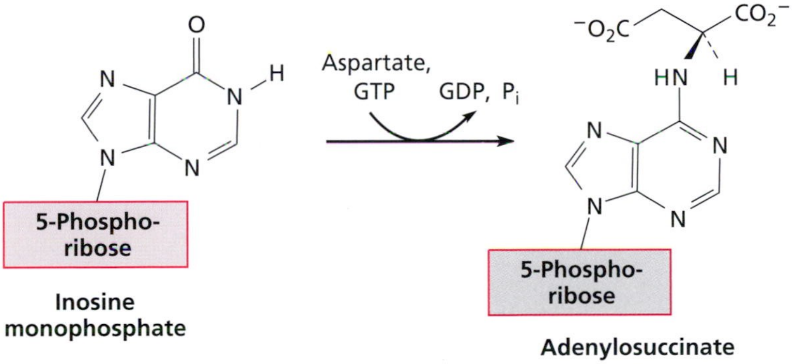

6.7 One route for the biosynthesis of NAD^+ involves the following reaction of quinolinate. Propose a mechanism for the reaction.

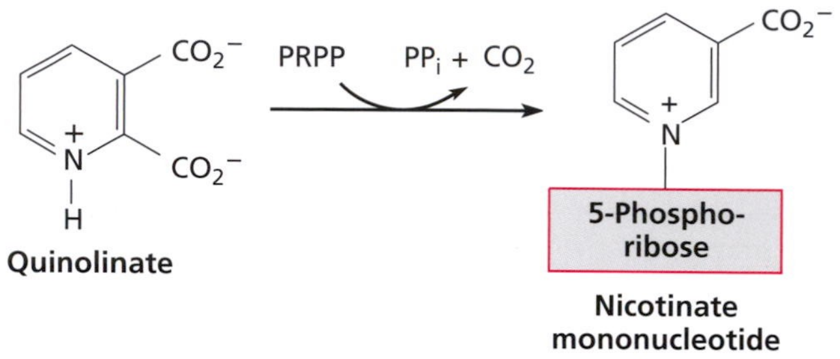

Quinolinate

PRPP → $PP_i + CO_2$

5-Phosphoribose

Nicotinate mononucleotide

6.8 Guanosine triphosphate is converted by GTP cyclohydrolase II into the following monophosphate. Propose a mechanism.

PPPOCH$_2$

Guanosine triphosphate

$2\ H_2O$ → PP_i, HCO_2^-

$^{2-}O_3POCH_2$

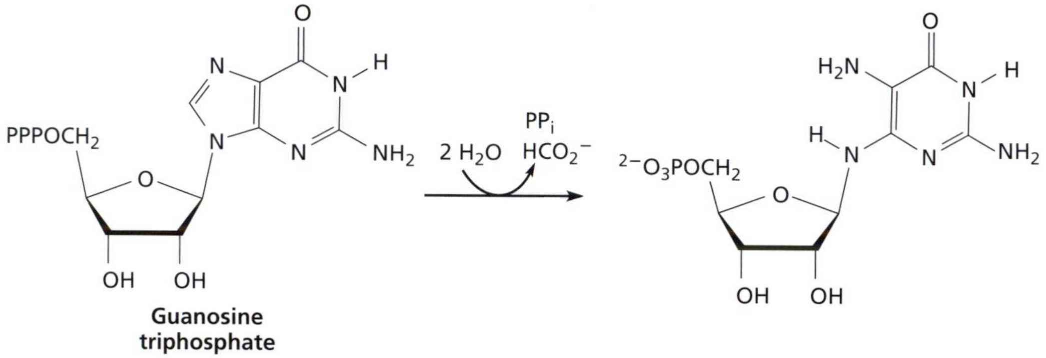

6.9 Guanosine triphosphate is converted by GTP cyclohydrolase I into dihydroneopterin triphosphate. Propose a mechanism.

PPPOCH$_2$

Guanosine triphosphate

$2\ H_2O$ → HCO_2^-

PPO

Dihydroneopterin triphosphate

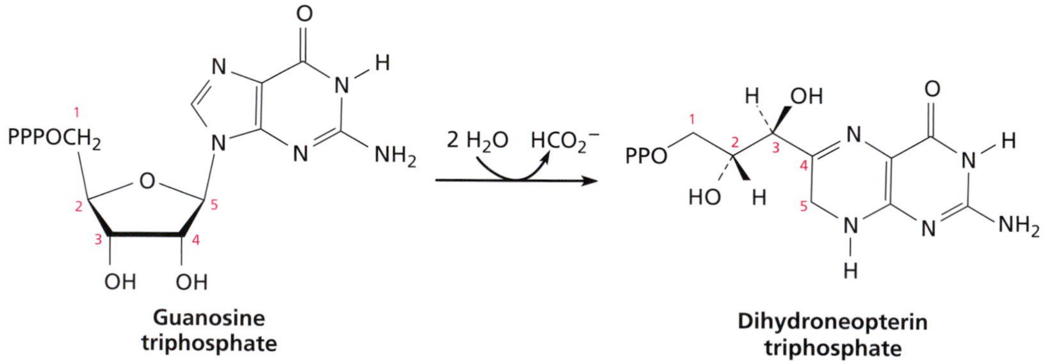

6.10 Retrieve the PDB coordinate file for dihydropyrimidine dehydrogenase (Figure 6.3), and display the structure using the Swiss PDB viewer. (The PDB code is 1GTH.) What is the distance between the C4 carbon of NADPH and the C6 carbon of the pyrimidine substrate? How is the hydride equivalent transferred over this long distance?

6.11 Retrieve the PDB coordinate file for thymidylate synthase (Figure 6.17), and display the structure using the Swiss PDB viewer. (The PDB code is 1B02.) Draw the structure of the adduct of 5-fluoro-2′-deoxyuridine-5′-monophosphate with 5,10-methylene-5,6,7,8-tetrahydrofolate, and propose a mechanism for its formation.

7 Biosynthesis of Some Natural Products

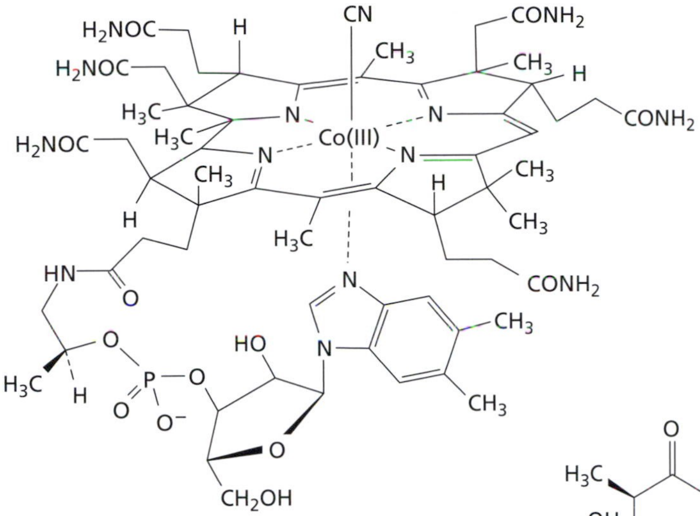

Cyanocobalamin
(Vitamin B_{12})

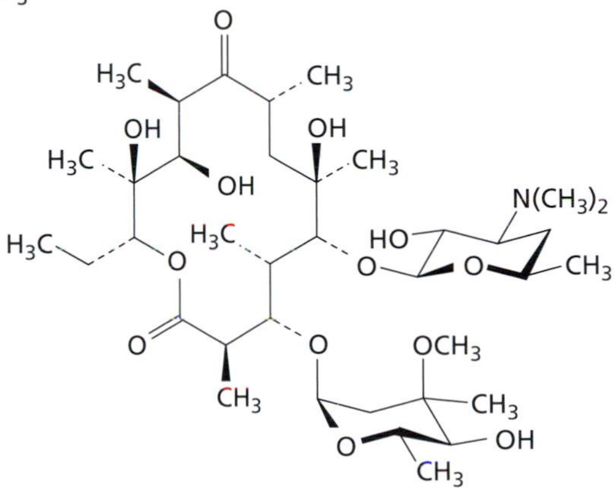

Important complex molecules such as vitamin B_{12} and the antibiotic erythromycin are biosynthesized using many of the same reactions we've seen up to this point.

References
Problems

In addition to the ubiquitous triacylglycerols, carbohydrates, proteins, and nucleic acids, living organisms also contain a vast diversity of other substances generally grouped under the heading "natural products." The term **natural product** really refers to *any* naturally occurring substance but is generally taken to mean a secondary metabolite—a small molecule that has no essential metabolic function in the producing organism and is not classified by structure. It's thought that the main function of most secondary metabolites is to increase the likelihood of an organism's survival by repelling or attracting other organisms.[1]

There is no rigidly defined scheme for classifying natural products—their immense diversity in structure, function, and biosynthesis is simply too great to allow them to fit neatly into a few simple categories. In practice, however, workers in the field often speak of five main classes of natural products: terpenoids, nonribosomal polypeptides, alkaloids, fatty acid–derived substances and polyketides, and enzyme cofactors. Representative structures from each class are shown in Figure 7.1.

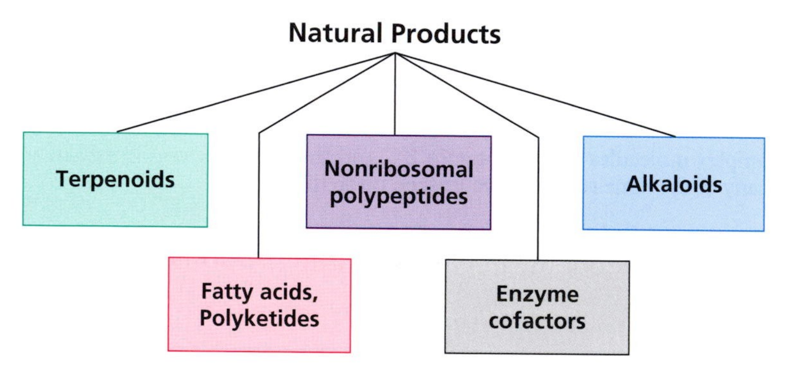

- *Terpenoids*, as discussed previously in Sections 2.2 and 3.5, are a vast group of substances derived biosynthetically from isopentenyl diphosphate. epiAristolochene (Section 3.5, Figure 3.28) and lanosterol (Section 3.6, Figure 3.33) are examples.
- *Nonribosomal polypeptides* are peptide-like compounds biosynthesized by a complex multifunctional synthetase without direct RNA transcription. We'll look at the penicillins as examples in Section 7.1.

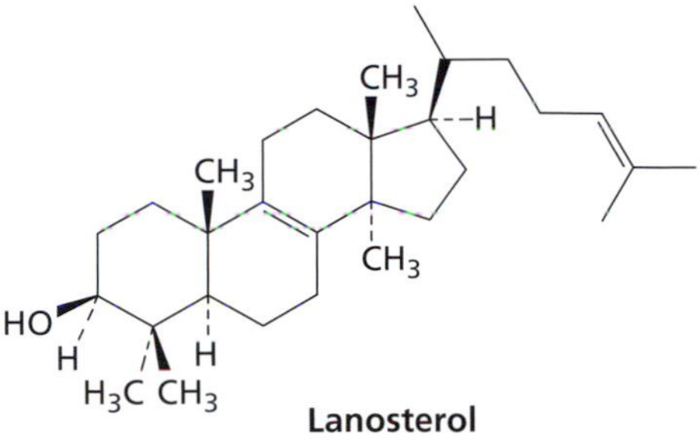

Lanosterol
(a terpenoid)

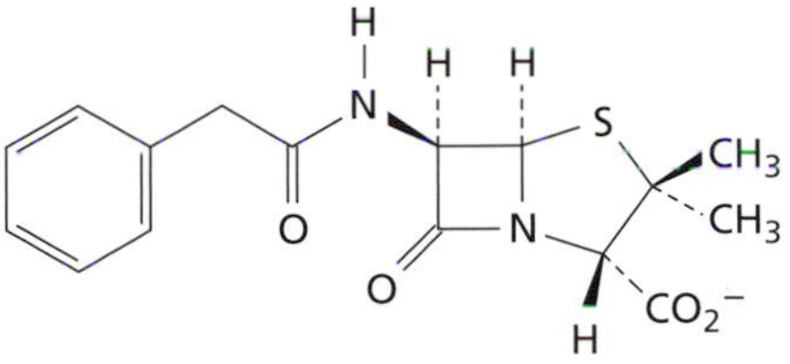

Benzylpenicillin
(a nonribosomal peptide)

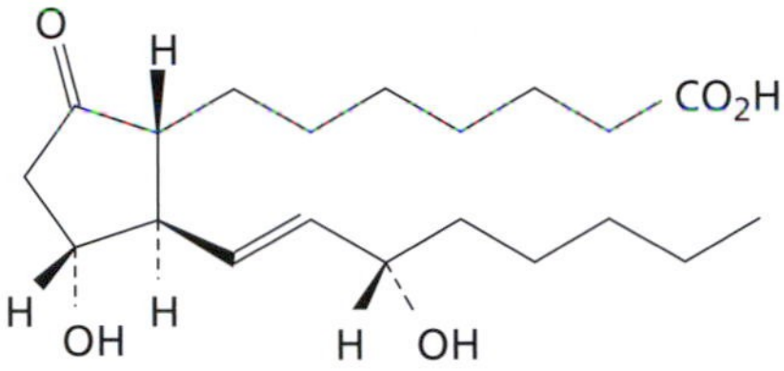

Prostaglandin E_1
(a fatty acid–derived eicosanoid)

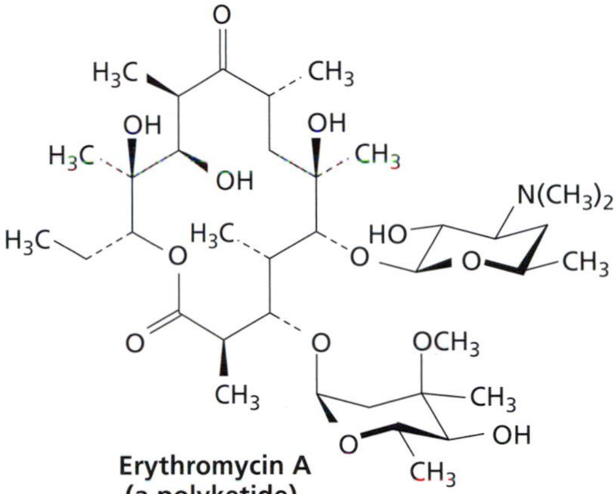

Erythromycin A
(a polyketide)

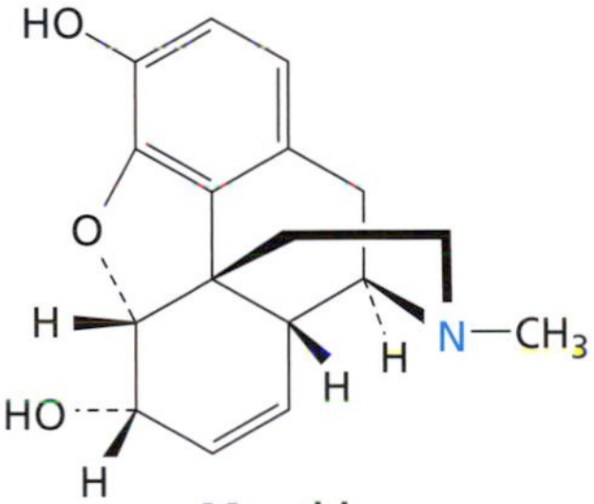

Morphine
(an alkaloid)

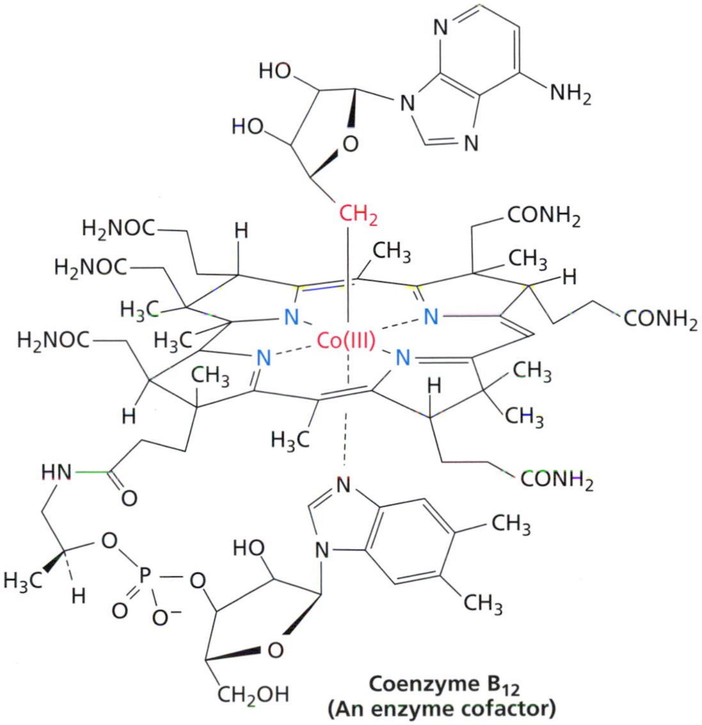

Coenzyme B_{12}
(An enzyme cofactor)

FIGURE 7.1 Structures of some representative natural products.

- *Alkaloids*, like terpenoids, are also a large and diverse class of compounds. They contain a basic amine group in their structure and are derived biosynthetically from amino acids. We'll look at morphine biosynthesis as an example in Section 7.2.
- *Fatty acid–derived substances* and *polyketides*, of which more than 10,000 are known, are biosynthesized by the joining together of simple acyl precursors such as acetyl CoA, propionyl CoA, and methylmalonyl CoA. Natural products derived from fatty acids generally have most of the oxygen atoms removed, but polyketides often have many oxygen substituents remaining on alternating carbons. We'll look at eicosanoid biosynthesis as an example of a fatty acid–derived natural-product (Section 7.3) and at erythromycin biosynthesis as an example of a polyketide (Section 7.4).
- *Enzyme cofactors* don't fit one of the more general categories and are usually classed separately. We'll look at coenzyme B_{12} biosynthesis as an example in Section 7.5.

All we can do in this chapter is to scratch the surface of natural-products chemistry. This chapter is not intended to be a comprehensive study; its purpose is to provide a brief introduction to an enormous, intellectually challenging, and immensely important area of modern biochemistry, perhaps tempting you to learn more on your own.

7.1 Biosynthesis of Penicillins and Cephalosporins

The story of penicillin has been told many times: In the late summer of 1928, the Scottish bacteriologist Alexander Fleming went on vacation, leaving in his lab a culture plate recently inoculated with the bacterium *Staphylococcus aureus.* While he was away, a remarkable chain of events occurred. First, a nine-day cold spell lowered the laboratory temperature to a point where the *Staphylococcus* on the plate could not grow. During this time, spores from a colony of the mold *Penicillium notatum* being grown on the floor below wafted up into Fleming's lab and landed in the culture plate. The temperature then rose, and both *Staphylococcus* and *Penicillium* began to grow. On returning from vacation, Fleming discarded the plate into a tray of antiseptic, intending to sterilize it. By chance, however, the plate did not sink deeply enough into the antiseptic, and when Fleming happened to glance at it a few days later, he noticed that the growing *Penicillium* mold appeared to dissolve the colonies of staphylococci.

Fleming realized that the *Penicillium* mold must be producing a chemical that killed the *Staphylococcus* bacteria, and he spent several years trying unsuccessfully to isolate the substance. Finally, in 1939, the Australian pathologist Howard Florey and the German refugee Ernst Chain managed to isolate the active substance, now called *benzylpenicillin*, or penicillin G. The dramatic ability of penicillin to cure bacterial infections in humans was soon demonstrated, and by 1943 it was being used on a large scale.

Penicillin G is but one member of a large class of *β-lactam antibiotics*, compounds with a four-membered lactam (cyclic amide) ring. The four-membered lactam ring is fused to a five-membered, sulfur-containing ring, and the carbon atom next to the lactam carbonyl group is bonded to an acylamino side chain, RCONH—, which can be varied in the laboratory to provide literally hundreds of penicillin analogs with different biological activity profiles. Further enzymatic transformations can also convert the penicillins into *cephalosporins*, another large class of β-lactam antibiotics characterized by the presence of a six-membered, sulfur-containing ring.

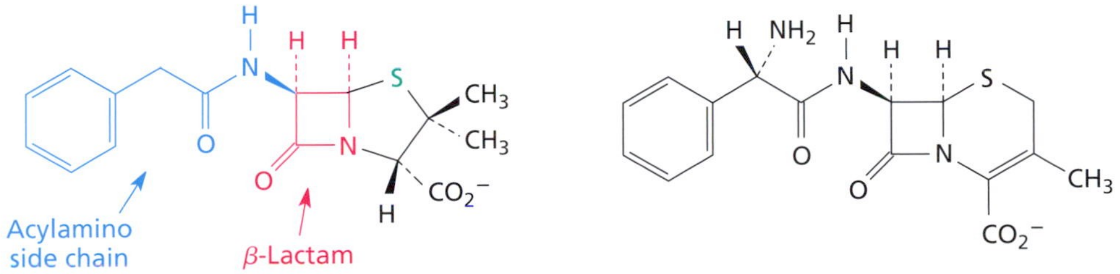

Benzylpenicillin (Penicillin G)

Cephalexin (a cephalosporin)

The biological activity of penicillins is due to the presence of the strained β-lactam ring, which reacts with and deactivates the transpeptidase needed to synthesize and repair bacterial cell walls. With the wall either incomplete or weakened, the bacterial cell ruptures and dies.

Penicillin is formed biosynthetically from three amino acids, L-cysteine, L-valine, and the nonprotein amino acid L-α-aminoadipate. Epimerization of L-valine to D-valine and condensation first gives the tripeptide L-δ-(α-aminoadipoyl)-L-cysteinyl-D-valine (ACV), and oxidative double cyclization yields isopenicillin N (IPN). Replacement of the L-δ-(α-aminoadipoyl) side chain by a variety of transamidation reactions then gives other penicillins. Cephalosporins are formed by epimerization of isopenicillin N to penicillin N, followed by expansion of the sulfur-containing ring to yield deacetoxycephalosporin C (Figure 7.2).

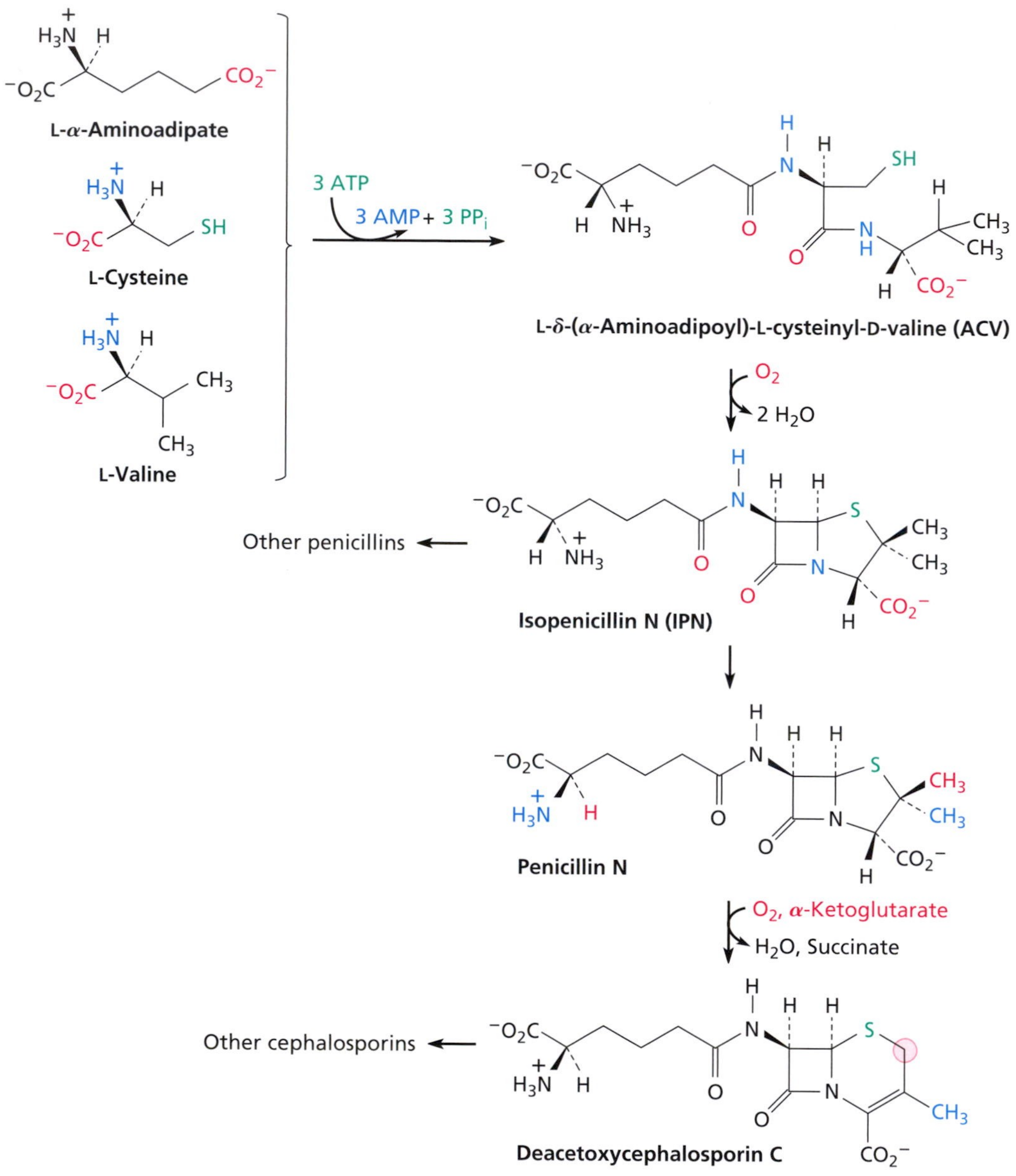

L-α-Aminoadipate

L-Cysteine

L-Valine

3 ATP

3 AMP + 3 PP_i

L-δ-(α-Aminoadipoyl)-L-cysteinyl-D-valine (ACV)

O_2

2 H_2O

Other penicillins ←

Isopenicillin N (IPN)

Penicillin N

O_2, α-Ketoglutarate

H_2O, Succinate

Other cephalosporins ←

Deacetoxycephalosporin C

FIGURE 7.2 An overview of penicillin and cephalosporin biosynthesis.

Penicillins

Both of the initial steps in penicillin synthesis—the amino acid couplings and the epimerization of L-valine to D-valine—occur on a single, multifunctional ACV synthetase.[2, 3] Present evidence suggests that the reaction takes place by the mechanism shown in Figure 7.3.

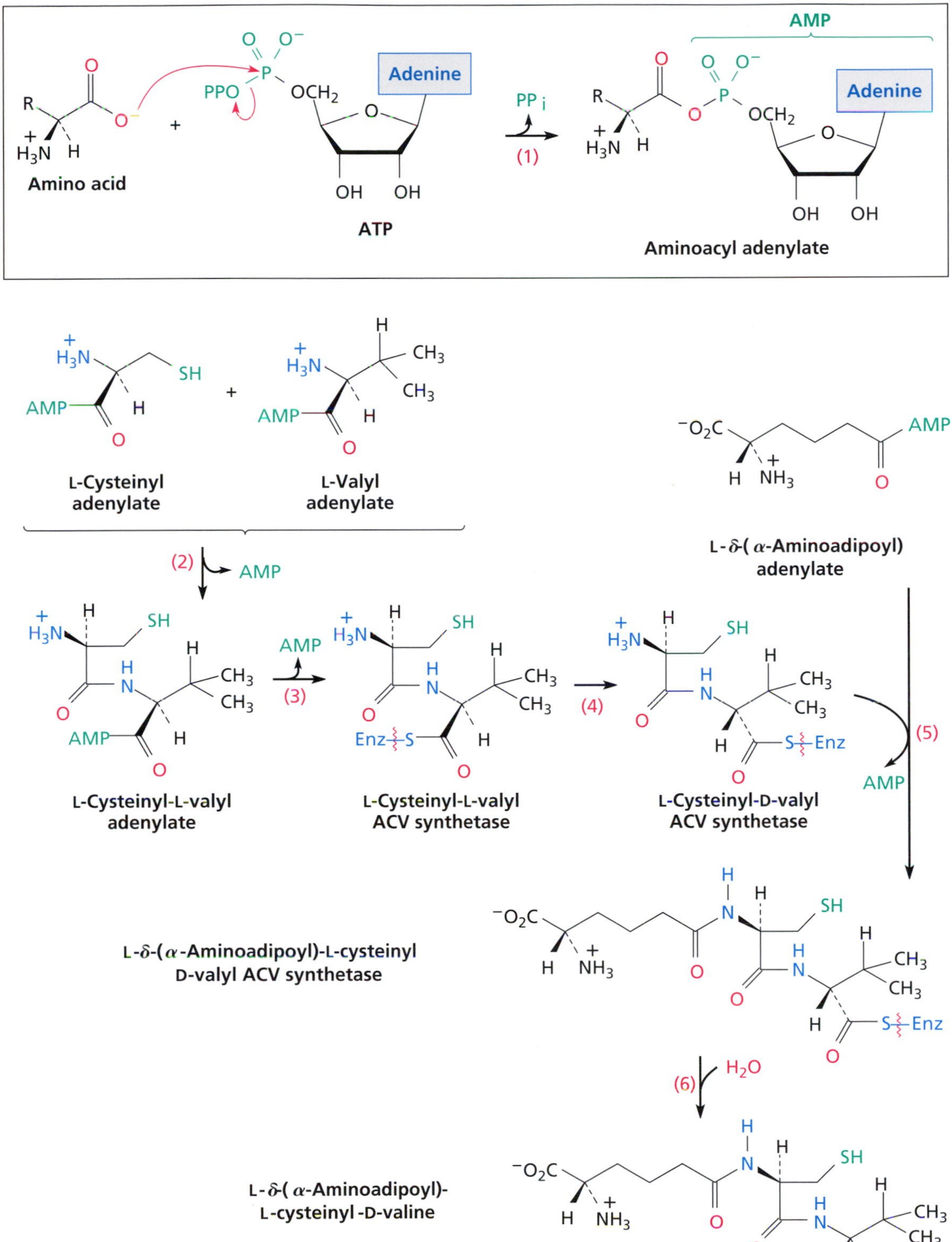

AMP

Adenine

Amino acid

ATP

PP_i

(1)

Aminoacyl adenylate

L-Cysteinyl adenylate

L-Valyl adenylate

L-δ-(α-Aminoadipoyl) adenylate

(2) AMP

L-Cysteinyl-L-valyl adenylate

(3) AMP

L-Cysteinyl-L-valyl ACV synthetase

(4)

L-Cysteinyl-D-valyl ACV synthetase

(5) AMP

L-δ-(α-Aminoadipoyl)-L-cysteinyl D-valyl ACV synthetase

(6) H_2O

L-δ-(α-Aminoadipoyl)-L-cysteinyl-D-valine

FIGURE 7.3 Mechanism of ACV biosynthesis from L-α-aminoadipate, L-cysteine, and L-valine.

Steps 1–2. Aminoacyl adenylate formation and coupling ACV biosynthesis begins with reaction of the three amino acids with ATP at different sites within the enzyme to give the corresponding aminoacyl adenylates. As usual, the reactions require Mg^{2+} as Lewis acid to neutralize the negative charges on the triphosphates. Cysteinyl and valyl adenylates then couple by nucleophilic acyl substitution of the valyl amino group onto the cysteinyl adenylate, giving L-cysteinyl-L-valyl adenylate.

Steps 3–4. Thioester formation and epimerization The relative timing of steps 3 and 4 is uncertain, but their combined result is to bind the L-cysteinyl-L-valyl adenylate to the enzyme and to epimerize the valyl residue from L to D. Binding the dipeptide to the enzyme occurs by nucleophilic acyl substitution on the acyl adenylate by the —SH group of a phosphopantetheine cofactor, with the phosphopantetheine itself bonded to an —OH group of a serine residue on the enzyme. Assuming that this binding step occurs first, it serves not only for translocation of the dipeptide intermediate to the next site in the enzyme but also to activate the valine carboxyl group for epimerization by making the α hydrogen more acidic and facilitating enolate formation.

Enz—SH / AMP / :B

L-Cysteinyl-L-valyl adenylate → **L-Cysteinyl-L-valyl ACV synthetase** ⇌ **L-Cysteinyl-D-valyl ACV synthetase**

Enz—SH = HSCH₂CH₂NHC(=O)CH₂CH₂NHC(=O)CH(OH)C(CH₃)₂CH₂OP(=O)(O⁻)O–Ser—

Phosphopantetheine

Steps 5–6. Coupling and hydrolysis Further coupling of L-cysteinyl-D-valyl-ACV synthetase by a nucleophilic acyl substitution reaction with L-δ-(α-aminoadipoyl) adenylate forms the ACV tripeptide, which is cleaved from the enzyme by hydrolysis.

Cyclization of the ACV tripeptide to give isopenicillin N is catalyzed by isopenicillin-N synthase (IPNS),[3, 4] a multifunctional, iron-dependent oxidase. As in the oxidation of phenylalanine to tyrosine (Section 5.3, Figure 5.28), an iron–oxo intermediate is involved, with the iron atom shuttling between various oxidation states. The mechanism is shown in Figure 7.4.

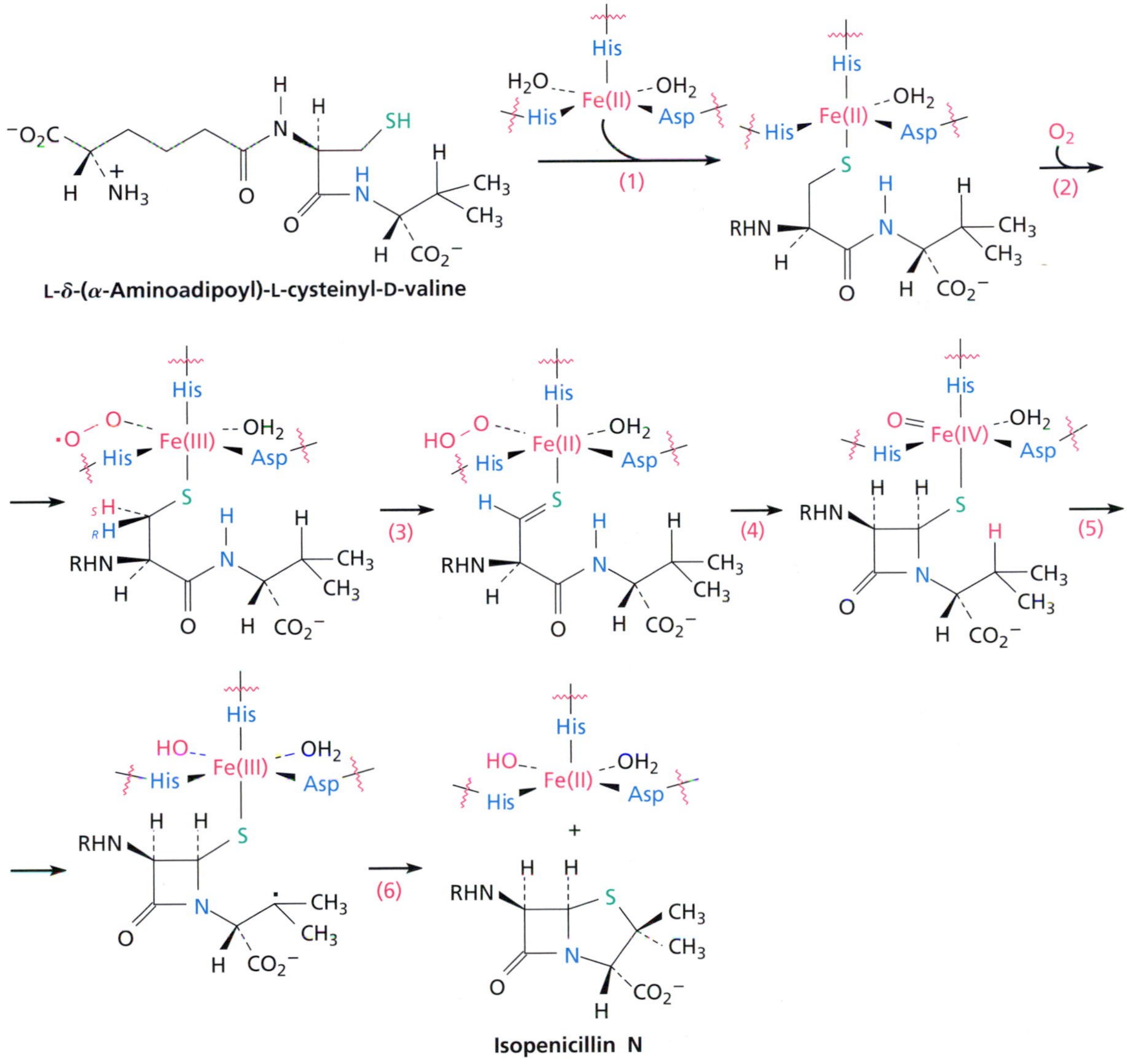

L-δ-(α-Aminoadipoyl)-L-cysteinyl-D-valine

Isopenicillin N

FIGURE 7.4 Pathway for isopenicillin N biosynthesis.

Steps 1–2. Formation and oxygenation of the iron complex The ACV tripeptide is converted to a thiolate ion, which complexes to an Fe(II) that is also ligated to two histidine residues and one aspartate residue in the enzyme. Reaction with O_2 forms a peroxy radical and oxidizes the iron to Fe(III). An X-ray crystal structure of the iron complex bonded to NO as a model for O_2 is shown in Figure 7.5.

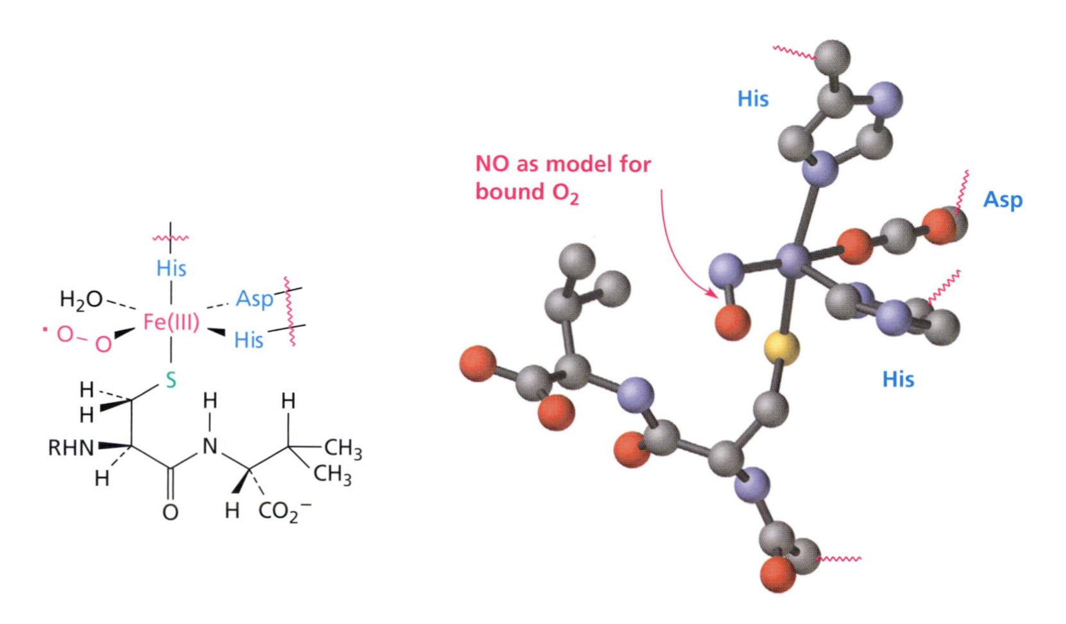

FIGURE 7.5 X-ray crystal structure of the IPNS iron complex bonded to NO as a model for O_2.

Steps 3–4. Oxidation and β-lactam formation The peroxy radical abstracts the *pro-S* hydrogen from the carbon atom adjacent to sulfur, forming a C=S double bond and reducing the iron to Fe(II). Nucleophilic addition to the C=S by the amide nitrogen atom from valine then forms the β-lactam ring, and expulsion of water oxidizes the iron atom to Fe(IV), giving an iron–oxo intermediate (Figure 7.6).

Steps 5–6. Oxidation and cyclization The iron–oxo intermediate abstracts a hydrogen atom from the valine residue, giving a radical that undergoes cycliza-

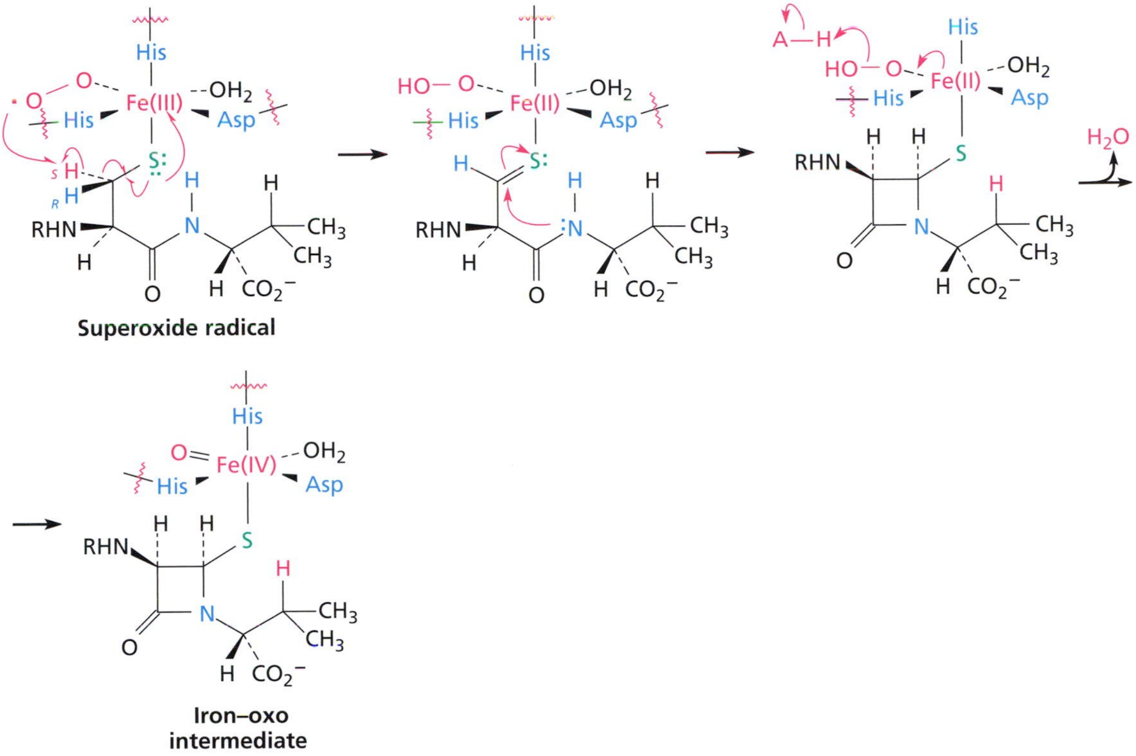

FIGURE 7.6 Mechanism of β-lactam formation.

tion by reaction on sulfur. At the same time, the iron atom is reduced back to Fe(II) and isopenicillin N is released from the enzyme.

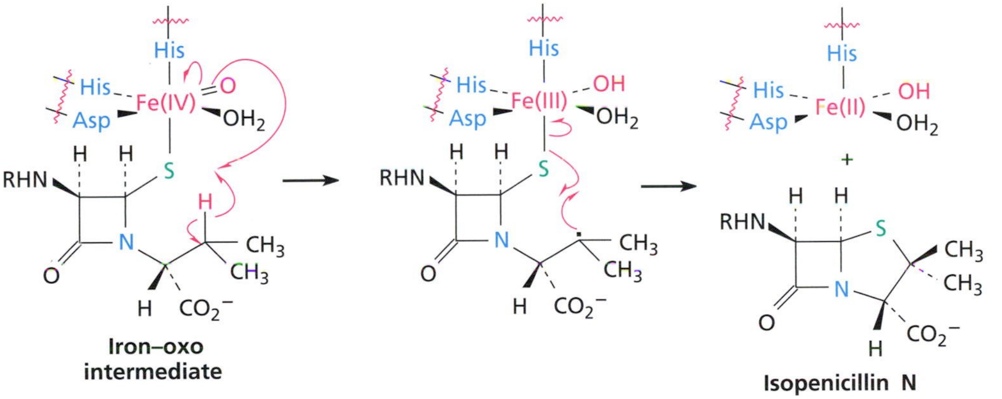

Cephalosporins

As noted earlier in this section, cephalosporins arise by an initial epimerization of isopenicillin N to penicillin N, followed by expansion of the sulfur-containing ring. Catalyzed by isopenicillin-N epimerase, the conversion of IPN to penicillin N is a PLP-dependent process that occurs by initial formation of a PLP–amino acid imine, followed by deprotonation of the α carbon, reprotonation, and PLP cleavage (Figure 7.7). Although cofactor-independent, base-catalyzed epimerization of amino acids can also occur (Section 5.5, Figure 5.33), the PLP-dependent reaction is much more common because PLP greatly increases the acidity of the α carbon (Section 5.3, Figure 5.14).

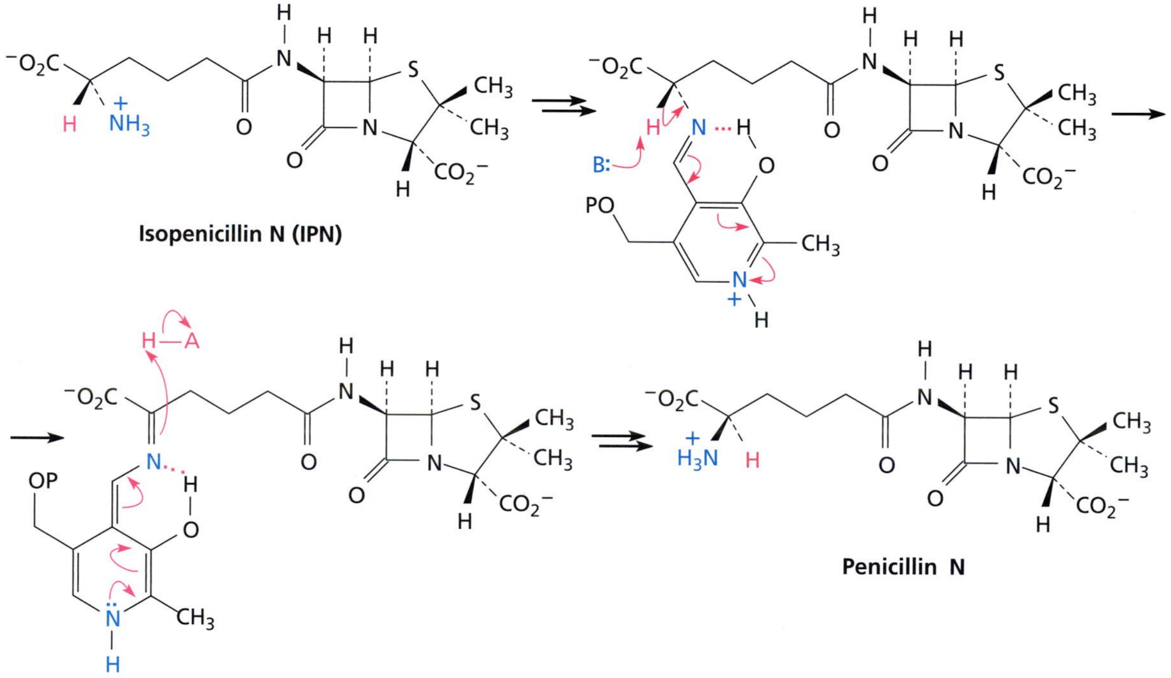

FIGURE 7.7 Mechanism of the PLP-dependent epimerization of isopenicillin N to penicillin N.

Ring expansion of penicillin N to give deacetoxycephalosporin C is catalyzed by deacetoxycephalosporin C synthase (DAOCS),[5, 6] as shown in Figure 7.8. DAOCS is a multifunctional, nonheme-iron, α keto acid–dependent oxidase (Section 5.3, Figure 5.29) that functions by a mechanism involving an iron–oxo intermediate. Formation of the iron–oxo intermediate proceeds by initial complexation of an Fe(II) species to α-ketoglutarate, followed by addition of O_2 and decarboxylation. One of the atoms in O_2 becomes the iron–oxo oxygen atom, and the other becomes a carboxyl oxygen atom in the succinate by-product released at the end. The α-ketoglutarate carboxyl is converted into CO_2.

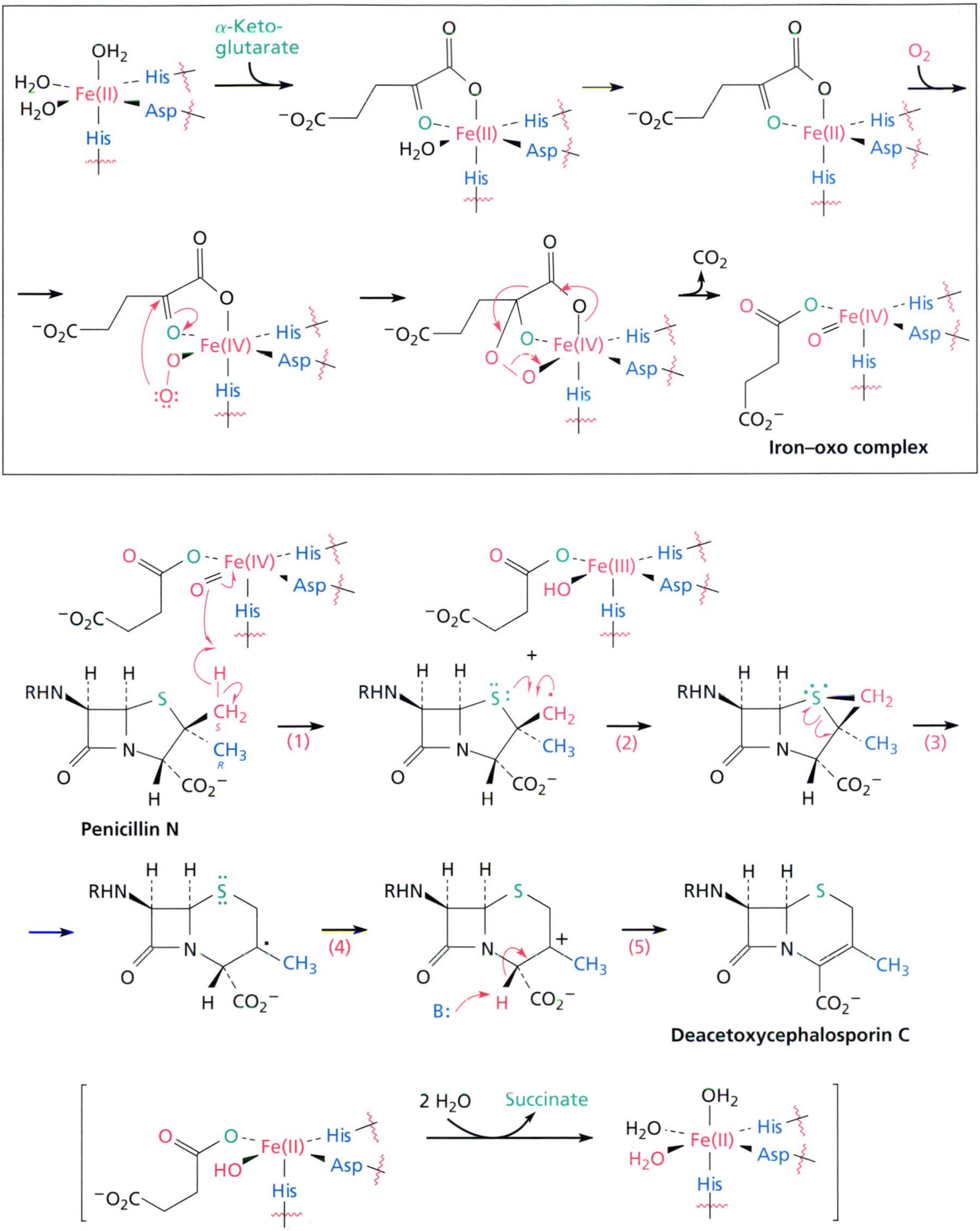

FIGURE 7.8 Mechanism of the biosynthesis of deacetoxycephalosporin C from penicillin N.

Once formed, the iron–oxo complex abstracts a hydrogen atom from the *pro-S* methyl group of penicillin N (Figure 7.8, step 1) to give a primary radical that cyclizes to a three-membered, sulfur-containing ring (step 2). Ring-opening then gives a tertiary radical (step 3), which is oxidized by the iron(III) complex to yield a tertiary carbocation (step 4). Loss of H^+ from the neighboring carbon gives deacetoxycephalosporin C (step 5).

7.2 Biosynthesis of Morphine

The opium poppy, *Papaver somniferum*, has been cultivated for more than 6000 years, and references to the so-called "plant of joy" appear in a Sumerian tablet dated to approximately 3400 B.C. Medical uses of the poppy have been known since the early 1500s when crude extracts, called *opium*, or *laudanum*, were used for the relief of pain. Morphine was the first pure compound to be isolated from opium, but its close relative codeine also occurs naturally. Codeine, which is the methyl ether of morphine and is converted to morphine in the body, is used in prescription cough medicines and as an analgesic. Heroin, another close relative of morphine, does not occur naturally but is synthesized by diacetylation of morphine.

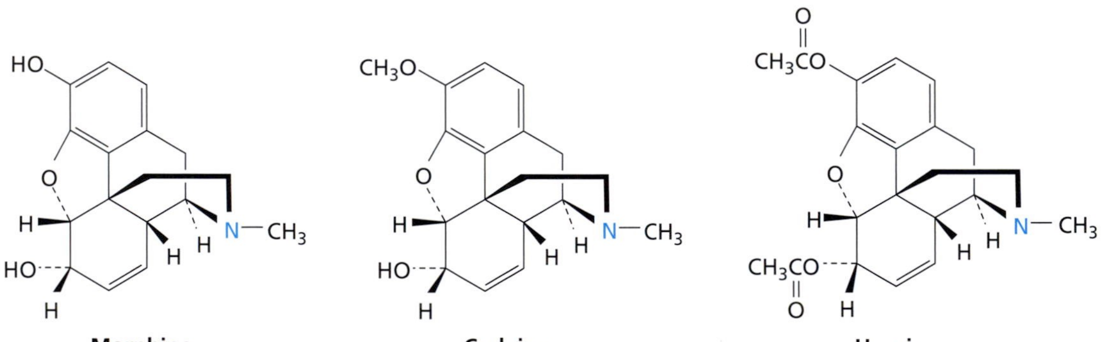

Morphine **Codeine** **Heroin**

Morphine and other natural products that contain a basic amine group are called **alkaloids**. Many thousands of alkaloids are known, with a diversity of structure rivaling that of the terpenoids (Section 3.5). Historically, the study of alkaloid chemistry, and of morphine in particular, was enormously influential in the development of organic chemistry as a science.

Morphine is biosynthesized from two tyrosine molecules.[7] One tyrosine is converted into dopamine, the second is converted into *p*-hydroxyphenylacetaldehyde, and the two are coupled. The pathway is summarized in Figure 7.9.

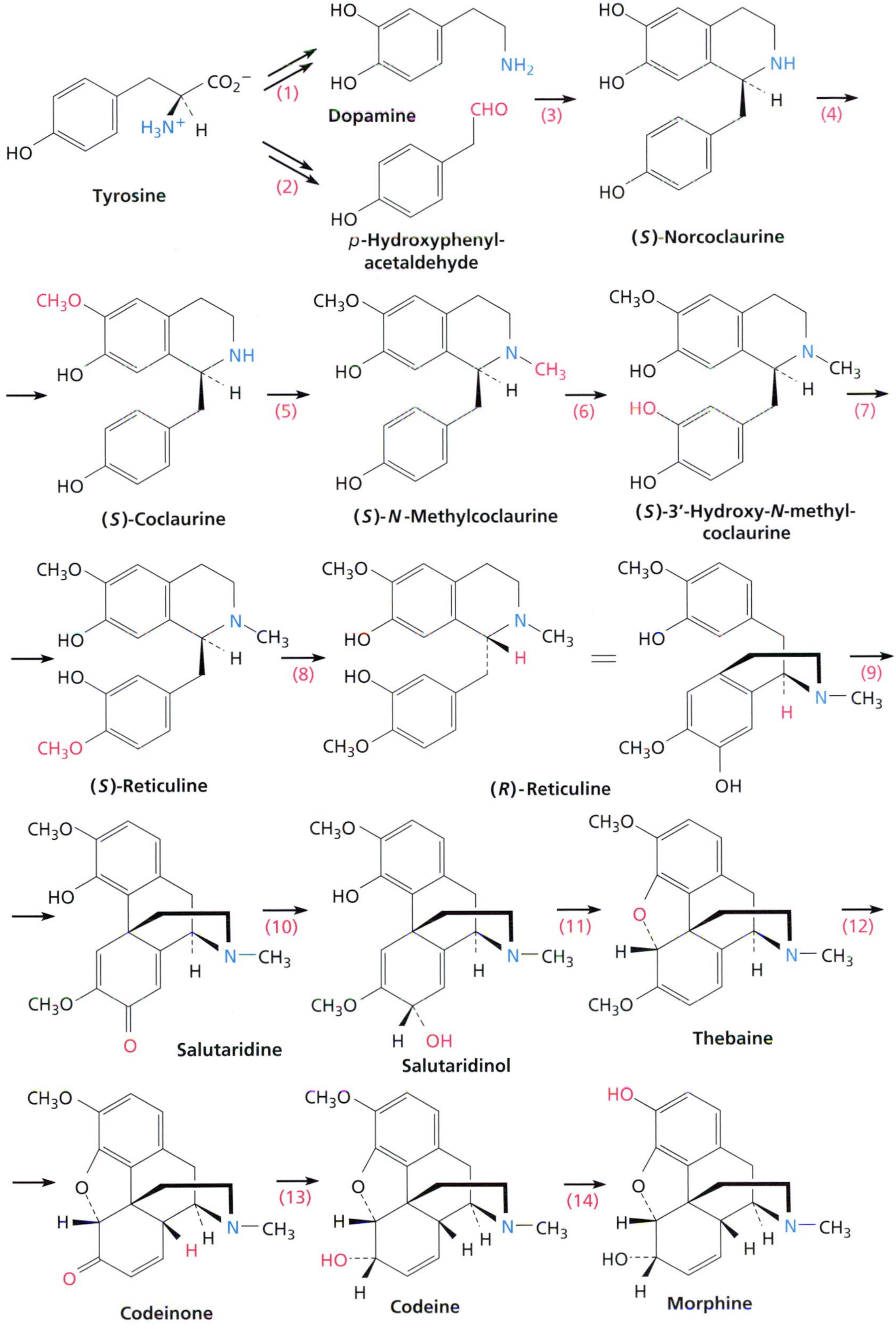

Tyrosine

(1)

(2)

Dopamine

p-Hydroxyphenylacetaldehyde

(3)

(S)-Norcoclaurine

(4)

(S)-Coclaurine

(5)

(S)-N-Methylcoclaurine

(6)

(S)-3′-Hydroxy-N-methylcoclaurine

(7)

(S)-Reticuline

(8)

(R)-Reticuline

(9)

Salutaridine

(10)

Salutaridinol

(11)

Thebaine

(12)

Codeinone

(13)

Codeine

(14)

Morphine

FIGURE 7.9 Pathway for the biosynthesis of morphine.

Step 1. Dopamine biosynthesis Dopamine is formed from tyrosine in two steps: an initial hydroxylation of the aromatic ring, followed by a PLP-dependent decarboxylation. The hydroxylation is catalyzed by tyrosine 3-monooxygenase, a tetrahydrobiopterin-containing enzyme whose mechanism[8] is thought to be analogous to that shown previously for phenylalanine hydroxylation (Figures 5.26 and 5.27). The decarboxylation is catalyzed by aromatic L-amino acid decarboxylase, a PLP-dependent enzyme whose mechanism is analogous to that of diaminopimelate decarboxylase (Figure 5.34).

Tetrahydrobiopterin
O_2
Dihydrobiopterin
H_2O
HO
H $\overset{+}{NH_3}$
CO_2^-
Tyrosine
HO
HO
H $\overset{+}{NH_3}$
CO_2^-
L-Dopa
(PLP-dependent)
CO_2
HO
HO
$\overset{+}{NH_3}$
Dopamine

Step 2. *p*-Hydroxyphenylacetaldehyde biosynthesis *p*-Hydroxyphenylacetaldehyde, the second tyrosine-derived precursor of morphine, is also formed in two steps: an initial PLP-dependent transamination with α-ketoglutarate to give *p*-hydroxyphenylpyruvate, followed by TPP-dependent decarboxylation of the α keto acid. The transamination is catalyzed by tyrosine transaminase and occurs by the mechanism previously discussed (Figure 5.1). The decarboxylation is catalyzed by 4-hydroxyphenylpyruvate decarboxylase and occurs by a mechanism analogous to that described previously for the formation of acetaldehyde from pyruvate (Section 4.3, Figure 4.8). The process is shown in Figure 7.10.

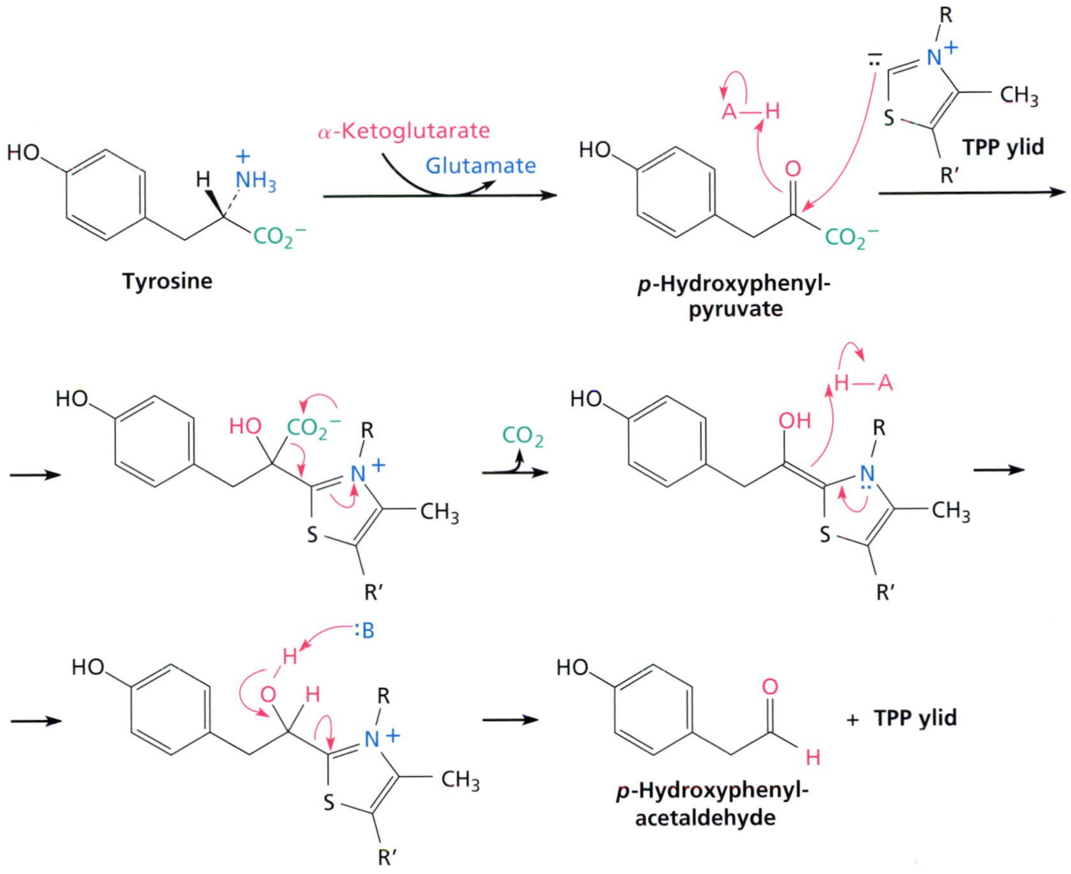

FIGURE 7.10 Mechanism of the TPP-dependent, nonoxidative decarboxylation of *p*-hydroxyphenylpyruvate to give *p*-hydroxyphenylacetaldehyde.

Step 3. Coupling Catalyzed by (*S*)-norcoclaurine synthase, the coupling of dopamine and *p*-hydroxyphenylacetaldehyde proceeds through formation of an intermediate iminium ion, followed by intramolecular electrophilic substitution of the aromatic ring (Figure 7.11).

Steps 4–6. Methylations and hydroxylation (*S*)-Norcoclaurine next undergoes two methylations and a hydroxylation to give (*S*)-3′-hydroxy-*N*-methylcoclaurine. Both methylations use *S*-adenosylmethionine (SAM) as the methyl donor, produce *S*-adenosylhomocysteine (SAH) as the by-product, and occur by the usual S_N2 substitution pathway (Section 5.3, Figure 5.21). The first methylation

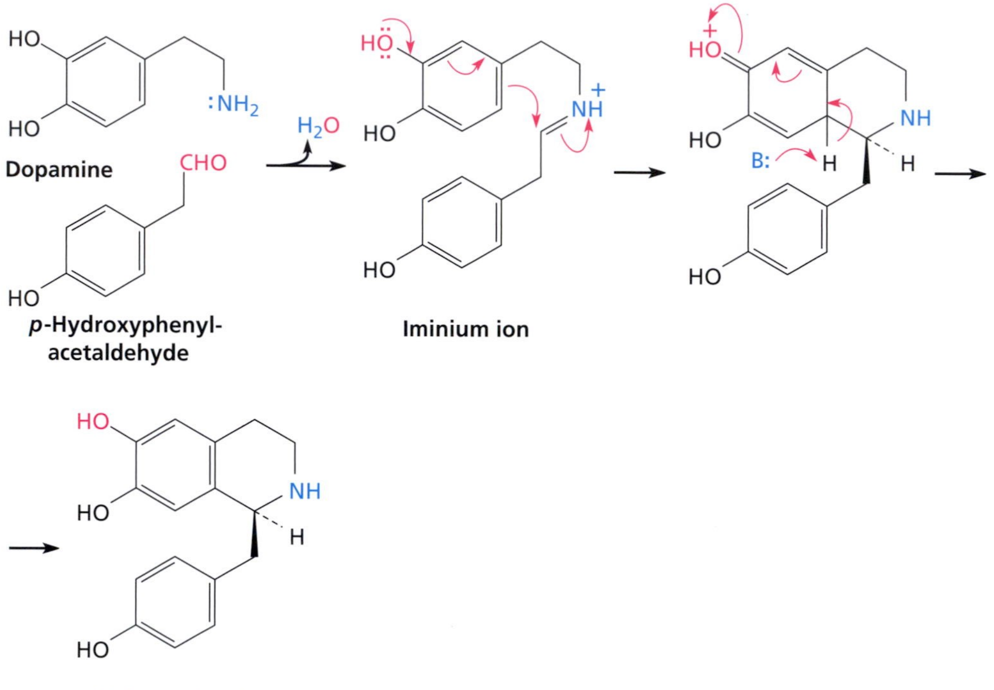

FIGURE 7.11 Mechanism of the coupling of dopamine and *p*-hydroxyphenylacetaldehyde to give (*S*)-norcoclaurine.

occurs on a phenol oxygen and is catalyzed by norcoclaurine 6-*O*-methyltransferase; the second methylation occurs on the amine nitrogen and is catalyzed by coclaurine-*N*-methyltransferase.

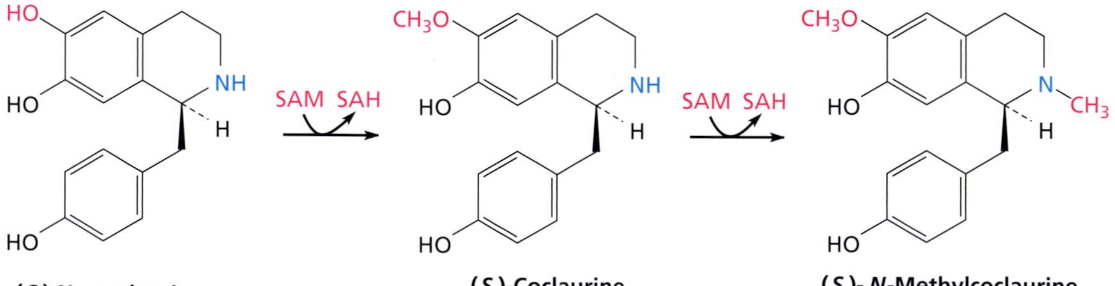

The hydroxylation of (*S*)-*N*-methylcoclaurine in step 6 to give (*S*)-3′-hydroxy-*N*-methylcoclaurine is superficially similar to the hydroxylation of kynurenine to give 3-hydroxykynurenine (Section 5.3, Figure 5.13) in that both require a flavin cofactor and NADPH as reducing agent. Unlike the enzyme in the kynurenine hydroxylation, however, that responsible for hydroxylation of *N*-methylcoclaurine is a *cytochrome P450* enzyme. These enzymes, of which more than 500 are known, contain an Fe(III)–heme cofactor ligated to a cysteine residue; they function to activate molecular oxygen for a variety of different oxidation processes.

As shown in Figure 7.12, activation of molecular oxygen by *N*-methylcoclaurine 3′-monooxygenase begins with reduction of Fe(III) in the heme to Fe(II).

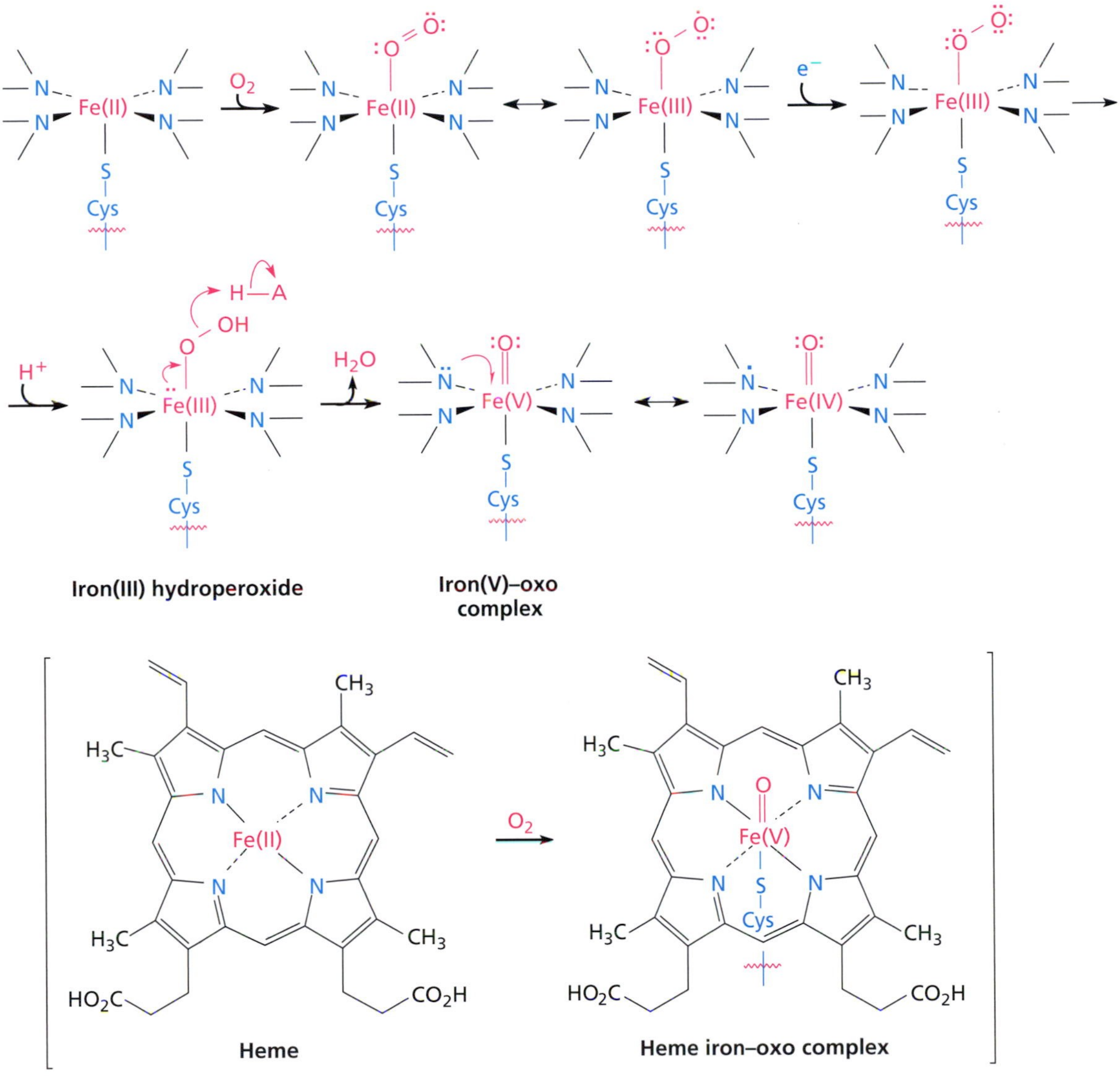

Iron(III) hydroperoxide

Iron(V)–oxo complex

Heme

Heme iron–oxo complex

FIGURE 7.12 The structure of heme and the mechanism of the activation of molecular oxygen by a cytochrome P450.

NADPH is the reducing agent, with participation by a flavin coenzyme in shuttling an electron. Molecular oxygen complexes to the Fe(II) center, a second electron is added, and protonation occurs to give an iron(III) hydroperoxide. Loss of water from the hydroperoxide then gives what can be formally written either as an iron(V)–oxo complex or, if an electron is transferred to iron from a ligand nitrogen atom, as an iron(IV) complex like those we've seen previously. Thus, the cytochrome P450 enzyme is a highly reactive oxidizing agent.

Details of the hydroxylation itself are not clear. The reaction may take place through an epoxide intermediate, or it may occur directly through an electrophilic aromatic substitution mechanism.

(*S*)-*N*-Methylcoclaurine

(*S*)-3′-Hydroxy-*N*-methylcoclaurine

Steps 7–8. Methylation and epimerization Methylation of a phenolic —OH group in (*S*)-3′-hydroxy-*N*-methylcoclaurine by SAM gives (*S*)-reticuline through the usual S_N2 pathway, and epimerization of the stereocenter forms (*R*)-reticuline. The epimerization is a two-step process, the first an oxidation of the tertiary amine to an intermediate iminium ion, and the second a hydride reduction of the iminium ion. The oxidation is catalyzed by an uncharacterized enzyme whose mechanism is not yet known. The reduction of the iminium ion is catalyzed by 1,2-dehydroreticulinium reductase, with NADPH as cofactor (Figure 7.13).

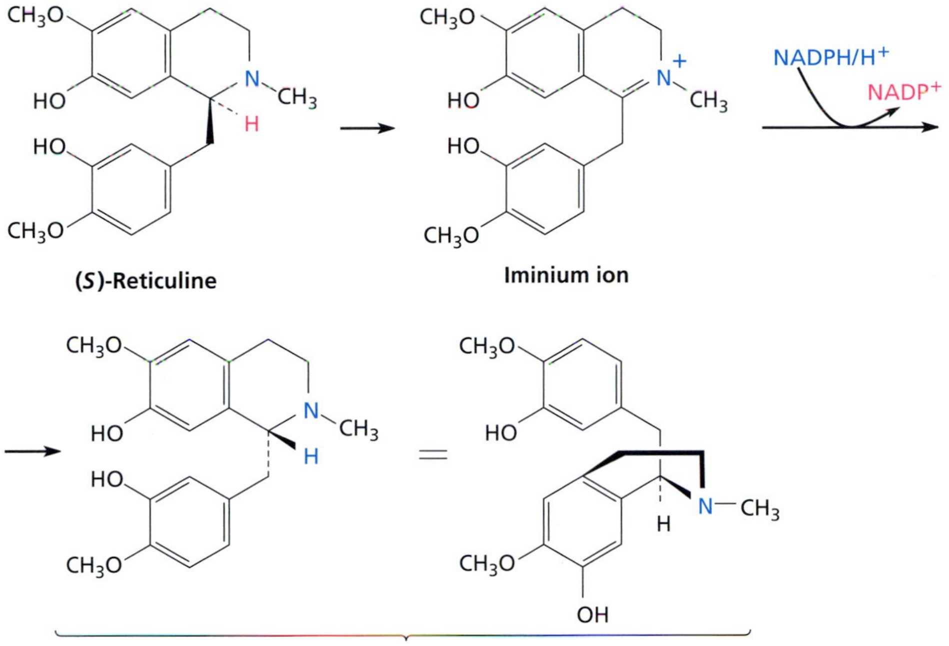

CH_3O HO HO CH_3O N CH_3 H

(S)-Reticuline

CH_3O HO HO CH_3O N^+ CH_3

Iminium ion

NADPH/H^+ $NADP^+$

CH_3O HO HO CH_3O N CH_3 H = CH_3O HO CH_3O OH H N—CH_3

(R)-Reticuline

FIGURE 7.13 Epimerization of (S)-reticuline to (R)-reticuline.

Step 9. Oxidative coupling (R)-Reticuline is converted into salutaridine by an oxidative coupling between the ortho position of one phenol ring and the para position of the other. The reaction is catalyzed by the cytochrome P450-dependent enzyme salutaridine synthase,[9] which contains an iron(V)–oxo heme complex like that in the hydroxylation of (S)-N-methylcoclaurine. Formation of the phenoxide ions and abstraction of a nonbonding electron from each oxygen atom occurs, followed by radical coupling and a keto–enol tautomerization to yield salutaridine (Figure 7.14).

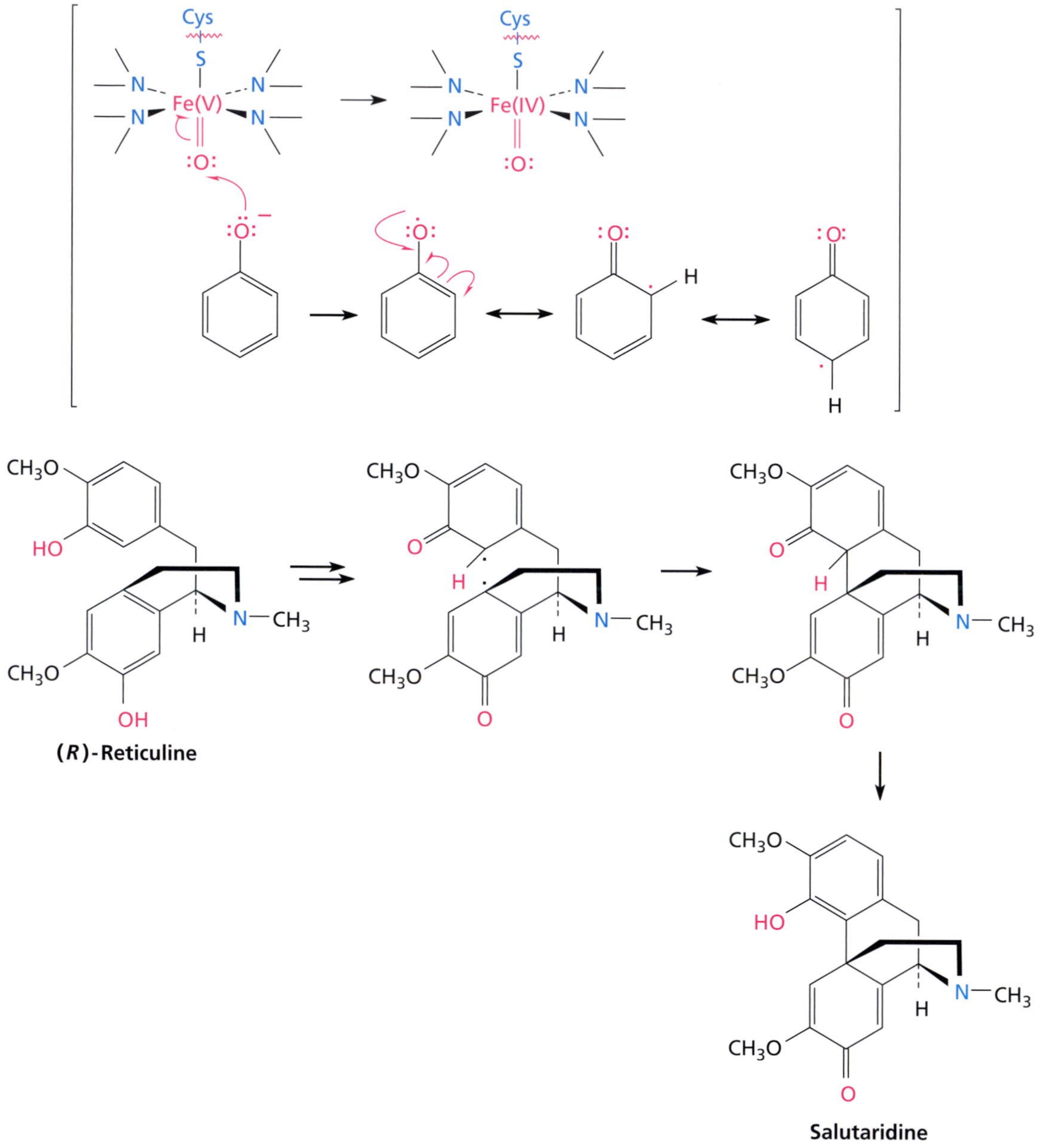

(*R*)-Reticuline

Salutaridine

FIGURE 7.14 Mechanism of the oxidative phenol coupling of (*R*)-reticuline to salutaridine.

Steps 10–11. Reduction and cyclization Reduction of salutaridine to salutaridinol is catalyzed by salutaridine reductase, with NADPH as cofactor. This alcohol is then acetylated by acetyl CoA in the presence of salutaridinol 7-*O*-acetyltransferase to give a doubly allylic acetate, which spontaneously cyclizes to thebaine (Figure 7.15). Because of the unusual stability of the doubly allylic cation intermediate, the cyclization is likely an S_N1 process.

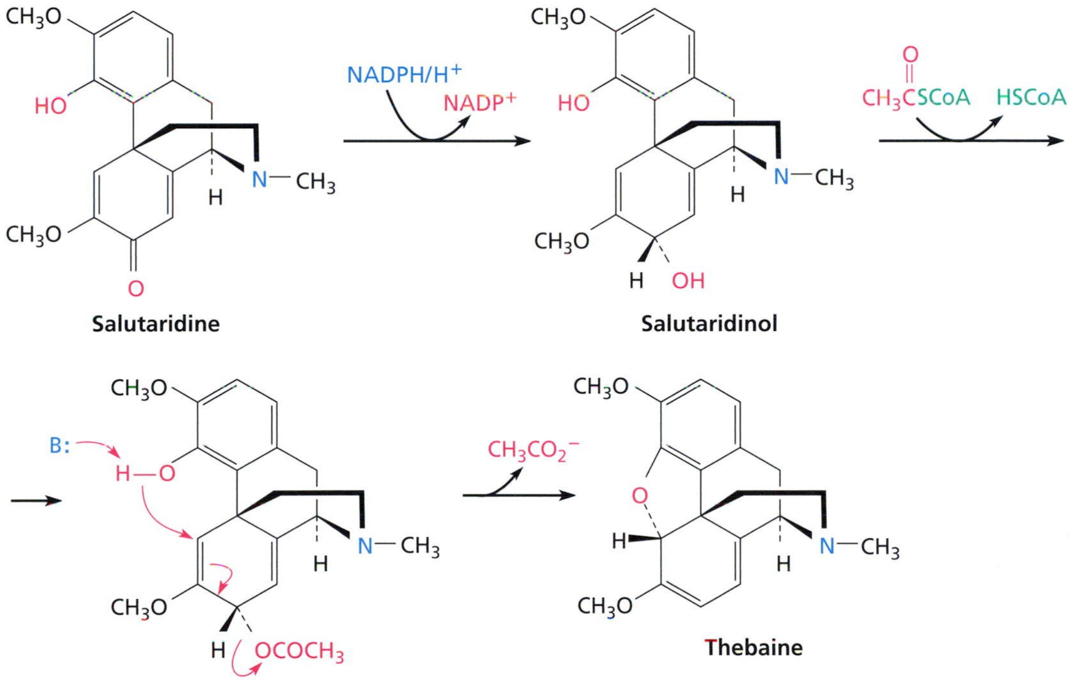

FIGURE 7.15 Mechanism of the formation of thebaine from salutaridine.

Steps 12–13. Demethylations and reduction The remaining steps in the biosynthesis of morphine involve two demethylation reactions and a reduction. The first demethylation is catalyzed by an uncharacterized, P450 monooxygenase that hydroxylates the $—OCH_3$ group to form $—OCH_2OH$, which then loses formaldehyde. A likely mechanism involves abstraction of a hydrogen from the methyl group by the Fe(V)–oxo complex, whose formation was shown in Figure 7.12. Transfer of an OH radical and reduction of the iron atom to Fe(III) then follow (Figure 7.16).

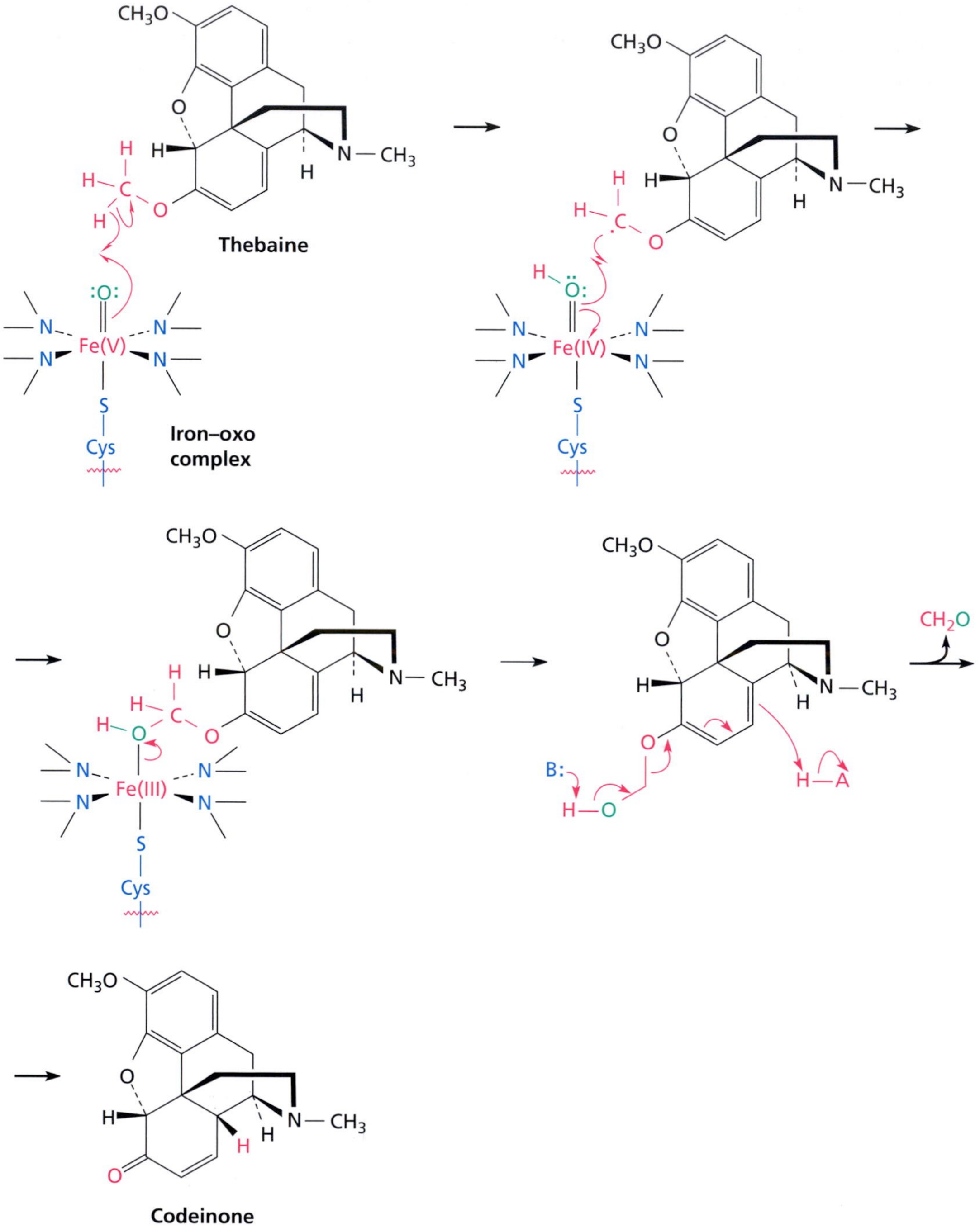

Thebaine

Iron–oxo complex

Codeinone

FIGURE 7.16 A mechanism for the demethylation of thebaine, catalyzed by a P450 enzyme.

The codeinone resulting from thebaine demethylation is reduced by codeinone reductase with NADPH as cofactor to give codeine, which is demethylated to yield morphine. As in the first demethylation, the P450 enzyme that catalyzes this final step is not yet characterized.

$NADPH/H^+$ → $NADP^+$

Codeinone → Codeine → Morphine

7.3 Biosynthesis of Prostaglandins and Other Eicosanoids

Prostaglandins (Section 2.2), thromboxanes, and leukotrienes make up a class of lipids called **eicosanoids** because they are derived biologically from 5,8,11,14-eicosatetraenoic acid, or arachidonic acid (Figure 7.17). Prostaglandins (PG) have a cyclopentane ring with two long side chains, thromboxanes (TX) have a six-membered, oxygen-containing ring, and leukotrienes (LT) are acyclic.

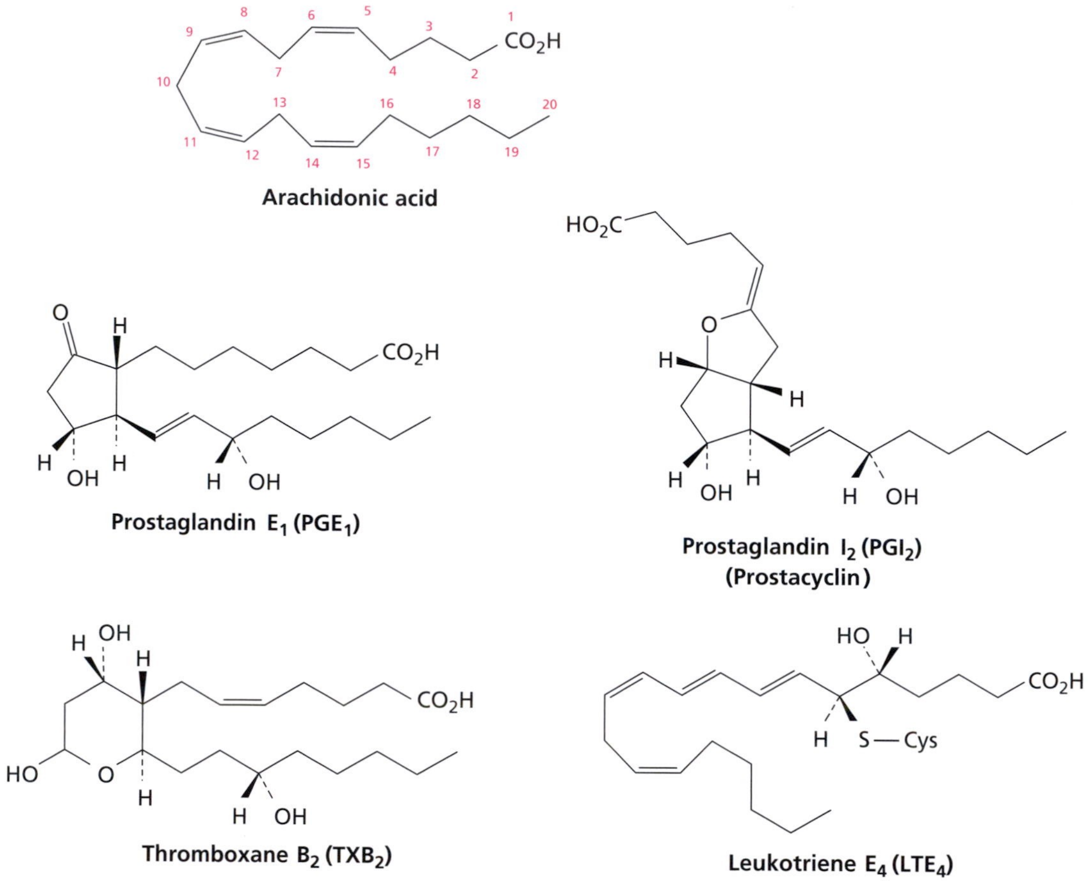

Arachidonic acid

Prostaglandin E_1 (PGE_1)

Prostaglandin I_2 (PGI_2) (Prostacyclin)

Thromboxane B_2 (TXB_2)

Leukotriene E_4 (LTE_4)

FIGURE 7.17 Structures of some representative eicosanoids. All are derived biologically from arachidonic acid.

Eicosanoids are named based on their ring system (PG, TX, or LT), substitution pattern, and number of double bonds. The various substitution patterns on the ring are indicated by letter as in Figure 7.18, and the number of double bonds is indicated by a subscript. Thus, PGE_1 is a prostaglandin with the "E" substitution pattern and one double bond. The numbering of the atoms in the various eicosanoids is the same as in arachidonic acid, starting with the —CO_2H carbon as C1, continuing around the ring, and ending with the —CH_3 carbon at the other end of the chain as C20.

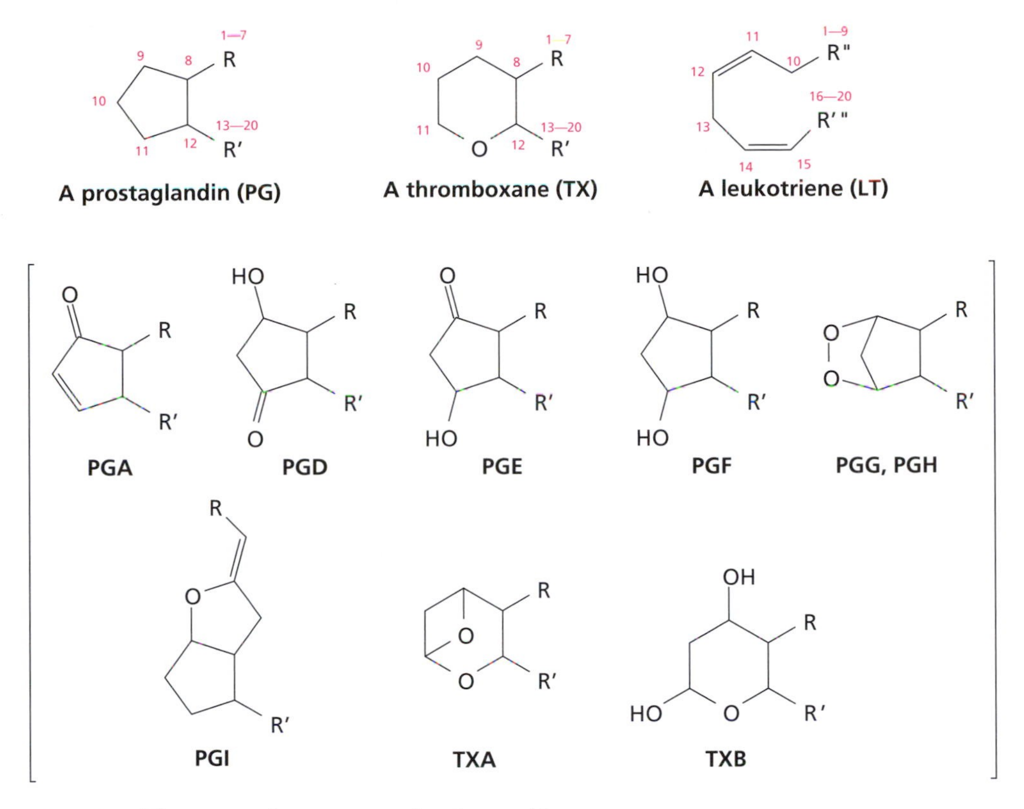

FIGURE 7.18 The nomenclature system for eicosanoids.

Eicosanoid biosynthesis begins with the conversion of arachidonic acid to PGH_2, catalyzed by the multifunctional PGH synthase (PGHS), also called cyclooxygenase (COX).[10–12] There are two distinct enzymes, PGHS-1 and PGHS-2 (or COX-1 and COX-2), both of which accomplish the same reaction and are often present in the same cell. The reason for the existence of two enzymes is not known, although they appear to function independently and may channel their products to different sites in the body. PGHS accomplishes two transformations at two sites, an initial reaction of arachidonic acid with 2 O_2 to yield PGG_2 at a cyclooxygenase site and a subsequent reduction of the hydroperoxide group (—OOH) to the alcohol PGH_2 at a peroxidase site. The mechanism is shown in Figure 7.19.

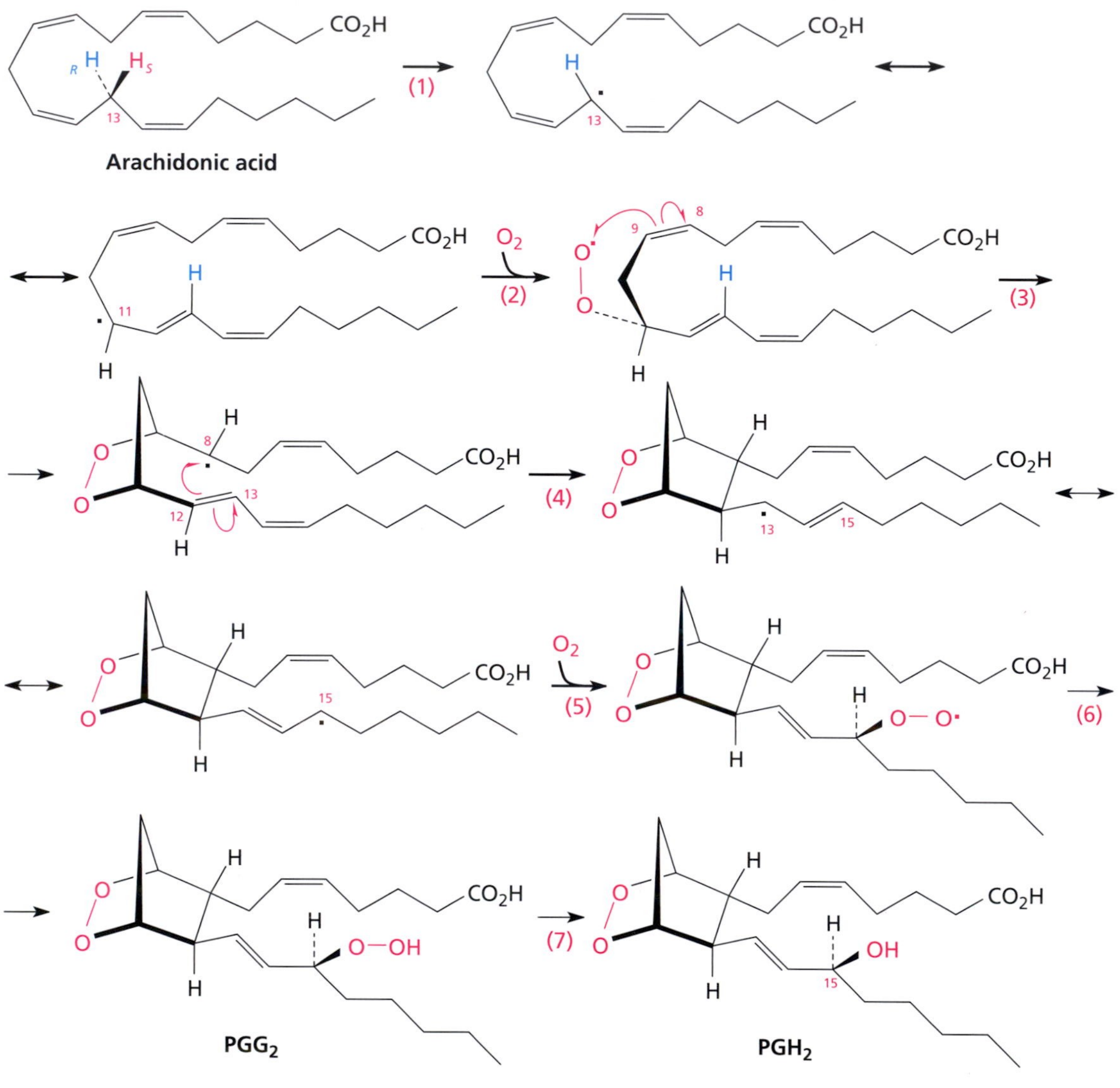

Arachidonic acid

PGG$_2$

PGH$_2$

FIGURE 7.19 Mechanism of the biosynthesis of PGH$_2$ from arachidonic acid.

Step 1. Radical formation Eicosanoid biosynthesis begins with formation of a radical by abstraction of the doubly allylic *pro-S* hydrogen on C13 of arachidonic acid. A tyrosyl radical (Tyr-385) on the enzyme carries out the actual abstraction but, as is often the case with radical oxygenations, an iron–oxo heme complex within the enzyme is involved in generating the radical. An oxidant, perhaps the peroxynitrite ion (ONO_2^-) formed by reaction of nitric oxide (NO) with oxygen, oxidizes the Fe(II) of heme to the corresponding Fe(IV)–oxo complex. Reaction of the complex with tyrosine by long-range electron transfer then forms the tyrosyl radical (Figure 7.20).

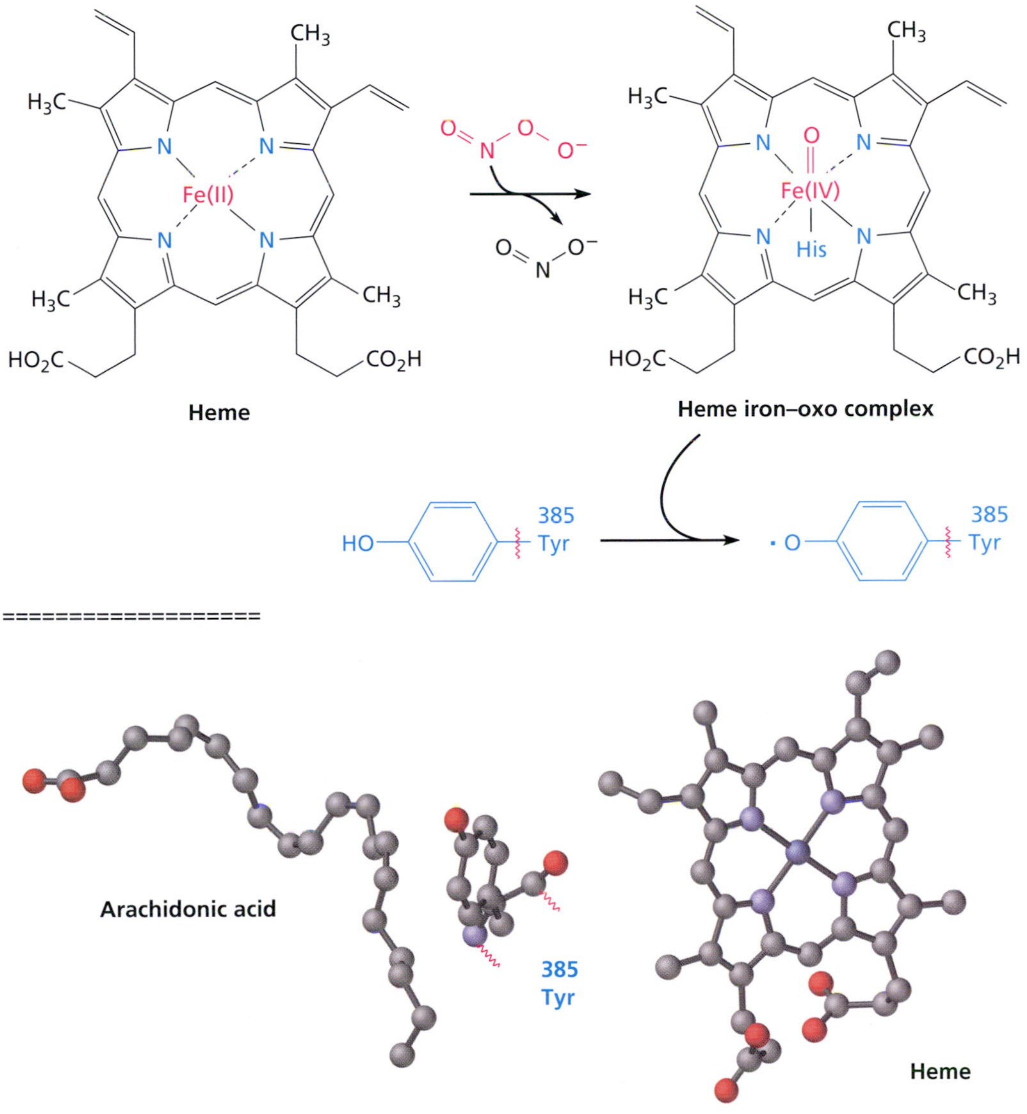

FIGURE 7.20 The structure of the heme iron–oxo complex in PGHS, along with an X-ray crystal structure of enzyme-bound arachidonic acid in the active site of PGHS.

Steps 2–6. Reaction with oxygen and cyclization The allylic radical produced by abstraction of the C13 hydrogen in step 1 has a resonance form that places the unpaired electron on C11. Reaction with molecular oxygen occurs at C11, forming a C—O bond and giving an oxygen radical (step 2) that adds to the double bond at C8–C9 (step 3). Further cyclization by reaction of the radical at C8 with the C12–C13 double bond gives another allylic radical at C13 (step 4), which reacts with O_2 at C15 through its resonance form (step 5). Finally, the peroxy radical abstracts the —OH hydrogen from Tyr-385, forming PGG_2 and regenerating the tyrosyl radical for the next turnover (step 6). Note that the overall conversion of the achiral arachidonic acid to PGG_2 results in the stereospecific formation of five chirality centers by a single enzyme!

Step 7. Reduction Following formation of PGG_2 at the cyclooxygenase site of PGHS, reduction of PGG_2 to PGH_2 occurs at the peroxidase site. Mechanistic details for reaction at the peroxidase site[13] are limited, but it appears that PGG_2 may be reduced by the heme Fe(II) cofactor, which is itself oxidized back to the heme iron–oxo complex in the process.

Further processing of PGH_2 leads to a variety of other eicosanoids. PGE_2, for instance, arises by an isomerization of PGH_2 catalyzed by PGE synthase (PGES).[14, 15] Glutathione (Section 5.3) is needed for enzyme activity, although it is not chemically changed during the isomerization and its role is not fully understood. One possibility is that the glutathione anion breaks the O—O bond in PGH_2 by nucleophilic attack on one of the oxygen atoms, giving a thioperoxy intermediate (R—S—O—R′) that loses glutathione to give the ketone.

RS⁻ H H CO_2H O O H OH H H—A PGH$_2$ → RS—O H H :B CO_2H H OH H H OH Thioperoxy intermediate ↓ RS⁻ O H CO_2H H OH H H OH PGE$_2$

7.4 Biosynthesis of Erythromycin

Polyketides constitute an immensely valuable class of natural products numbering over 10,000 compounds. Commercially important polyketides include antibiotics (erythromycin A, tetracycline) and immunosuppressants (rapamycin, FK506), as well as anticancer (doxorubicin, adriamycin), antifungal (amphotericin B), antiparasitic (avermectin), and cholesterol-lowering (lovastatin) agents. It has been estimated that the sales of these and other polyketide pharmaceuticals total over $15 billion per year.

Polyketides are biosynthesized by the joining together of the simple acyl CoA's acetyl CoA, propionyl CoA, methylmalonyl CoA, and (less frequently) butyryl CoA.[16–19] The key carbon–carbon bond-forming step in each joining is a Claisen condensation (Section 1.7). Once the carbon chain is assembled and released from the enzyme, further transformations take place to give the final product. Erythromycin A, for instance, is biosynthesized from one propionate and six methylmalonate units by the pathway outlined in Figure 7.21. Following initial assembly of the acyl units into the macrocyclic lactone 6-deoxyerythronolide B, two hydroxylations, two glycosylations, and a final methylation complete the biosynthesis.[17]

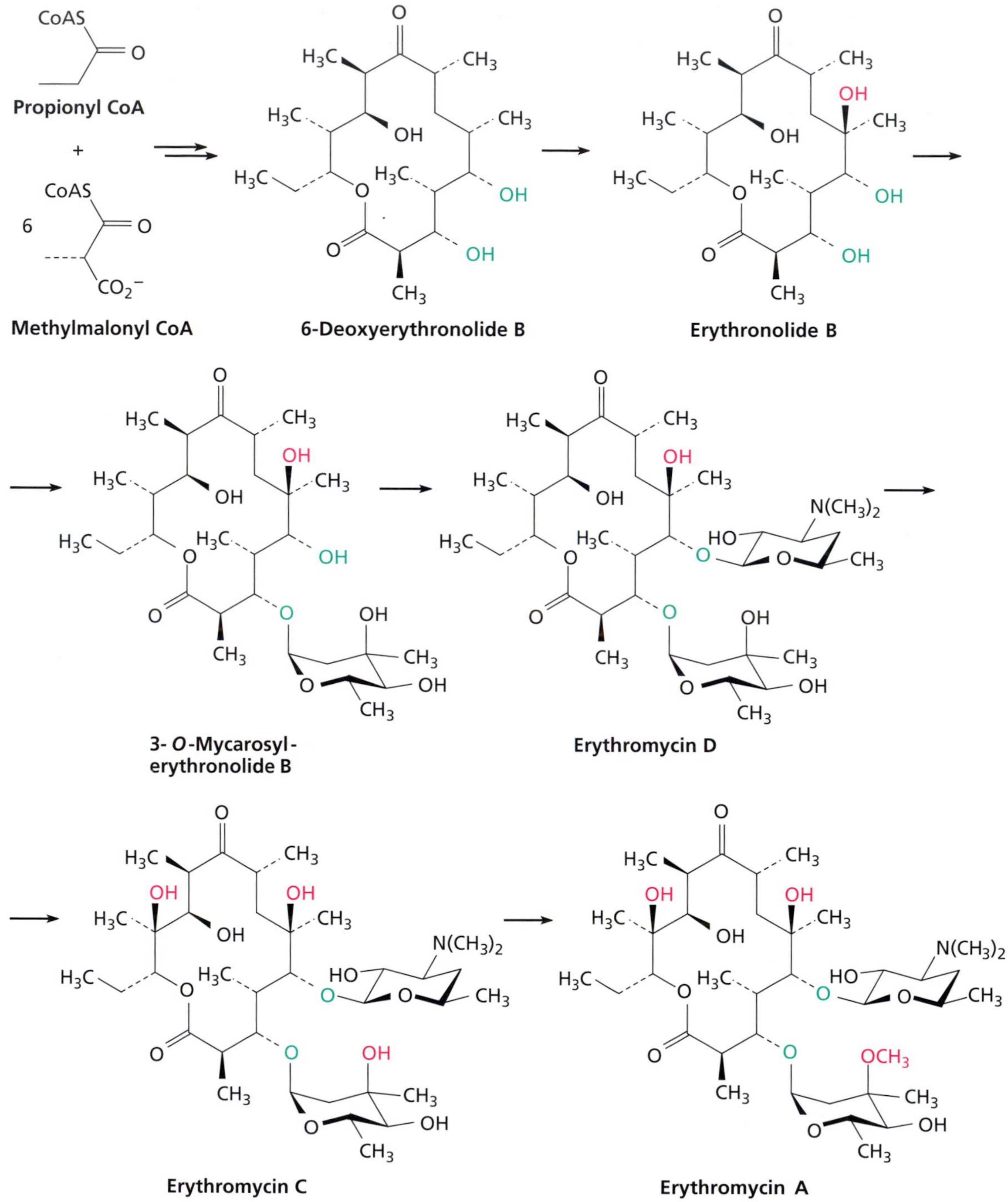

Propionyl CoA

+

6 **Methylmalonyl CoA**

6-Deoxyerythronolide B

Erythronolide B

3-*O*-Mycarosyl-erythronolide B

Erythromycin D

Erythromycin C

Erythromycin A

FIGURE 7.21 Pathway for the biosynthesis of erythromycin A. One propionate and six methylmalonate units are first assembled into the macrocyclic lactone 6-deoxyerythronolide B, which is then hydroxylated, glycosylated by two different sugars, hydroxylated again, and finally methylated.

The initial assembly of acyl CoA precursors to build a polyketide carbon chain is carried out by a polyketide synthase, or PKS. The erythromycin PKS is a massive synthase of greater than 2 million molecular weight and containing more than 20,000 amino acids. Furthermore, it is a *homodimer*, meaning that it consists of two identical protein chains twisted around each other, with each chain containing all the enzymes necessary for constructing the polyketide chain. Both chains contain many separate enzyme domains, each of which is a folded, globular region inside the PKS that catalyzes a specific biosynthetic step. The domains are grouped into modules, where each module carries out the sequential addition and processing of an acyl CoA to the growing polyketide. In addition, adjacent modules form three larger enzymes (DEBS 1, DEBS 2, and DEBS 3) that are linked by peptide spacers. As shown in Figure 7.22, the erythromycin PKS consists of an initial *loading module* to attach the first acyl group, six *extension modules* to add six further acyl groups, and an *ending module* to cleave the thioester bond and release the polyketide. In this instance, the ending module also catalyzes cyclization to give a macrocyclic lactone.

The loading module has two domains: an acyl transfer (AT) domain and an acyl carrier protein (ACP) domain. The AT selects the first acyl CoA (propionyl CoA in the case of erythromycin) and catalyzes its transfer to the adjacent ACP, which binds it and holds it for further reaction. Each extension module in a PKS has a minimum of three domains: an AT, an ACP, and a ketosynthase (KS). Polyketide chain extension occurs when the AT selects a new acyl CoA, transfers it to the ACP, and the KS then catalyzes a Claisen condensation reaction to extend the chain. In addition to the three minimum domains, some extension modules also contain a ketoreductase (KR) to reduce a ketone carbonyl group and produce an alcohol, a dehydratase (DH) to dehydrate the alcohol and produce a C=C bond, and an enoyl reductase (ER) to reduce the C=C bond. Finally, the ending domain is a thioesterase (TE), which releases the product by catalyzing a lactonization.

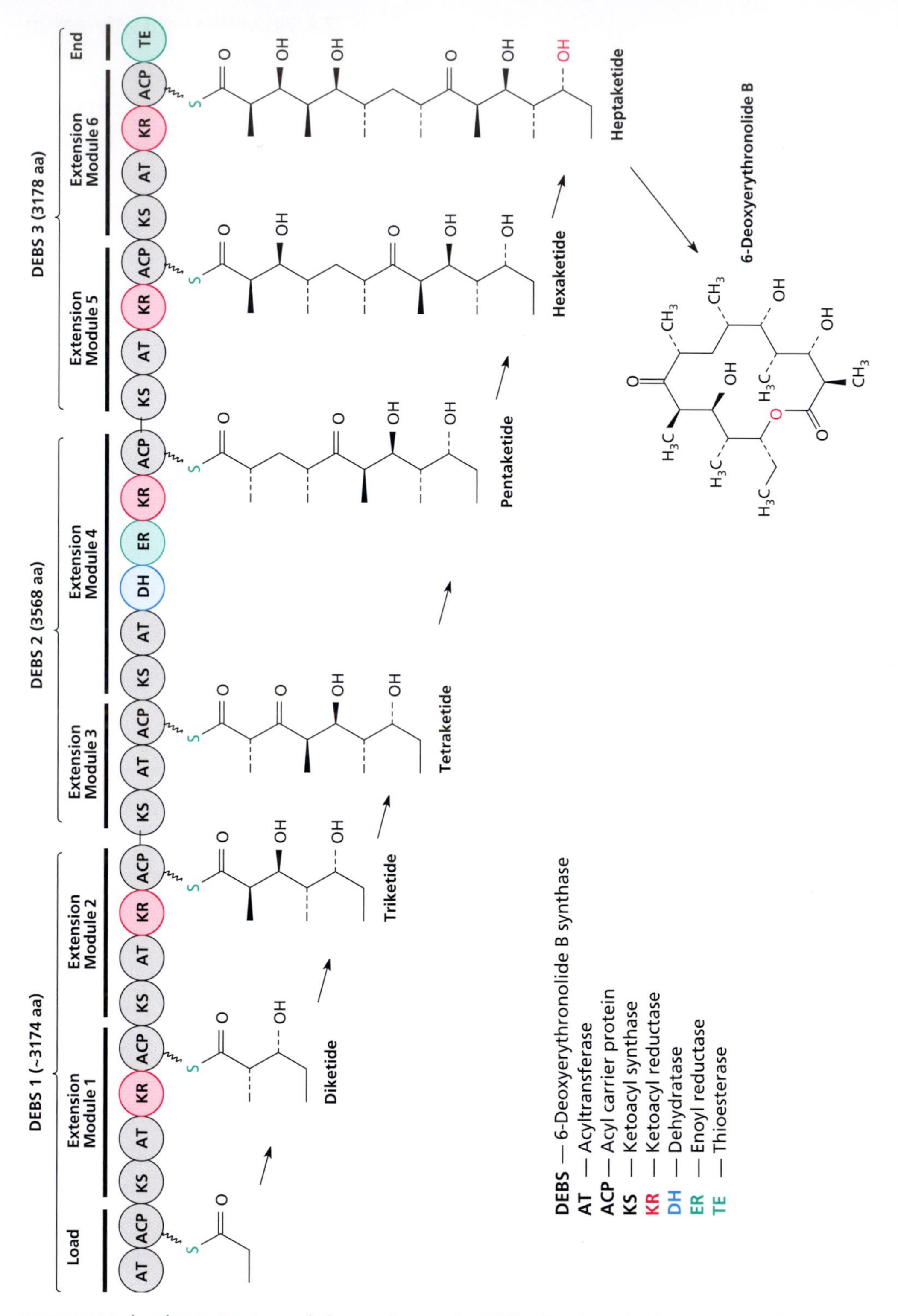

FIGURE 7.22 A schematic view of the erythromycin PKS, showing the locations of the enzyme domains within the loading module and the six extension modules. The figure is explained in detail in the text.

Step 1. Loading In the erythromycin PKS, loading occurs when the loading AT domain binds a propionyl CoA using a thioester bond to the —SH of a cysteine residue. The AT then transfers the propionyl group to the adjacent ACP. Each ACP in the synthase contains a phosphopantetheinyl group bonded to a serine hydroxyl, and bonding of the acyl group to the enzyme occurs by thioester formation with the phosphopantetheine —SH (Figure 7.23). The phosphopantetheine acts in effect as a long, flexible arm attached to the enzyme to allow movement of the acyl group from one catalytic domain to another.

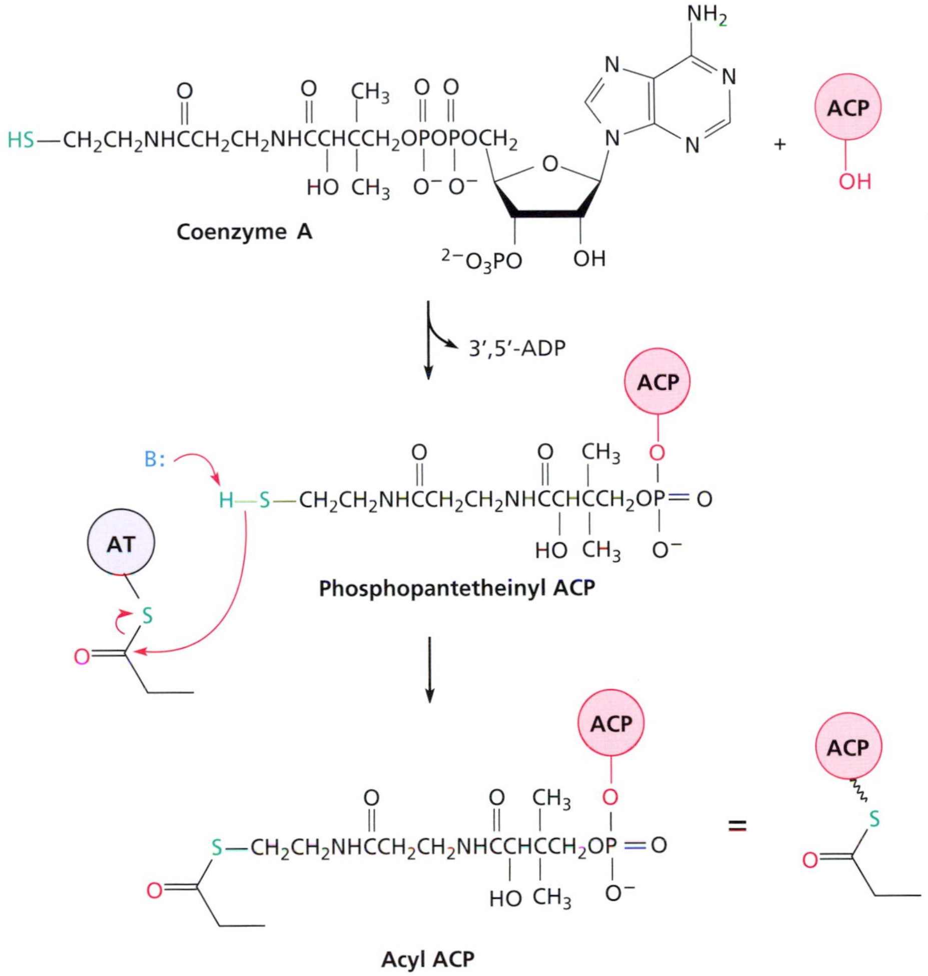

FIGURE 7.23 Formation of an acyl ACP during polyketide biosynthesis. The phosphopantetheine, symbolized by a zigzag line between S and ACP, acts as a long, flexible arm to allow the acyl group to move from one catalytic domain to another.

Step 2. Chain extension Following loading of the first propionyl group (Figure 7.24, step i), polyketide chain extension begins when the ACP transfers the propionyl group to the ketosynthase of module 1 (KS1), again forming a thioester bond to a cysteine residue (step ii). At the same time, the AT and ACP of module 1 load a (2*S*)-methylmalonyl CoA onto the thiol terminus of the ACP1 phosphopantetheine (step iii), and KS1 catalyzes a decarboxylative Claisen condensation

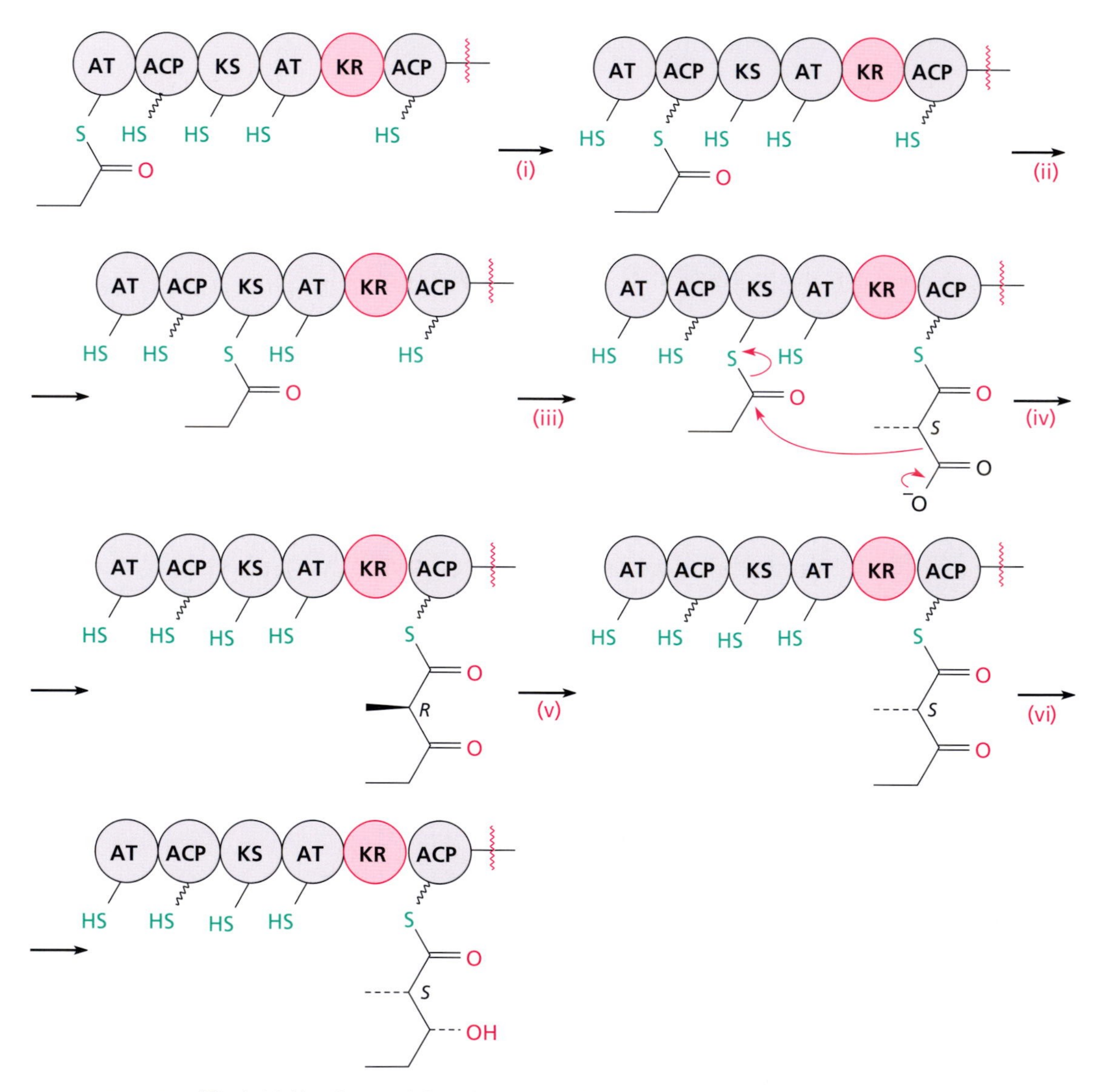

FIGURE 7.24 The initial loading and first chain-extension cycle catalyzed by the erythromycin PKS.

to form an enzyme-bound β-keto thioester (step iv). The condensation occurs with inversion of configuration at the methyl-bearing chirality center, but the initially formed diketide undergoes epimerization to give a product with apparent retention of configuration (step v). Finally, KR1 reduces the ketone to a β-hydroxy thioester by transfer of the *pro-S* hydrogen from NADPH as cofactor (step vi).[20] Module 1 is now finished, so the diketide is transferred to KS2 for another chain extension.

Step 3. Other modifications The reactions catalyzed by extension modules 2, 5, and 6 are similar to those of module 1, although the stereochemistries of the Claisen condensation and reduction steps may differ. The reactions in modules 3 and 4, however, are different. Module 3 lacks a KR domain, so no reduction occurs and the tetraketide product contains a ketone carbonyl group (Figure 7.22). Module 4 contains a KR and two additional enzyme domains, so it catalyzes two additional reactions. Following the KS4 decarboxylative Claisen condensation of (2*S*)-methylmalonate with a tetraketide and KR4 reduction of the resultant pentaketide to an alcohol, a dehydratase (DH) converts the pentaketide alcohol to an α,β-unsaturated thioester. The double bond is then reduced by an enoyl reductase (ER) domain (Figure 7.25).

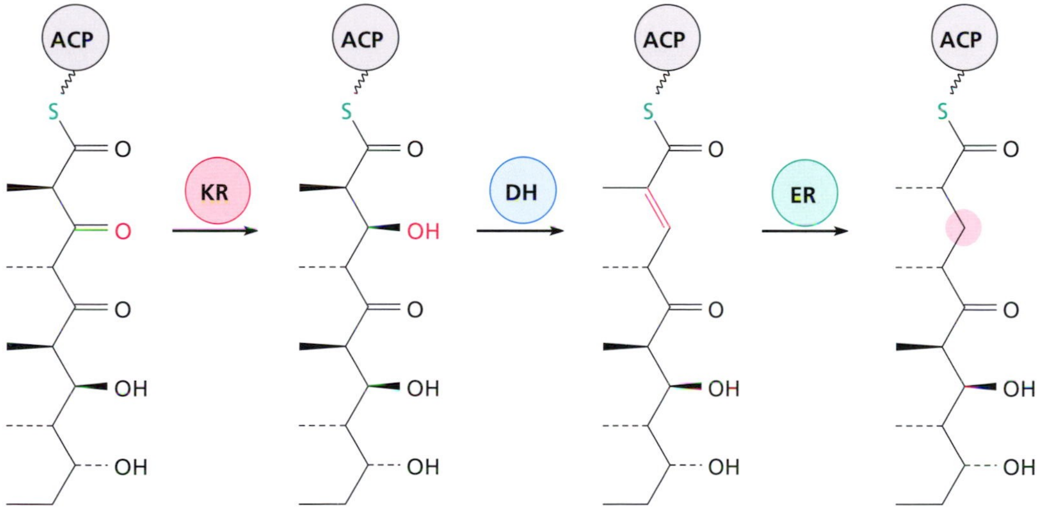

FIGURE 7.25 Additional processing of the pentaketide intermediate in module 4 removes a carbonyl group by a reduction–dehydration–reduction sequence.

Note that the complete sequence of reactions—Claisen condensation, ketone reduction, dehydration, and double-bond reduction—is identical to that found in fatty-acid biosynthesis (Section 3.4, Figure 3.12). All fatty-acid synthases have the same set of AT, ACP, KS, KR, DH, and ER components as the polyketide synthases.

Step 4. Lactonization Release of 6-deoxyerythronolide B from the PKS is catalyzed by the ending thioesterase module. A serine residue on the TE module first carries out a nucleophilic acyl substitution on the ACP-bound heptaketide, and the acyl enzyme that results undergoes lactonization. A histidine residue in the TE acts as base to catalyze nucleophilic acyl substitution of the serine ester by the C13—OH group (Figure 7.26).

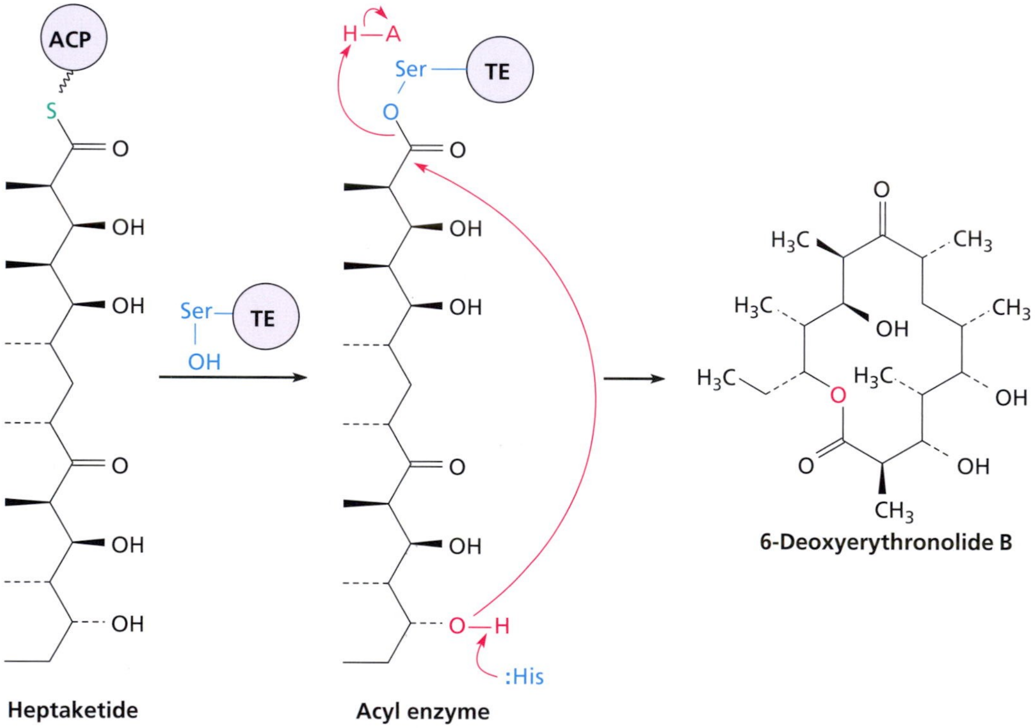

FIGURE 7.26 Release of 6-deoxyerythronolide from the PKS occurs by lactonization of an acyl enzyme, formed by reaction of a serine residue in the TE module with the heptaketide.

Following its release from the PKS, 6-deoxyerythronolide B is hydroxylated at C6 with retention of configuration to give erythronolide B. The reaction is catalyzed by a P450 hydroxylase[21, 22] analogous to that involved in morphine biosynthesis (Section 7.2, Figures 7.12 and 7.16). L-Mycarose is then attached to the C3 hydroxyl group by a TDP-L-mycarosyltransferase (Figure 7.27). The glycosylating agent is thymidyl diphosphomycarose, which is biosynthesized from glucose by the erythromycin-producing organism.[17] The reaction is probably an S_N1-like process that proceeds by initial formation of the glycosyl carbocation.

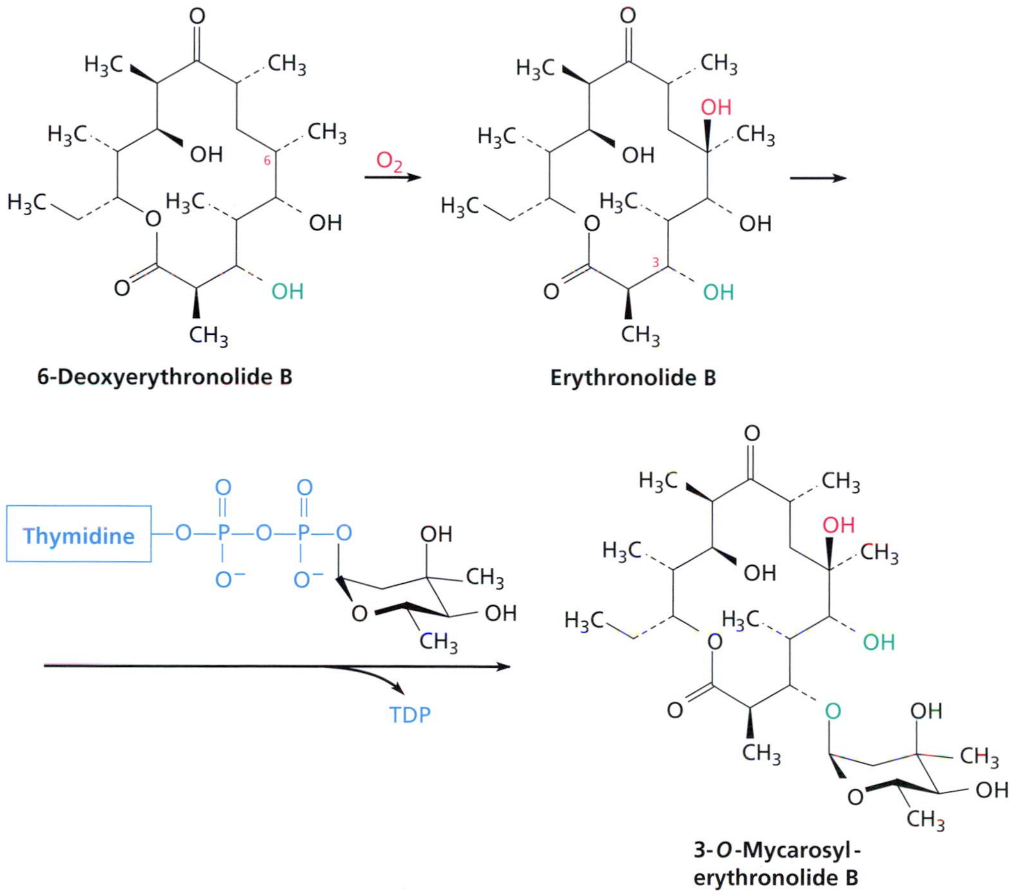

6-Deoxyerythronolide B

O_2

Erythronolide B

Thymidine

TDP

3-*O*-Mycarosyl-erythronolide B

FIGURE 7.27 Initial hydroxylation and glycosylation of 6-deoxyerythronolide B.

The final steps in erythromycin A biosynthesis are a further glycosylation, a further hydroxylation, and a methylation (Figure 7.28). As in the attachment of mycarose, the attachment of the amino sugar D-desosamine also takes place by transfer from a thymidyl diphosphosugar in a reaction catalyzed by TDP-D-desosamine glycosyltransferase. C12 hydroxylation by another P450 enzyme occurs with retention of configuration to give erythromycin C, and methylation of the C3"hydroxyl group of the mycarose unit by reaction with *S*-adenosyl-methionine (Section 5.3, Figure 5.21) gives erythromycin A. As with TDP-mycarose, the TDP-desosamine needed for glycosylation is biosynthesized from glucose.

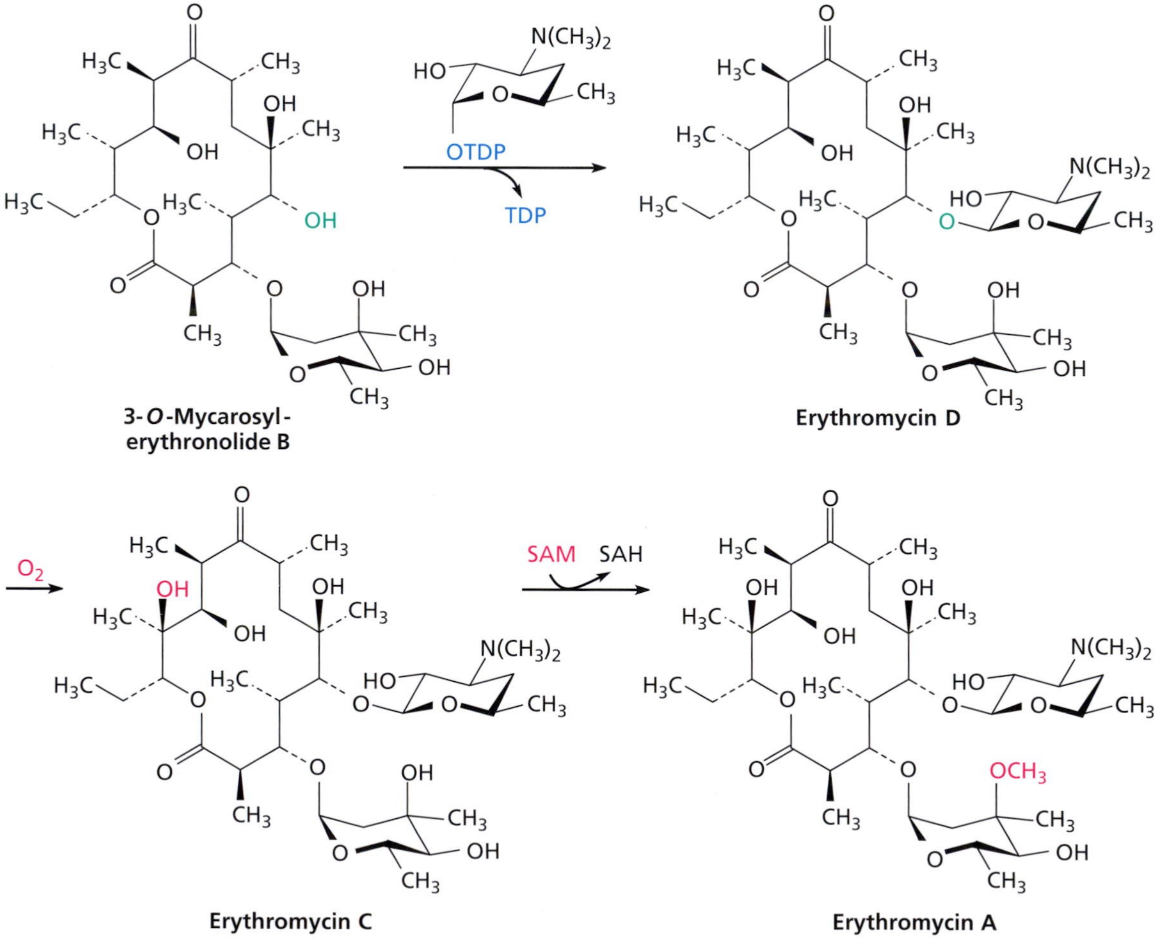

FIGURE 7.28 Final steps in the biosynthesis of erythromycin A.

7.5 Biosynthesis of Coenzyme B_{12} and Other Tetrapyrroles

Tetrapyrroles are a group of natural products that contain four pyrrole rings linked by bridging atoms into either a linear or macrocyclic arrangement. Linear tetrapyrroles, such as bilirubin, the principal pigment in bile, are systematically named as substituted bilanes. Macrocyclic tetrapyrroles, such as coenzyme B_{12} (Figure 5.37), heme (Figure 7.12), and chlorophyll are systematically named as substituted porphyrins or corrins (Figure 7.29). Numbering begins at the upper

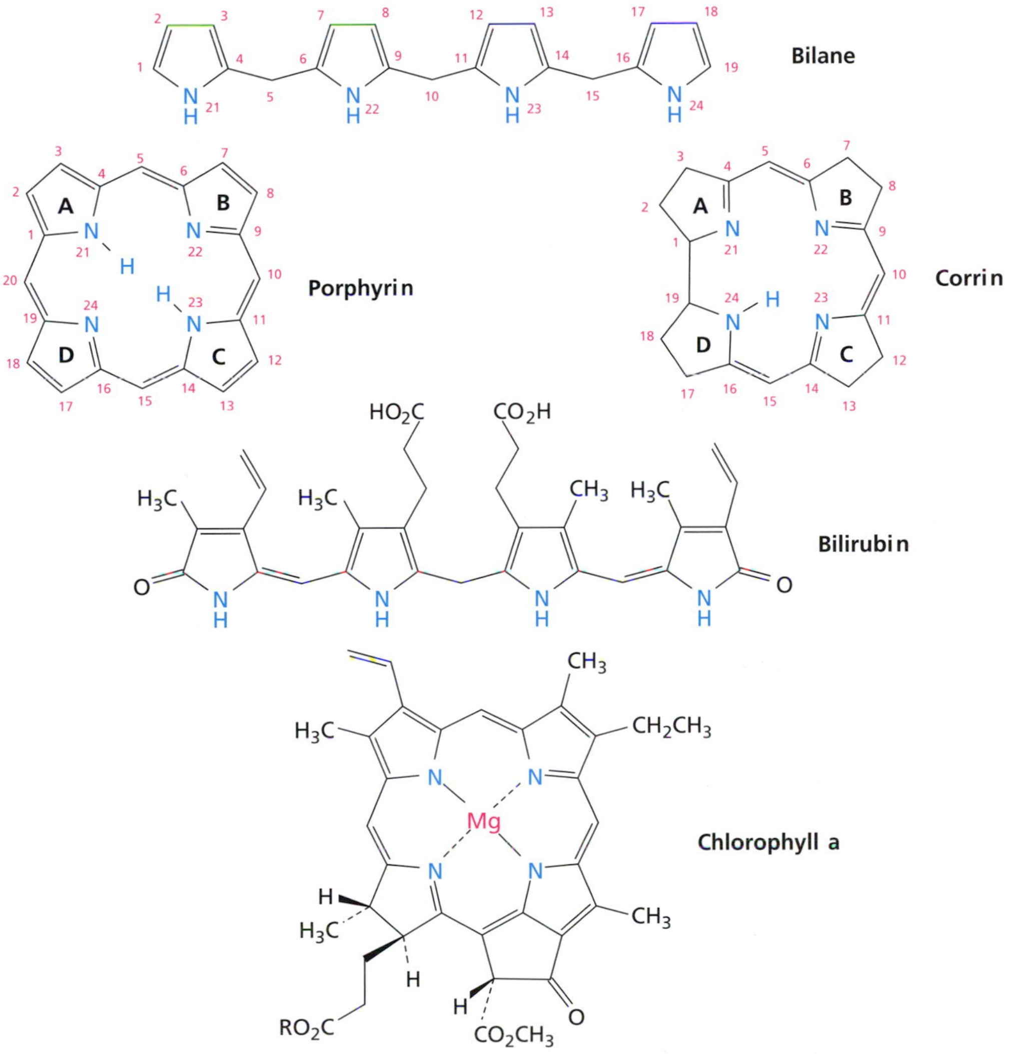

FIGURE 7.29 Nomenclature, numbering, and examples of some tetrapyrroles.

left and continues in a clockwise direction. Note that in both bilanes and corrins, there is no position numbered "20."

Uro'gen III

All tetrapyrroles, whether linear or macrocyclic, are formed biosynthetically from glycine and succinyl CoA by pathways that use the same opening steps and give the monopyrrole porphobilinogen (PBG) as the intermediate. The process is shown in Figure 7.30.

Steps 1–4. Condensation and decarboxylation Glycine and succinyl CoA undergo Claisen condensation and subsequent decarboxylation to give 5-aminolevulinate (ALA). The reaction is catalyzed by the PLP-dependent 5-aminolevulinate synthase[23, 24] and occurs by formation of a glycine–PLP imine that is deprotonated by removal of the *pro-R* hydrogen (step 1). The imine that results undergoes a Claisen condensation reaction with succinyl CoA (step 2), followed by decarboxylation of the β keto acid intermediate (step 3) and release of 5-aminolevulinate (step 4).

Steps 5–9. Condensation and cyclization Following the formation of 5-aminolevulinate from glycine and succinyl CoA, two molecules of ALA condense to give porphobilinogen in a reaction catalyzed by porphobilinogen synthase (PBGS).[25] The mechanism is not completely understood, but it's thought that both ALA molecules are bound by Schiff base linkages to lysine residues in the enzyme. Nucleophilic addition of the $—NH_2$ group of one ALA to the iminium ion of the other (step 5) is followed by a cyclization (step 6) and elimination of a lysine residue (step 7). Elimination of the second lysine residue then occurs (step 8), and tautomerization (step 9) yields the pyrrole.

Condensation of four porphobilinogen molecules to give a linear tetrapyrrole is catalyzed by porphobilinogen deaminase, also called hydroxymethylbilane synthase.[26, 27] The reaction requires a dipyrromethane cofactor that is covalently bonded to the sulfur of a cysteine residue in the enzyme and acts as a primer for the sequential additions of the four porphyrinogens.

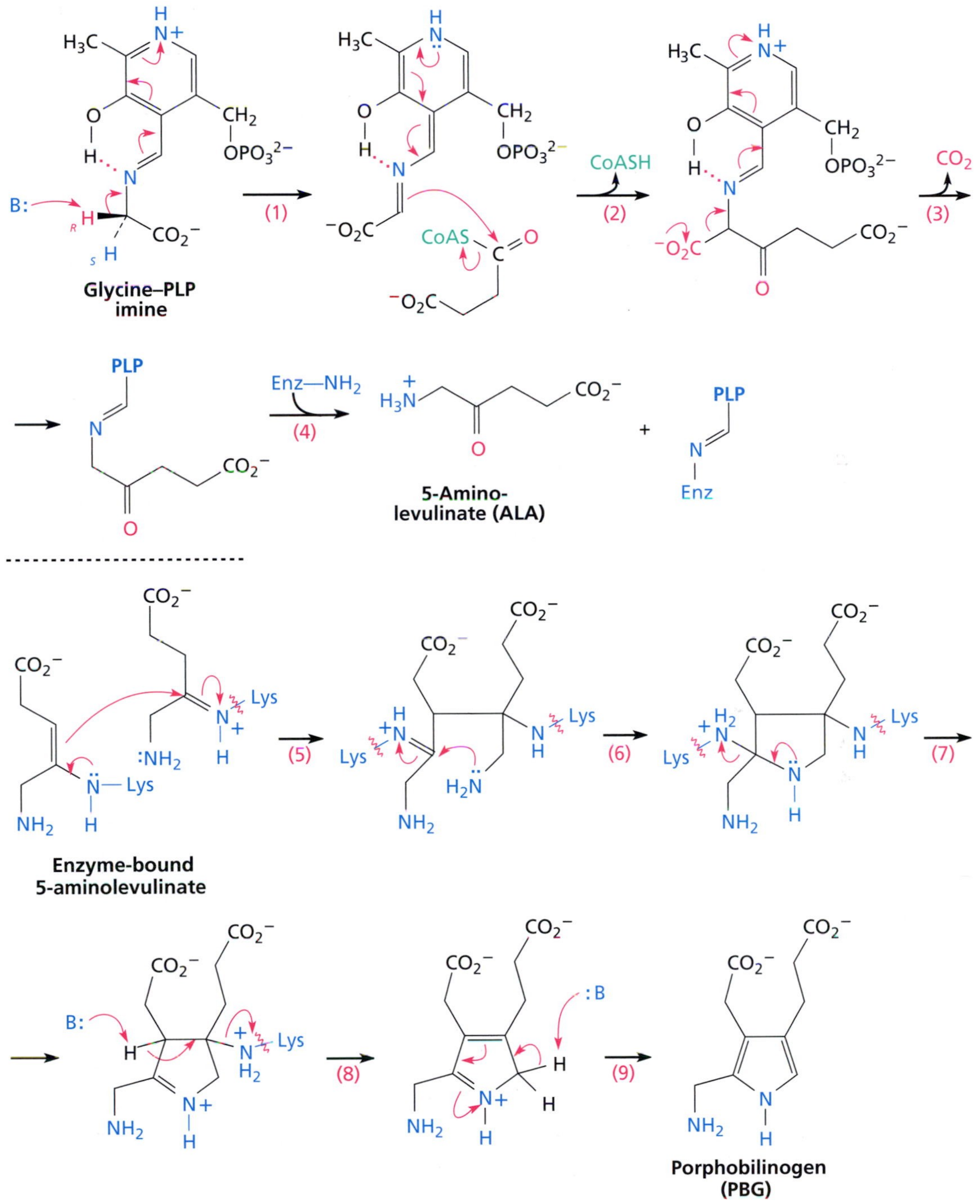

FIGURE 7.30 Mechanism of the biosynthesis of the tetrapyrrole precursor porphobilinogen from glycine and succinyl CoA.

As shown in Figure 7.31, the porphobilinogen additions are thought to take place by sequential electrophilic aromatic substitution reactions onto the growing chain. Porphobilinogen loses ammonia to form a cation (step 1), which has a resonance structure placing the positive charge on the side-chain carbon atom. Sequential electrophilic substitutions to the terminal pyrrole group on the growing chain lead ultimately to a hexapyrrole (steps 2–3), which undergoes cleavage by a retro-electrophilic aromatic substitution (step 4). That is, H^+ as the electrophile adds to the second pyrrole ring, regenerating the dipyrromethane cofactor plus a tetrapyrrole. Addition of water to the cation then yields hydroxymethylbilane (step 5).

The cyclization of hydroxymethylbilane to give the cyclic tetrapyrrole uroporphyrinogen III, usually abbreviated uro'gen III, is somewhat surprising in that a skeletal rearrangement occurs during the reaction (Figure 7.32). As a result, the cyclization product has a different relationship among its acetic acid (A) and propionic acid (P) side chains than the starting hydroxymethylbilane does. Beginning at the upper left (ring A) of the molecule as drawn in Figure 7.32, hydroxymethylbilane has an alternating A–P–A–P–A–P–A–P arrangement of side chains, but uro'gen III has an A–P–A–P–A–P–P–A arrangement.

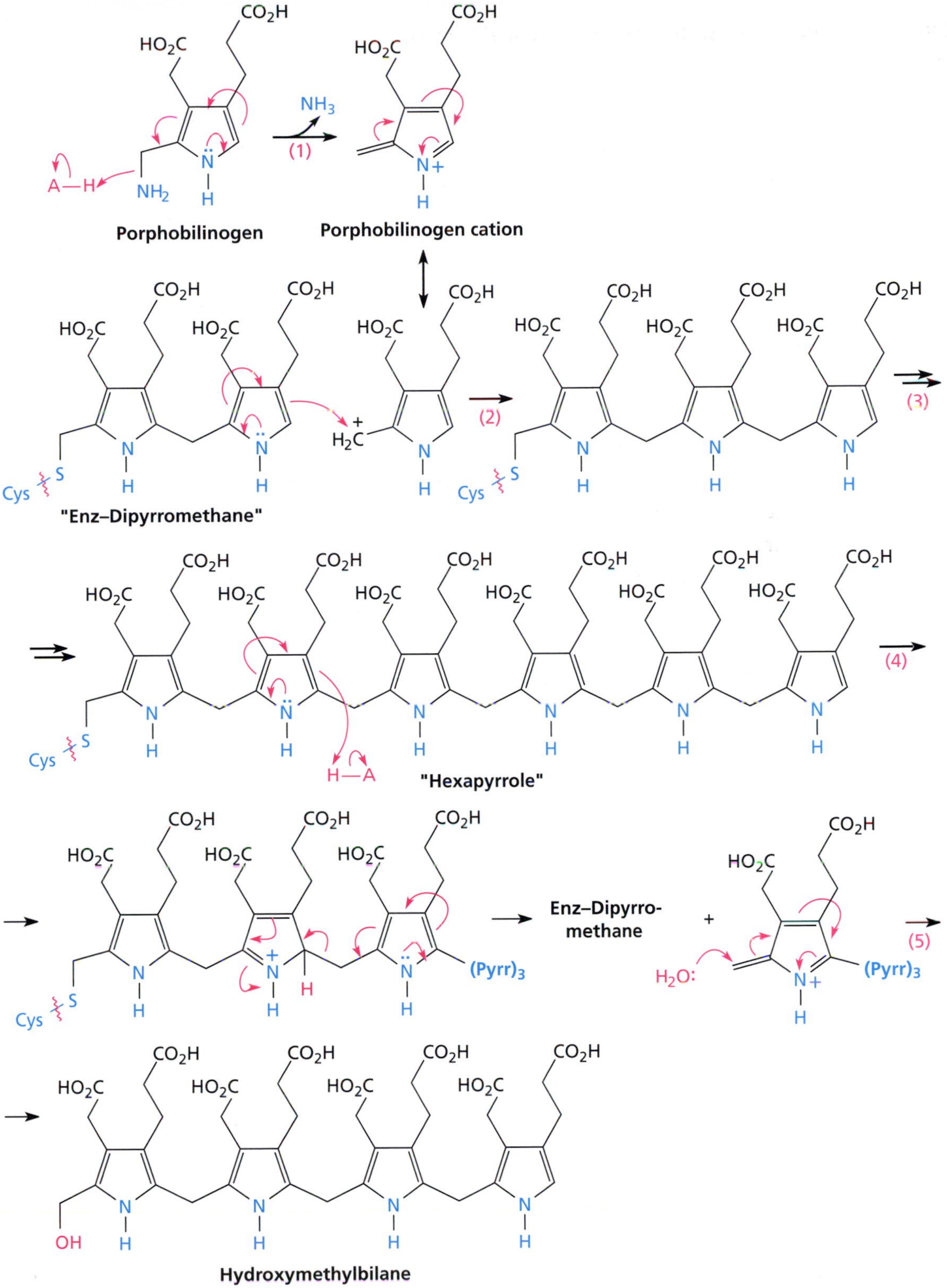

Porphobilinogen

Porphobilinogen cation

"Enz–Dipyrromethane"

"Hexapyrrole"

Enz–Dipyrromethane + (Pyrr)$_3$

Hydroxymethylbilane

FIGURE 7.31 Biosynthesis of the tetrapyrrole hydroxymethylbilane by sequential electrophilic aromatic substitution reactions.

Catalyzed by uroporphyrinogen III synthase, the cyclization of hydroxymethylbilane occurs by an initial protonation and loss to water to give a cation, which adds to the pyrrole ring at the other terminus. Rather than add to the unsubstituted position next to the propionic acid side chain, however, addition occurs at the substituted position next to acetic acid, generating a spirocyclic intermediate (Figure 7.32, step 1). Ring-opening yields an acyclic cation similar to those involved in hydroxymethylbilane synthesis (step 2), and reclosure on the opposite side of the pyrrole ring gives uro'gen III (step 3).

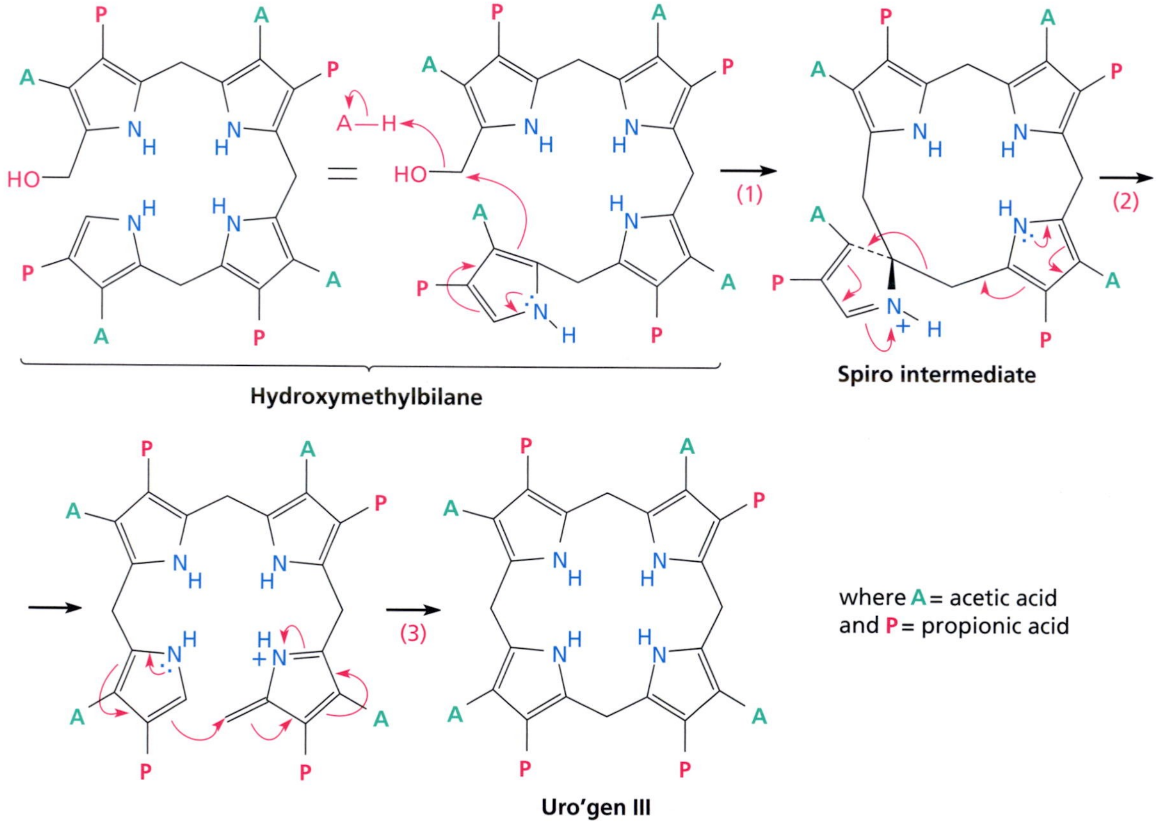

Hydroxymethylbilane

Spiro intermediate

Uro'gen III

where A = acetic acid
and P = propionic acid

FIGURE 7.32 Mechanism of the cyclization of hydroxymethylbilane to uroporphyrinogen III (uro'gen III).

Heme

Heme, the oxygen-carrying pigment in blood, is biosynthesized from uro'gen III by the four-step reaction sequence summarized in Figure 7.33. First, the four acetic acid side chains of uro'gen III are converted by decarboxylation into

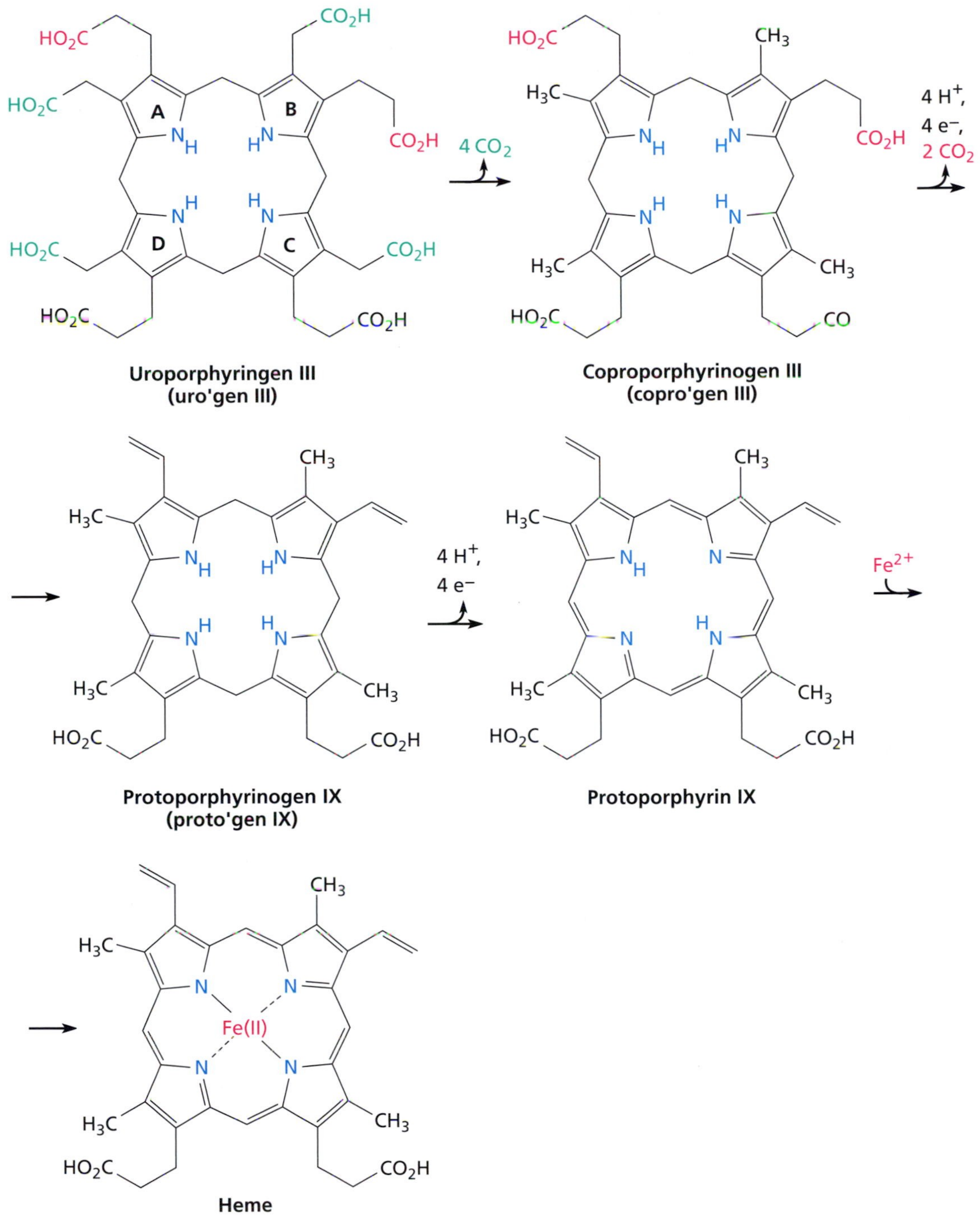

Uroporphyringen III
(uro'gen III)

$4\ CO_2$

Coproporphyrinogen III
(copro'gen III)

$4\ H^+$, $4\ e^-$, $2\ CO_2$

Protoporphyrinogen IX
(proto'gen IX)

$4\ H^+$, $4\ e^-$

Protoporphyrin IX

Fe^{2+}

Heme

FIGURE 7.33 Biosynthesis of heme from uro'gen III.

—CH_3 groups, yielding coproporphyrinogen III (copro'gen III). Two of the four propionic acid side chains are then oxidatively decarboxylated, converting them into vinyl groups and giving protoporphyrinogen IX (proto'gen IX). Next, the bridging —CH_2— groups are dehydrogenated, the double bonds tautomerize to give protoporphyrin IX, and an iron(II) atom is then inserted into the center of the macrocycle by a ferrochelatase to produce heme.

The conversion of uro'gen III to copro'gen III is catalyzed by uroporphyrinogen decarboxylase[28] and is thought to occur by the mechanism shown in Figure 7.34. Protonation of the D ring of uro'gen III (Figure 7.33) by a tyrosine residue in the enzyme gives a cation (step 1), which loses CO_2 with the iminium ion of the pyrrole ring acting as the electron acceptor (step 2). Protonation of the side-chain carbon (step 3) and abstraction of H^+ from the ring (step 4) complete the first decarboxylation. Repetition of the process three more times in the ring sequence A → B → C yields copro'gen III.

Oxidative decarboxylation of two propionic acid side chains to form vinyl groups is catalyzed by coproporphyrinogen oxidase and is not well understood mechanistically. Similarly, the mechanism of the dehydrogenation of the bridging methylene groups, catalyzed by protoporphyrinogen oxidase, is not yet known.

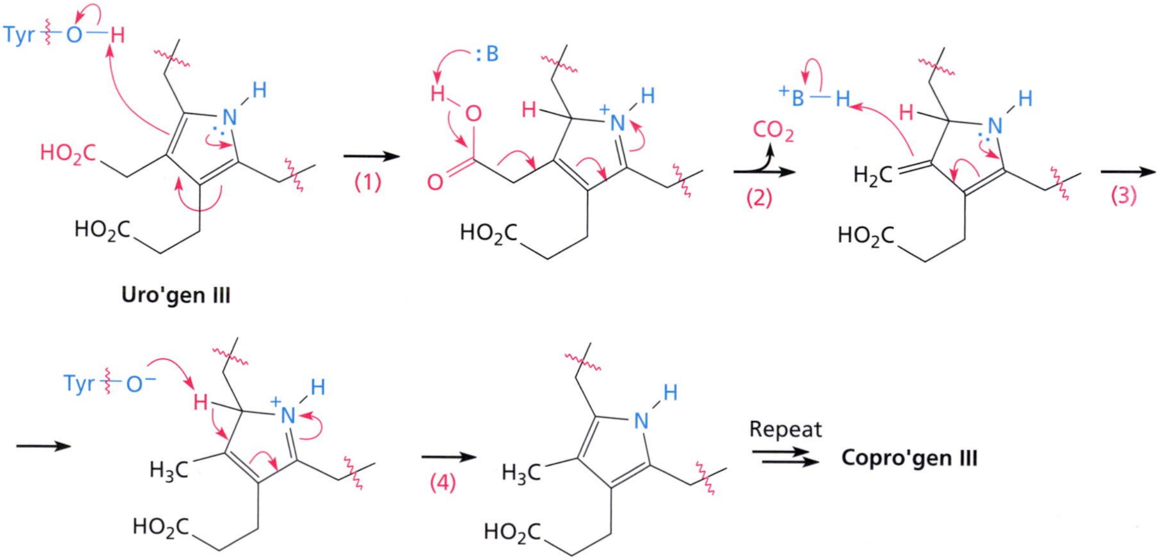

FIGURE 7.34 Proposed mechanism for the fourfold decarboxylation of uro'gen III to copro'gen III.

More is known, however, about the final step of heme biosynthesis, insertion of the iron(II) atom into the center of the macrocycle. The iron insertion is catalyzed by ferrochelatase[29] and appears to involve a domelike distortion of the planar protoporphyrin IX ring, with the Fe^{2+} entering from one face while 2 H^+ ions depart from the opposite face.

Coenzyme B_{12}

Coenzyme B_{12}, or adenosylcobalamin, is the most structurally complex of all enzyme cofactors, and the elucidation of its biosynthetic pathway represents an outstanding achievement in natural-products chemistry.[30–32] (*Vitamin* B_{12}, as opposed to coenzyme B_{12}, has a cyano ligand bonded to the central cobalt in place of adenosyl.) Note that coenzyme B_{12} has the contracted *corrin* ring system (Figure 7.29) rather than the larger porphyrin ring system as in heme. Corrins arise from porphyrins by a ring-contraction reaction and, as with other tetrapyrroles, uro'gen III is the precursor.

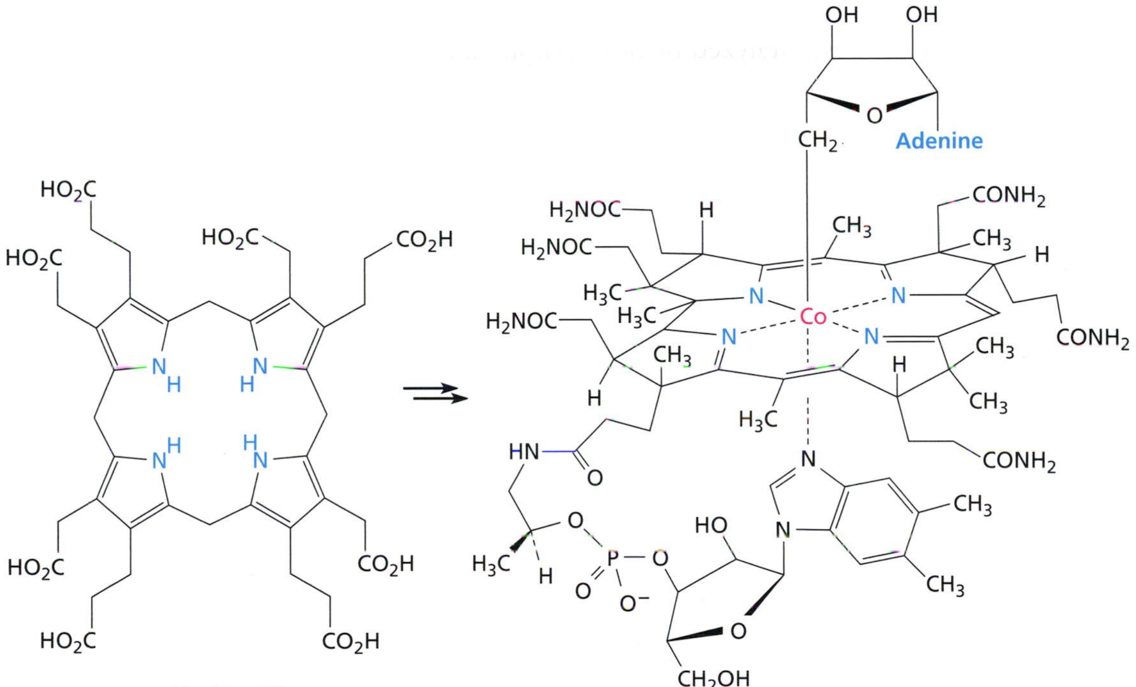

Uro'gen III

Coenzyme B_{12} — Adenosylcobalamin

The biosynthetic conversion of uro'gen III into coenzyme B_{12} begins with three sequential methylation reactions, all involving *S*-adenosylmethionine as the methyl donor. As shown in Figure 7.35, uro'gen III first reacts with SAM at C2

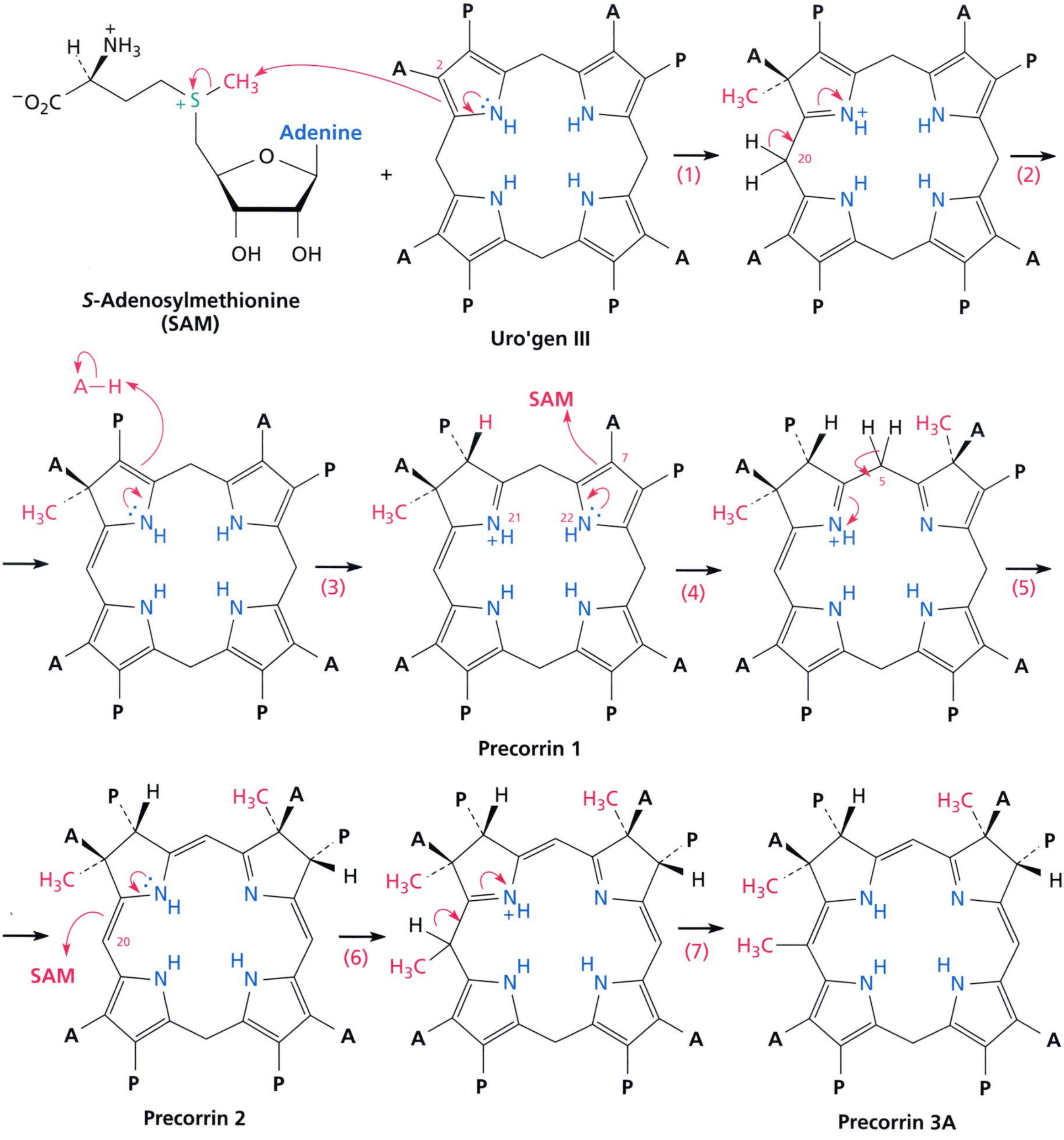

FIGURE 7.35 Mechanism of the conversion of uro'gen III to precorrin 3A by three successive methylations with *S*-adenosylmethionine.

to give a cation (step 1), which loses a proton from C20 on the neighboring bridge (step 2). The enamine that results tautomerizes to an imine, yielding precorrin 1 (step 3). A second methylation then occurs on C7 (step 4), followed by loss of H^+ from C5 to yield precorrin 2 (step 5). Finally, a third methylation takes place on C20 (step 6), and loss of a proton gives precorrin 3A (step 7). The first two methylation reactions are catalyzed by *S*-adenosylmethionine–uroporphyrinogen III methyltransferase, and the third is catalyzed by *S*-adenosylmethionine–precorrin 2 methyltransferase.

Following the first three methylation steps, an oxidation, a ring contraction, five additional methylations, and a decarboxylation convert precorrin 3A into hydrogenobyrinic acid. The overall pathway is shown in Figure 7.36.

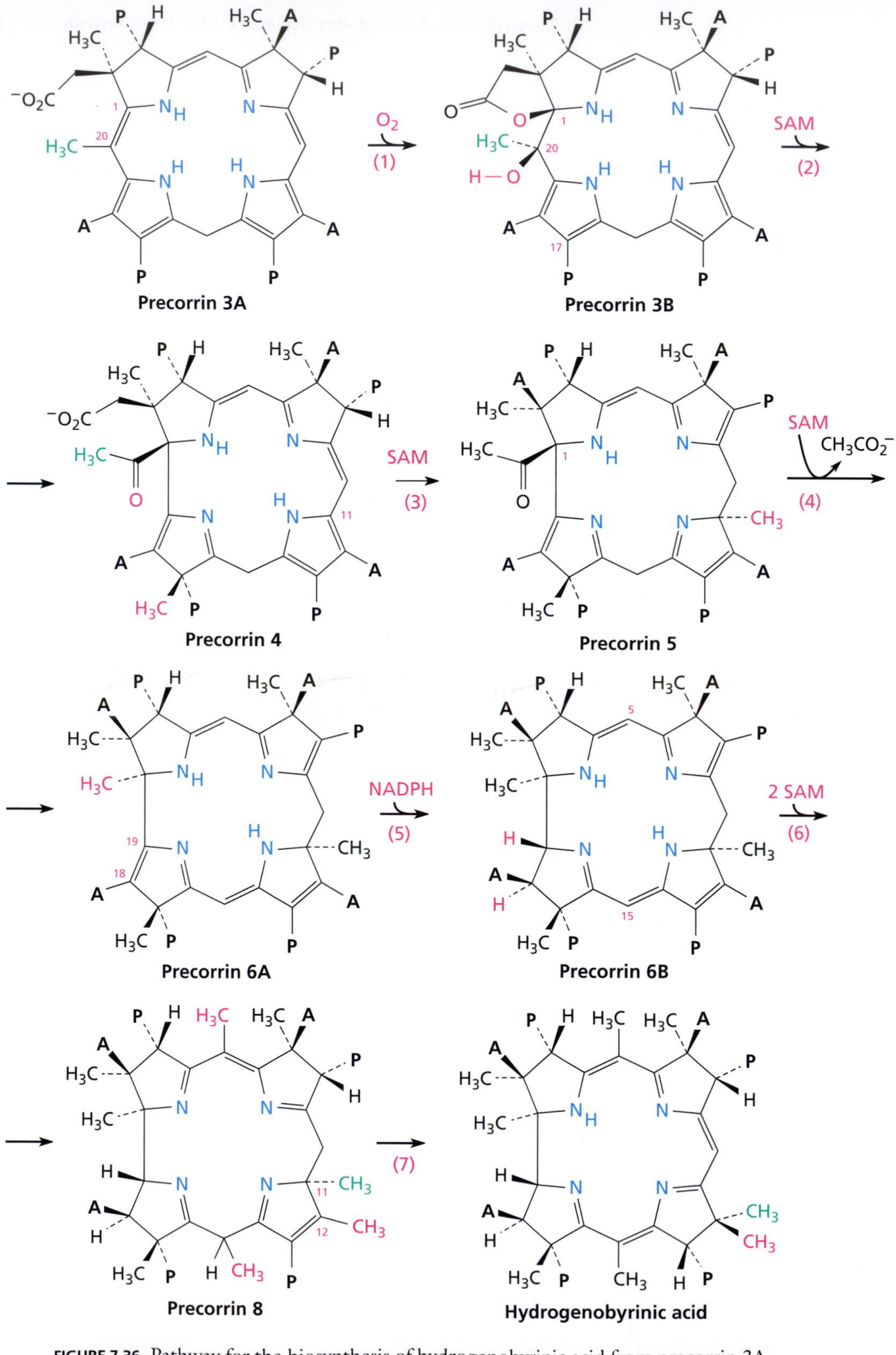

Precorrin 3A

O_2 (1)

Precorrin 3B

SAM (2)

Precorrin 4

SAM (3)

Precorrin 5

SAM, $CH_3CO_2^-$ (4)

Precorrin 6A

NADPH (5)

Precorrin 6B

2 SAM (6)

Precorrin 8

(7)

Hydrogenobyrinic acid

FIGURE 7.36 Pathway for the biosynthesis of hydrogenobyrinic acid from precorrin 3A.

Steps 1–2. Oxidation, rearrangement, and methylation Oxidation of precorrin 3A to precorrin 3B requires molecular oxygen in most organisms and is catalyzed by a nonheme monooxygenase that contains an iron–sulfur cluster. Mechanistic details of this step are not known, although an iron–oxo intermediate may be involved (Figure 7.37, step i). Addition of the side-chain carboxylate at C2 to an

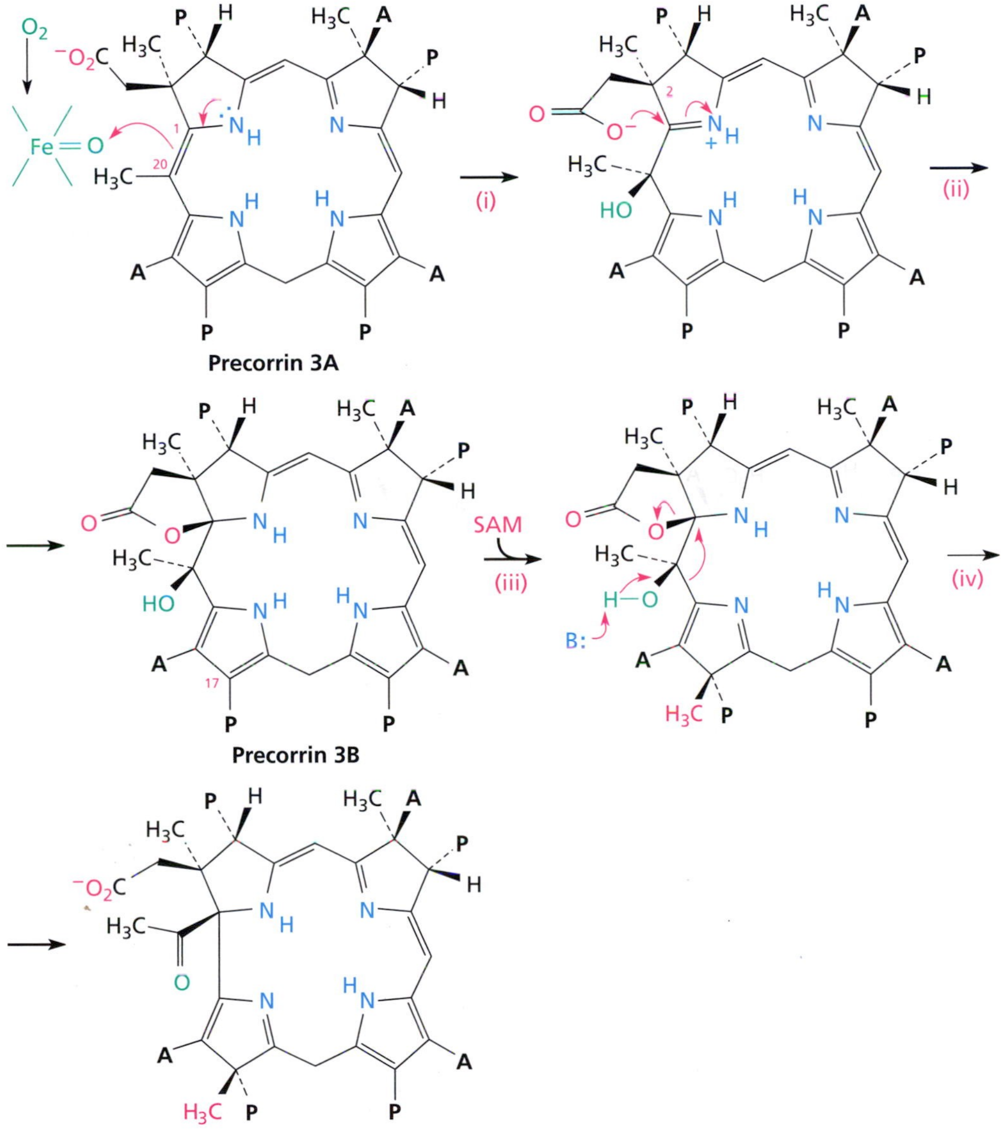

FIGURE 7.37 A possible mechanism for the oxidation and ring contraction of precorrin 3A to precorrin 4.

iminium ion gives the hydroxy lactone precorrin 3B (step ii), and methylation by SAM at C17 (step iii) forms an intermediate that undergoes a pinacol-like rearrangement to accomplish the ring contraction (step iv). Both of the latter two steps are catalyzed by the same methyltransferase.

Steps 3–4. Methylation and loss of acetic acid The methylation of precorrin 4 at C11 to yield precorrin 5 occurs by the same mechanism as previous methylations in a methyltransferase-catalyzed reaction. Precorrin 5 then loses acetic acid and undergoes methylation at C1 to give precorrin 6A in a process catalyzed by a different, bifunctional methyltransferase. As shown in Figure 7.38, the mechanism of the reaction is thought to involve initial tautomerization (step ii), followed by nucleophilic addition of water to the acetyl side chain and loss of acetic acid by a retro-Claisen reaction (step iii). Methylation at C1 (step iv) and another tautomerization (step v) yield precorrin 6A.

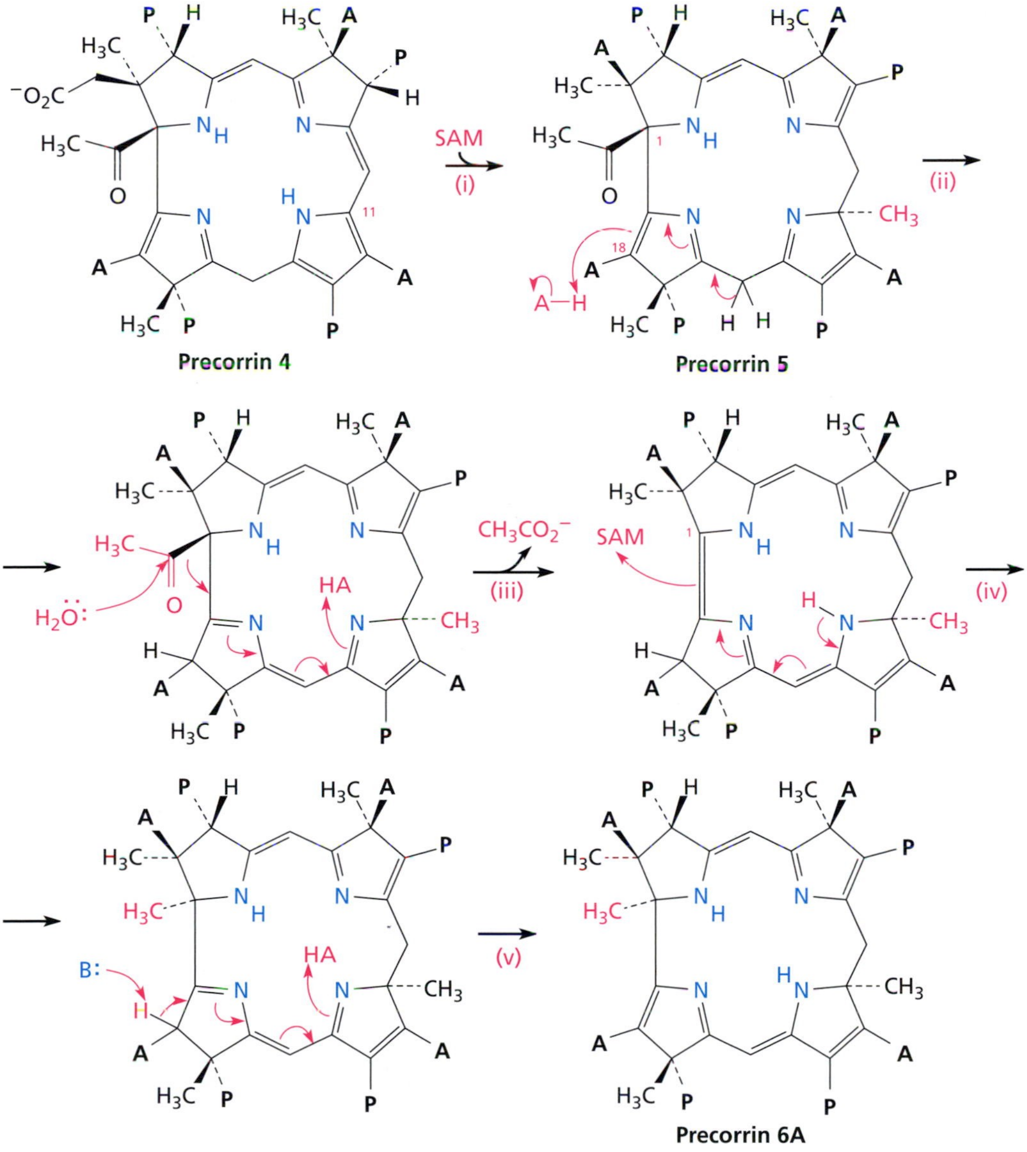

FIGURE 7.38 A possible mechanism for the loss of acetic acid from precorrin 5.

Step 5. Reduction Precorrin 6A is reduced to precorrin 6B in an NADPH-dependent reaction. Protonation of C18 gives an iminium ion, which undergoes hydride addition from NADPH at C19 (Figure 7.39).

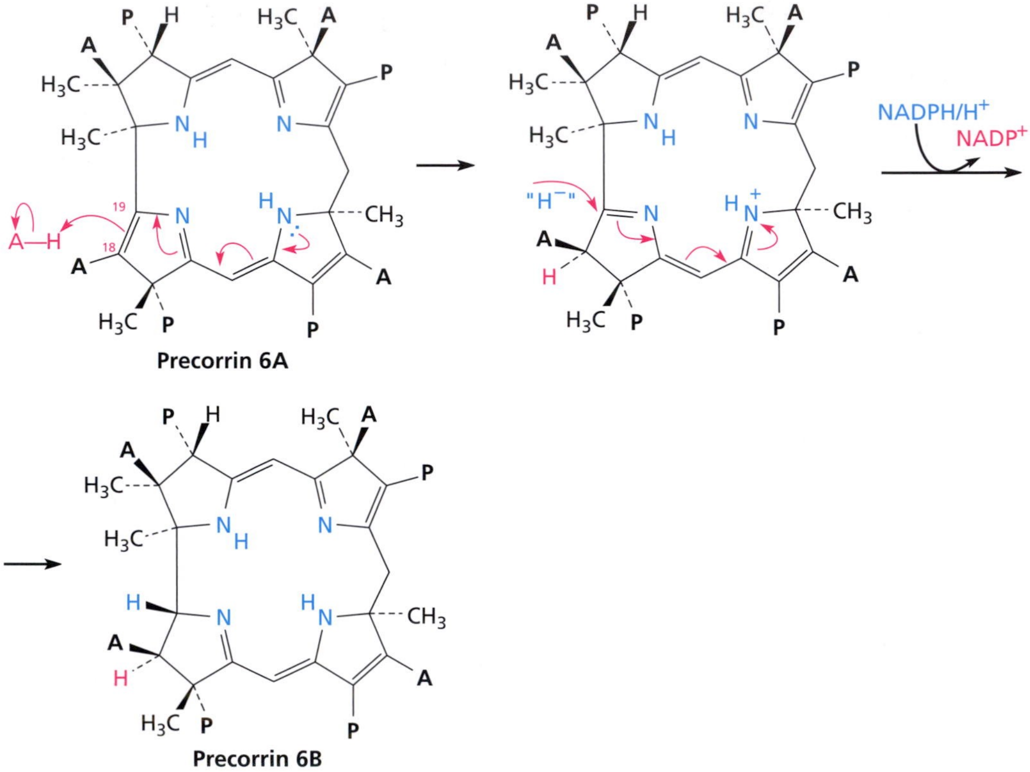

Precorrin 6A

NADPH/H$^+$ NADP$^+$

Precorrin 6B

FIGURE 7.39 Mechanism of the reduction of precorrin 6A to precorrin 6B.

Step 6. Methylations and decarboxylation Precorrin 6B next undergoes two additional SAM-dependent methylations, at C5 and C15 (Figure 7.40), followed by bond tautomerization. The acetic acid side chain at C12 then decarboxylates by a mechanism similar to that by which uro'gen III decarboxylates in heme biosynthesis (Figure 7.34), producing precorrin 8. All steps are catalyzed by the multifunctional precorrin 8 synthase.

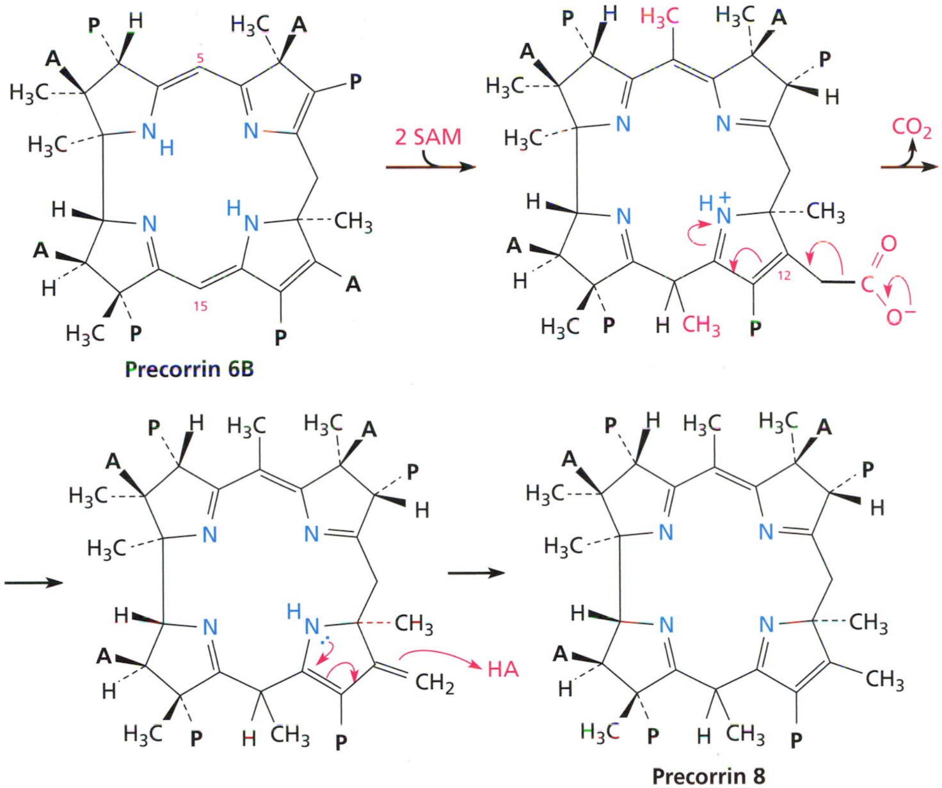

FIGURE 7.40 Mechanism of the methylation and decarboxylation of precorrin 6B to precorrin 8.

Step 7. Methyl shift Protonation of the C-ring nitrogen atom in precorrin 8 gives a carbocation (Figure 7.41, step i) that allows migration of the methyl group from C11 to C12 to occur (step ii). Tautomerization of the double bonds (step iii) then gives hydrogenobyrinic acid (HBA). The transformation is catalyzed by precorrin 8 mutase.[33]

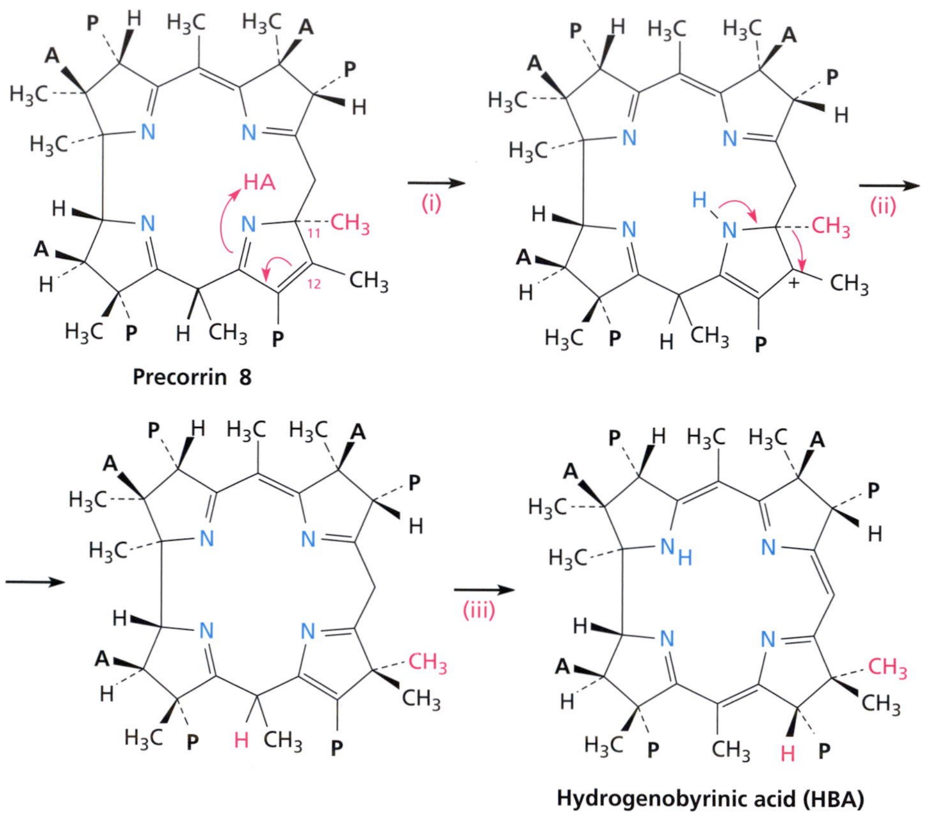

FIGURE 7.41 Mechanism of the conversion of precorrin 8 to hydrogenobyrinic acid.

With the appropriately substituted corrin ring constructed, the next phase in coenzyme B_{12} biosynthesis involves introduction of the central cobalt atom and modification of the various side chains (Figure 7.42). Hydrogenobyrinic acid is first converted to the corresponding *a,c*-diamide (step 1) in a reaction catalyzed by HBA *a,c*-diamide synthetase. Glutamine and ATP are required as cofactors, and the mechanism is presumably similar to that by which asparagine is biosynthesized from aspartate (Section 5.4). The diamide then has a cobalt(II) atom inserted by a cobaltochelatase (step 2) in a reaction analogous to the Fe(II) insertion that occurs in heme biosynthesis.

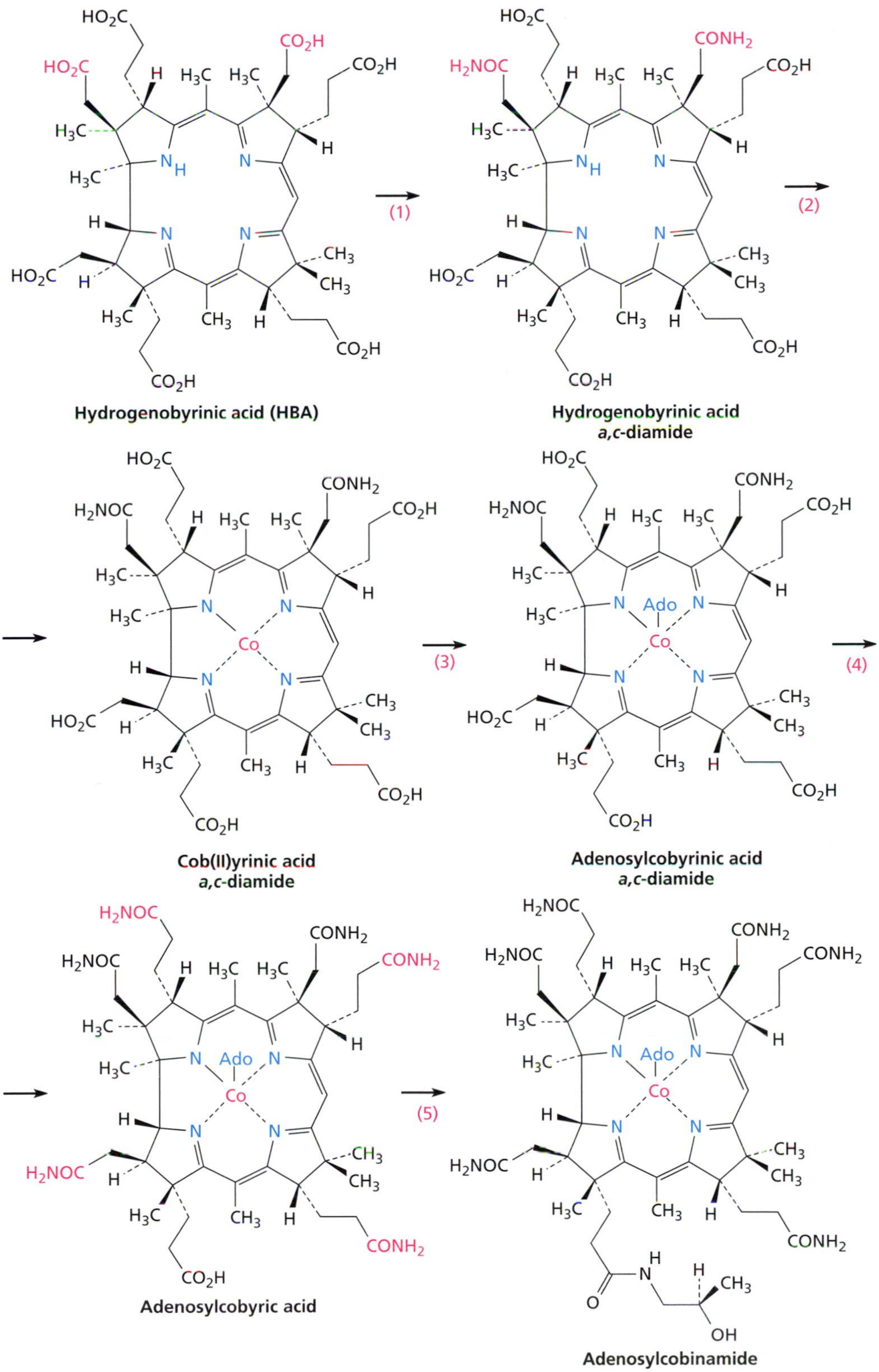

Hydrogenobyrinic acid (HBA)

(1)

Hydrogenobyrinic acid *a,c*-diamide

(2)

Cob(II)yrinic acid *a,c*-diamide

(3)

Adenosylcobyrinic acid *a,c*-diamide

(4)

Adenosylcobyric acid

(5)

Adenosylcobinamide

FIGURE 7.42 Pathway for the conversion of hydrogenobyrinic acid to adenosyl cobinamide.

The cobalt atom of cob(II)yrinic acid *a,c*-diamide is next reduced to the Co(I) oxidation state, and an adenosyl (Ado) ligand is added by cob(I)alamin adenosyl transferase (step 3). Four further side-chain amidations, all catalyzed by the same amidase enzyme, then lead to adenosylcobyric acid (step 4), which is attached to (*R*)-1-amino-2-propanol at the sole remaining carboxyl group to give adenosylcobinamide (step 5). Details of this step are not known, although the necessary aminopropanol is thought to arise from decarboxylation of threonine.

The final steps in B_{12} biosynthesis, shown in Figure 7.43, involve construction of the extended nucleotide side chain that acts as a ligand for the cobalt atom. Adenosylcobinamide is first converted to a guanosine diphosphate derivative, adenosyl-GDP-cobinamide, in a two-step process catalyzed by a bifunctional synthetase. The substrate reacts with ATP to give a monophosphate, which in turn reacts with GTP to complete the process. Substitution of guanosine monophosphate by α-ribazole then expels GMP and yields coenzyme B_{12}.

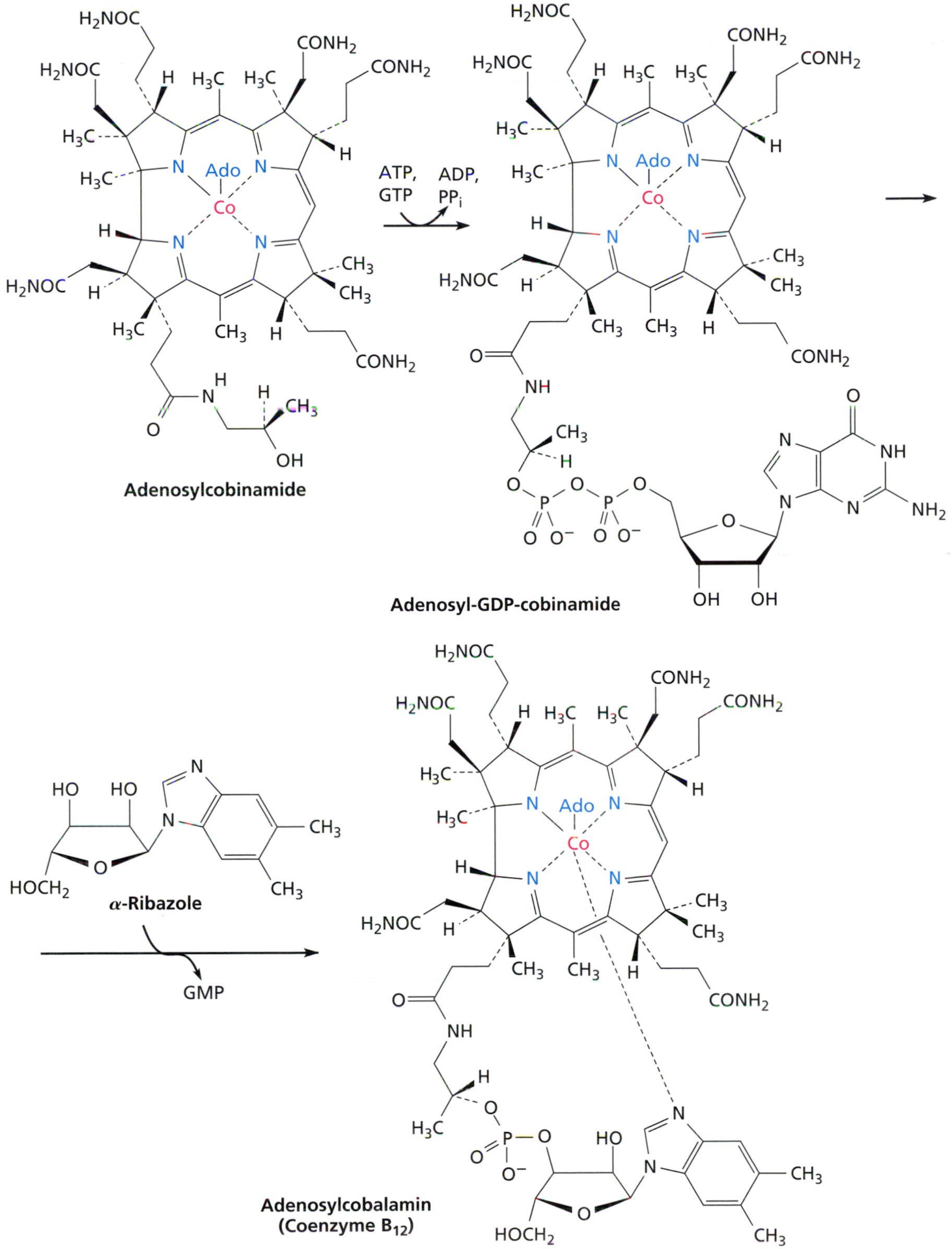

Adenosylcobinamide

ATP, GTP → ADP, PP_i

Adenosyl-GDP-cobinamide

α-Ribazole → GMP

Adenosylcobalamin (Coenzyme B_{12})

FIGURE 7.43 Final steps in the biosynthesis of coenzyme B_{12}.

The α-ribazole needed for the final step in B_{12} biosynthesis is itself synthesized by reaction of 5,6-dimethylbenzimidazole with nicotinic acid mononucleotide, followed by dephosphorylation. The substitution probably occurs by an S_N1 mechanism, similar to that seen previously for inverting glycosidases (Section 4.1, Figure 4.2). In aerobic organisms, the 5,6-dimethylbenzimidazole is derived from riboflavin, although the details of that process are not yet clear.[34]

Nicotinic acid mononucleotide + Dimethylbenzimidazole → α-Ribazole

References

1. Williams, D. H.; Stone, M. J.; Hauck, P. R.; Rahman, S. K., "Why Are Secondary Metabolites (Natural Products) Biosynthesized?" *J. Nat'l Prod.*, **1989**, *52*, 1189–1208.
2. Baldwin, J. E.; Byford, M. F.; Shiau, C.-Y.; Schofield, C. J., "The Mechanism of ACV Synthetase," *Chem. Rev.*, **1997**, *97*, 2631–2649.
3. Roach, P. L.; Clifton, I. J.; Hensgens, C. M. H.; Shibta, N.; Schofield, C. J.; Hajdu, J.; Baldwin, J. E., "Structure of Isopenicillin N Synthase Complexed with Substrate, and the Mechanism of Penicillin Formation," *Nature*, **1997**, *387*, 827–830.
4. Burzlaff, N. I.; Rutledge, P. J.; Clifton, I. J.; Hensgens, C. M.; Pickford, M.; Adlington, R. M.; Roach, P. L.; Baldwin, J. E., "The Reaction Cycle of Isopenicillin N Synthase Observed by X-ray Diffraction," *Nature*, **1999**, *401*, 721–724.
5. Valegard, K.; Terwisscha v.-S., A. C.; Lloyd, M. D.; Hara, T.; Ramaswamy, S.; Perrakis, A.; Thompson, A.; Lee, H.-J.; Baldwin, J. E.; Schofield, C. J.; Hajdu, J.; Andersson, I., "Structure of a Cephalosporin Synthase," *Nature*, **1998**, *394*, 805–809.
6. Lloyd, M. D.; Lee, H.-J.; Harlos, K.; Zhang, Z.-H.; Baldwin, J. E.; Schofield, C. J.; Charnock, J. M.; Garner, C. D.; Hara, T.; Terwisscha van Scheltinga, A. C.; Valegard, K.; Viklund, J. A. C.; Hajdu, J.; Andersson, I.; Danielsson, A.; Bhikhabhai, R., "Studies on the Active Site of Deacetoxycephalosporin C Synthase," *J. Mol. Biol.*, **1999**, *287*, 943–960.

7. Novak, B. H.; Hudlicky, T.; Reed, J. W.; Mulzer, J.; Trauner, D., "Morphine Synthesis and Biosynthesis: An Update," *Curr. Org. Chem.*, **2000**, *4*, 343–362.
8. Fitzpatrick, P. F., "Mechanism of Aromatic Amino Acid Hydroxylation," *Biochemistry*, **2000**, *42*, 14083–14091.
9. Gerardy, R.; Zenk, M. H., "Formation of Salutaridine from (*R*)-Reticuline by a Membrane-Bound Cytochrome P-450 Enzyme from *Papaver somniferum*," *Phytochemistry*, **1992**, *32*, 79–86.
10. Malkowski, M. G.; Ginell, S. L.; Smith, W. L.; Garavito, R. M., "The Productive Conformation of Arachidonic Acid Bound to Prostaglandin Synthase," *Science*, **2000**, *289*, 1933–1938.
11. Kiefer, J. R.; Pawlitz, J. L.; Moreland, K. T.; Stegeman, R. A.; Hood, W. F.; Gierse, J. K.; Stevens, A. M.; Goodwin, D. C.; Rowlinson, S. W.; Marnet, L. J.; Stallings, W. C.; Kurumbal, R. G., "Structural Insights into the Stereochemistry of the Cyclooxygenase Reaction," *Nature*, **2000**, *405*, 97–101.
12. van der Donk, W. A.; Tsai, A.-L.; Kulmacz, R. J., "The Cyclooxygenase Reaction Mechanism," *Biochemistry*, **2002**, *41*, 15451–15458.
13. Seibold, S. A.; Ball, T.; Hsi, L. C.; Mills, D. A.; Abeysinghe, R. D.; Micielli, R.; Rieke, C. J.; Cukier, R. I.; Smith, W. L., "Histidine-386 and Its Role in Cyclooxygenase and Peroxidase Catalysis by Prostaglandin–Endoperoxide H Synthases," *J. Biol. Chem.*, **2003**, *278*, 46163–46170.
14. Thoren, S.; Weinander, R.; Saha, S.; Jegerschoeld, C.; Pettersson, P. L.; Samuelsson, B.; Hebert, H.; Hamberg, M.; Morgenstern, R.; Jakobsson, P.-J., "Human Microsomal Prostaglandin E Synthase-1: Purification, Functional Characterization, and Projection Structure Determination," *J. Biol. Chem.*, **2003**, *278*, 22199–22209.
15. Ouellet, M.; Falgueyret, J.-P.; Hien E. P.; Pen, A.; Mancini, J. A.; Riendeau, D.; Percival, M. D., "Purification and Characterization of Recombinant Microsomal Prostaglandin E Synthase-1," *Protein Exp. Pur.*, **2002**, *26*, 489–495.
16. Staunton, J.; Weissman, K. J., "Polyketide Biosynthesis: A Millennium Review," *Nat. Prod. Rep.*, **2001**, *18*, 380–416.
17. Rawlings, B. J., "Type I Polyketide Biosynthesis in Bacteria (Part A—Erythromycin Biosynthesis)," *Nat. Prod. Rep.*, **2001**, *18*, 190–227.
18. Rawlings, B. J., "Type I Polyketide Biosynthesis in Bacteria (Part B)," *Nat. Prod. Rep.*, **2001**, *18*, 231–281.
19. Khosla, C.; Gokhale, R. S.; Jacobsen, J. R.; Cane, D. E., "Tolerance and Specificity of Polyketide Synthases," *Ann. Rev. Biochem*,, **1999**, *68*, 219–253.
20. McPherson, M.; Khosla, C.; Cane, D. E., "Erythromycin Biosynthesis: The β-Ketoreductase Domains Catalyze the Stereospecific Transfer of the 4-*pro-S* Hydride of NADPH," *J. Am. Chem. Soc.*, **1998**, *120*, 3267–3268.
21. Cupp-Vickery, J. R.; Han, O.; Hutchinson, R.; Poulos, T. L., "Substrate-Assisted Catalysis in Cytochrome P450eryF," *Nature Struct. Biol.*, **1996**, *3*, 632–637.
22. Guallar, V.; Harris, D. L.; Batista, V. S.; Miller, W. H., "Proton-Transfer Dynamics in the Activation of Cytochrome P450eryF," *J. Amer. Chem. Soc.*, **2002**, *124*, 1430–1437.

23. Shoolingin-Jordan, P. M.; Al-Daihan, S.; Alexeev, D.; Baxter, R. L.; Bottomley, S. S.; Kahari, I. D.; Roy, I.; Sarwar, M.; Sawyer, L.; Wang, S.-F., "5-Aminolevulinic Acid Synthase: Mechanism, Mutations and Medicine," *Biochim. Biophys. Acta*, **2003**, *1647*, 361–366.
24. Zhang, J.; Ferreira, G. C., "Transient State Kinetic Investigation of 5-Aminolevulinate Synthase Reaction Mechanism," *J. Biol. Chem.*, **2000**, *277*, 44660–44669.
25. Goodwin, C. E.; Leeper, F. J., "Stereochemistry and Mechanism of the Conversion of 5-Aminolevulinic Acid into Porphobilinogen Catalyzed by Porphobilinogen Synthase," *Org. Biomol. Chem.*, **2003**, *1*, 1443–1446.
26. Shoolingin-Jordan, P. M.; Al-Dbass, A.; McNeill, L. A.; Sarwar, M.; Butler, D., "Human Porphobilinogen Deaminase Mutations in the Investigation of the Mechanism of Dipyrromethane Cofactor Assembly and Tetrapyrrole Formation," *Biochem. Soc. Trans.*, **2003**, *31*, 731–735.
27. Shoolingin-Jordan, P. M., "Structure and Mechanism of Enzymes Involved in the Assembly of the Tetrapyrrole Macrocycle," *Biochem. Soc. Trans.*, **1998**, *26*, 326–336.
28. Martins, B. M.; Grimm, B.; Mock, H.-P.; Huber, R.; Messerschmidt, A., "Crystal Structure and Substrate Binding Modeling of the Uroporphyrinogen-III Decarboxylase from *Nicotiana tabacum*. Implications for the Catalytic Mechanism," *J. Biol. Chem.*, **2001**, *276*, 44108–44116.
29. Dailey, H. A.; Dailey, T. A.; Wu, C.-K.; Medlock, A. E.; Wang, K.-F.; Rose, J. P.; Wang, B.-C., "Ferrochelatase at the Millennium: Structures, Mechanisms, and [2Fe–2S] Clusters," *Cell. Mol. Life Sci.*, **2000**, *57*, 1909–1926.
30. Scott, A. I.; Roessner, C. A.; Santander, P. J., "Genetic and Mechanistic Exploration of the Two Pathways of Vitamin B_{12} Biosynthesis," Editors: Kadish, K. M.; Smith, K. M.; Guilard, R., *Porphyrin Handbook*, **2003**, *12*, 211–228. Elsevier Science, San Diego, CA.
31. Battersby, A. R., "Tetrapyrroles: The Pigments of Life," *Nat. Prod. Rep.*, **2000**, *17*, 507–526.
32. Eschenmoser, A., "Vitamin B_{12}: Experiments Concerning the Origin of Its Molecular Structure," *Angew. Chem. Int., Ed. Engl.*, **1988**, *27*, 5–39.
33. Shipman, L. W.; Li, D.; Roessner, C. A.; Scott, A. I.; Sacchettini, J. C., "Crystal Structure of Precorrin-8x Methyl Mutase," *Structure*, **2001**, *9*, 587–596.
34. Maggio-Hall, L. A.; Dorrestein, P. C.; Escalante-Semerena, J. C.; Begley, T. P., "Formation of the Dimethylbenzimidazole Ligand of Coenzyme B_{12} under Physiological Conditions by a Facile Oxidative Cascade," *Org. Letters*, **2003**, *5*, 2211–2213.

Problems

7.1 Show the mechanism of the PLP-dependent epimerization of isopenicillin N to penicillin N, a step in cephalosporin biosynthesis.

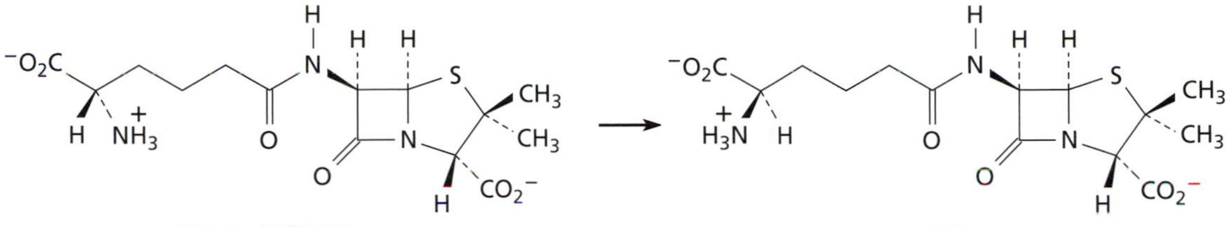

Isopenicillin N → **Penicillin N**

7.2 Assign *R* or *S* stereochemistry to the five chirality centers formed in the conversion of arachidonic acid to PGG_2.

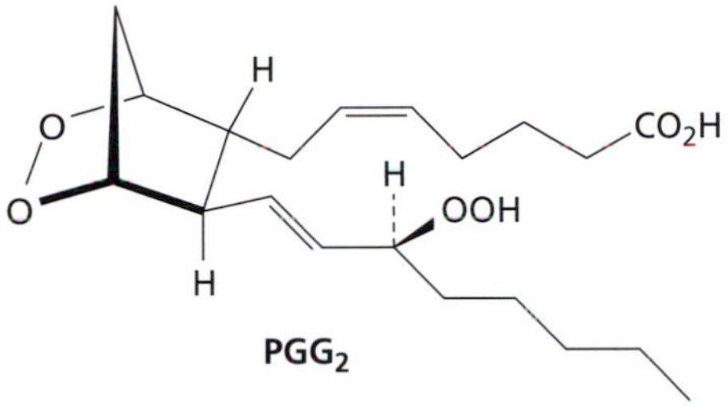

PGG_2

7.3 Assign *R* or *S* stereochemistry to the chirality centers in 6-deoxyerythronolide B (Figure 7.21).

7.4 Does the ketone-reduction step catalyzed by KR1 in erythromycin biosynthesis take place on the *re* face or the *si* face of the substrate? What about the reduction catalyzed by KR2? (See Figure 7.22).

7.5 When the enoyl reductase domain (ER4) in the erythromycin PKS is deactivated by gene mutation, all further steps still occur normally. What is the structure of the lactone that results?

7.6 Show the mechanism of the PLP-dependent formation of 5-aminolevulinate from glycine and succinyl CoA, the first step in tetrapyrrole biosynthesis.

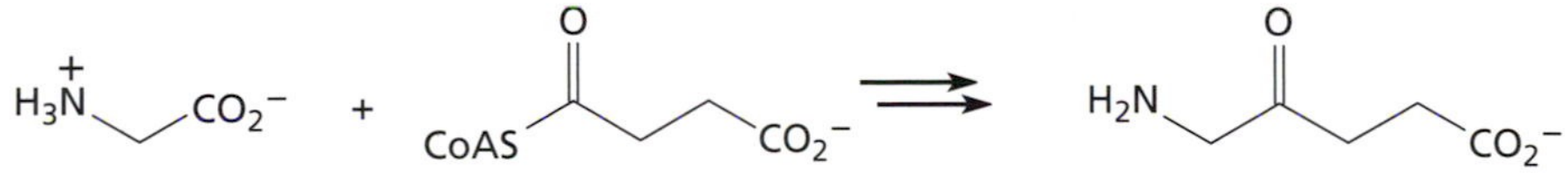

Glycine + **Succinyl CoA** ⇌ **5-Aminolevulinate**

7.7 Show the mechanism of the addition of a porphobilinogen molecule to the growing chain in hydroxymethylbilane biosynthesis.

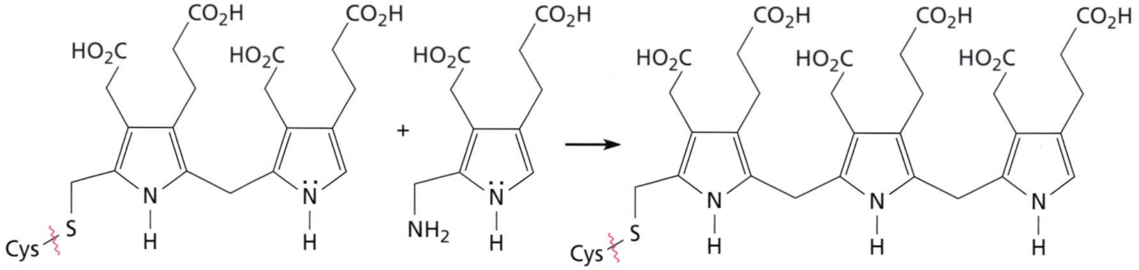

7.8 Show the mechanism of the twofold methylation of precorrin 6B, a step in coenzyme B_{12} biosynthesis (Figure 7.40).

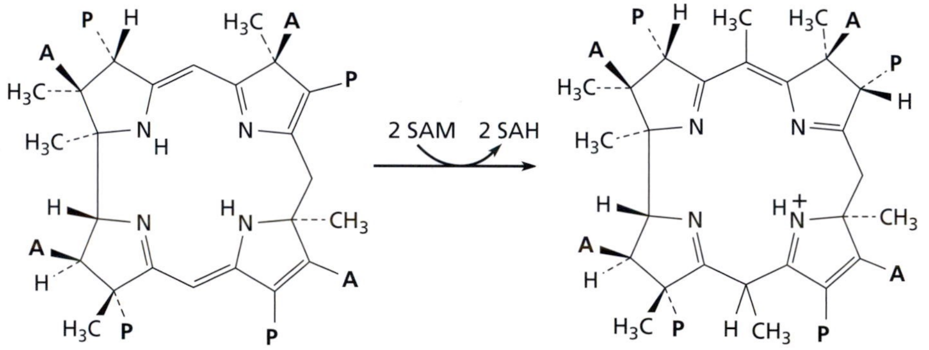

2 SAM 2 SAH

Precorrin 6B

7.9 Does the methyl group that migrates in the conversion of precorrin 8 to hydrogenobyrinic acid have *pro-R* or *pro-S* stereochemistry in the product?

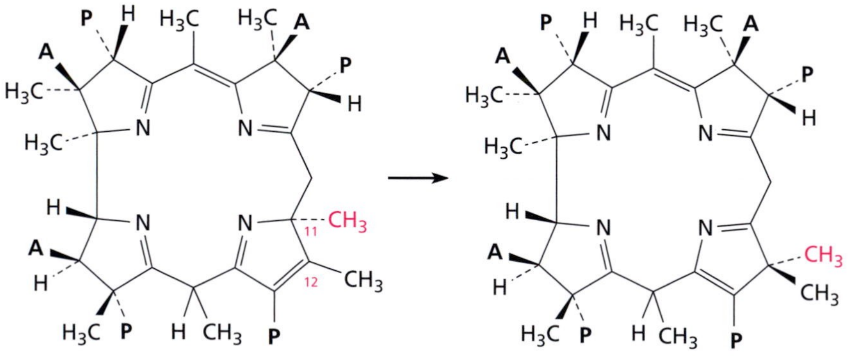

Precorrin 8

7.10 Show the mechanism of the final bond tautomerizations in hydrogenobyrinic acid biosynthesis (Figure 7.41).

Hydrogenobyrinic acid (HBA)

7.11 The alkaloid berbamunine is biosynthesized from (R)-N-methylcoclaurine in two steps, an initial epimerization to (S)-N-methylcoclaurine followed by a coupling of the two epimers. Review the morphine biosynthesis in Figure 7.9, and propose mechanisms for both steps.

(*S*)-*N*-Methylcoclaurine

(*R*)-*N*-Methylcoclaurine

Berbamunine

7.12 Ethylene is biosynthesized by a pathway that begins with the conversion of *S*-adenosylmethionine into 1-aminocyclopropanecarboxylate. What cofactor is likely to be needed for this reaction? Propose a mechanism.

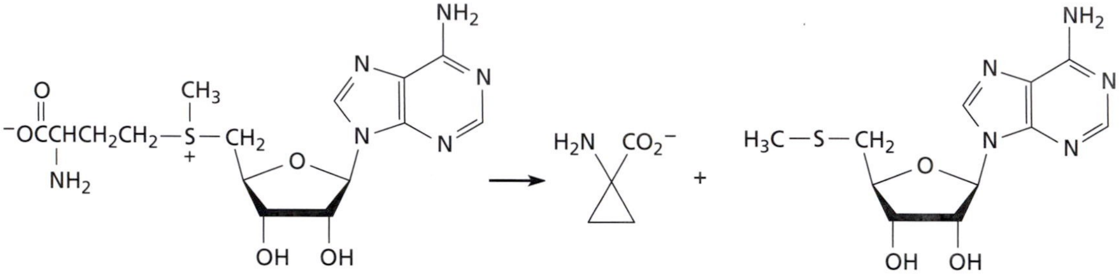

S-Adenosylmethionine

7.13 Propose a biosynthetic pathway for the following conversion, and identify any cofactors that are likely to be involved:

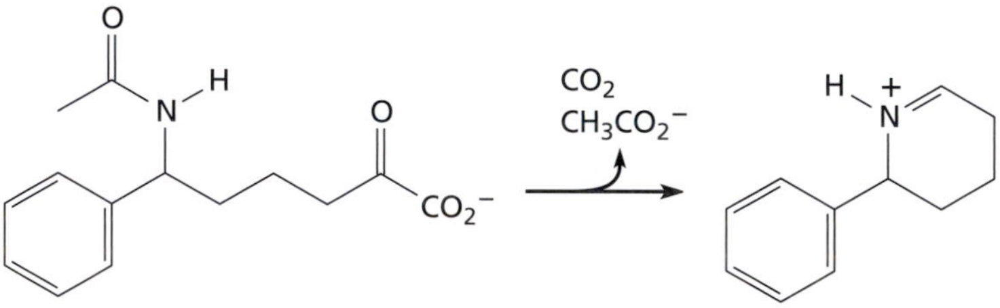

7.14 The following reaction is observed on treatment of deacetoxycephalosporin C with an α keto acid–dependent, nonheme iron monooxygenase. Propose a mechanism.

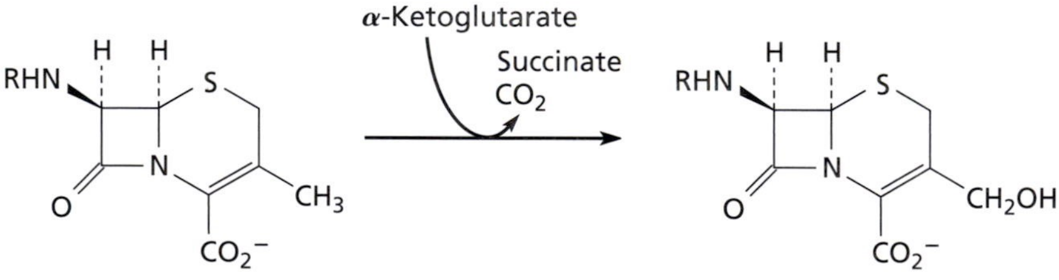

Deacetoxycephalosporin C

7.15 Pyridoxal phosphate is biosynthesized by a pathway that includes the following step. Propose a mechanism.

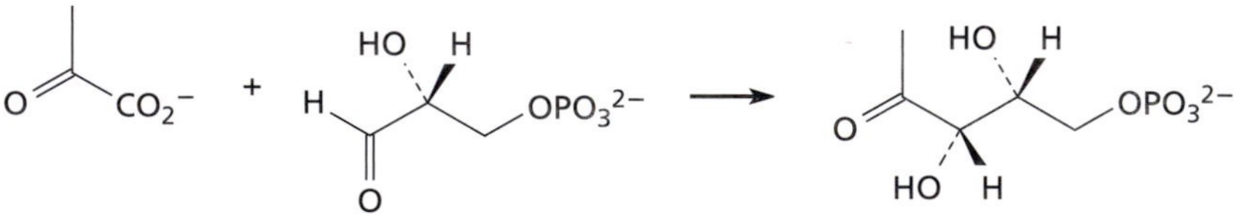

7.16 Two of the final steps in the biosynthesis of pyridoxal phosphate are shown. Identify any cofactors involved, and propose mechanisms for both steps.

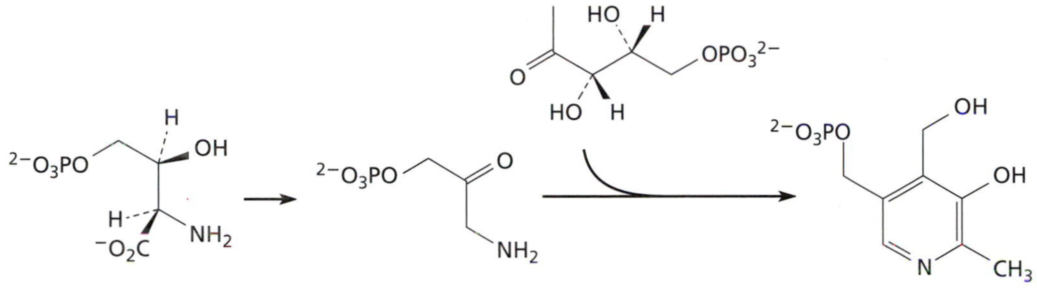

7.17 Retrieve the PDB coordinate file for isopenicillin N synthase (Figure 7.5), and display the structure using the Swiss PDB viewer. (The PDB code is 1BLZ.)

(a) In the mechanism outlined in Figure 7.6, an amide nitrogen adds to a thioaldehyde. Suggest a possible active-site base for amide deprotonation.

(b) What is the separation between the amide nitrogen and the thiocarbonyl carbon?

(c) What is the driving force for the formation of this strained ring?

7.18 Retrieve the PDB coordinate file for prostaglandin H synthase (Figure 7.20), and display the structure using the Swiss PDB viewer. (The PDB code is 1DIY.)

(a) Which carbon of arachidonic acid is closest to the phenoxy radical of Tyr-385? Is this consistent with the mechanistic proposal in Figure 7.19?

(b) The electron transfer from Tyr-485 to the porphyrin occurs by a long-distance electron-transfer mechanism. What is the shortest distance between the π system of the porphyrin and the π system of Tyr-385?

8 A Summary of Biological Transformations

In writing this book, we chose to organize our coverage by biological pathway rather than by type of transformation or reaction mechanism. We did this because we believe it's easier to understand why transformations take place as they do when they're viewed in the context of an overall biochemical scheme rather than in isolation. Now that we've seen examples of almost all the common biological transformations, however, it's worthwhile to go back and organize those transformations by reaction type. That way, you'll know what kind of mechanism to expect when you see a common sort of transformation in an unfamiliar pathway or biosynthetic scheme.

Hydrolyses, Esterifications, Thioesterifications, and Amidations

Amide and ester hydrolyses occur by several mechanisms, but the most common is through formation of an acyl enzyme intermediate by action of a catalytic triad of amino acid residues. A serine residue in the enzyme does a nucleophilic acyl substitution on the ester or amide, and the acyl enzyme is then hydrolyzed. The hydrolysis of triacylglycerols discussed in Section 3.1 and shown in Figure 3.1 is an example.

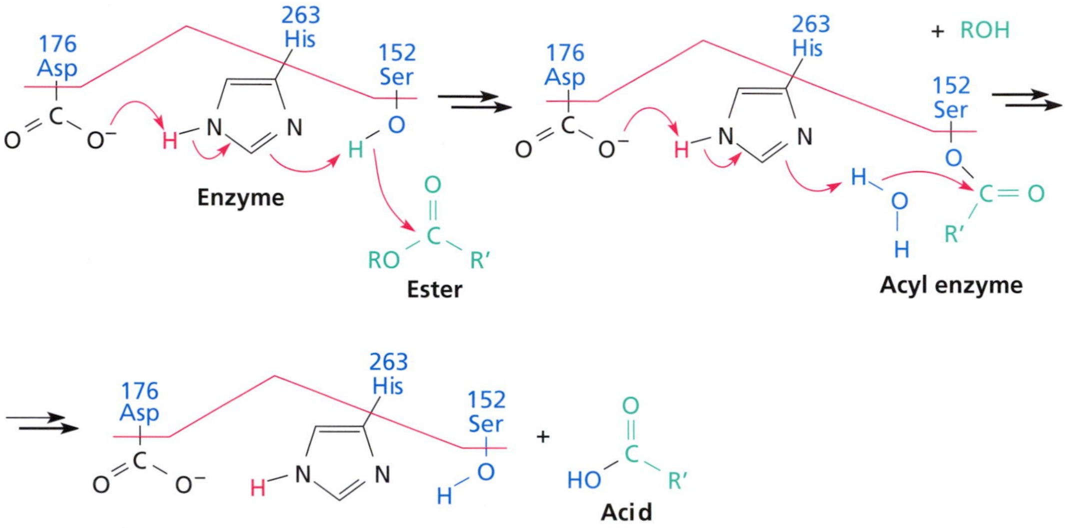

176 Asp, 263 His, 152 Ser, Enzyme, Ester, + ROH, Acyl enzyme, Acid

Esters are formed from alcohols by several mechanisms, frequently involving an intermediate thioester. The biosynthesis of acylglycerols discussed in Section 3.1 is an example.

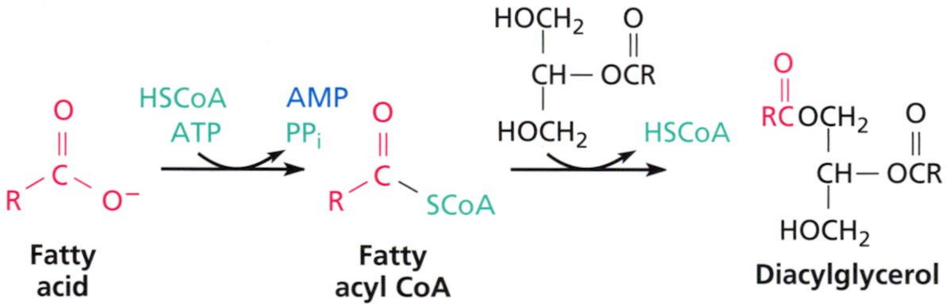

Fatty acid → (HSCoA, ATP; AMP, PP_i) Fatty acyl CoA → (HOCH$_2$–CH(OCOR)–CH$_2$OH; HSCoA) Diacylglycerol

Primary amides are often formed by nucleophilic acyl substitution of ammonia on an acyl adenylate or acyl phosphate, with the ammonia produced by hydrolysis of glutamine. The biosyntheses of asparagine and glutamine are examples (Section 5.4). Secondary amides are formed similarly.

Aspartate → (ATP, PP_i) → (AMP) → Asparagine

Glutamate → (ATP, ADP) → (P_i) → Glutamine

Carbonyl Condensations

A Claisen condensation reaction between two esters or thioesters, frequently through an acyl enzyme intermediate, is commonly used to form β keto esters. The synthesis of acetoacetyl CoA in step 1 of the mevalonate pathway for terpenoid biosynthesis is an example (Section 3.5 , Figure 3.17).

Acetyl CoA → + CoASH → Acetoacetyl CoA

The Claisen condensation reaction is reversible, so β keto esters can be cleaved. Step 4 of the β-oxidation pathway for fatty-acid catabolism is an example (Section 3.3, Figure 3.10).

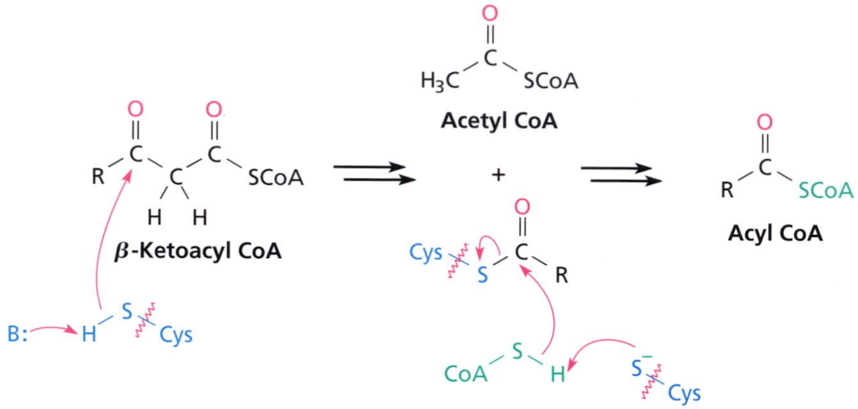

Aldol condensation of a ketone is mechanistically similar to a Claisen condensation and yields a β hydroxy ketone or ester. The addition of acetyl CoA to oxaloacetate in the first step of the citric acid cycle is an example (Section 4.4, Figure 4.12).

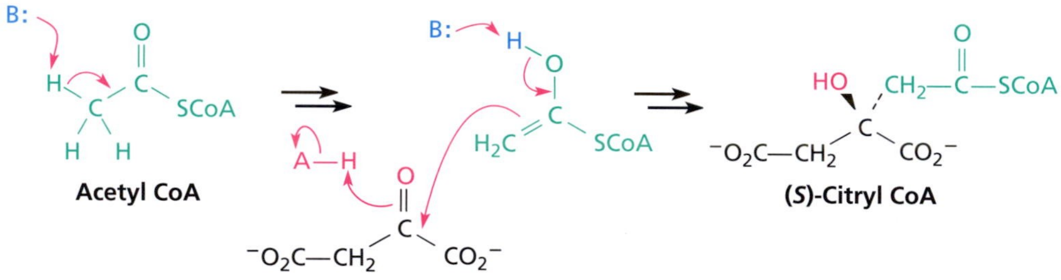

The aldol condensation reaction is reversible, so a β hydroxy carbonyl compound can split into two pieces. The cleavage of fructose 1,6-bisphosphate in step 4 of glycolysis is an example (Section 4.2, Figure 4.5). Note that the ketone is activated by conversion to an iminium ion prior to cleavage.

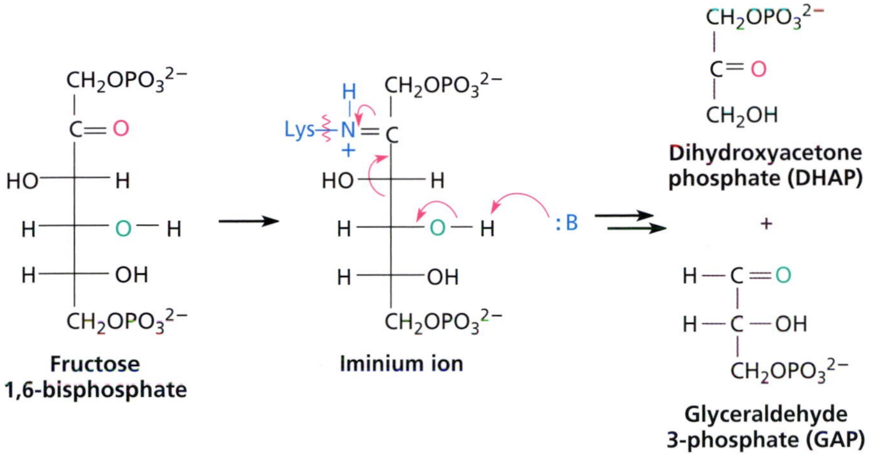

Fructose 1,6-bisphosphate → Iminium ion ⇌ Dihydroxyacetone phosphate (DHAP) + Glyceraldehyde 3-phosphate (GAP)

Carboxylations and Decarboxylations

Carboxylation α to a carbonyl group in bacteria and animals is usually carried out by carboxybiotin. The formation of malonyl CoA from acetyl CoA in step 3 of fatty-acid biosynthesis is an example (Section 3.4, Figure 3.14).

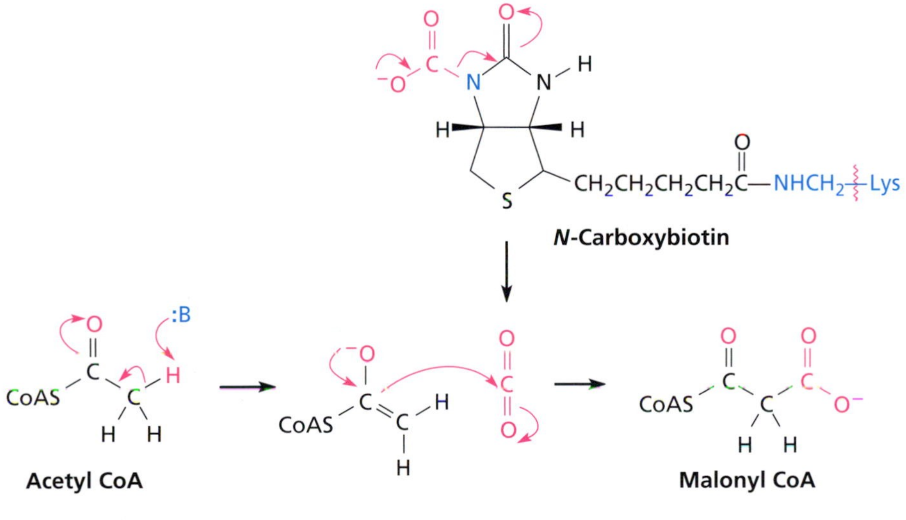

N-Carboxybiotin

Acetyl CoA → Malonyl CoA

In plants, which have access to higher concentrations of CO_2, direct carboxylation can occur, as in the carboxylation of ribulose 1,5-bisphosphate catalyzed by Rubisco in step 1 of the Calvin cycle (Section 4.7, Figure 4.22).

Decarboxylations occur by several different mechanisms but have as their common theme the presence of an electron-accepting group two carbons away from the carboxylate. Frequently, the substrate is a β keto acid, as occurs when oxalosuccinate is converted to α-ketoglutarate in step 3 of the citric acid cycle (Section 4.4).

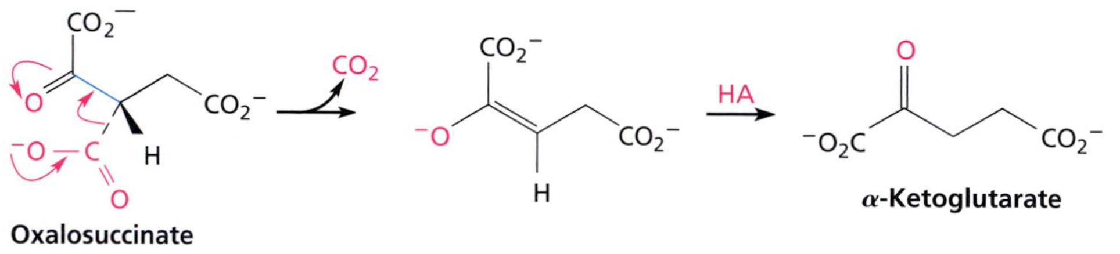

Oxalosuccinate → **α-Ketoglutarate**

Alternatively, the electron acceptor may be a leaving group such as phosphate ion. An example occurs in the conversion of 3-phosphomevalonate 5-diphosphate to isopentenyl diphosphate in step 4 of the mevalonate pathway for terpenoid biosynthesis (Section 3.5, Figure 3.20).

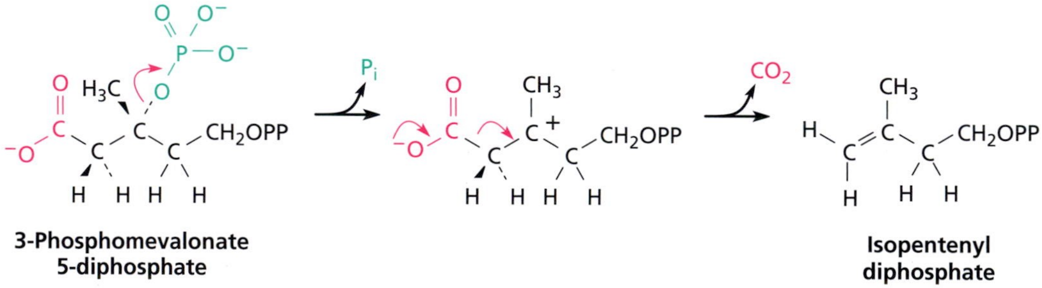

3-Phosphomevalonate 5-diphosphate → **Isopentenyl diphosphate**

α Keto acids decarboxylate by a multistep, thiamin-dependent pathway that begins with nucleophilic addition of thiamin diphosphate (TPP) ylid to the ketone. Protonation of the decarboxylated intermediate, followed by loss of TPP ylid, gives an aldehyde. The decarboxylation of pyruvate to yield acetaldehyde (Section 4.3, Figure 4.8) and the decarboxylation of *p*-hydroxyphenylpyruvate in the opening steps of morphine biosynthesis (Section 7.2, Figure 7.10) are examples.

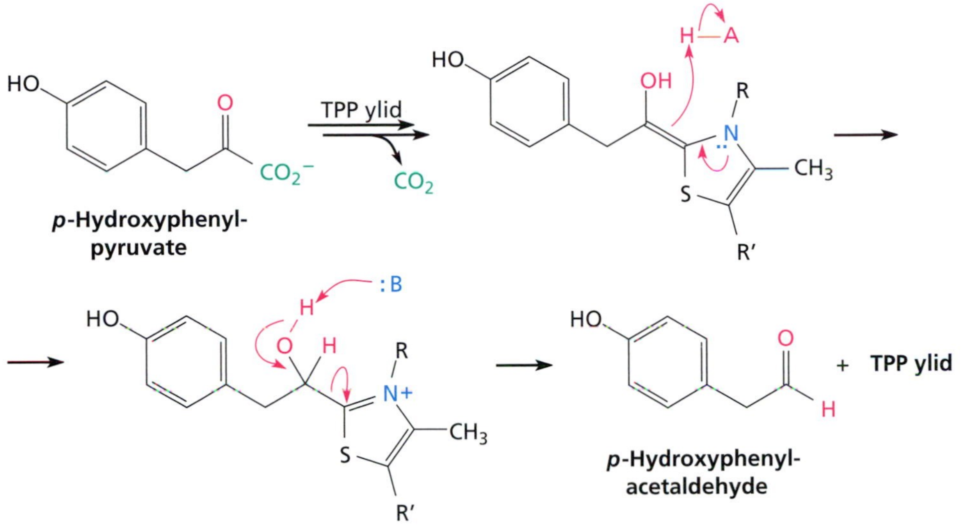

HO TPP ylid CO_2^- CO_2 OH H—A R N CH_3 S R'

p-Hydroxyphenyl-pyruvate

:B H HO O H R N+ CH_3 S R' HO O H + **TPP ylid**

p-Hydroxyphenyl-acetaldehyde

Alternatively, the initial TPP addition product can be oxidized by reaction with lipoamide to yield a thioester, resulting in an overall *oxidative* decarboxylation. The conversion of pyruvate to acetyl CoA is an example (Section 4.3, Figure 4.9).

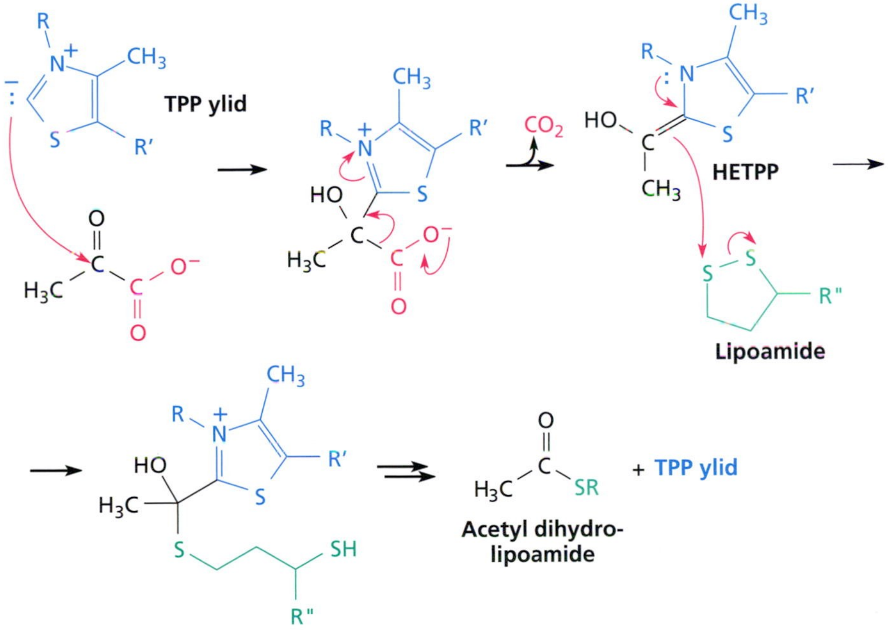

R N^+ CH_3 S R' **TPP ylid** O C O^- H_3C C O

CH_3 R N+ R' HO S C H_3C C O^- O CO_2

CH_3 R N R' HO C S CH_3 **HETPP** S S R" **Lipoamide**

CH_3 R + N R' HO H_3C S S SH R"

O C H_3C SR + **TPP ylid**

Acetyl dihydro-lipoamide

Yet another common mechanism for decarboxylation is the PLP-dependent pathway used by amino acids. The decarboxylation of *meso*-2,6-diaminopimelate that occurs in lysine biosynthesis is an example (Section 5.5, Figure 5.34).

PLP–Diaminopimelate imine → (CO_2) → … → **Lysine**

Aminations and Deaminations

Aminations and deaminations are generally carried out by a PLP-dependent transamination mechanism operating in either a forward or reverse direction (Section 5.1, Figure 5.1). In the amination direction, an *α* keto acid reacts with PMP to give an amino acid. In the deamination direction, an amino acid reacts with PLP to give an *α* keto acid.

***α* Amino acid** ⇌ **PLP–amino acid imine** ⇌ ***α* Keto acid imine** ⇌ (H_2O) **Pyridoxamine phosphate (PMP)** + ***α* Keto acid**

Deamination can also be carried out oxidatively by reaction of an amino acid with NAD^+ to yield an imine that is hydrolyzed. The conversion of glutamate to α-ketoglutarate discussed in Section 5.1 is an example.

Glutamate —(NAD^+ → NADH)→ α-Iminoglutarate —(H_2O → NH_4^+)→ α-Ketoglutarate

1-Carbon Transfers

The most common 1-carbon transfer is the S_N2 methylation reaction carried out by *S*-adenosylmethionine (SAM). The methylation of uro'gen III to precorrin 1, a step in coenzyme B_{12} biosynthesis, is an example (Section 7.5, Figure 7.35).

S-Adenosylmethionine (SAM) + Uro'gen III →

Derivatives of tetrahydrofolate are also found as 1-carbon donors. 5,10-Methylenetetrahydrofolate can donate a CH_2 group, as in the synthesis of thymidine monophosphate from deoxyuridine monophosphate (Figure 8.1 and Section 6.4, Figure 6.16).

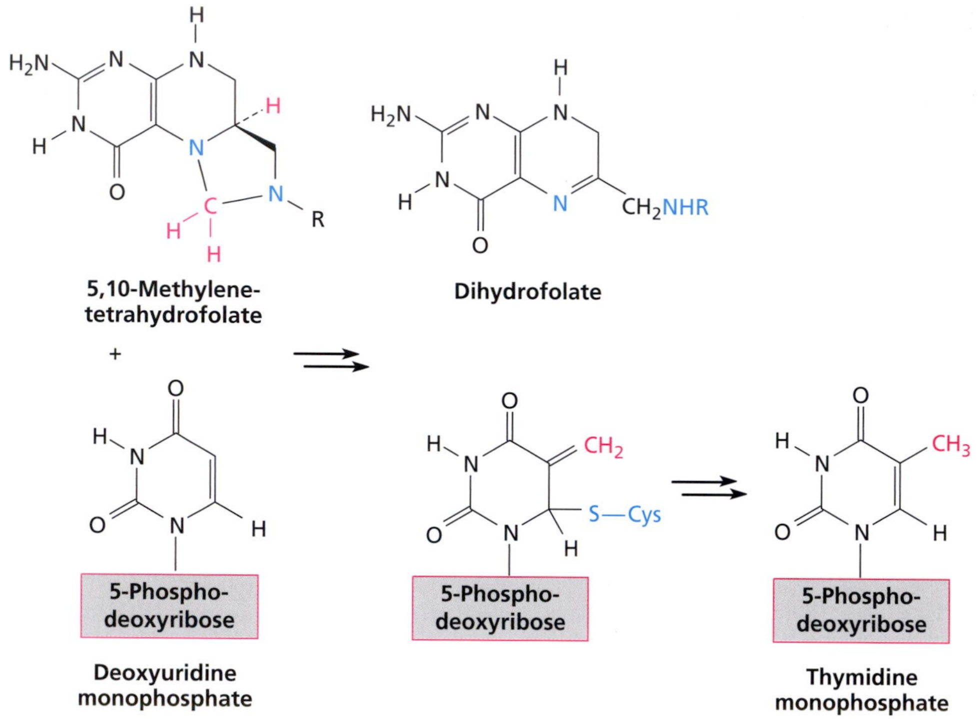

5,10-Methylene-tetrahydrofolate + Deoxyuridine monophosphate ⇌ Dihydrofolate + Thymidine monophosphate

FIGURE 8.1 Synthesis of thymidine monophosphate from deoxyuridine monophosphate.

Alternatively, 10-formyltetrahydrofolate can donate a formyl group, as occurs in step 3 of the biosynthesis of inosine (Section 6.3, Figure 6.10).

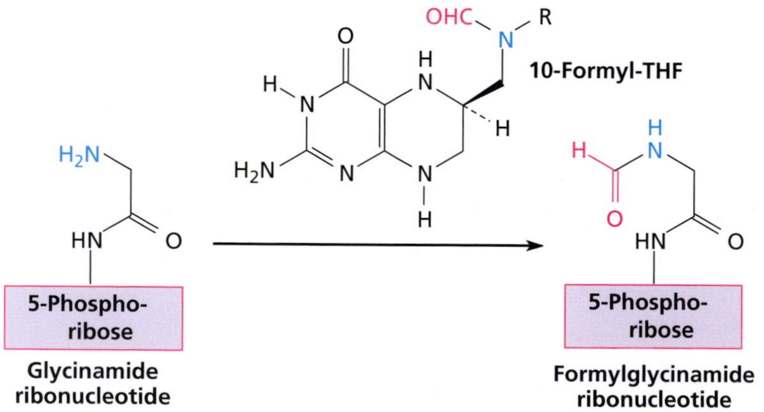

Glycinamide ribonucleotide + 10-Formyl-THF → Formylglycinamide ribonucleotide

Rearrangements

Rearrangements of many mechanistic types occur. One example is the acyloin rearrangement of a hydroxy ketone to an isomeric hydroxy ketone, as occurs in the DXP pathway for terpenoid biosynthesis (Section 3.5, Figure 3.23).

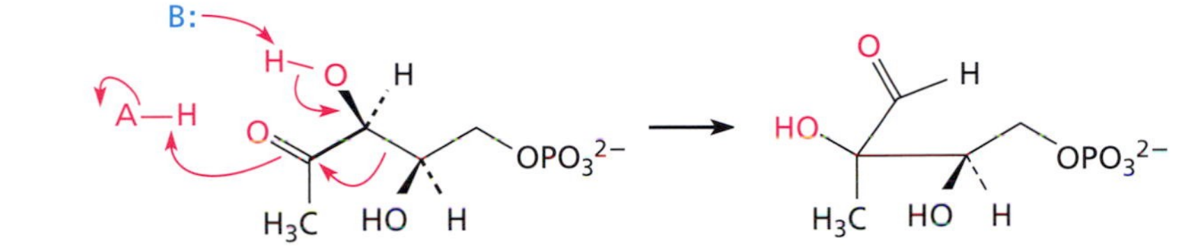

1-Deoxy-D-xylulose 5-phosphate

Pinacol rearrangement of a diol or diol monoester to form a ketone is closely related to the acyloin rearrangement. An example occurs in the conversion of precorrin 3B to precorrin 4 (Section 7.5, Figure 7.37).

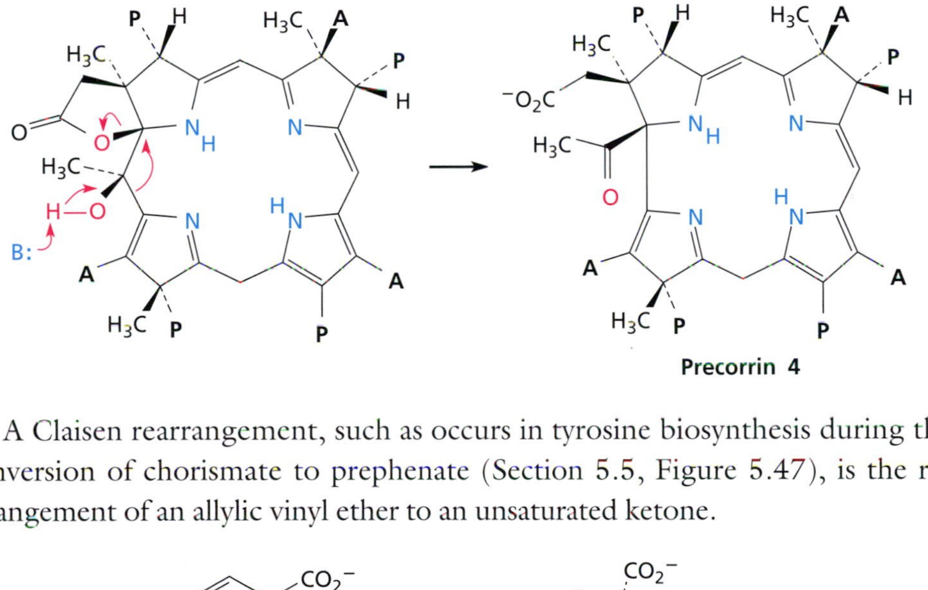

Precorrin 4

A Claisen rearrangement, such as occurs in tyrosine biosynthesis during the conversion of chorismate to prephenate (Section 5.5, Figure 5.47), is the rearrangement of an allylic vinyl ether to an unsaturated ketone.

Chorismate **Prephenate**

Enzymes containing coenzyme B_{12} can catalyze rearrangements that involve the exchange of a hydrogen atom with a group on the neighboring carbon. An example occurs during threonine metabolism when methylmalonyl CoA rearranges to succinyl CoA (Section 5.3, Figure 5.9).

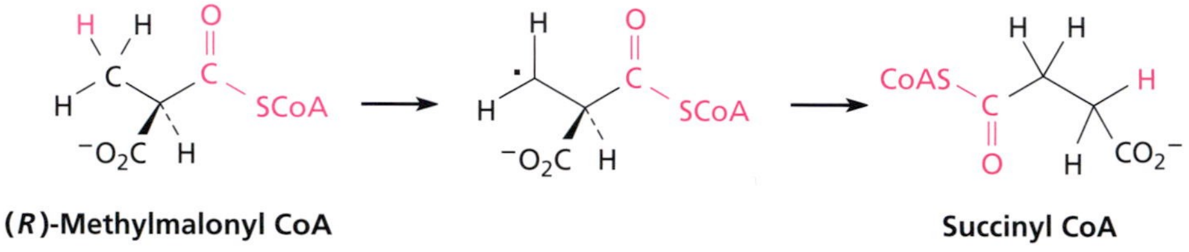

(R)-Methylmalonyl CoA → → **Succinyl CoA**

Isomerizations and Epimerizations

Isomerizations involving carbonyl compounds frequently occur through enol intermediates. The interconversion of glucose 6-phosphate with fructose 6-phosphate during step 2 of glycolysis is an example (Section 4.2, Figure 4.4).

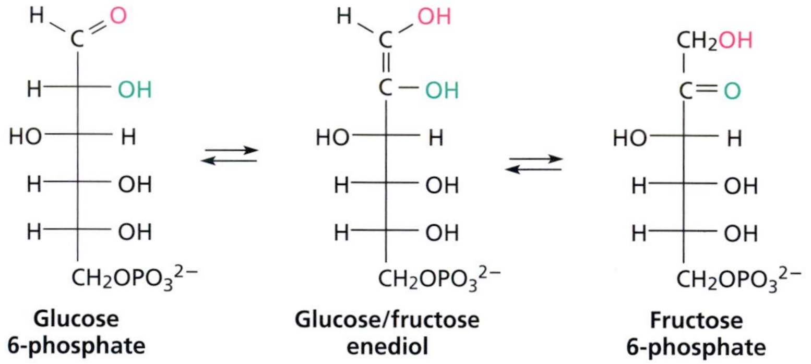

Glucose 6-phosphate ⇌ **Glucose/fructose enediol** ⇌ **Fructose 6-phosphate**

Epimerization of the chiral center next to a carbonyl group can occur by several mechanisms. Often, a direct deprotonation–reprotonation path is followed, as in the epimerization of ribulose 5-phosphate to xylulose 5-phosphate in step 4 of the pentose phosphate pathway (Section 4.6).

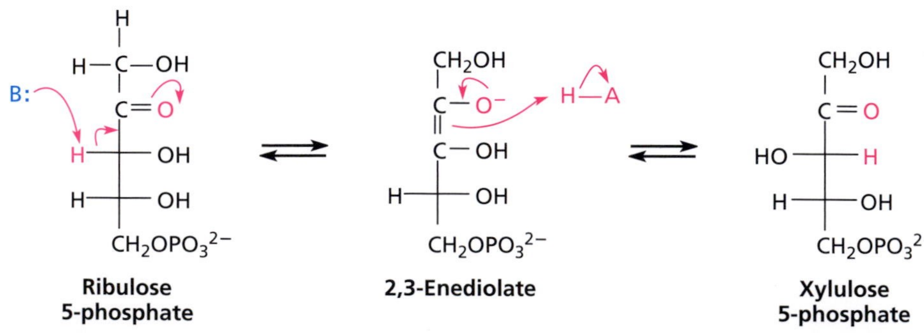

Ribulose 5-phosphate ⇌ **2,3-Enediolate** ⇌ **Xylulose 5-phosphate**

Alternatively, a PLP-dependent path is often followed for epimerization of amino acids. Formation of the PLP–imine makes the α hydrogen more acidic and more easily removed. An example occurs in the isomerization of isopenicillin N to penicillin N (Figure 8.2 and Figure 7.7, Section 7.1).

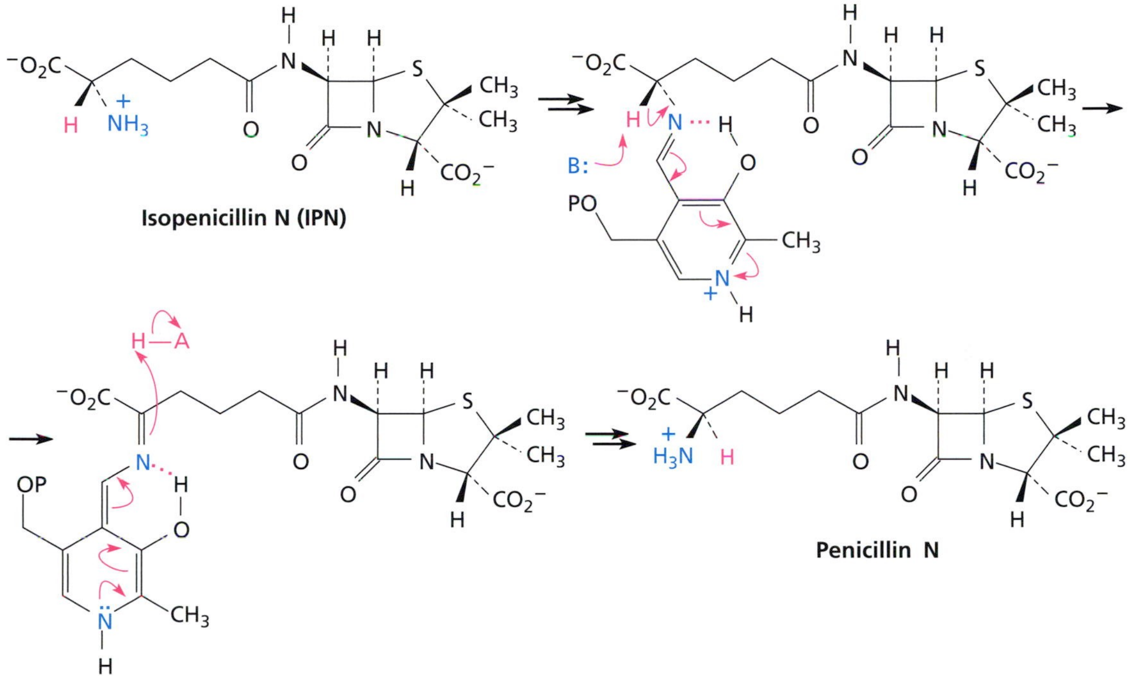

Isopenicillin N (IPN)

Penicillin N

FIGURE 8.2 The isomerization of isopenicillin N to penicillin N.

Oxidations and Reductions of Carbonyl Compounds

Oxidation of an alcohol to give an aldehyde or ketone generally occurs by hydride transfer from the alcohol to either NAD^+ or $NADP^+$ (or, less frequently, FAD). The oxidation of *sn*-glycerol 3-phosphate to dihydroxyacetone phosphate that occurs during glycerol catabolism is an example (Section 3.2, Figure 3.5).

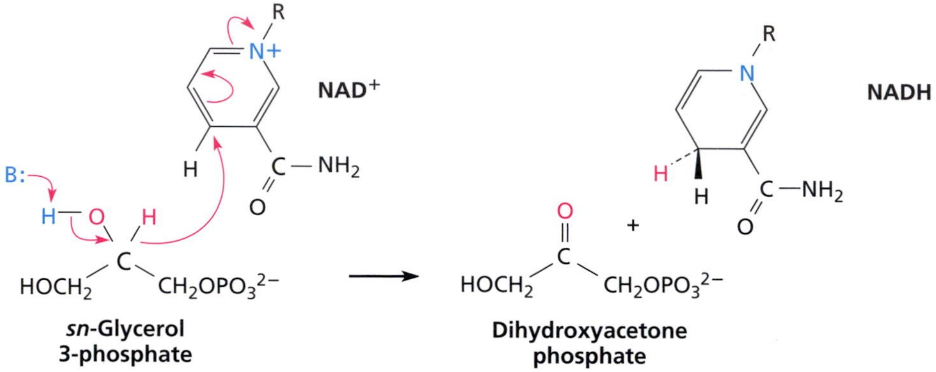

NAD^+ NADH

sn-Glycerol 3-phosphate → Dihydroxyacetone phosphate

Oxidation of an aldehyde to a carboxylic acid or ester generally involves formation of an intermediate hydrate or hemithioacetal, which is oxidized by NAD^+ by the usual hydride-transfer mechanism. An example is the oxidation of glyceraldehyde 3-phosphate in step 6 of glycolysis (Section 4.2, Figure 4.7).

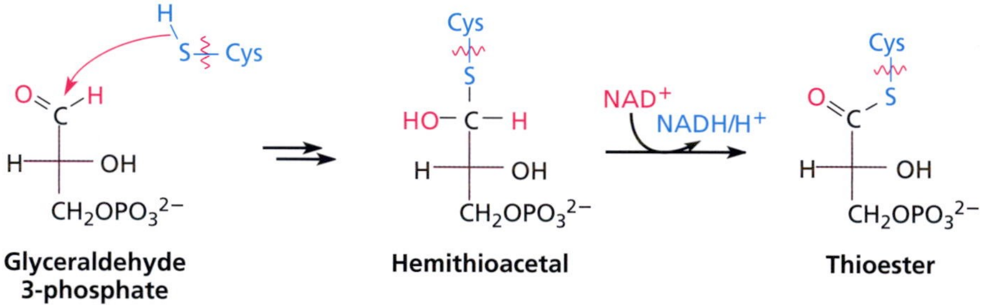

Glyceraldehyde 3-phosphate ⇌ Hemithioacetal → (NAD^+ → NADH/H^+) Thioester

Carbonyl compounds are dehydrogenated by FAD to α,β-unsaturated products in a mechanistically complex reaction that may either involve radicals or a hydride transfer. An example occurs in step 1 of the fatty-acid β-oxidation pathway (Section 3.3).

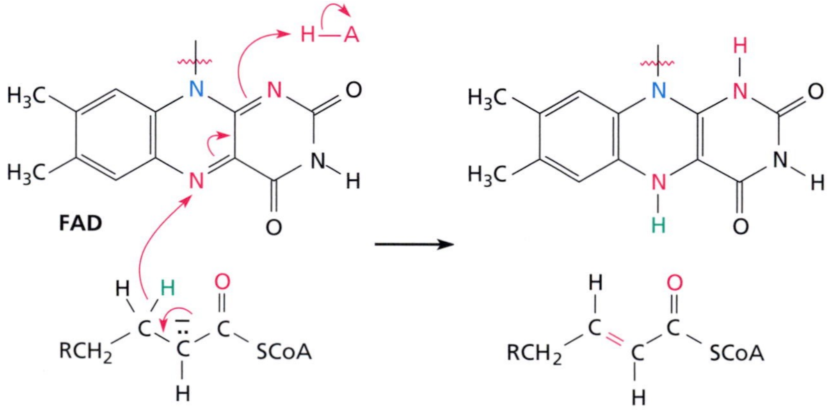

FAD

The reduction of a carbonyl compound to an alcohol is the reverse of alcohol oxidation and occurs by hydride transfer from NADH or NADPH. Reduction of acetoacetyl ACP to β-hydroxybutyryl ACP in step 6 of fatty-acid biosynthesis is an example (Section 3.4, Figure 3.15).

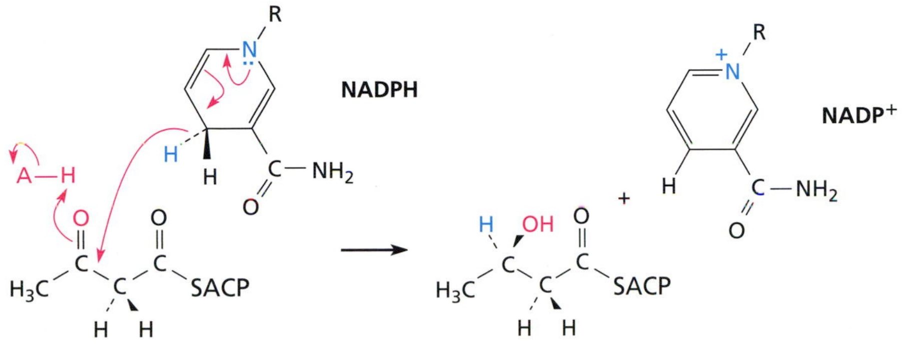

NADPH NADP$^+$

Acetoacetyl ACP β-Hydroxybutyryl ACP

α,β-Unsaturated carbonyl compounds are reduced by a conjugate addition reaction of hydride ion to the β carbon, followed by protonation on the α carbon. The reduction of crotonyl ACP that occurs in step 8 of fatty-acid biosynthesis (Section 3.4) is an example.

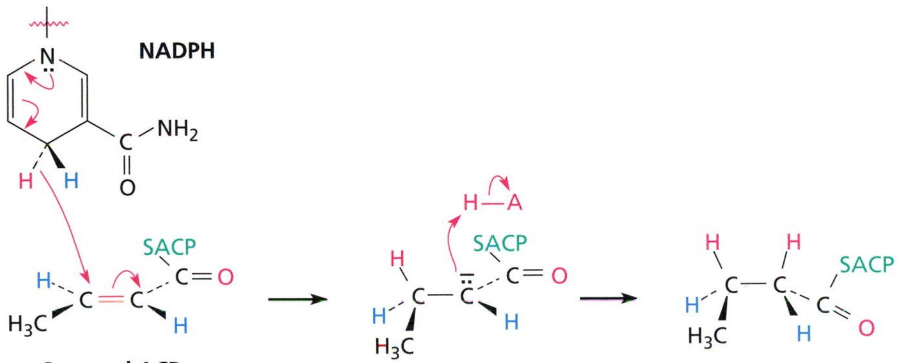

NADPH

Crotonyl ACP Butyryl ACP

Hydroxylations and Other Oxidations via Metal Complexes

An enzyme that uses molecular oxygen as the oxidizing agent and incorporates only one of the O_2 atoms into the product is called a *monooxygenase*. Many monooxygenases make use of an iron–oxo cofactor, which can be formed in several different ways. In the hydroxylation of phenylalanine to give tyrosine (Section 5.3, Figures 5.26 and 5.27), for instance, an iron(IV)–oxo complex is formed by reaction of 5,6,7,8-tetrahydrobiopterin with O_2 and an Fe(II) precursor (Figure 8.3).

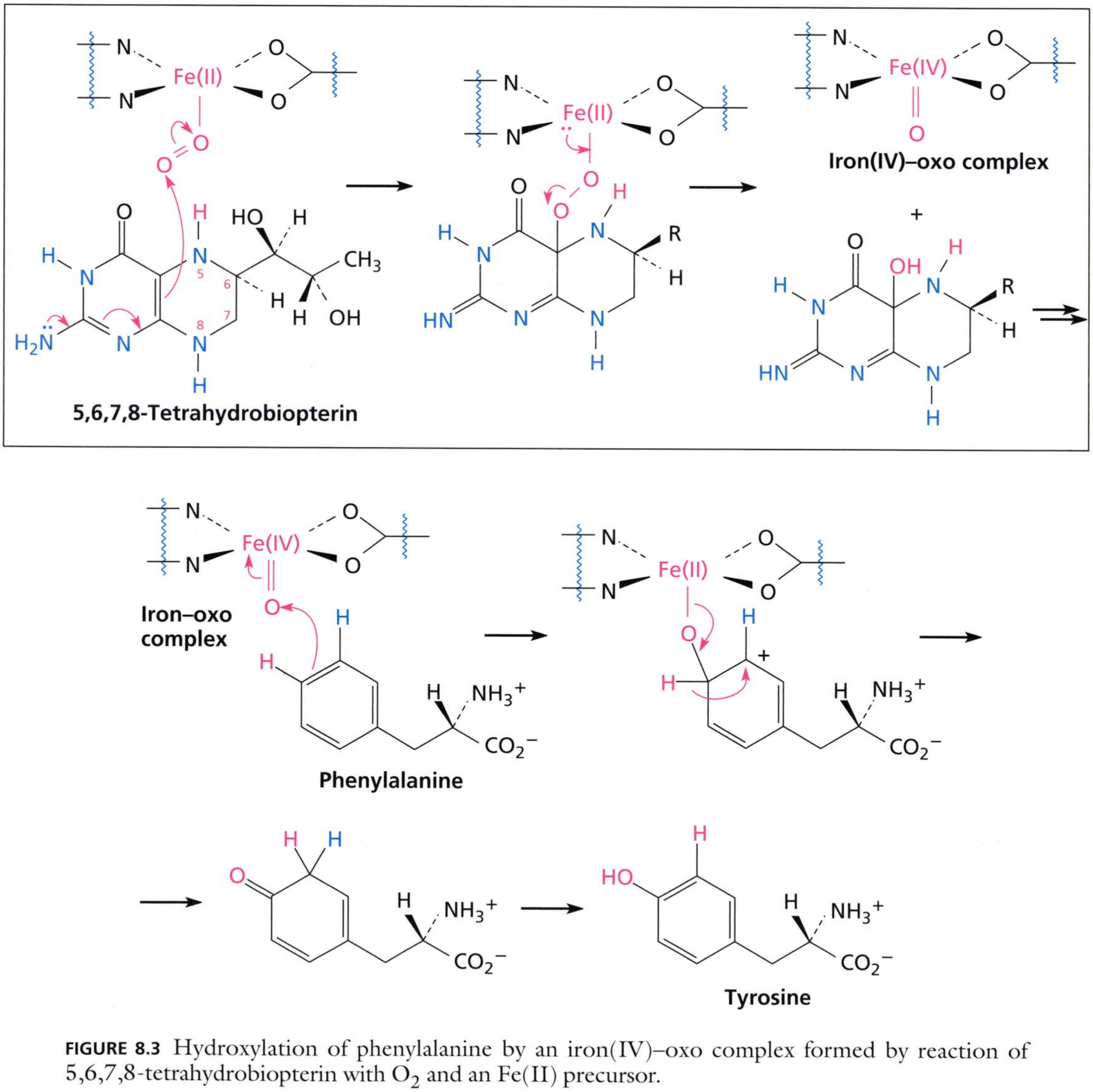

5,6,7,8-Tetrahydrobiopterin

Iron(IV)–oxo complex

Iron–oxo complex

Phenylalanine

Tyrosine

FIGURE 8.3 Hydroxylation of phenylalanine by an iron(IV)–oxo complex formed by reaction of 5,6,7,8-tetrahydrobiopterin with O_2 and an Fe(II) precursor.

Iron(IV)-oxo complexes are frequently involved in radical oxidations, such as occurs during the ring-closure steps in the biosynthesis of penicillin (Section 7.1, Figures 7.4 and 7.5). In this instance, the iron–oxo complex is formed by loss of hydroxide ion from an iron(II) hydroperoxide intermediate (Figure 8.4).

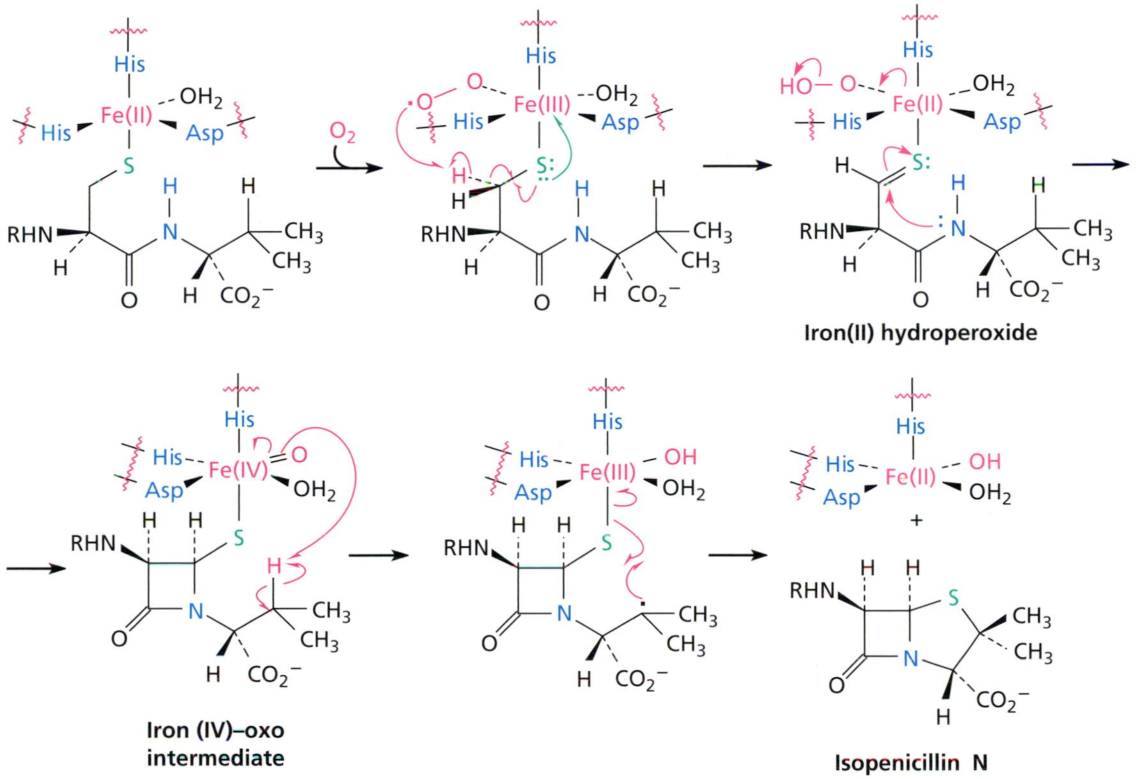

Iron(II) hydroperoxide

Iron (IV)–oxo intermediate

Isopenicillin N

FIGURE 8.4 Formation and reaction of an iron(IV)–oxo complex during penicillin biosynthesis.

Yet another method for forming an iron(IV)–oxo complex is by initial complexation of a nonheme Fe(II) species to α-ketoglutarate, followed by addition of O_2 and decarboxylation. Such α keto acid–dependent oxidases also require ascorbate cofactor and are frequently used for oxidation of unactivated C—H bonds, as in the biosynthesis of deacetoxycephalosporin C from penicillin N (Section 7.1, Figure 7.8). The process is shown in Figure 8.5.

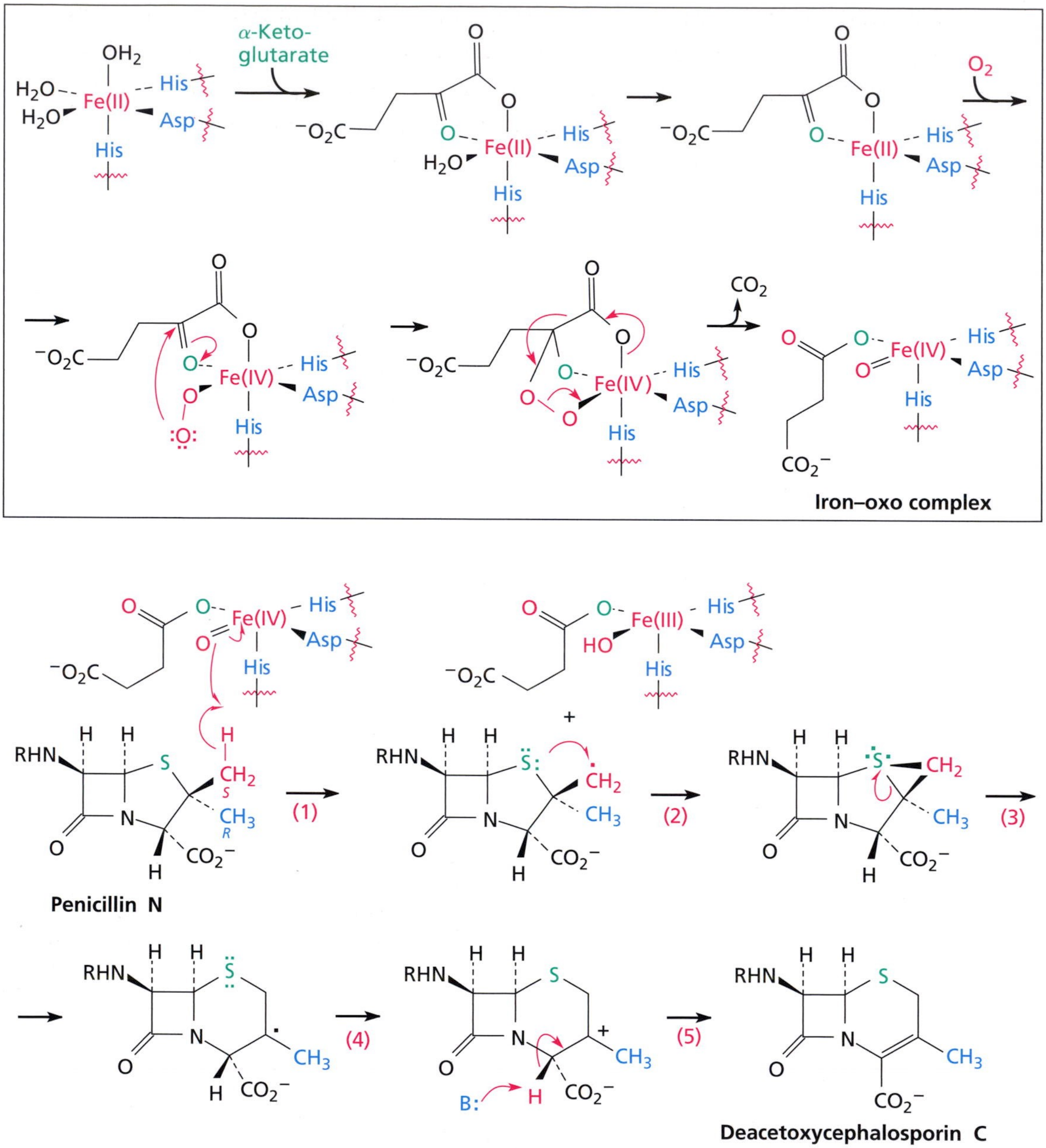

OH$_2$
H$_2$O-- Fe(II) -- His
H$_2$O Asp
His
α-Keto-
glutarate
O
$^-$O$_2$C
O
O-- Fe(II) -- His
H$_2$O Asp
His
O
$^-$O$_2$C
O
O-- Fe(II) -- His
Asp
His
O$_2$

O
$^-$O$_2$C
O
O-- Fe(IV) -- His
O Asp
:O:
His
O
$^-$O$_2$C
O
O-- Fe(IV) -- His
O Asp
O
His
CO$_2$
O
O-- Fe(IV) -- His
O= Asp
His
CO$_2^-$
Iron–oxo complex

O
O-- Fe(IV) -- His
O= Asp
$^-$O$_2$C
His
O
O-- Fe(III) -- His
HO Asp
$^-$O$_2$C
His
+

H H
RHN
S
H
CH$_2$
S
CH$_3$
R
N
O
H
CO$_2^-$
Penicillin N
(1)
H H
RHN
S:
CH$_2$
CH$_3$
N
O
H
CO$_2^-$
(2)
H H
RHN
S
CH$_2$
CH$_3$
N
O
H
CO$_2^-$
(3)

H H
RHN
S
N
O
CH$_3$
H CO$_2^-$
(4)
H H
RHN
S
N
+
CH$_3$
O
B: H CO$_2^-$
(5)
H H
RHN
S
N
CH$_3$
O
CO$_2^-$
Deacetoxycephalosporin C

FIGURE 8.5 Oxidation of an unactivated C—H bond by an iron(IV)–oxo complex formed by an α-ketoglutarate–dependent enzyme.

Finally, there are a large number of cytochrome P450 enzymes, which contain an Fe(III)–heme cofactor ligated to a cysteine residue. As shown in Figure 8.6, their function is to activate molecular oxygen for a variety of different oxidation processes, such as the oxidation of an unactivated C—H bond that occurs in the demethylation of thebaine (Section 7.2, Figure 7.16).

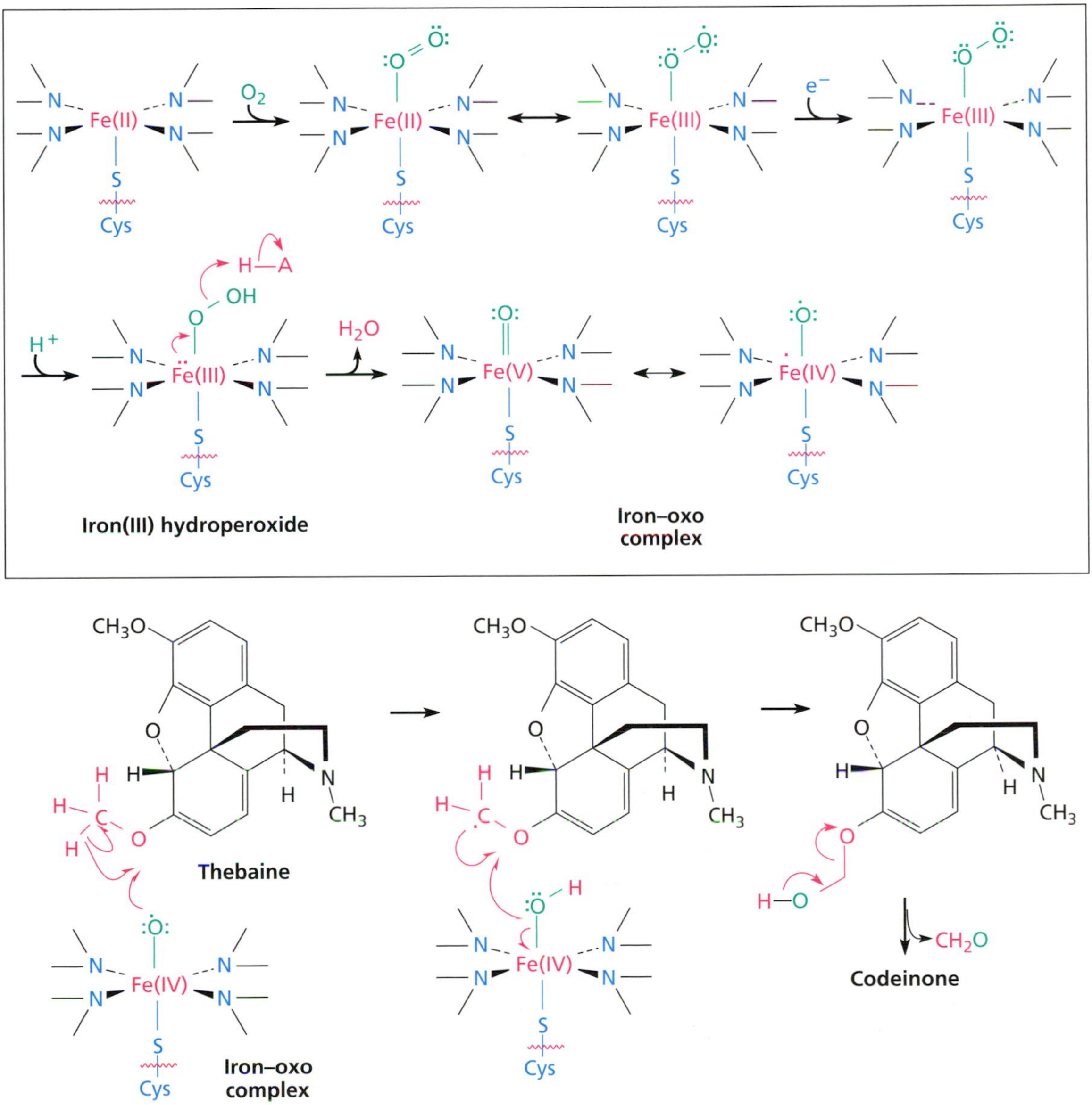

FIGURE 8.6 Oxygen activation and a subsequent hydroxylation reaction catalyzed by a P450 enzyme.

Iron(IV) complexes are also found in reactions catalyzed by *dioxygenases*, enzymes that incorporate both of the O_2 atoms in the product. An example is the conversion of homogentisate to 4-maleylacetoacetate that occurs in step 4 of tyrosine catabolism (Section 5.3, Figure 5.30).

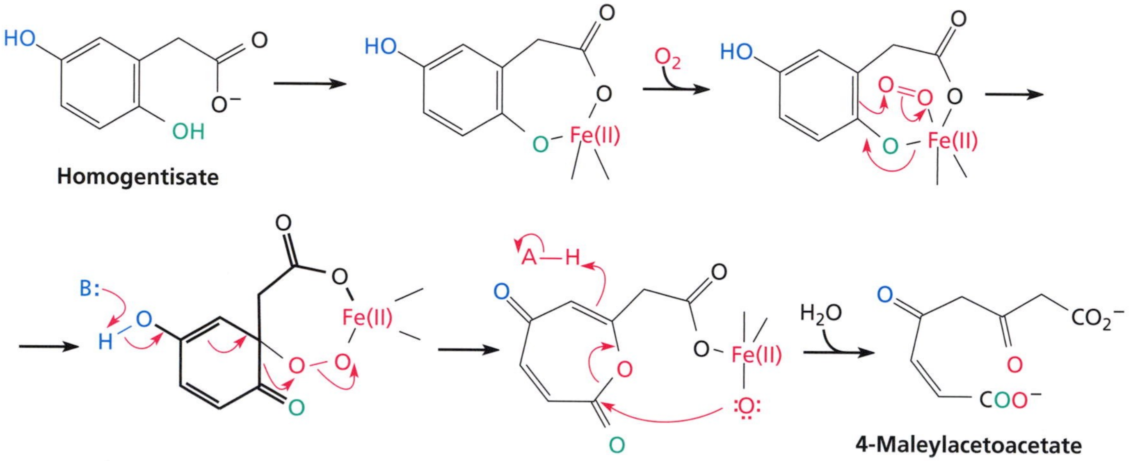

Homogentisate

4-Maleylacetoacetate

Other oxo–metal complexes are also used for oxygenations. The oxidation of hypoxanthine to xanthine during adenosine catabolism, for example, uses a molybdenum–oxo species (Section 6.1, Figure 6.5).

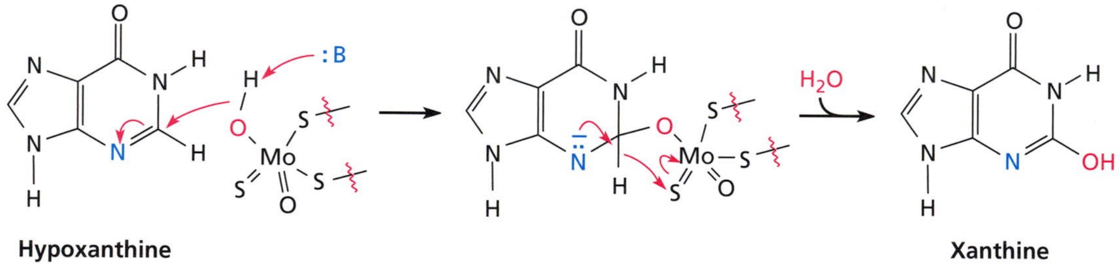

Hypoxanthine

Xanthine

Appendix A
Visualizing Protein Structures Using the Swiss PDB Viewer

As we remarked in the brief note at the end of Section 3.1, computer-based visualization programs are essential for examining the three-dimensional structures of proteins, and coordinate data for more than 22,000 substances are currently available from the Protein Data Bank (PDB). Learning how to access and use this information will greatly enhance your understanding of protein structure and enzymatic catalysis. The following steps show how to find, download, and view an enzyme using the Swiss PDB Viewer.

Downloading a PDB coordinate file

Let's use alcohol dehydrogenase as the example of an enzyme we want to study. To download the coordinates for alcohol dehydrogenase, go to the Protein Data Bank at http://www.rcsb.org/pdb/, whose homepage is shown in Figure A1. Type "alcohol dehydrogenase" into the search window, and click on "Search." A list of 89 structures will appear (Figure A2). We'll look only at the second one on the list, with a PDB code of 1A71. Click on the small panel with the curved blue arrow to the right of the 1A71 code, and the PDB file will be downloaded to your desktop with the filename 1A71.pdb.gz.

RCSB Home | wwPDB Home | Contact Us | Help

Did you find what you wanted?

Welcome to the PDB, the single worldwide repository for the processing and distribution of 3-D biological macromolecular structure data.

ABOUT PDB | NEW FEATURES | USER GUIDES | FILE FORMATS | DATA UNIFORMITY | STRUCTURAL GENOMICS | SOFTWARE | PUBLICATIONS | EDUCATION

Search the Archive

Enter a PDB ID or keyword Query Tutorial

Search

PDB ID Authors Full Text Search
match exact word remove similar sequences

QuickSearch! search Web pages and structures
SearchLite keyword search form with examples
SearchFields customizable search form
Status Search find entries awaiting release

News Complete News Newsletter pdb-l Archive Subscribe

14-Sep-2004
"The Impact of Structural Genomics on the Protein Data Bank" Published
A paper describing structural genomics' effects on the PDB's data pipeline, data capture, and target tracking has been published in American Journal of PharmacoGenomics. [MORE...]

PDB Mirrors

Please bookmark a mirror site
San Diego Supercomputer Center, UCSD*
Rutgers University*
Center for Advanced Research in Biotechnology, NIST*
Cambridge Crystallographic Data Centre, UK
National University of Singapore
Osaka University, Japan
Max Delbrück Center for Molecular Medicine, Germany

OCA / PDB Lite MORE...

**RCSB partner*

In citing the PDB please refer to:

H.M. Berman, J. Westbrook, Z. Feng, G. Gilliland, T.N. Bhat, H. Weissig, I.N. Shindyalov, P.E. Bourne: The Protein Data Bank. *Nucleic Acids Research*, **28** pp. 235-242 (2000)

ABOUT PDB | NEW FEATURES | USER GUIDES | FILE FORMATS | DATA UNIFORMITY | STRUCTURAL GENOMICS | SOFTWARE | PUBLICATIONS | EDUCATION

FIGURE A1 The PDB homepage.

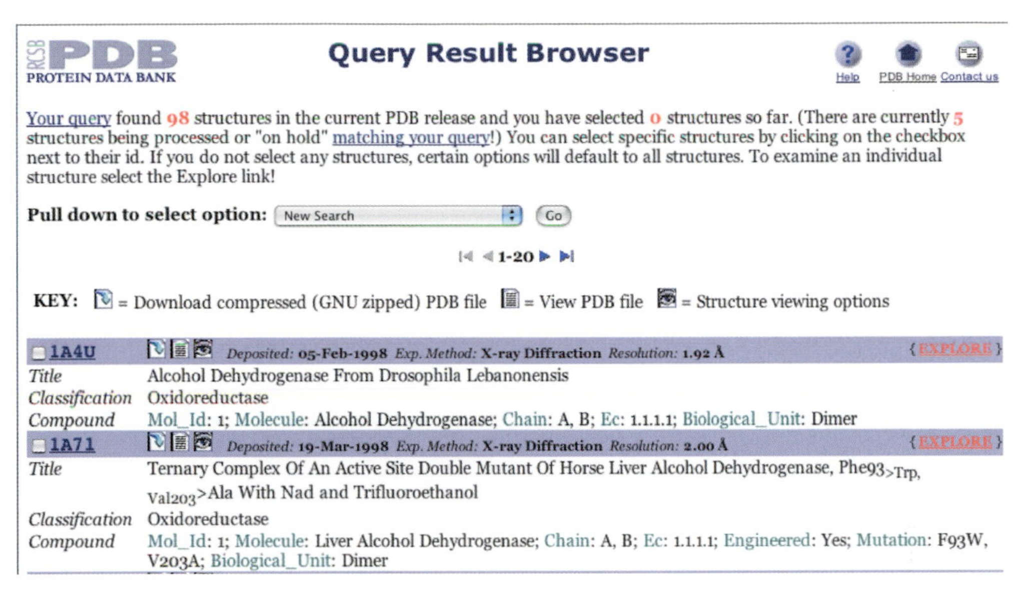

FIGURE A2 A partial list of alcohol dehydrogenase structures.

Downloading the Swiss PDB viewer

To download the Swiss PDB viewer, go to http://us.expasy.org/spdbv/. Click on "Download," select the Mac or PC version of the program, and a file called spdbv37_sea will appear on your desktop. Decompress this file to give a folder called SPDBV37, which contains the SwissPdbViewer program. From within SwissPdbViewer, open the alcohol dehydrogenase PDB file, click "OK" on each of the warning windows, and the structure of the enzyme will appear as in Figure A3.

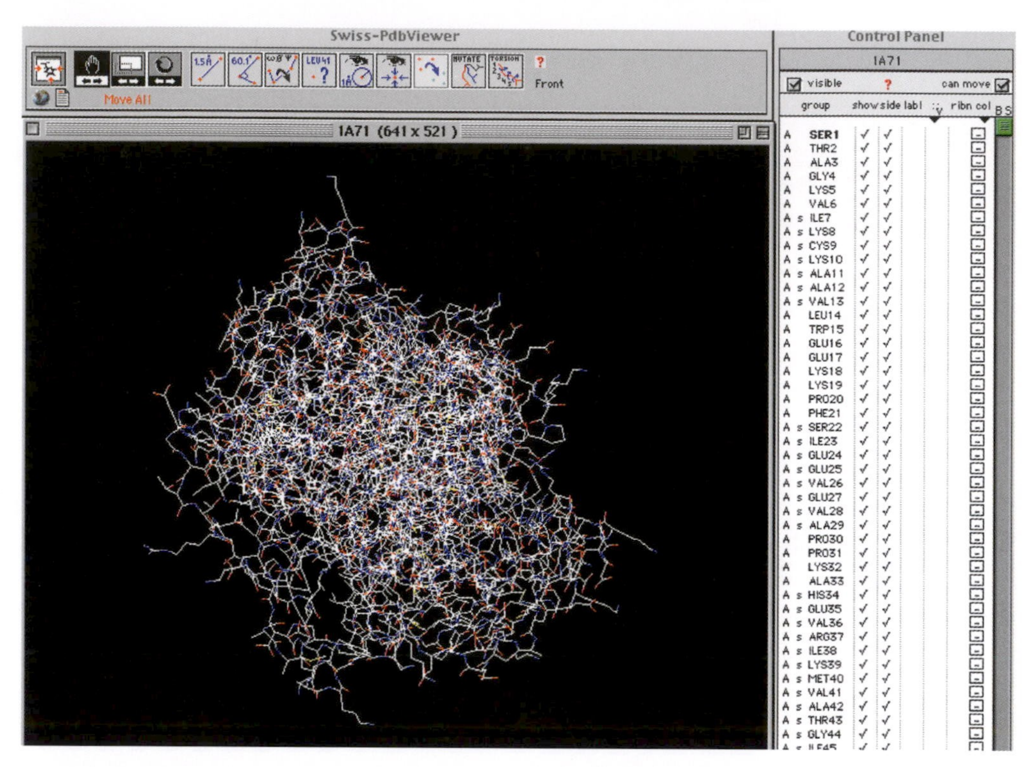

FIGURE A3 The structure of alcohol dehydrogenase shown using the Swiss PDB Viewer.

In addition to the structure window, two other windows will appear on your desktop (Figure A4). The Tools window gives you a set of tools to manipulate your structure, and the Control Panel gives you a list of all of the amino acids and small molecules bound to the protein—substrate, product, cofactor, metal ions, water, and inhibitor if present. If either of these windows are not open, click on the "Wind" menu and select the window that needs to be opened. The Control Panel also shows checkmarks that allow you to hide any of the residues on the protein ("Show" column) as well as a box on the far right that allows you to change the color of any residue ("Ribn col" column).

The most striking feature of the enzyme structure at this stage is its complexity, so it will be difficult to extract useful information unless you simplify the structure by getting rid of amino acids that aren't relevant to the questions you want to ask. Fortunately, this editing is easy in the Swiss PDB viewer, as we can illustrate by focusing on the active site of alcohol dehydrogenase.

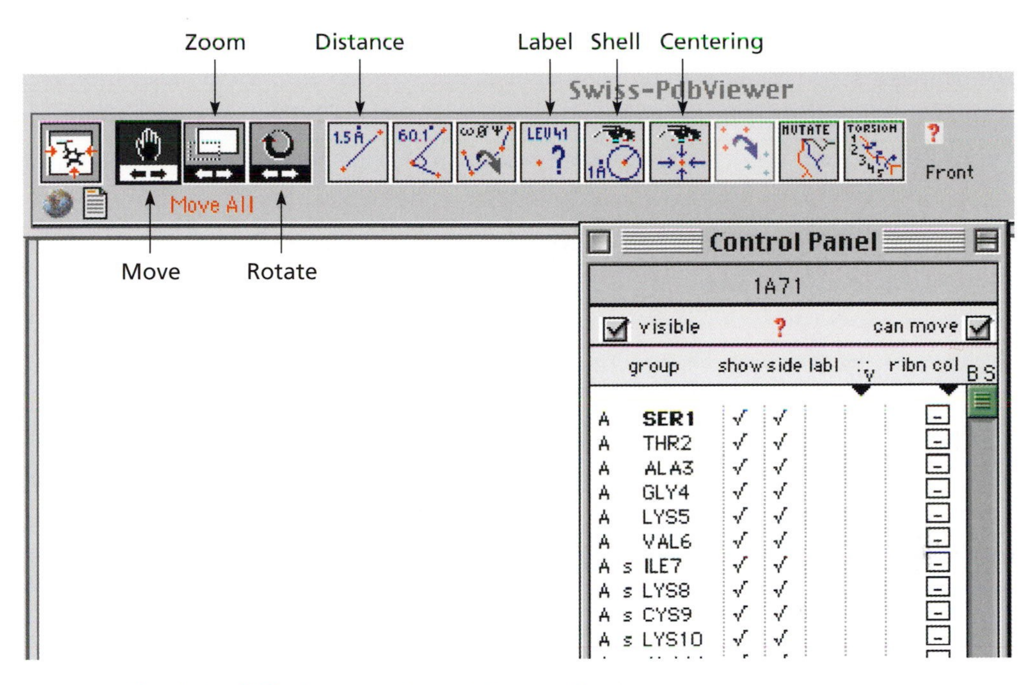

FIGURE A4 The Swiss PDB viewer toolbar and Control Panel.

Isolating the active site of alcohol dehydrogenase

The easiest way to isolate the active site of an enzyme is to locate a small molecule bound at the active site or find a previously identified active-site residue. For alcohol dehydrogenase, locate the NAD cofactor in the Control Panel. (Protein-bound small molecules are usually listed between the amino acids and the water molecules.) Click on the NAD color box, and select red. You should then be able to see the red NAD in the structure. Center the NAD on the screen by clicking first on the "Centering" tool and then on the atom that you want to fix in the center of the screen. You can now use the "Zoom" tool to zoom in on the active site.

The structure is still quite complex, so you need to use the "Shell" tool to achieve further simplification by isolating a shell of residues around an atom. First click on the Shell tool, and then click on C4 of NAD. Select "Display only groups that are within," type "6" in the open window, and click OK. The structure now shows only the residues that are within a 6 Å shell of the C4 carbon of NAD and is sufficiently simple that you can work with it. (You can restore the complete structure by choosing a large radius in the Shell tool, and you can change the color of the NAD back to CPK colors by clicking on the red NAD color box and selecting "Cancel" in the resulting window.)

At this point, you can rotate and zoom in or out using the "Rotate" and "Zoom" tools, and you can view the structure in stereo by selecting "Stereo" from the Display menu. This is a very useful tool, and it's well worth the effort to train your eyes to see in stereo. (A useful web-based tutorial for learning how to see stereoviews can be found at http://www.usm.maine.edu/~rhodes/0Help/StereoView.html#con.) Deselecting "Stereo" in the Display menu returns the structure to the mono view. It may seem like a lot of steps to this point, but with a little practice you can isolate the active site of any enzyme in a few minutes.

Exploring the active site of alcohol dehydrogenase

Once you've located the active site, you can explore it in many different ways.

Labeling atoms You can identify any atom in the structure using the "Label" tool. First click on the tool, then click on the atom of interest and a label will appear identifying the atom. For example, if you click on the yellow sulfur atoms, they will be identified at Cys-46 and Cys-174. Sometimes it's easier to locate a residue by clicking on the checkmark in the "Show" column of the Control Panel. For example, if you click repeatedly on the ETF1 checkmark, the active-site trifluoroethanol will appear and disappear. The structure can rapidly become crowded with labels, but you can remove them all by going to the Display menu and selecting "Labels," followed by "Clear user labels."

Measuring the distance between atoms You can determine the distance between any two atoms using the "Distance" tool. Click on the tool, then click on atom 1 followed by atom 2, and the distance between the atoms will appear on the structure.

Identification of hydrogen bonds Hydrogen bonds often play an important part in substrate binding and activation. To visualize them using the Swiss PDB viewer, select "Compute hydrogen bonds" in the Tools menu and all hydrogen bonds will appear as dotted green lines in the structure. They can be turned on or off by selecting or deselecting the "Show hydrogen bonds" option in the Display menu.

Problems

Here are a few problems to help you practice what you've learned.

1. Open the alcohol dehydrogenase structure (PDB code = 1A71) using the Swiss PDB viewer.

(a) How many subunits are in the enzyme?

(b) What is the secondary structure of residues 324–337?

(c) What is the secondary structure of residues 42–45 and 370–373?

(d) Which residue is hydrogen-bonded to the amide carbonyl group of NAD?

(e) What is the length of this hydrogen bond?

(f) What is the distance between the two Zn^{2+} ions in each subunit?

(g) Which Zn^{2+} ion is involved in catalysis? Explain.

(h) What are the ligands for the Zn^{2+} ion involved in catalysis? What are the ligands for the other zinc ion?

2. Open the DNA structure (PDB code = 113D) using the Swiss PDB viewer.

(a) Write the sequence, and locate the two GT mismatches.

(b) Write the structure of the hydrogen-bonded GT mismatched base pair.

(c) Compare the mismatch with the Watson–Crick AT and GC base pairs.

3. Although tRNA is made up of the A/U/C/G bases, many of these bases have undergone additional modification to give the active form of tRNA. To explore one of these modifications, open the phenylalanine tRNA structure (PDB code = 1EHZ) using the Swiss PDB viewer.

(a) What is the structure of residue 39?

(b) What residue is hydrogen-bonded to residue 39?

(c) What base is located immediately 3′ and 5′ to residue 39?

(d) Identify two other modified bases.

4. Open the phenylalanine tRNA structure (PDB code = 1EHZ) using the Swiss PDB viewer.

(a) Locate the phenylalanine anticodon (GAA).

(b) Are any of the bases in the anticodon loop modified? If so, draw the structure of each modified base.

(c) Locate the amino acid attachment site at the 3′ end of the tRNA.

(d) What is the distance between the amino acid attachment site and the anticodon loop? Indicate the two atoms that you used for your distance measurement.

Appendix B
The KEGG and BRENDA Databases

In this book, we have described most of the basic principles of the reactions found in living systems, and you are now prepared to begin exploring the much larger world of metabolism. A useful starting point for retrieving information on biosynthetic pathways is the KEGG database—the Kyoto Encyclopedia of Genes and Genomes maintained by the Kanehisa Laboratory of Kyoto University Bioinformatics Center. Let's illustrate the use of this database using cysteine metabolism as our example.

First, go to the KEGG database at http://www.genome.ad.jp/kegg, and click on the "KEGG Table of Contents" at the bottom of the page (Figure B1).

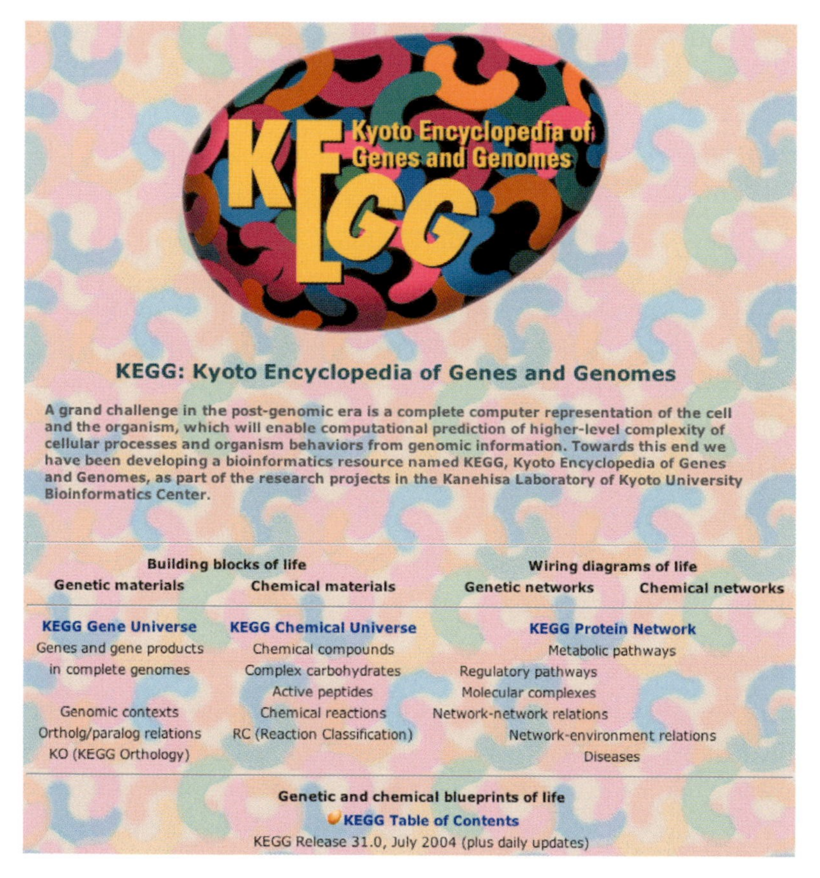

FIGURE B1 The homepage of the KEGG database.

In the Table of Contents window, click on the first item, "KEGG pathway database," to see a list of metabolic pathways. Scroll down through this list to Section 1.5: Amino acid metabolism (Figure B2).

1.5 Amino Acid Metabolism	
Glutamate metabolism	Ortholog, Aminotransferases
Alanine and aspartate metabolism	Ortholog
Glycine, serine and threonine metabolism	Ortholog
Methionine metabolism	Ortholog
Cysteine metabolism	Ortholog
Valine, leucine and isoleucine degradation	Ortholog
Valine, leucine and isoleucine biosynthesis	Ortholog
Lysine biosynthesis	Ortholog
Lysine degradation	
Arginine and proline metabolism	Ortholog
Histidine metabolism	Ortholog
Tyrosine metabolism	Ortholog
Phenylalanine metabolism	Ortholog
Tryptophan metabolism	
Phenylalanine, tyrosine and tryptophan biosynthesis	Ortholog
Urea cycle and metabolism of amino groups	Ortholog
1.6 Metabolism of Other Amino Acids	
beta-Alanine metabolism	Ortholog
Taurine and hypotaurine metabolism	
Aminophosphonate metabolism	
Selenoamino acid metabolism	

FIGURE B2 A KEGG list of metabolic pathways for amino acids.

Click on "Cysteine metabolism," and a window showing the metabolic pathway in a compressed form will appear. To obtain the structure of any compound on the map, click on the circle near the compound name and the structure will appear on a separate window. To obtain additional information on any enzyme, click on the box containing the Enzyme Commission (EC) number for that enzyme and a new window will open giving information on the enzyme. In this way you can construct the entire biosynthetic pathway from L-serine and sulfate to L-cysteine (Figure B3).

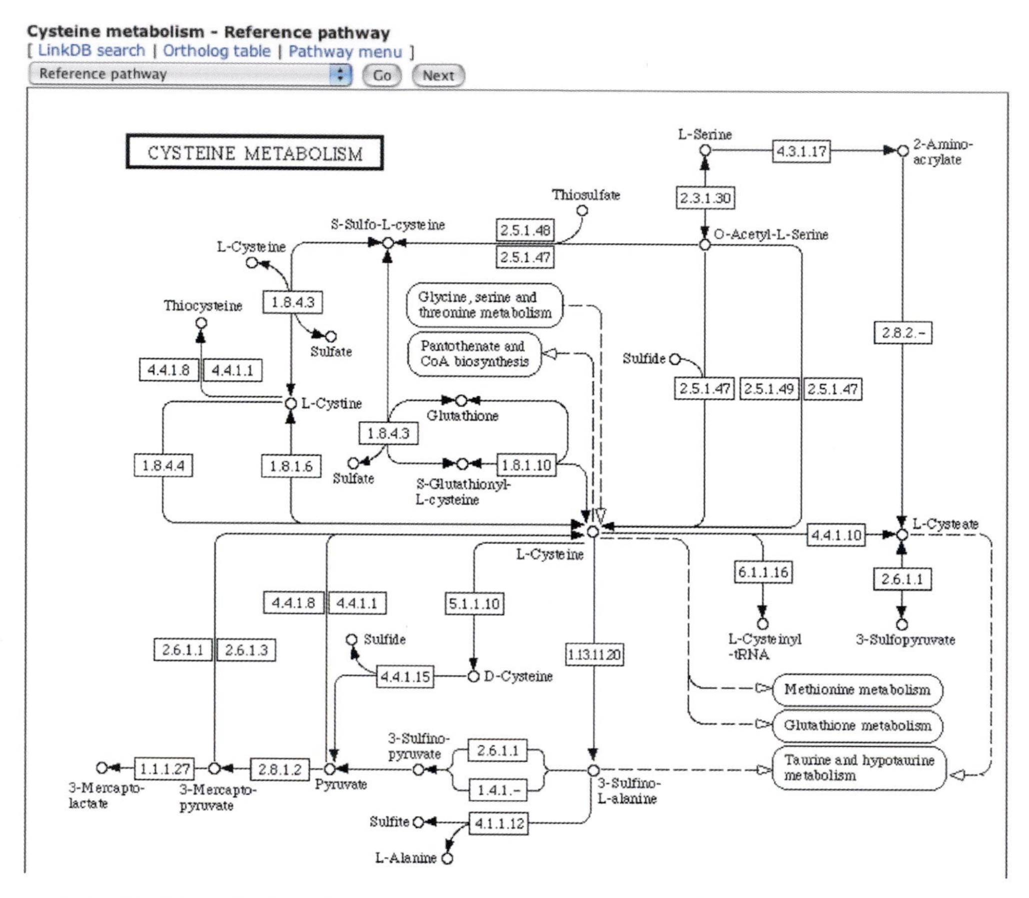

FIGURE B3 The biosynthetic pathway for L-cysteine as shown in the KEGG database.

Further information on specific enzymes can be obtained from the BRENDA database, which is maintained by the Institute of Biochemistry at the University of Cologne and can be accessed at http://www.brenda.uni-koeln.de/ (Figure B4). We'll use serine *O*-acetyltransferase, one of the enzymes on the cysteine biosynthesis pathway, to illustrate the use of the database.

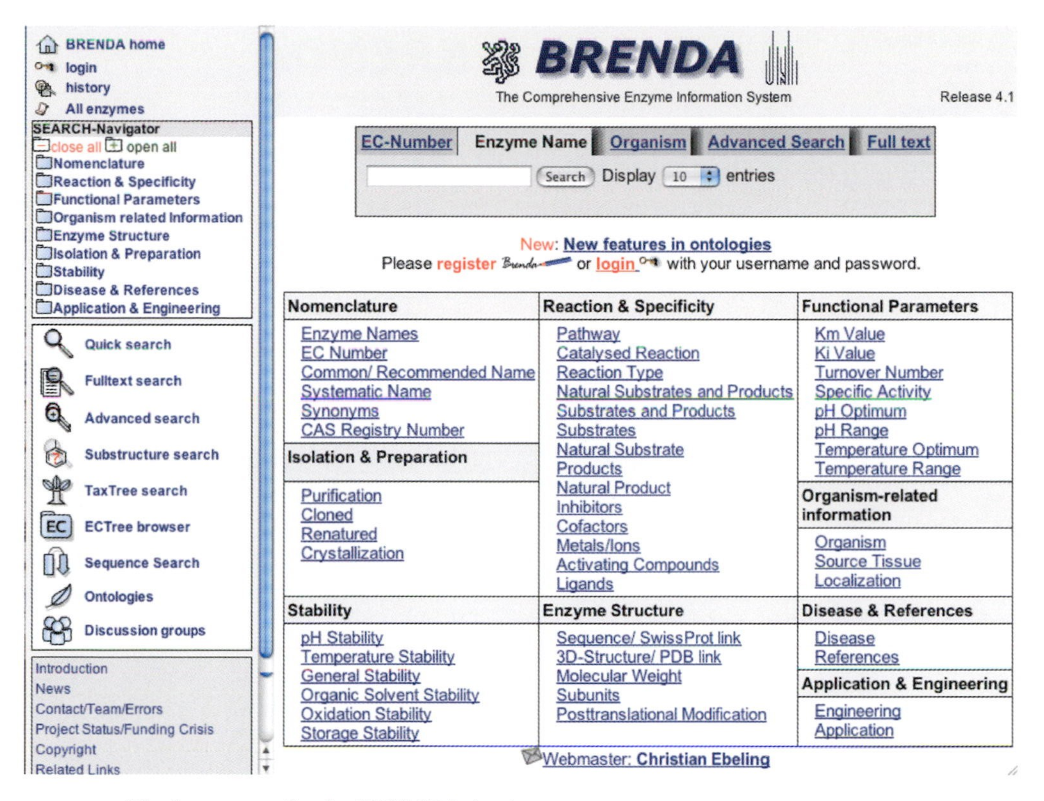

FIGURE B4 The homepage for the BRENDA database.

Type "serine *O*-acetyltransferase" or its EC number (2.3.1.30) in to the search window, and click on "Search." This will generate an output describing two families of the enzyme (Figure B5). Click on one of these and you'll get a window giving extensive information on the enzyme, including sources, properties, inhibitors, kinetic parameters, and so forth.

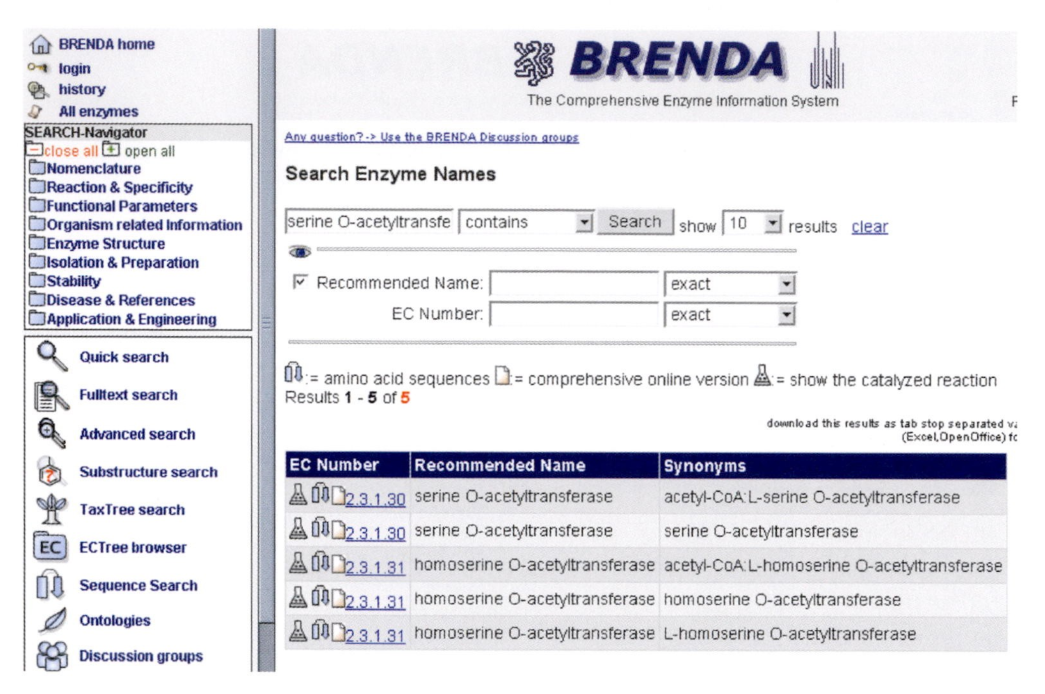

FIGURE B5 Search results for serine *O*-acetyltransferase in the BRENDA database.

By using the KEGG and BRENDA databases, you can quickly assemble information on a wide variety of biosynthetic pathways and on the enzymes involved. It's important to realize, however, that neither of these databases is fully current, so you should update any information you retrieve by carrying out a search of the current literature.

Appendix C
Answers to End-of-Chapter Problems

These answers to end-of-chapter problems are somewhat abbreviated and don't show all the mechanistic details. They do, however, cover the important points about each answer, leaving you to fill in the rest.

Chapter 1 Common Mechanisms in Biological Chemistry

1.1 (a), (c), (g), (h) can behave either as an acid or a base.

1.2 1-Butene is more nucleophilic.

1.3 (a) < (e) < (c) < (b) < (d)

1.4 (b) < (a) < (d) < (c)

1.5 The product of protonation on the double-bonded oxygen is stabilized by two resonance forms.

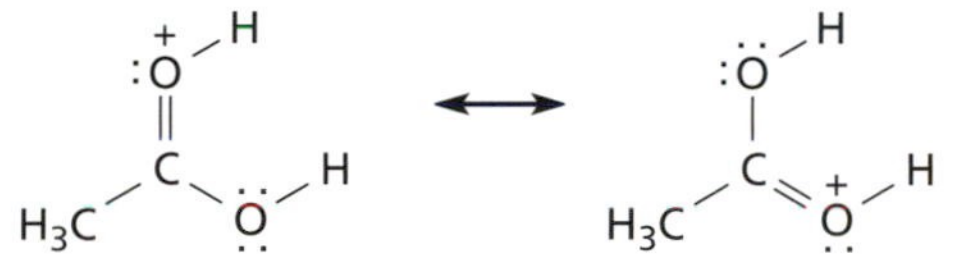

1.6 The product of protonation on the double-bonded nitrogen is stabilized by three resonance forms.

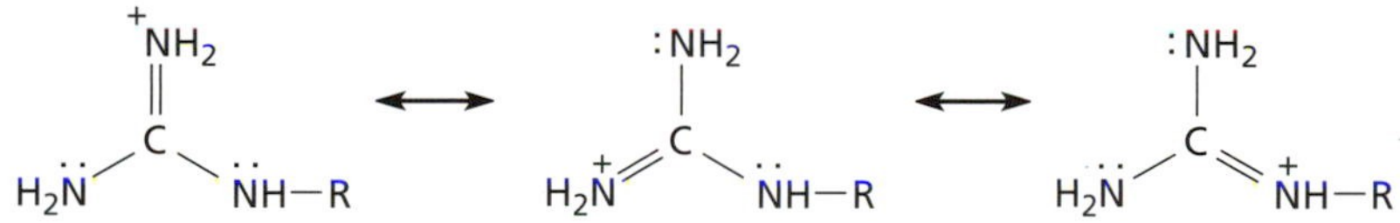

1.7

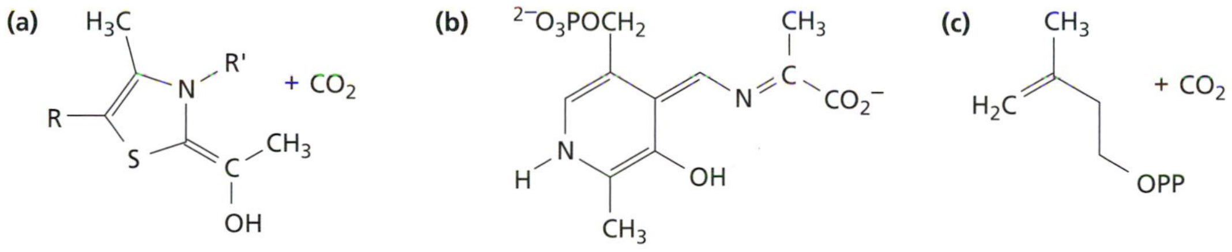

(a) $+ CO_2$

(b)

(c) $+ CO_2$

1.8

(a)

$$R{-}CH(OH){-}CH_2{-}\overset{O}{\overset{\|}{C}}OCH_3 \;\; (:B) \longrightarrow R{-}\overset{O}{\overset{\|}{C}}H + {}^{-}{:}CH_2{-}\overset{O}{\overset{\|}{C}}OCH_3 + BH^+$$

(b)

$$R{-}\overset{O}{\overset{\|}{C}}{-}SCoA \;+\; H{-}\underset{R''}{\underset{|}{\overset{H}{\overset{|}{C}}}}{-}\overset{O}{\overset{\|}{C}}{-}R' \;\; (:B) \longrightarrow R{-}\underset{SCoA}{\underset{|}{\overset{O^-}{\overset{|}{C}}}}{-}CHR''{-}\overset{O}{\overset{\|}{C}}{-}R' + BH^+ \longrightarrow R{-}\overset{O}{\overset{\|}{C}}{-}CHR''{-}\overset{O}{\overset{\|}{C}}{-}R' + CoAS^-$$

(c)

H_3C, CO_2CH_3, CH_3, H—A → + → + → H_3C, CO_2CH_3, CH_3, + →

B: H CO_2CH_3, H_3C, +, CH_3 → H_3C, CO_2CH_3, CH_3 + BH^+

1.9

$$RCH_2CH_2\overset{O}{\overset{\|}{C}}CH_2\overset{O}{\overset{\|}{C}}SCoA \;\; (B{:}\; H{-}SCoA) \longrightarrow RCH_2CH_2\underset{SCoA}{\underset{|}{\overset{O^-}{\overset{|}{C}}}}{-}CH_2\overset{O}{\overset{\|}{C}}SCoA \;\; (H{-}A) \longrightarrow RCH_2CH_2\overset{O}{\overset{\|}{C}}SCoA + CH_3\overset{O}{\overset{\|}{C}}SCoA$$

1.10

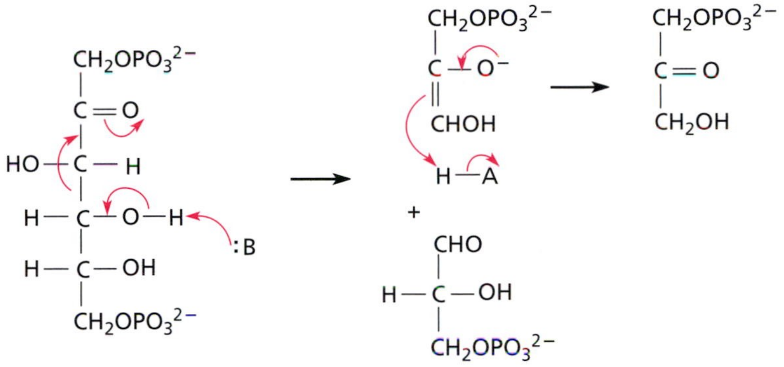

1.11

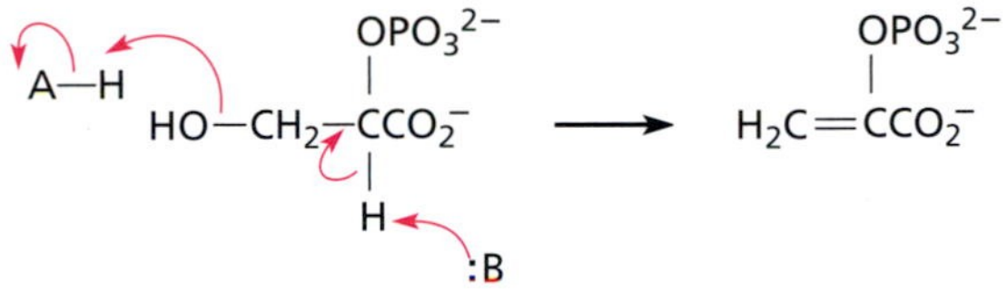

1.12

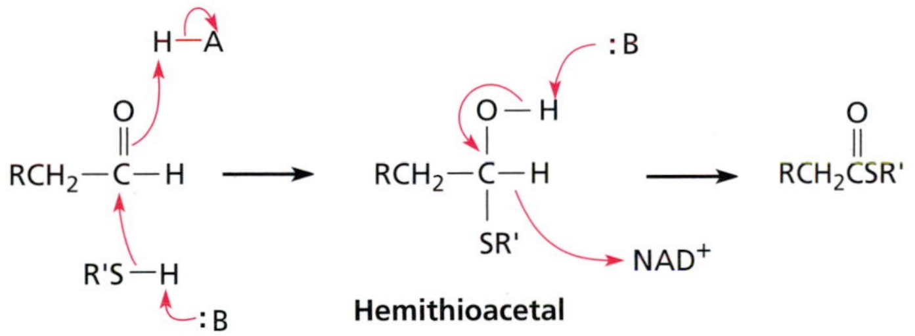

Hemithioacetal

1.13

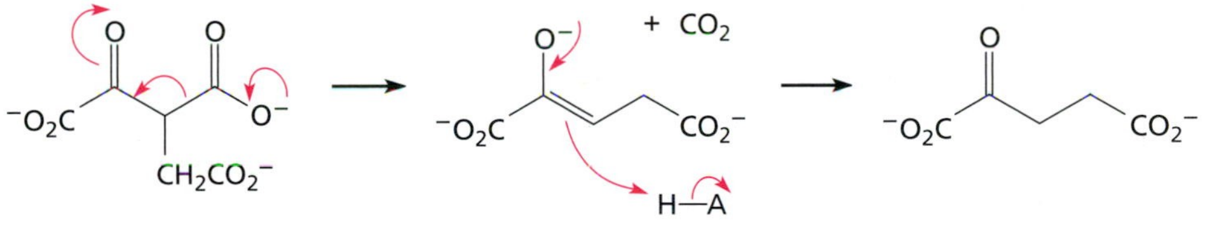

1.14

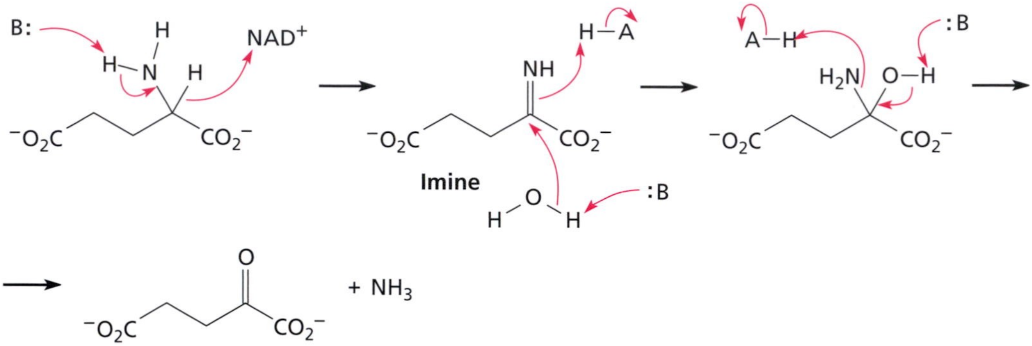

1.15

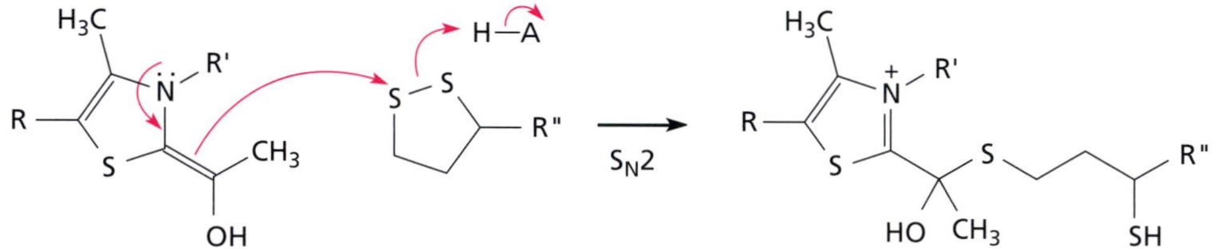

1.16 Inversion of configuration implies an S_N2 reaction.

Chapter 2 Biomolecules

2.1

(a)

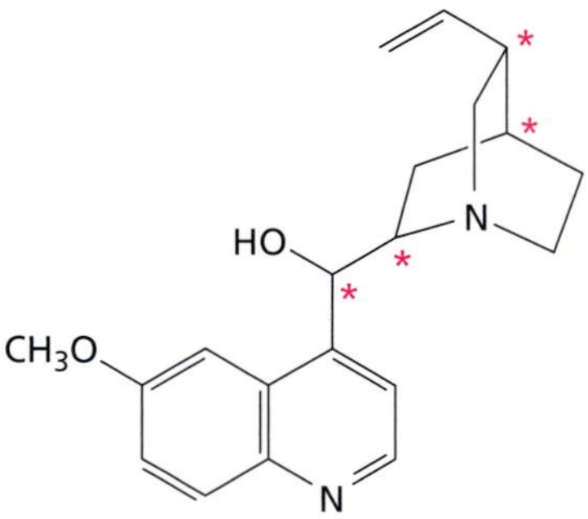

(b)

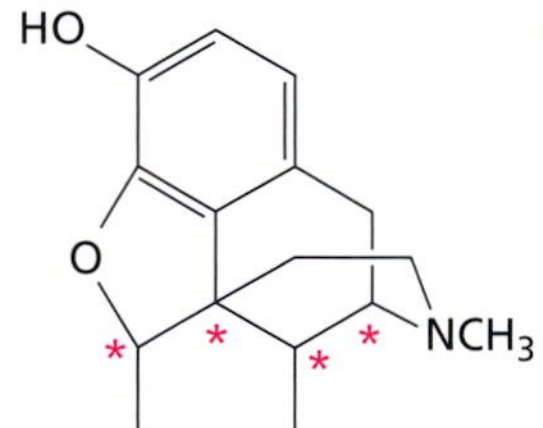

(c)

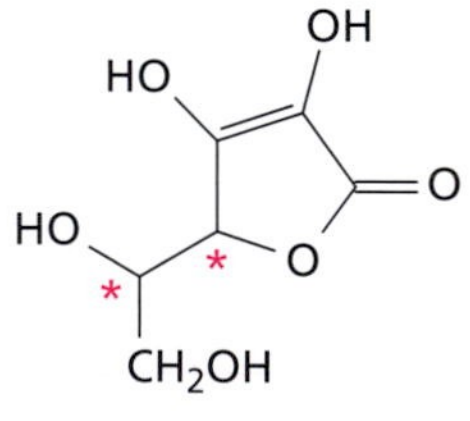

(d)

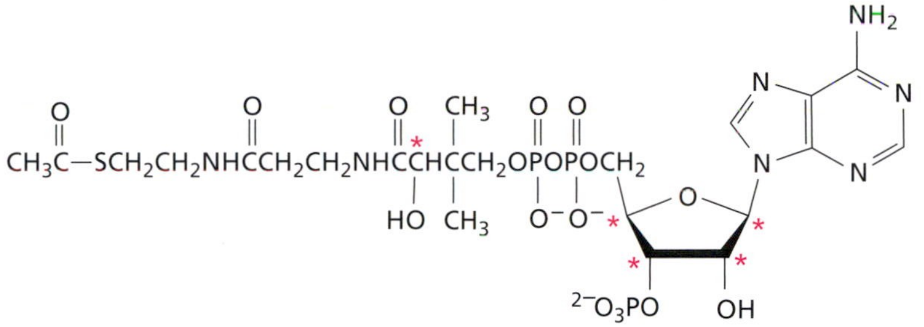

NH2
N N
O O O CH3 O O
CH3C—SCH2CH2NHCCH2CH2NHCCHCCH2OPOPOCH2 N N
HO CH3 O⁻ O⁻ O
2⁻O3PO OH

2.2

O
CH2OH
OH CH3
H OH
CH3 H
H H
O

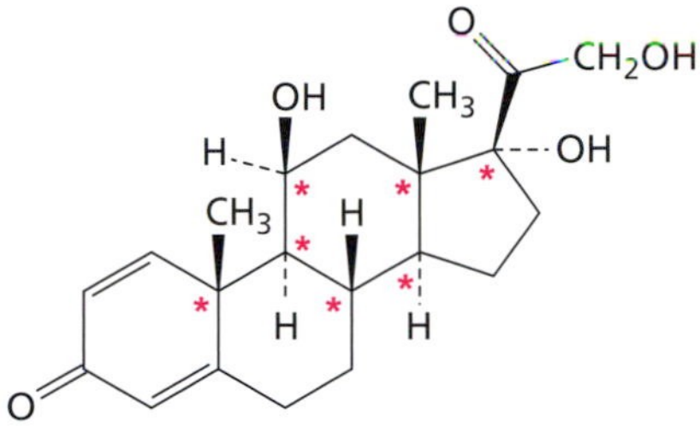

2.3 (**a**) meso compound; (**b**) epimers; (**c**) enantiomers

2.4

(a)
H H
H3C S S OH
R H
CH(CH3)2

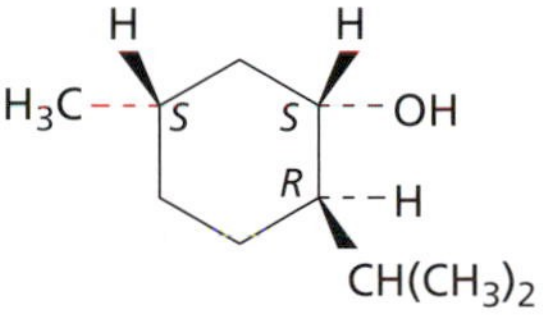

(b)
O
H—N N—H
H R S H
H
S S CH2CH2CH2CH2CO2⁻

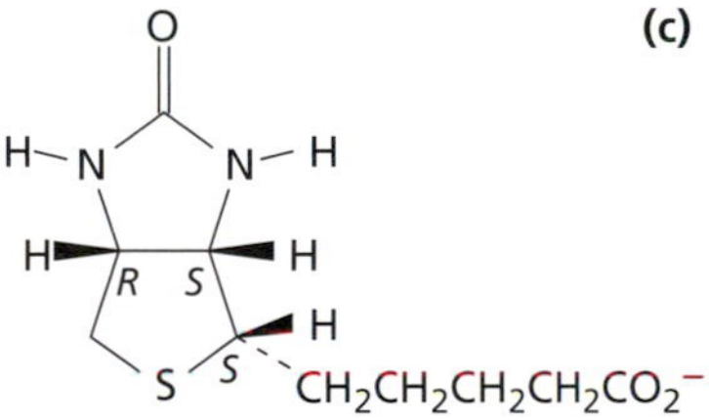

(c)
HO CH2OH
R R O
HO R OH
R R NH2

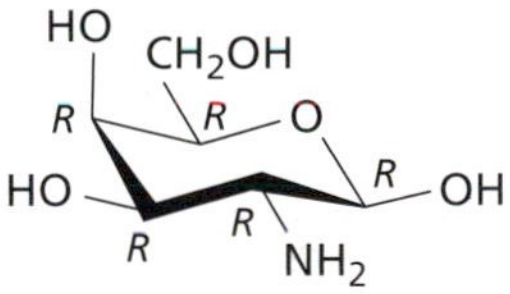

(d)
O H
R CO2H
R
R R S
HO H
H H HO H

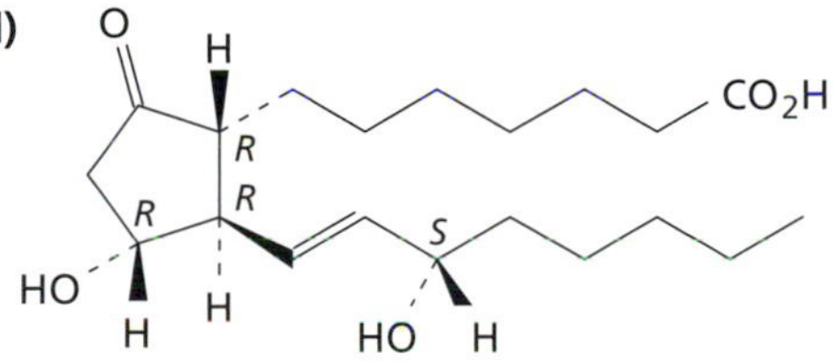

2.5

HO H
R S CH3
H NHCH3

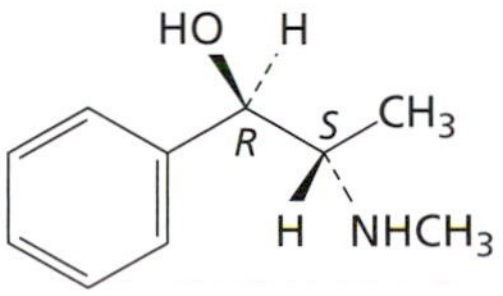

2.6

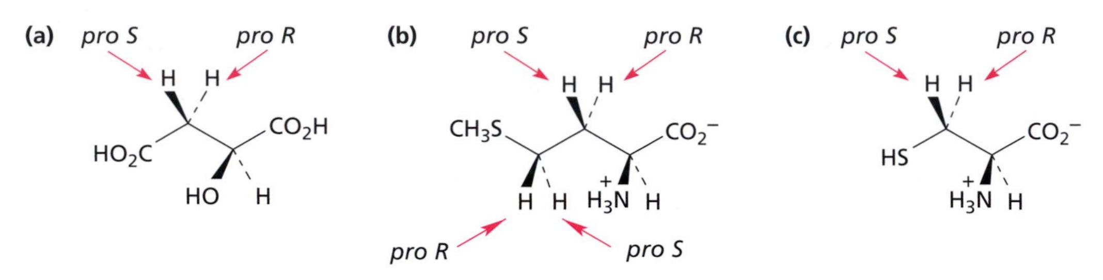

2.7

(a) re, si — H_3C–C(=O)–CO_2^-

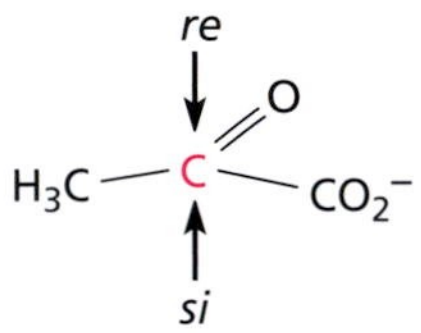

(b) si, re — H–C–CH_3, ^-O_2C–C, H

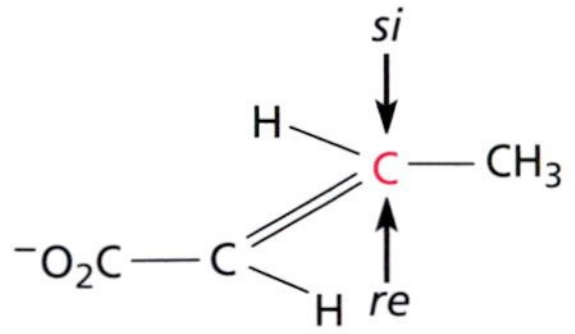

2.8

(a) H, H_3C, OH, *S*, CO_2^-

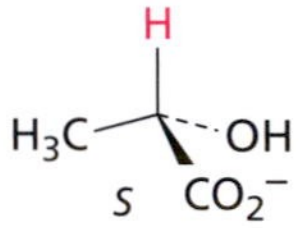

(b) OH, *S*, H, CH_3, ^-O_2C, H, H

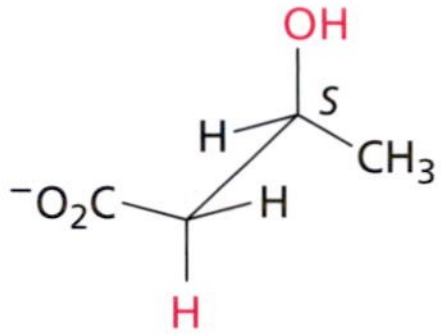

2.9

HO CO_2^-, ^-O_2C, CO_2^-, *pro R* ⟶ CO_2^-, ^-O_2C, CO_2^-

2.10 *anti* Addition of H_2O occurs initially from the *re* face.

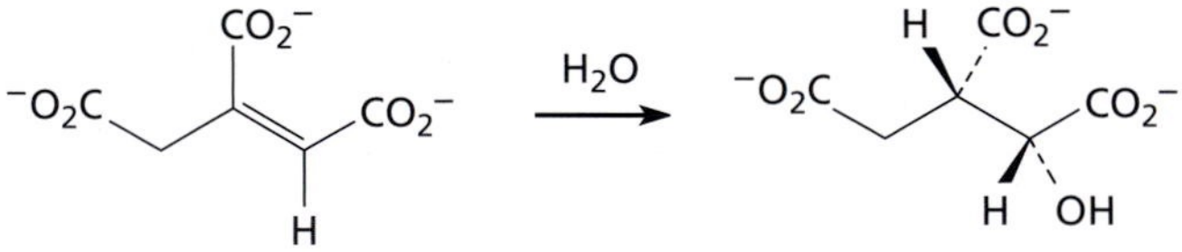

CO_2^-, ^-O_2C, CO_2^-, H $\xrightarrow{H_2O}$ H CO_2^-, ^-O_2C, CO_2^-, H OH

2.11

equatorial

Cholestanol

axial

Coprostanol

2.12

2.13

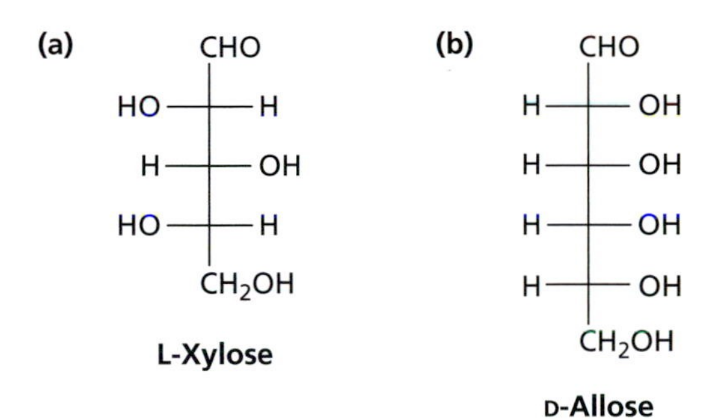

2.14

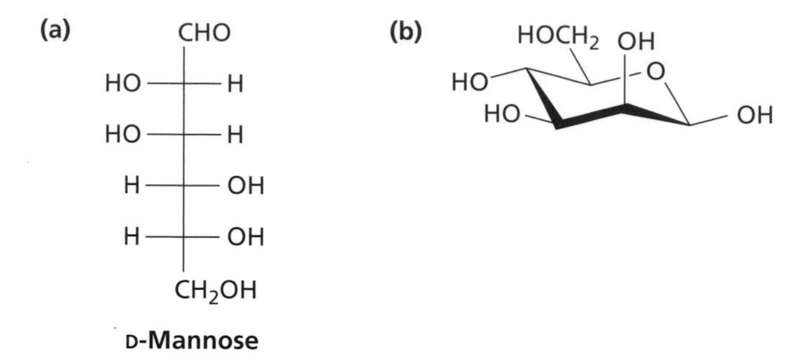

2.15

(a)

CHO
HO — H
H — OH
H — OH
H — OH
CH_2OH

(b)

CH_2OH
C=O
H — OH
H — OH
H — OH
CH_2OH

2.16

OH
O
CH_2OH
OH OH

2.17 Cys-Lys-Phe-Asp

2.18

$H_3\overset{+}{N}$—CH—C(=O)—NH—CH—C(=O)—N—CH—C(=O)—NH—CH—C(=O)—NH—CH—CO_2^-

$CHCH_3$ / CH_3 (Val); CH_2 / CH_2 / CO_2^- (Glu); ring (Pro); CH_3 (Ala); CH_2SH (Cys)

Val **Glu** **Pro** **Ala** **Cys**

2.19 In peptides, proline does not have an N—H and cannot form hydrogen bonds.

2.20 The double-bonded nitrogen is more basic because its protonated form is stabilized by resonance.

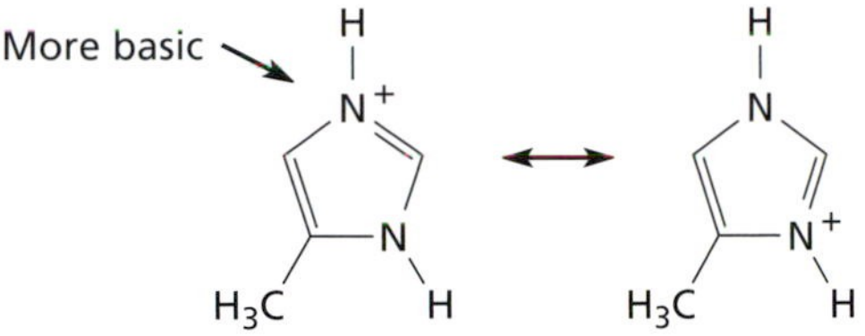

More basic

2.21

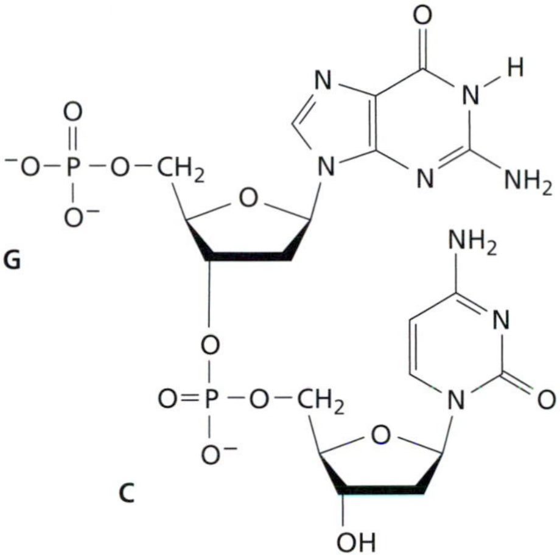

G

C

2.22 (**a**) oxidase; (**b**) dehydrase; (**c**) carboxylase

Chapter 3 Lipid Metabolism

3.1

(a) $CH_3CH_2CH_2CH{=}CHC(=O)SCoA$

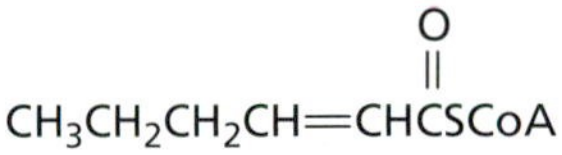

(b) $CH_3CH_2CH_2CH(OH)CH_2C(=O)SCoA$

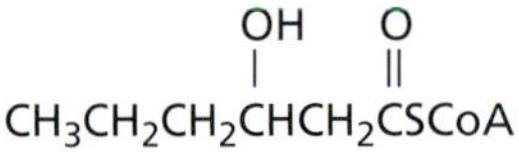

(c) $CH_3CH_2CH_2C(=O)CH_2C(=O)SCoA$

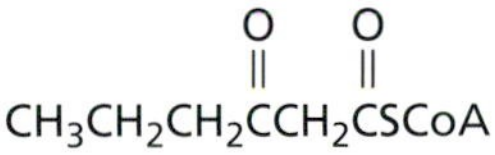

3.2

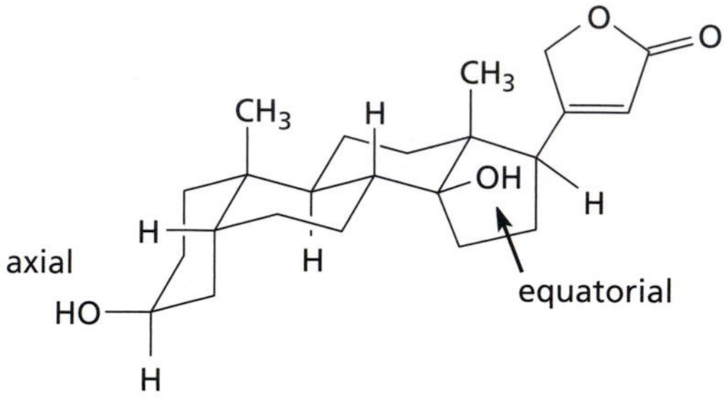

CH_3 H CH_3 O O

H OH H

axial H equatorial

HO

H

3.3

S $CH_2OPO_3^{2-}$

HO — H

CH_2OH

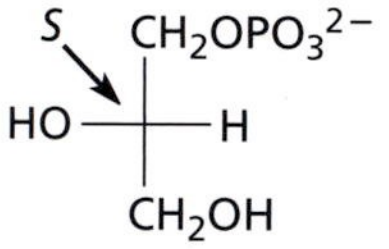

***sn*-Glycerol 1-phosphate**

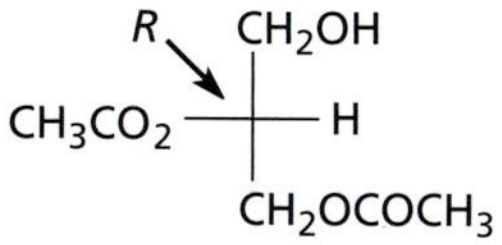

R CH_2OH

CH_3CO_2 — H

CH_2OCOCH_3

***sn*-Glycerol 2,3-diacetate**

3.4

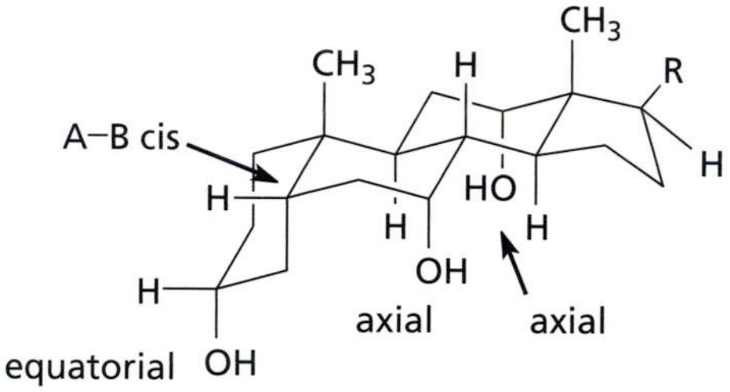

CH_3 H CH_3 R

A–B cis H

H HO H H

H OH

axial axial

equatorial OH

3.5

(a) * * OPP

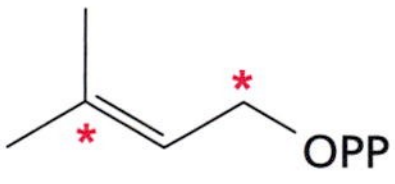

(b) * * * *

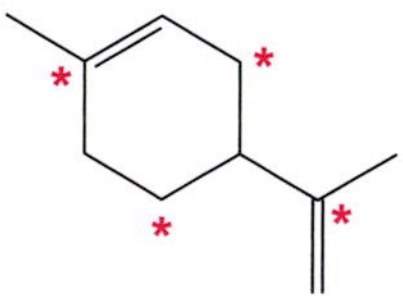

(c) * * * H_3C * H_2C * *

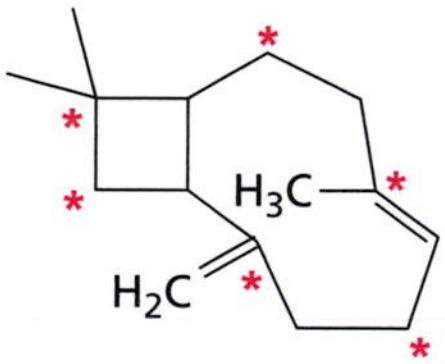

3.6

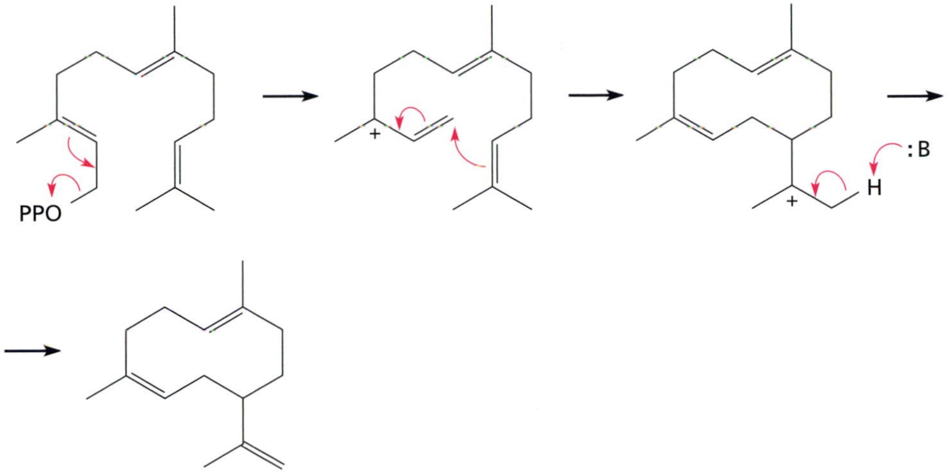

PPO

:B

H

3.7

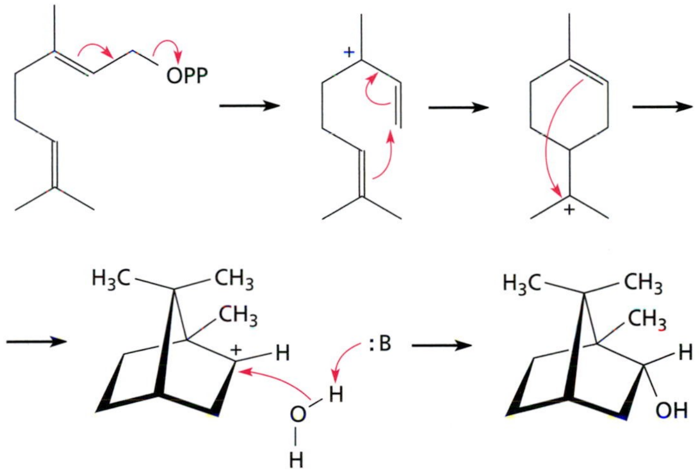

OPP

H_3C CH_3 CH_3 H :B H O H

H_3C CH_3 CH_3 H OH

3.8

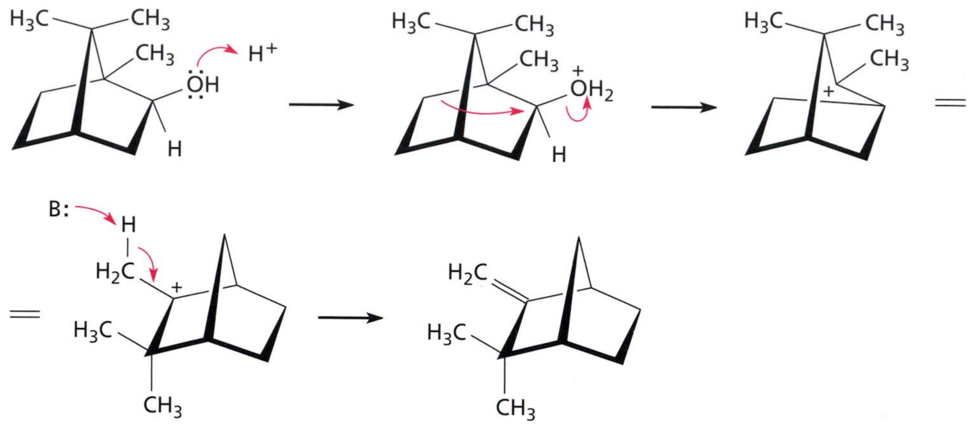
H3C
CH3
CH3
ÖH
H+
H
+
OH2
B:
H2C
H3C
CH3

3.9

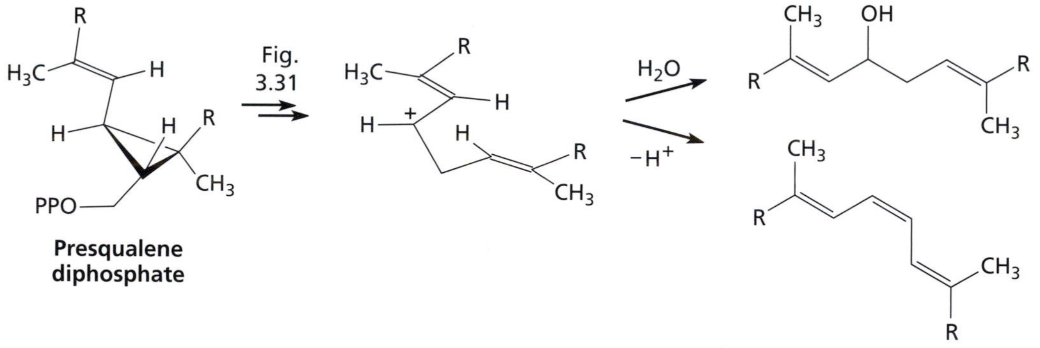
R
H3C
H
R
PPO
CH3
Presqualene
diphosphate
Fig.
3.31
+
H2O
−H+
OH

3.10

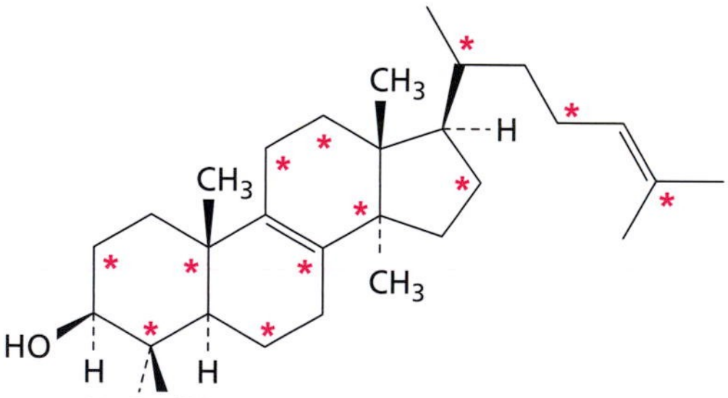
CH3
H
HO
H3C CH3

3.11

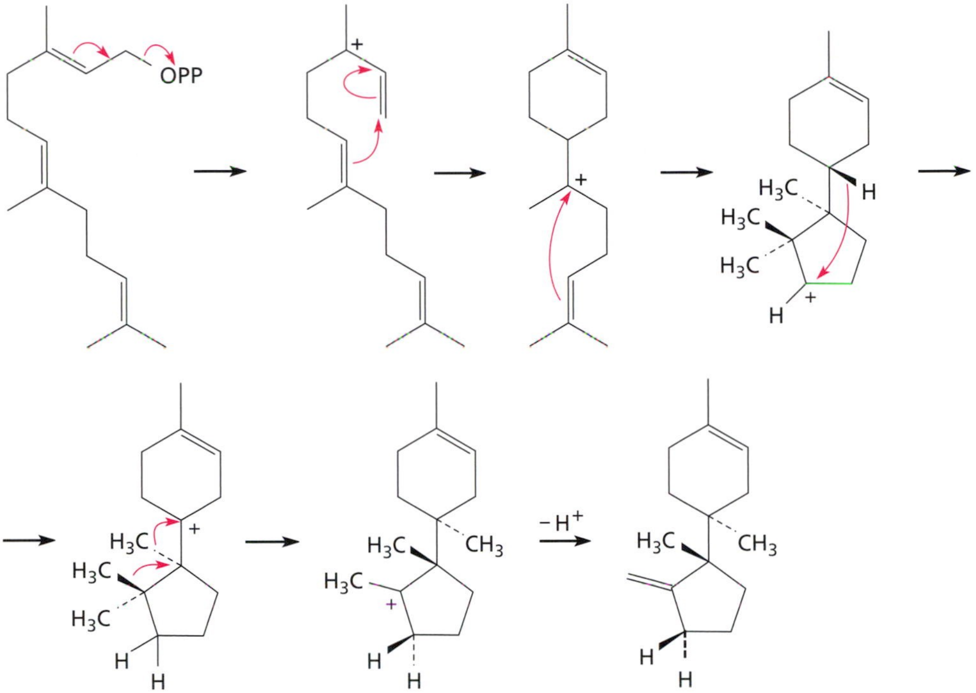

OPP

H_3C H_3C H_3C H H

H_3C H_3C H_3C H H

H_3C CH_3 H_3C H H

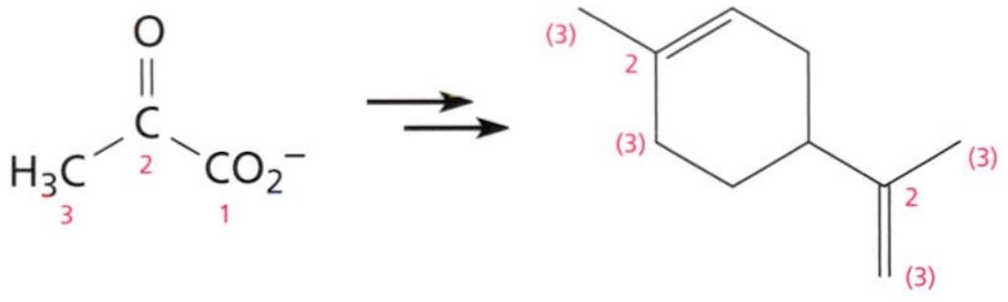

$-H^+$

H_3C CH_3 H H

3.12

O C 2 H_3C CO_2^- 3 1

(3) 2 (3) (3) 2 (3)

3.13

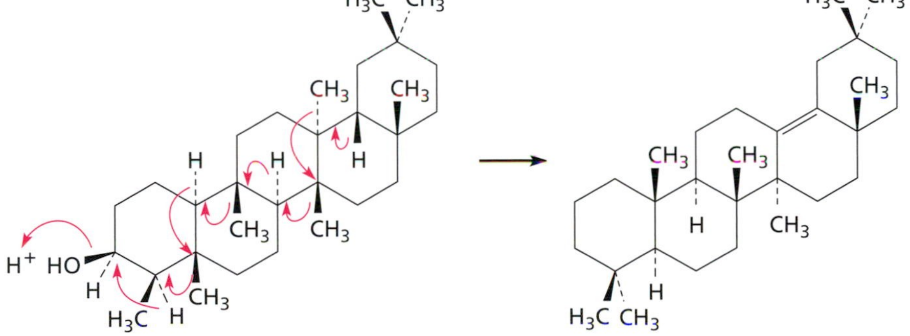

H_3C CH_3 CH_3 CH_3 H H H CH_3 CH_3 H^+ HO H H_3C H CH_3

H_3C CH_3 CH_3 CH_3 CH_3 H CH_3 H H_3C CH_3

3.14 (**a**) 3.05 Å; (**b**) Leu-153 and Phe-77; (**c**) 2.8 Å; (**d**) His-151 is too far from Ser-152 (5.17 Å) and also too far from Asp-176 (5.87 Å).

3.15 (**a**) Yes, 3.05 Å; (**b**) No, it is 4.2 Å away, so a bond rotation is needed to swing it into position; (**c**) Hydrogen-bond to the C2 ribose —OH group (distance = 2.89 Å) or to the backbone NH of Glu-376 (distance = 2.87 Å).

Chapter 4 Carbohydrate Metabolism

4.1 (**a**) ATP; (**b**) thiamin diphosphate; (**c**) biotin

4.2

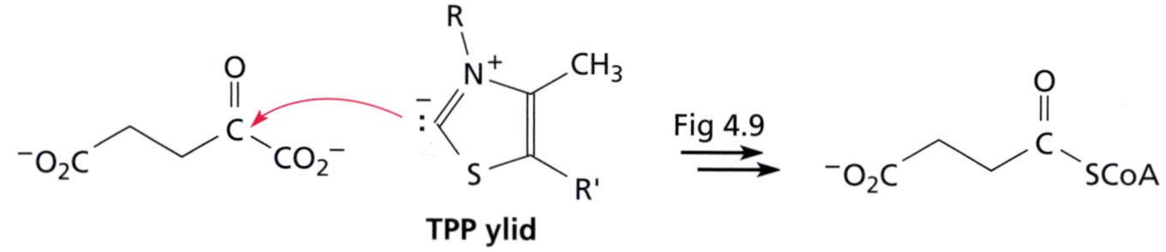

$^{-}O_2C$–CH_2CH_2–C(=O)–CO_2^- + TPP ylid $\xrightarrow{\text{Fig 4.9}}$ $^{-}O_2C$–CH_2CH_2–C(=O)–SCoA

TPP ylid

4.3

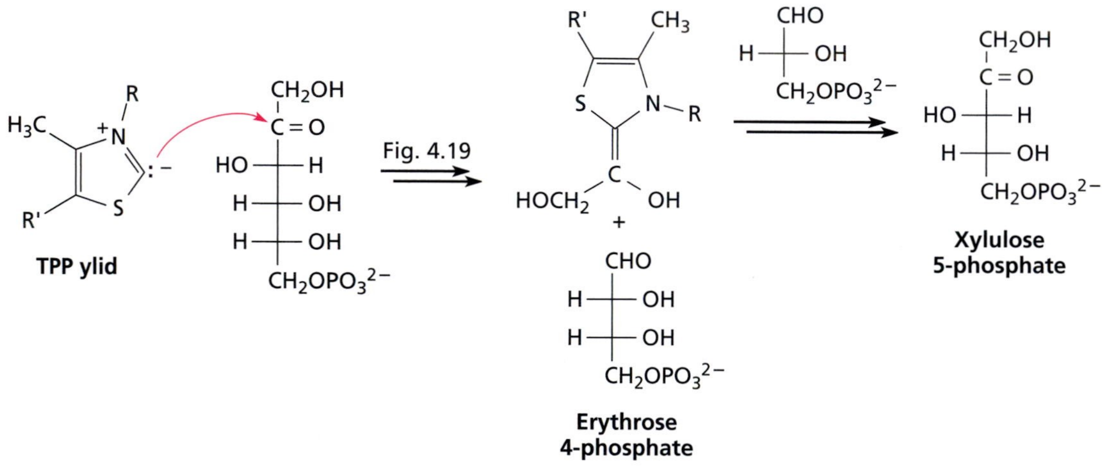

TPP ylid + xylulose-type ketose (CH_2OH, C=O, HO–H, H–OH, H–OH, $CH_2OPO_3^{2-}$) $\xrightarrow{\text{Fig. 4.19}}$ TPP adduct ($HOCH_2$–C–OH) + Erythrose 4-phosphate (CHO, H–OH, H–OH, $CH_2OPO_3^{2-}$)

Adduct + CHO, H–OH, $CH_2OPO_3^{2-}$ ⟶ Xylulose 5-phosphate (CH_2OH, C=O, HO–H, H–OH, $CH_2OPO_3^{2-}$)

TPP ylid

Erythrose 4-phosphate

Xylulose 5-phosphate

4.4

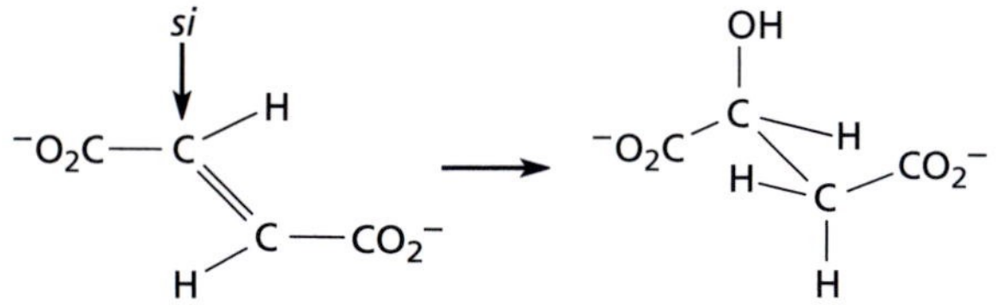

si

$^{-}O_2C$–CH=CH–CO_2^- ⟶ $^{-}O_2C$–CH(OH)–CH_2–CO_2^-

4.5

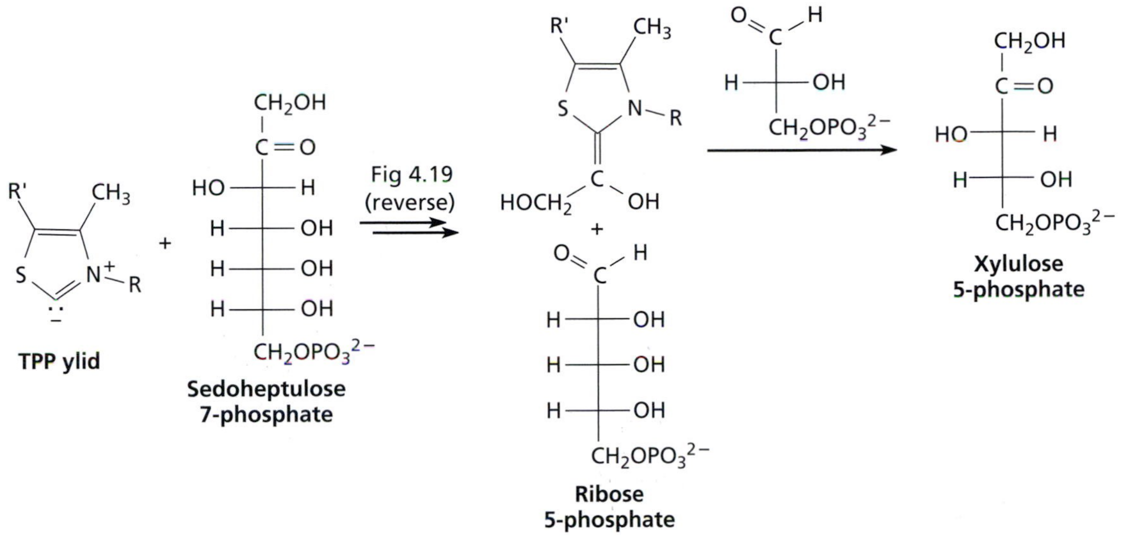

R' CH_3
S N^+ R

TPP ylid

+

CH_2OH
C=O
HO H
H OH
H OH
H OH
$CH_2OPO_3^{2-}$

Sedoheptulose 7-phosphate

Fig 4.19 (reverse)

R' CH_3
S N R
C
$HOCH_2$ OH

+

O C H
H OH
H OH
H OH
$CH_2OPO_3^{2-}$

Ribose 5-phosphate

O C H
H OH
$CH_2OPO_3^{2-}$

CH_2OH
C=O
HO H
H OH
$CH_2OPO_3^{2-}$

Xylulose 5-phosphate

4.6

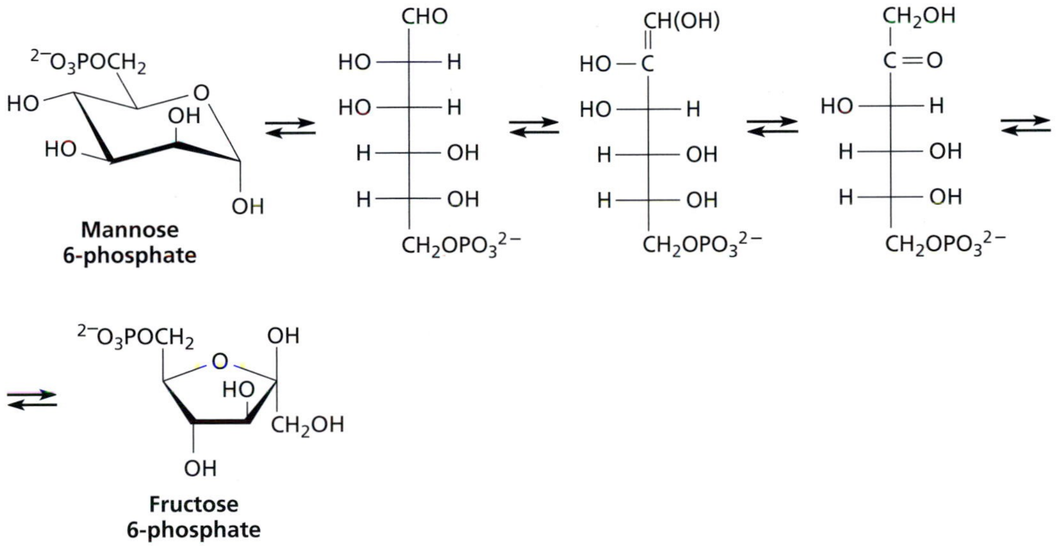

$^{2-}O_3POCH_2$
O
HO OH
HO
OH

Mannose 6-phosphate

⇌

CHO
HO H
HO H
H OH
H OH
$CH_2OPO_3^{2-}$

⇌

CH(OH)
HO—C
HO H
H OH
H OH
$CH_2OPO_3^{2-}$

⇌

CH_2OH
C=O
HO H
H OH
H OH
$CH_2OPO_3^{2-}$

⇌

⇌

$^{2-}O_3POCH_2$ OH
O
HO
CH_2OH
OH

Fructose 6-phosphate

4.7

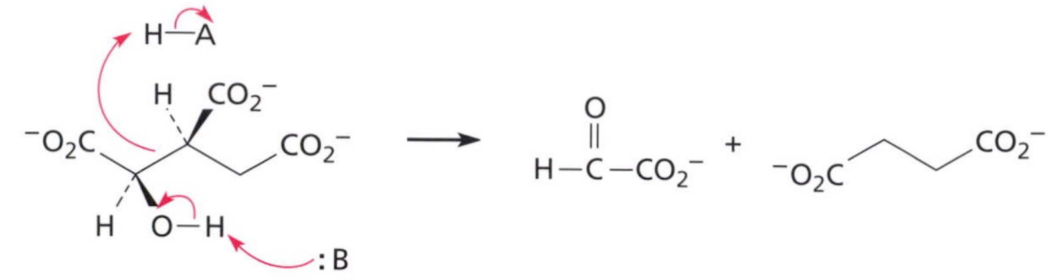

H—A

H CO_2^-

^-O_2C CO_2^- → H—C(=O)—CO_2^- + ^-O_2C CO_2^-

H O—H

:B

4.8

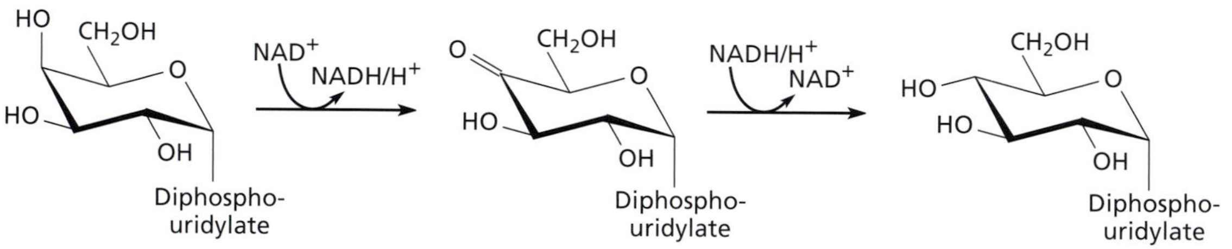

HO CH_2OH O HO OH Diphospho-uridylate

NAD^+ NADH/H^+ →

O CH_2OH O HO OH Diphospho-uridylate

NADH/H^+ NAD^+ →

CH_2OH O HO HO OH Diphospho-uridylate

4.9

B: CO_2^-

H—OH

A—H HO—H

H—OH

H—OH

$CH_2OPO_3^{2-}$

→

CO_2^-

C—OH

C—H

H—OH

H—OH

$CH_2OPO_3^{2-}$

⇌

CO_2^-

C=O

H—H

H—OH

H—OH

$CH_2OPO_3^{2-}$

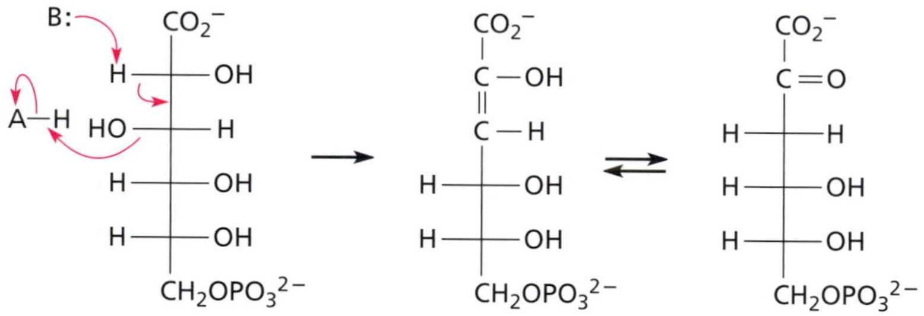

Chapter 5 Amino Acid Metabolism

5.1 The mechanism is the exact reverse of that shown in Figure 5.1.

5.2 *pro-R*

5.3

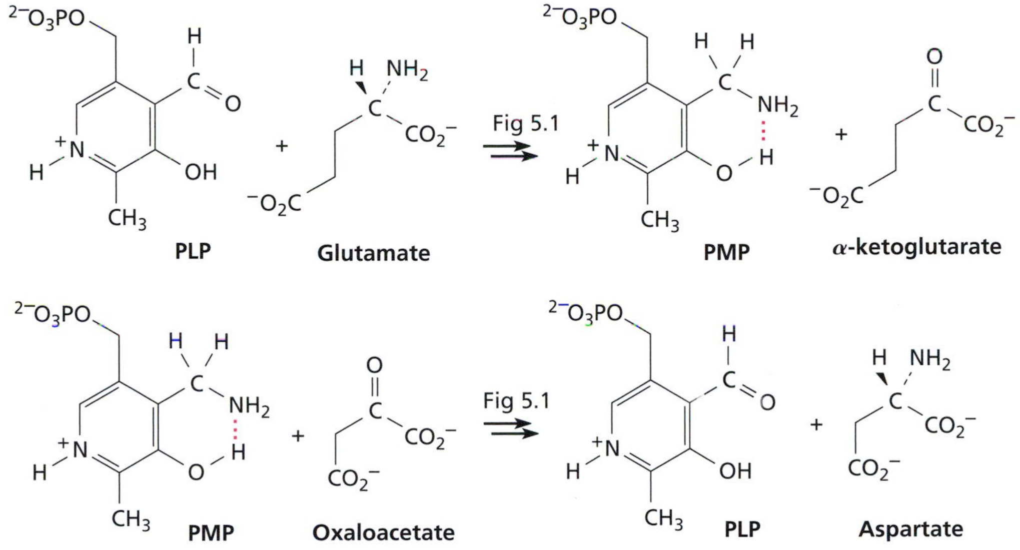

PLP + **Glutamate** $\overset{\text{Fig 5.1}}{\rightleftharpoons}$ **PMP** + **α-ketoglutarate**

PMP + **Oxaloacetate** $\overset{\text{Fig 5.1}}{\rightleftharpoons}$ **PLP** + **Aspartate**

5.4

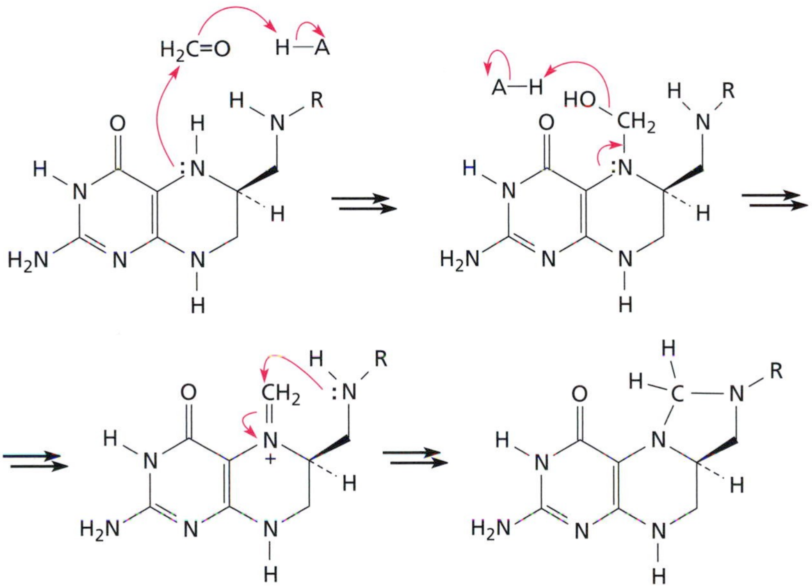

$H_2C{=}O$ H—A

A—H HO CH_2

5.5

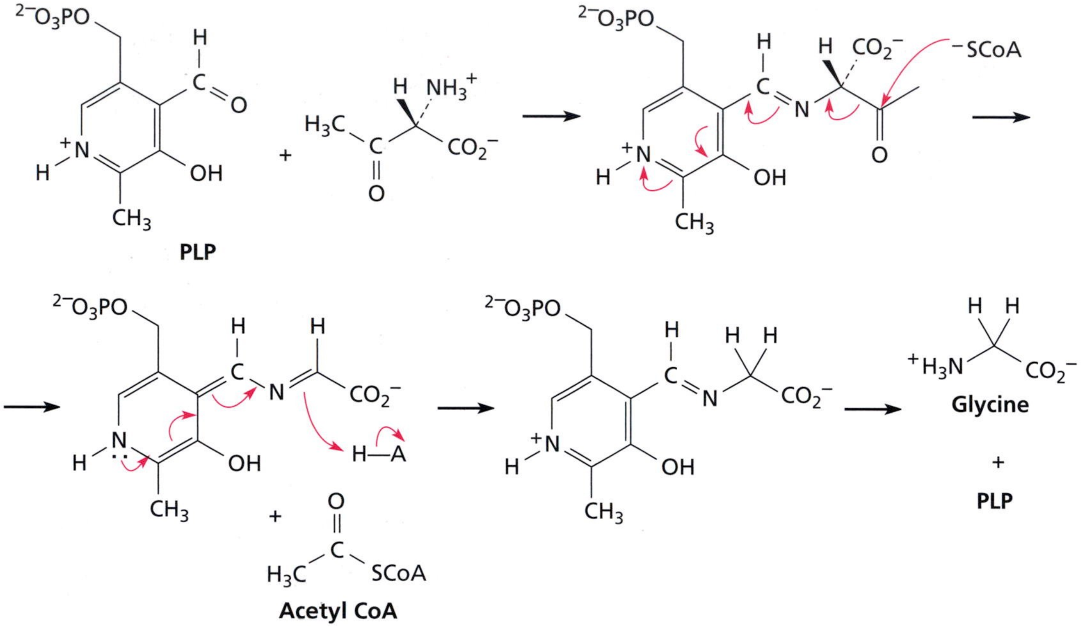

$^{2-}O_3PO$; H ; C ; O ; ^{+}N ; H ; OH ; CH_3 ; **PLP** ; + ; H ; NH_3^+ ; H_3C ; C ; O ; CO_2^- ; → ; $^{2-}O_3PO$; H ; C ; N ; H ; CO_2^- ; ^-SCoA ; O ; ^{+}N ; H ; OH ; CH_3 ; →

→ ; $^{2-}O_3PO$; H ; C ; N ; H ; CO_2^- ; N ; H ; OH ; CH_3 ; H—A ; + ; O ; C ; H_3C ; SCoA ; **Acetyl CoA** ; → ; $^{2-}O_3PO$; H ; C ; N ; H H ; CO_2^- ; ^{+}N ; H ; OH ; CH_3 ; → ; H H ; ^+H_3N ; CO_2^- ; **Glycine** ; + ; **PLP**

5.6

H ; CH_2OH ; O ; N ; H ; N: ; H ; H ; O ; CH_3 ; NH—Enz ; Enz ; O ; H—A ; → ; H ; CH_2—OH ; O ; N ; H ; :B ; N ; OH ; H—A ; H ; Enz ; Enz ; CH_3 ; → ; B: ; H ; CH_2—OH ; H—A ; O ; N ; N ; H ; Enz ; Enz ; CH_3

↓

4-Methylidene-imidazol-5-one

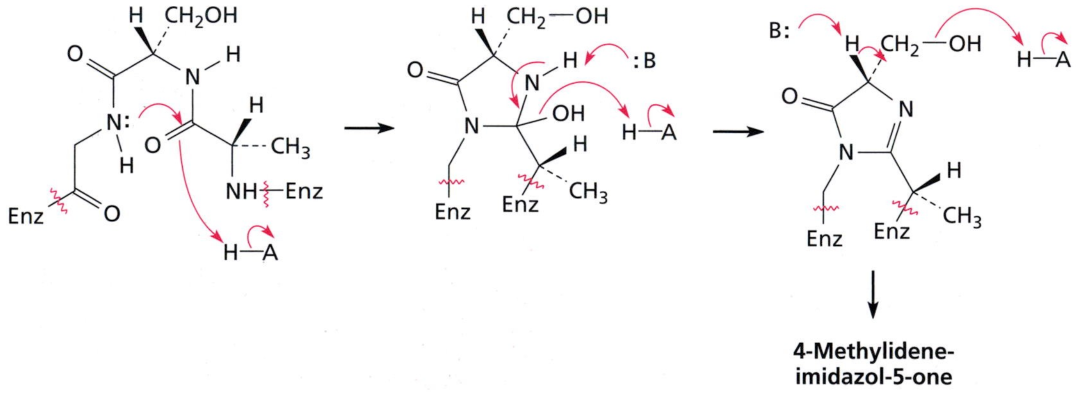

5.7

N-Carboxybiotin ⟶ CO_2 + Biotin

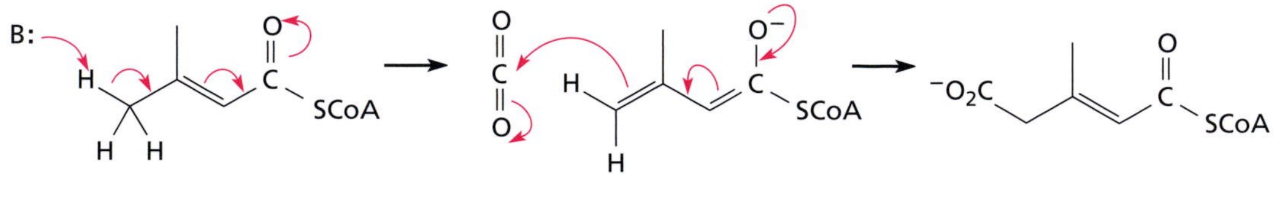

B: ; H ; H H ; O ; C ; SCoA ; → ; O ; C ; O ; H ; H ; O^- ; C ; SCoA ; → ; ^-O_2C ; O ; C ; SCoA

5.8

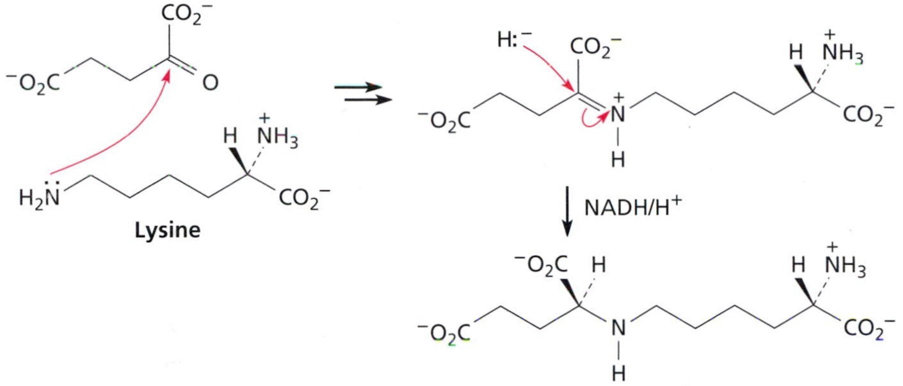

^-O_2C CO_2^- O H $\overset{+}{N}H_3$ $H_2\ddot{N}$ CO_2^-

Lysine

H:$^-$ CO_2^- H $\overset{+}{N}H_3$ ^-O_2C $\overset{+}{N}$ CO_2^- H

NADH/H^+

^-O_2C H H $\overset{+}{N}H_3$ ^-O_2C N CO_2^- H

5.9

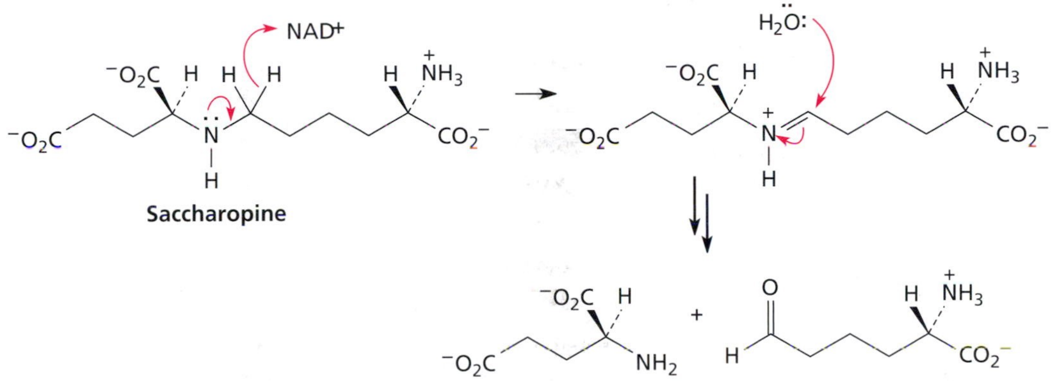

NAD^+ ^-O_2C H H H H $\overset{+}{N}H_3$ ^-O_2C $\ddot{N}$ CO_2^- H

Saccharopine

$H_2\ddot{O}$: ^-O_2C H H $\overset{+}{N}H_3$ ^-O_2C $\overset{+}{N}$ CO_2^- H

^-O_2C H + O H $\overset{+}{N}H_3$ ^-O_2C NH_2 H CO_2^-

5.10

O R H N N H HN N N H :B H—A

H—A O R H N N H H H_2N N N H :B

O H H N N R H H_2N N N

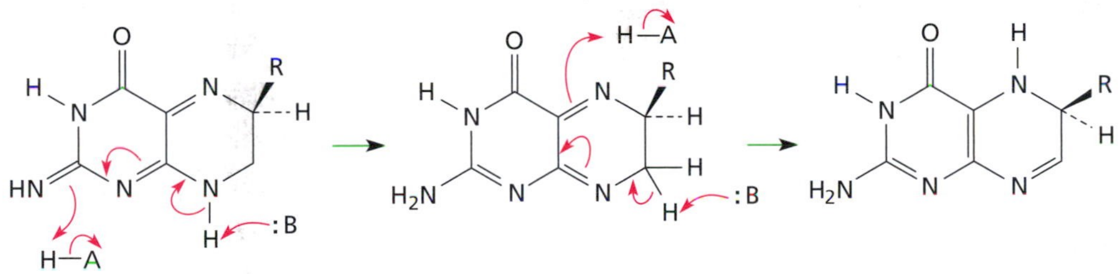

5.11

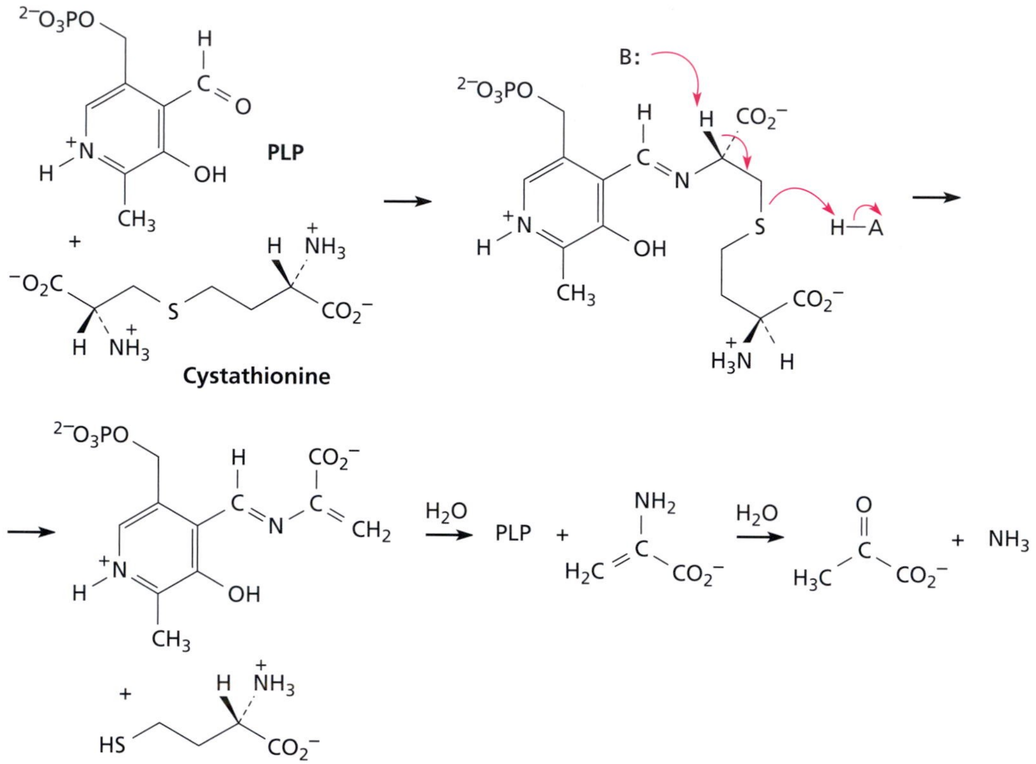

PLP + Cystathionine → → PLP + $H_2C{=}C(NH_2)CO_2^-$ → (H_2O) $H_3C\text{–}CO\text{–}CO_2^-$ + NH_3

5.12

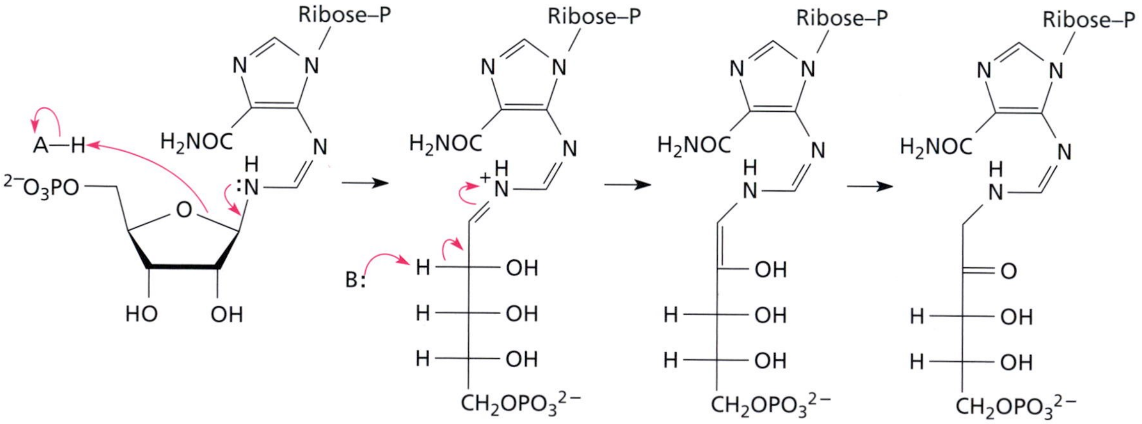

5.13

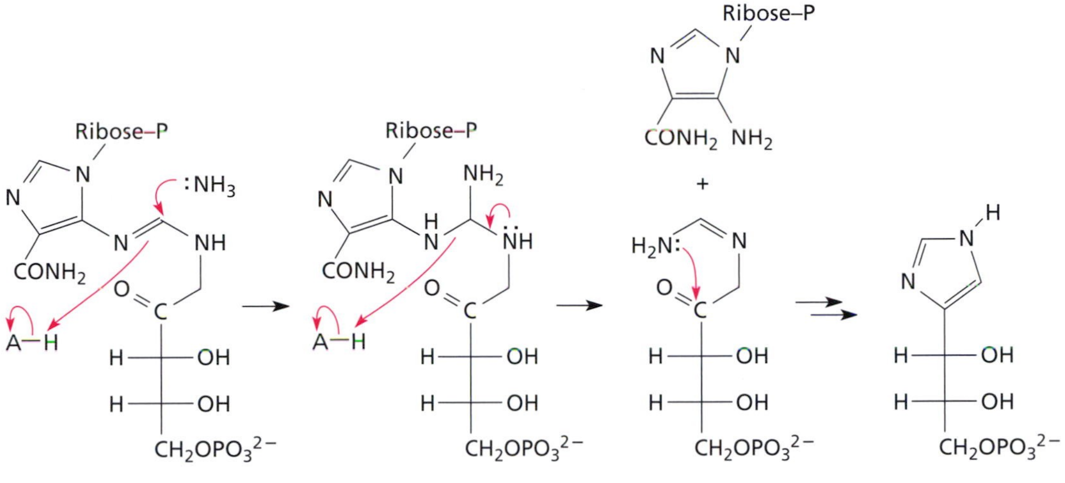

Ribose–P
N
N
:NH$_3$
N
NH
CONH$_2$
O
C
A–H
H–OH
H–OH
$CH_2OPO_3^{2-}$
→
Ribose–P
N
N
NH_2
H
N
NH
CONH$_2$
O
C
A–H
H–OH
H–OH
$CH_2OPO_3^{2-}$
→
Ribose–P
N
N
CONH$_2$ NH$_2$
+
H$_2$N:
N
O
C
H–OH
H–OH
$CH_2OPO_3^{2-}$
⇌
H
N
N
H–OH
H–OH
$CH_2OPO_3^{2-}$

5.14 The decarboxylation requires PLP as coenzyme and occurs by a mechanism analogous to that shown in Figure 5.34.

Chapter 6 Nucleotide Metabolism

6.1

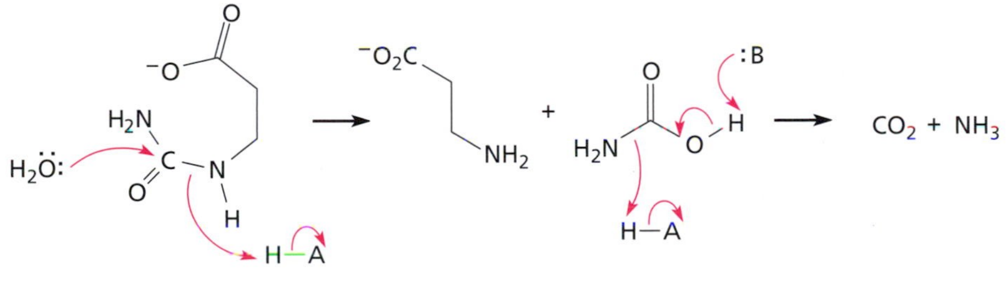

O
$^-$O
H$_2$N
H$_2$Ö:
C
N
O
H
H–A
→
^-O_2C
NH$_2$
+
O
:B
H$_2$N
O
H
H–A
→
CO_2 + NH_3

6.2

CO_2^-
O=C
H
S–Enz
H
:B
H–A
→
B:
H
O–C
CO_2^-
S
H
Enz
NAD$^+$
→
B:
H
S–CoA
CO_2^-
O=C
S
Enz
H–A
→
CO_2^-
O=C
SCoA

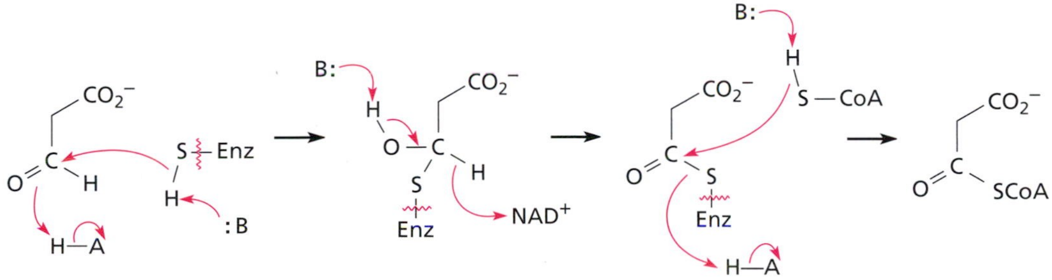

6.3

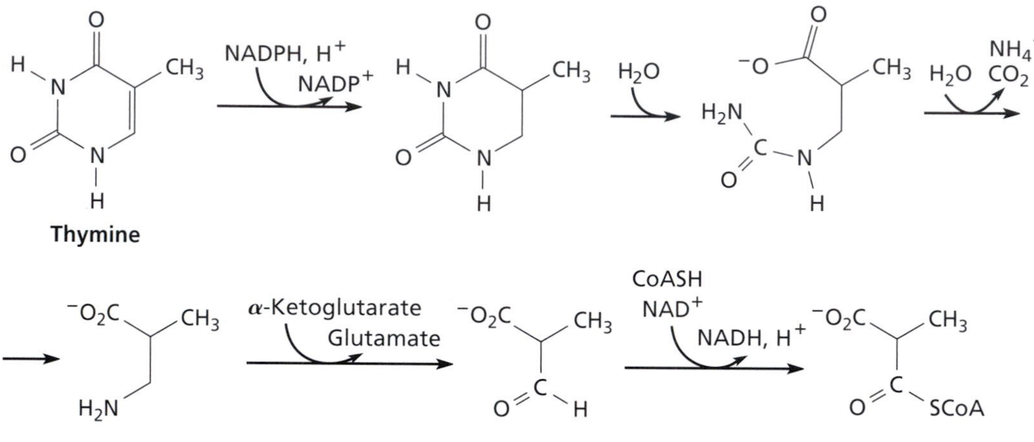

NADPH, H^+ NADP$^+$ H_2O H_2O NH_4^+ CO_2

Thymine

α-Ketoglutarate Glutamate CoASH NAD$^+$ NADH, H^+

6.4

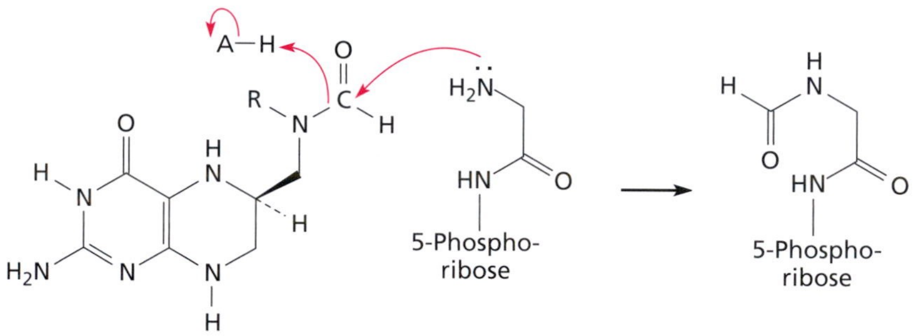

5-Phospho-ribose

5-Phospho-ribose

6.5

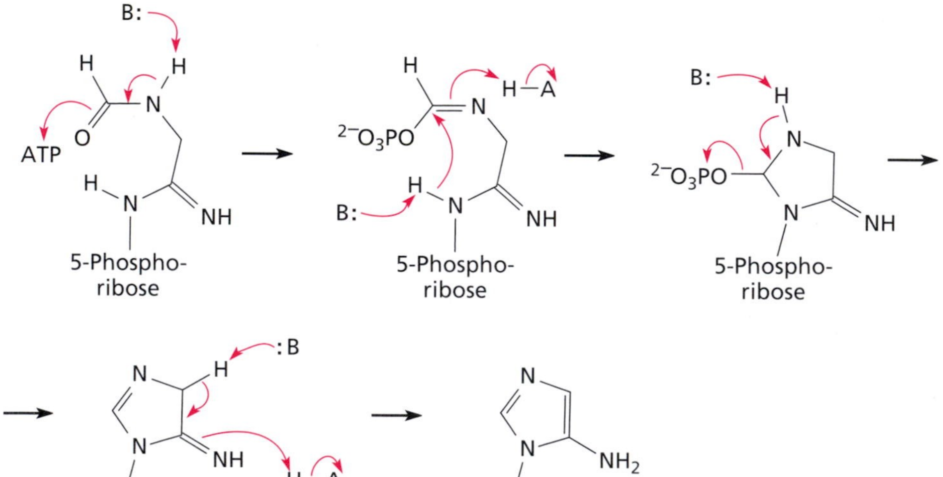

ATP

5-Phospho-ribose

5-Phospho-ribose

5-Phospho-ribose

5-Phospho-ribose

5-Phospho-ribose

6.6

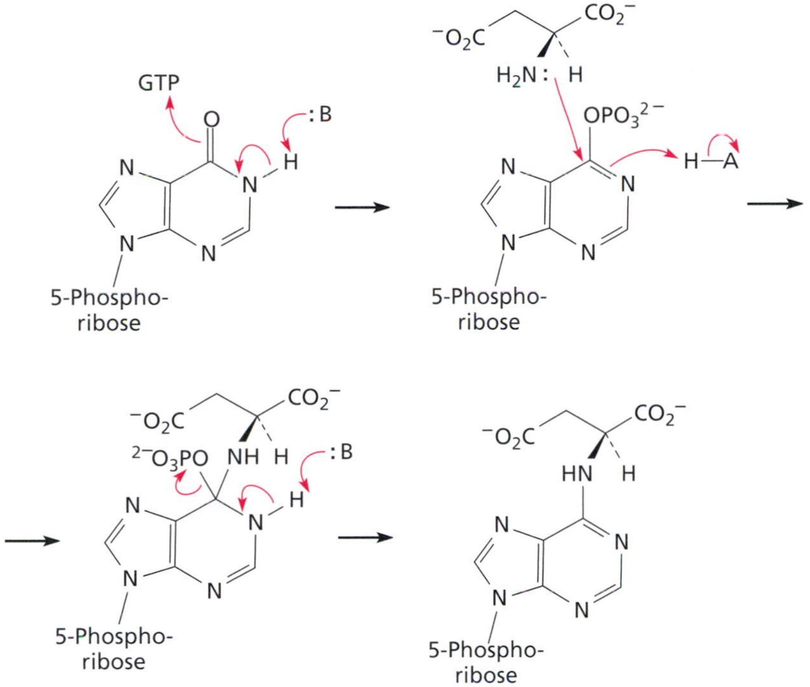

GTP
O
:B
N
N
H
N
N
5-Phospho-
ribose

$^{-}O_2C$
CO_2^-
H_2N: H
OPO_3^{2-}
N
N
H—A
N
N
5-Phospho-
ribose

$^{-}O_2C$
CO_2^-
$^{2-}O_3PO$ NH H
:B
N
N
H
N
N
5-Phospho-
ribose

$^{-}O_2C$
CO_2^-
HN H
N
N
N
N
5-Phospho-
ribose

6.7

$^{2-}O_3POCH_2$
O
OPP
OH OH

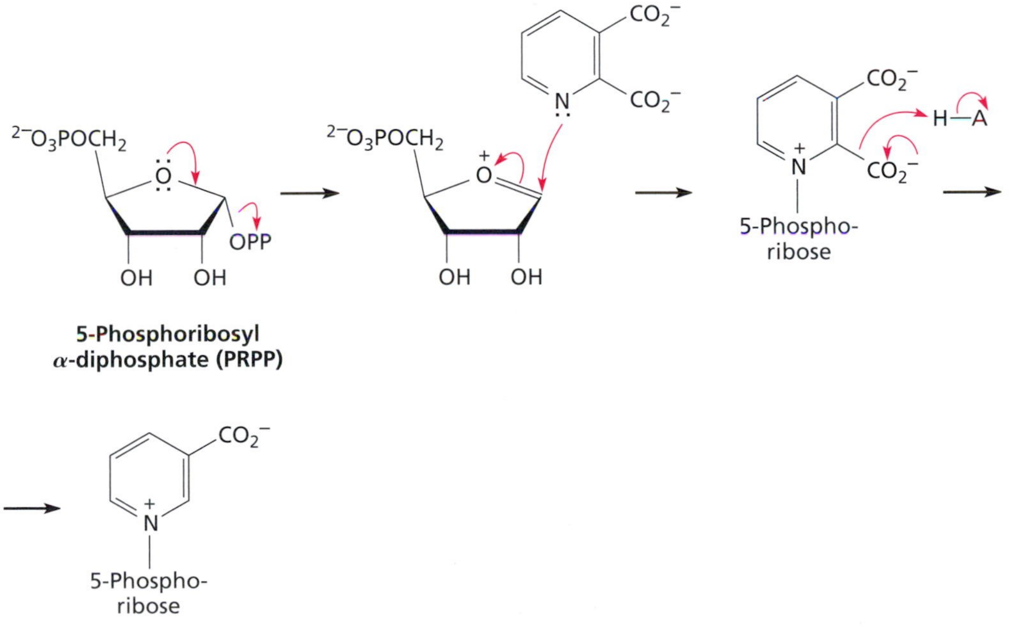

5-Phosphoribosyl
α-diphosphate (PRPP)

CO_2^-
CO_2^-
N
$^{2-}O_3POCH_2$
O
OH OH

CO_2^-
H—A
N
CO_2^-
5-Phospho-
ribose

CO_2^-
N
5-Phospho-
ribose

6.8

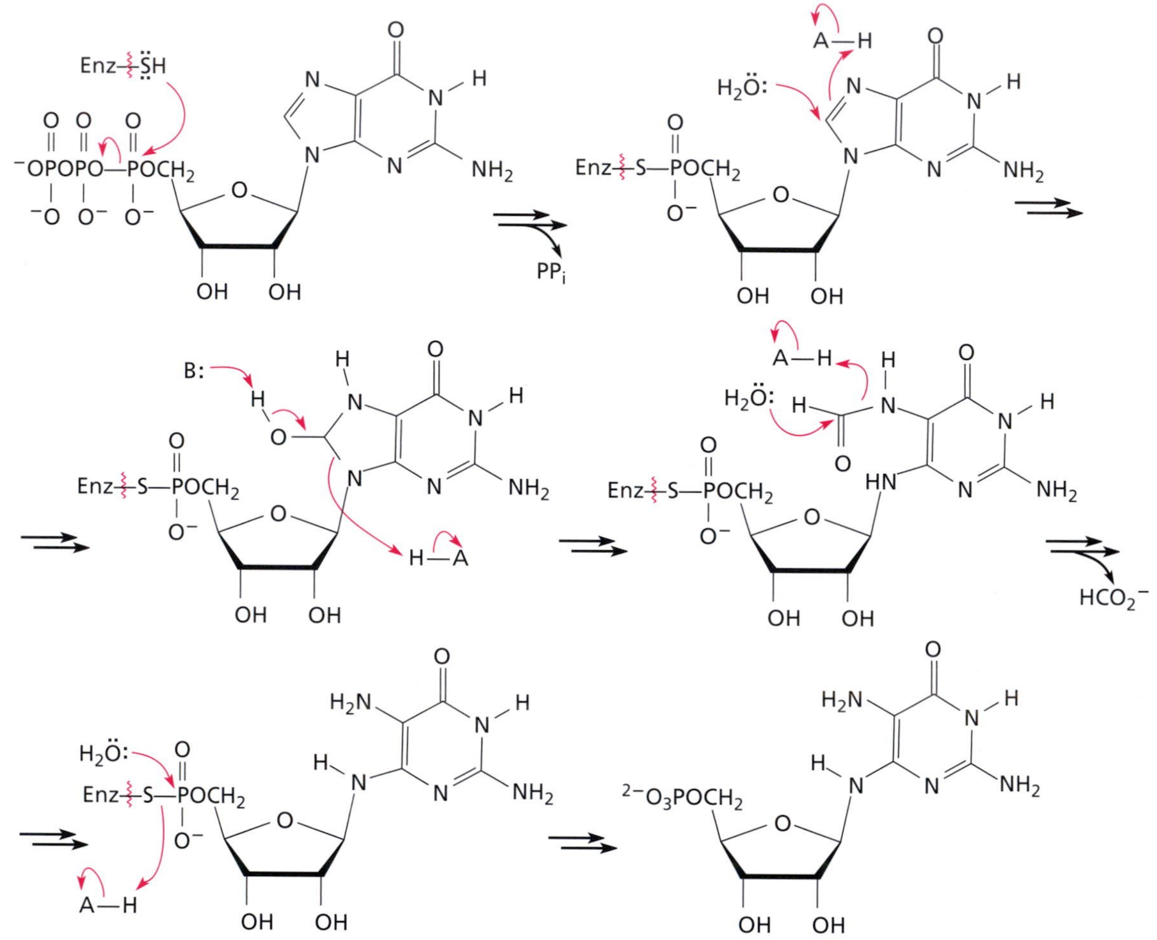

Enz—SH
$^{-}$OPOPO—POCH$_2$
O
N
N
H
NH$_2$
OH OH
PP$_i$
A—H
H$_2$Ö:
Enz—S—POCH$_2$
O$^-$
B:
H
HCO$_2^-$
H$_2$N
HN
$^{2-}$O$_3$POCH$_2$

6.9

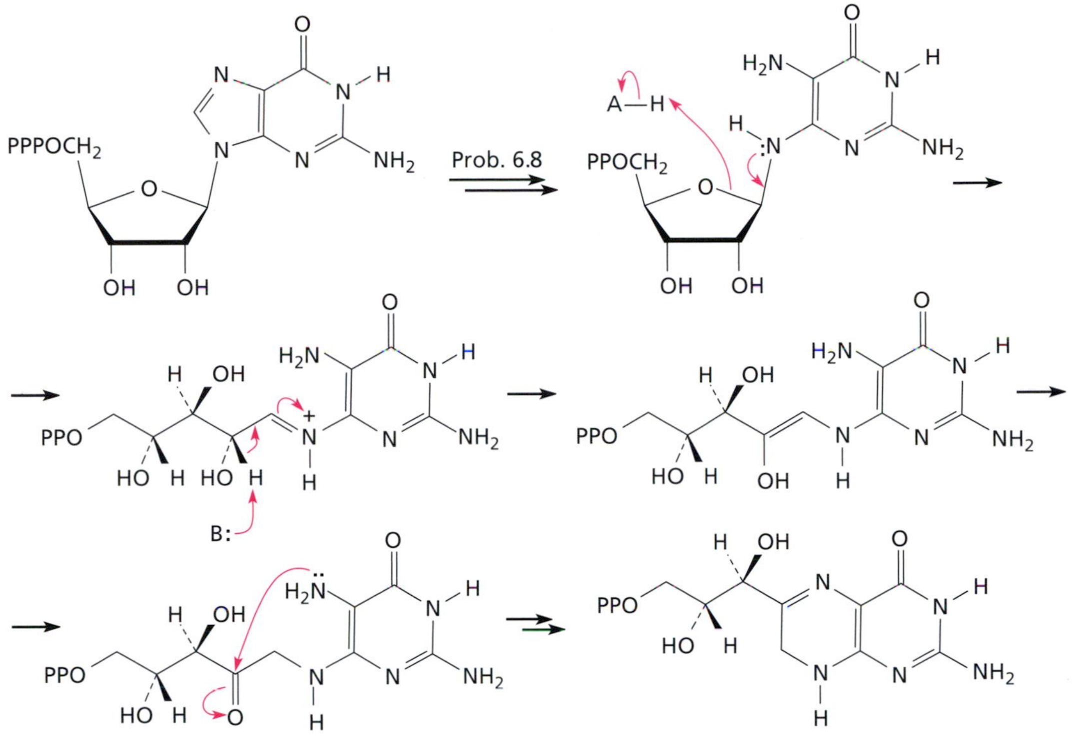
PPPOCH2
Prob. 6.8
PPOCH2
A—H
B:
PPO

6.10 55.91 Å; A long-range transfer of electrons is mediated by two iron–sulfur clusters.

6.11

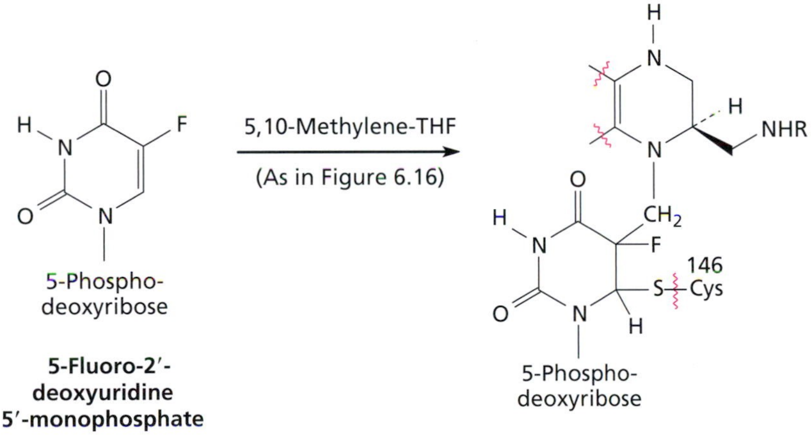
5,10-Methylene-THF
(As in Figure 6.16)
5-Phospho-deoxyribose
5-Fluoro-2′-deoxyuridine 5′-monophosphate
NHR
146
Cys
5-Phospho-deoxyribose

Chapter 7 Biosynthesis of Some Natural Products

7.1

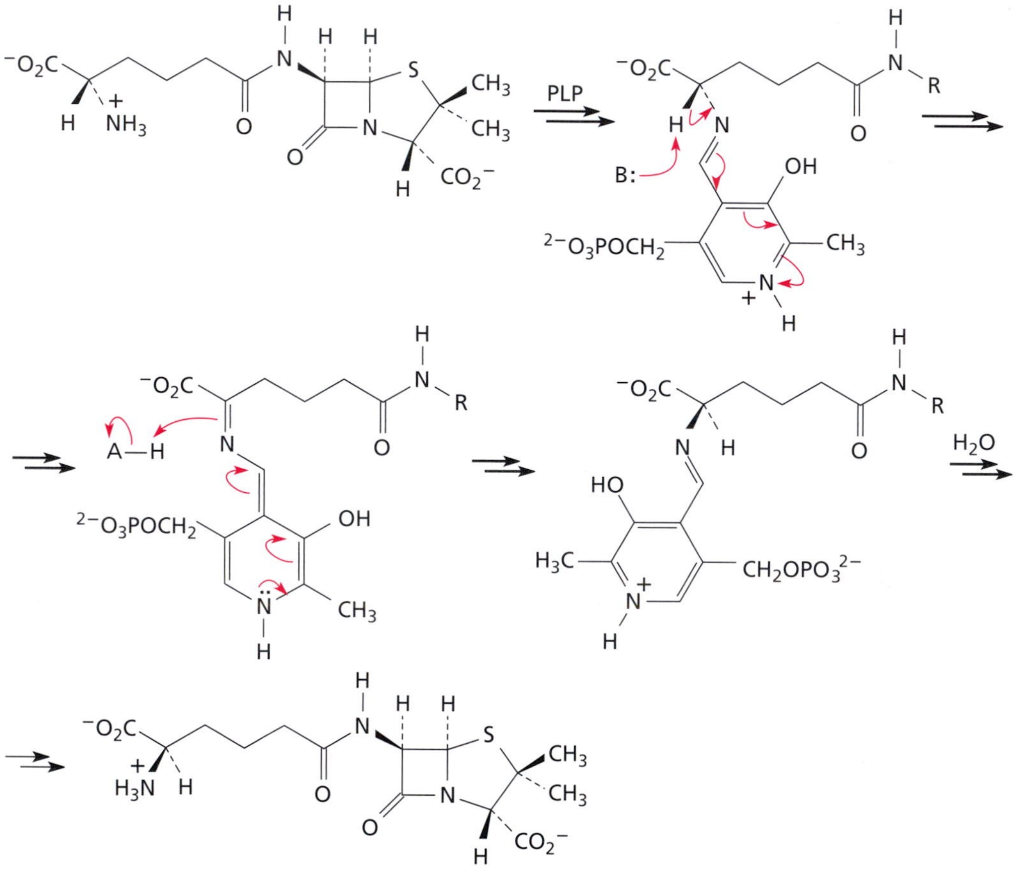

7.2

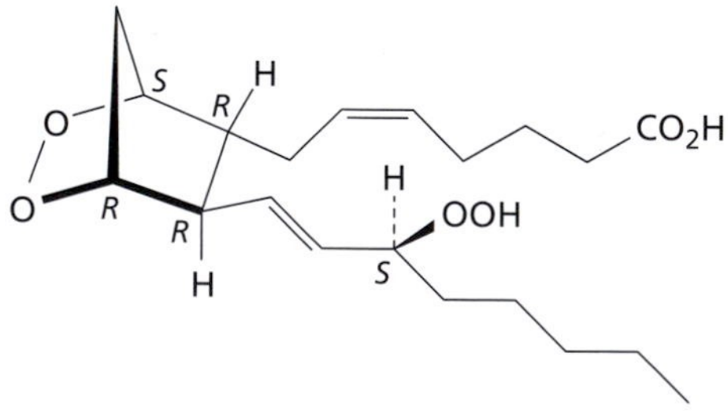

7.3 2*R*, 3*S*, 4*R*, 5*S*, 6*S*, 8*R*, 10*R*, 11*S*, 12*S*, 13*R*

7.4 *si* face for KR1; *re* face for KR2

7.5

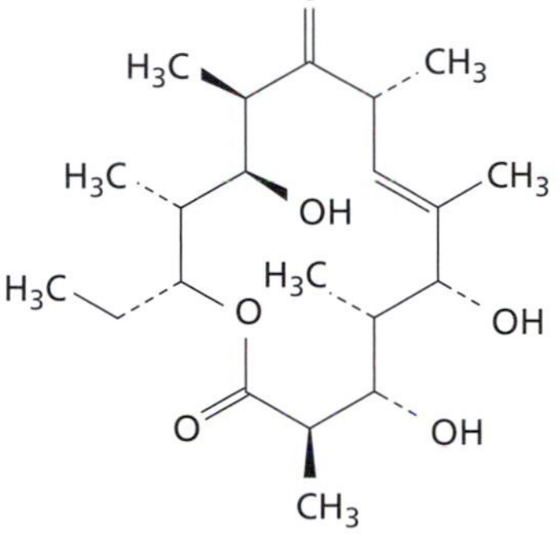

O
H_3C CH_3
H_3C OH CH_3
H_3C O H_3C OH
O OH
CH_3

7.6

H
H_3C N^+
HO CH_2
OPO_3^{2-}
N
B: H CO_2^-
H

Glycine–PLP imine

H
H_3C N:
HO CH_2
OPO_3^{2-}
N
^-O_2C CoAS C=O
^-O_2C

CoASH

H
H_3C N^+
HO CH_2
OPO_3^{2-}
N
^-O_2C CO_2^-
O H—A

H
H_3C N^+
HO CH_2
OPO_3^{2-}
N
CO_2^-
O

Enz—NH_2

H_3N CO_2^-
O

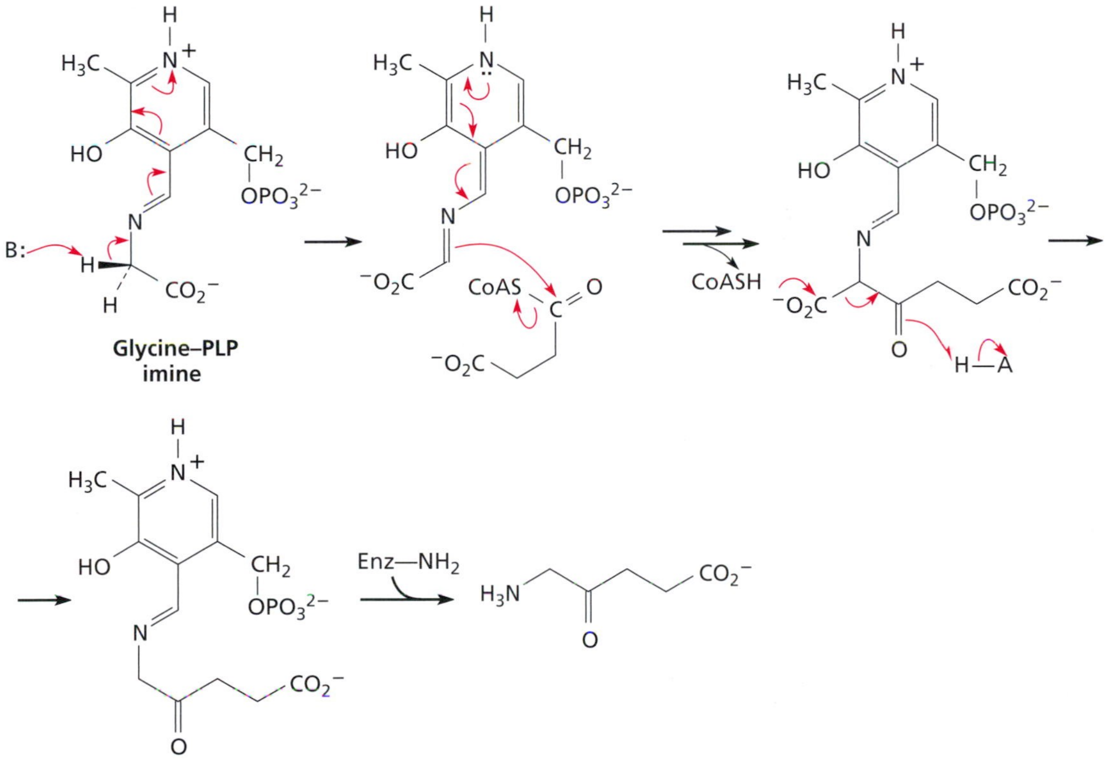

7.7 See Figure 7.31.

7.8

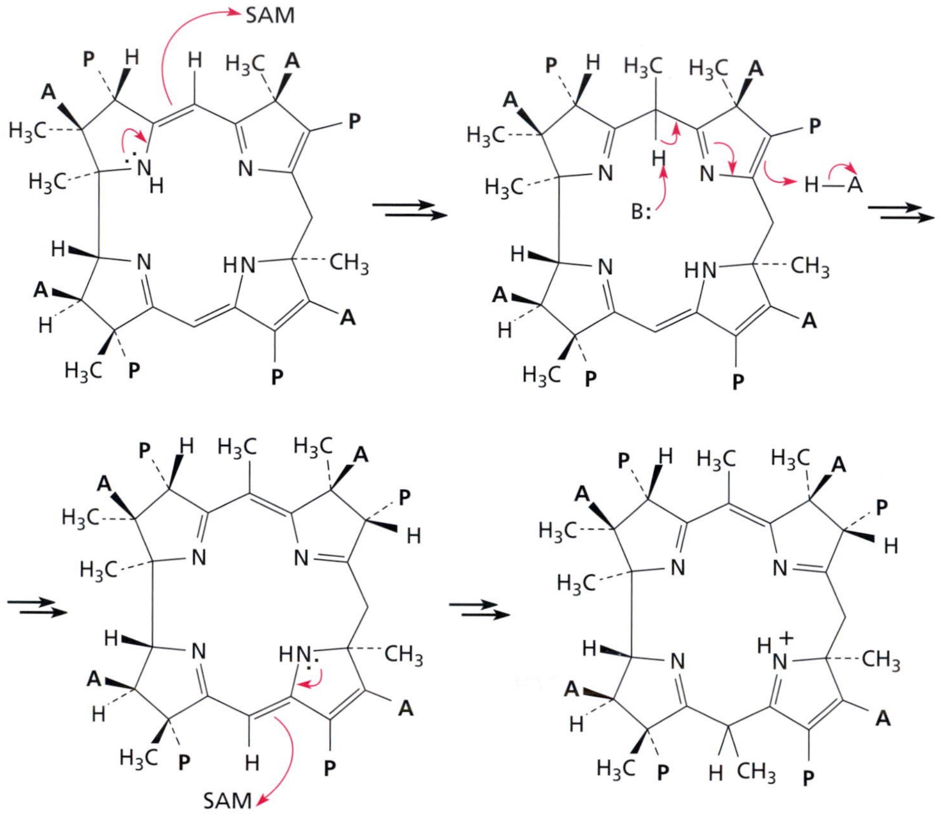

7.9 *pro-R*

7.10

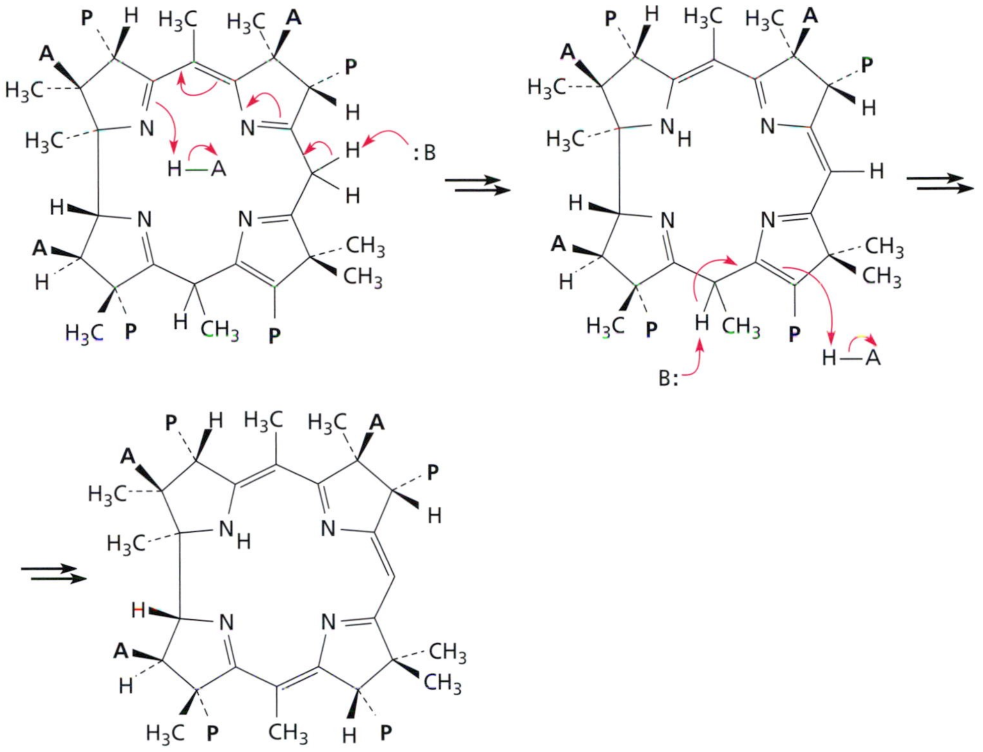

7.11

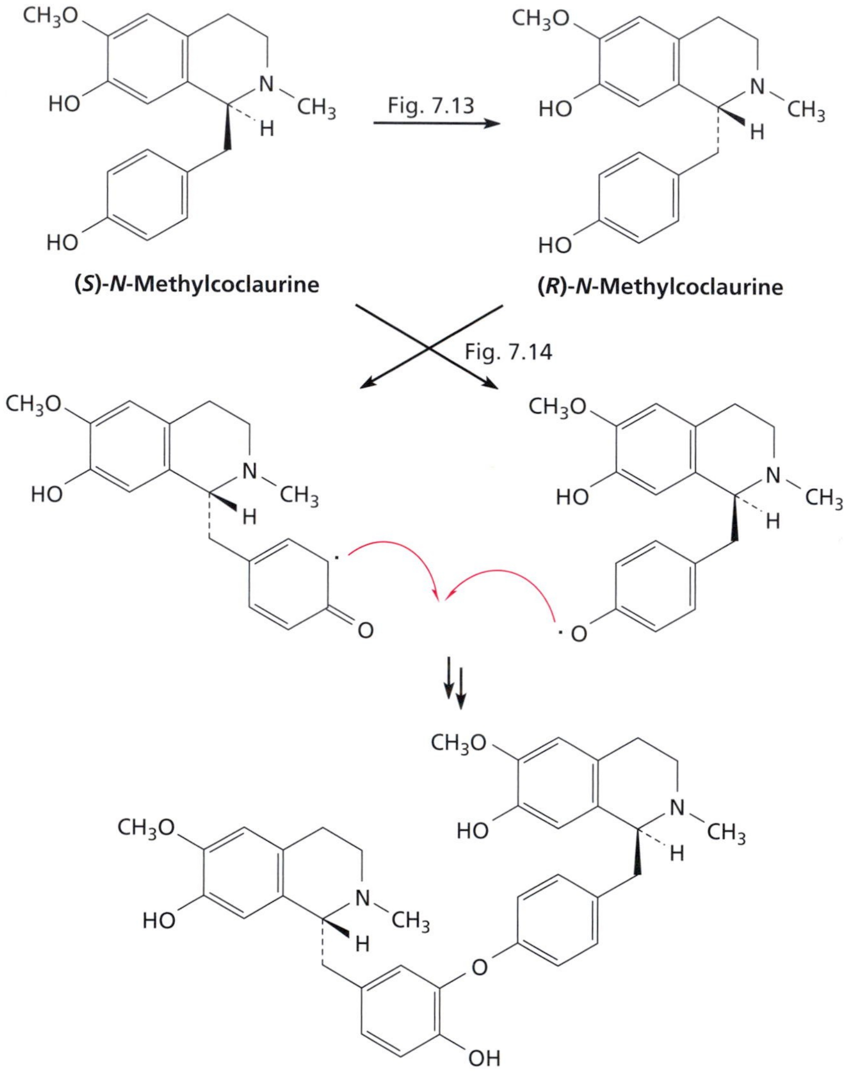

CH$_3$O, HO, HO, N, CH$_3$, H

(*S*)-*N*-Methylcoclaurine

Fig. 7.13

CH$_3$O, HO, HO, N, CH$_3$, H

(*R*)-*N*-Methylcoclaurine

Fig. 7.14

CH$_3$O, HO, N, CH$_3$, H, O

CH$_3$O, HO, N, CH$_3$, H, ·O

CH$_3$O, HO, N, CH$_3$, H

CH$_3$O, HO, N, CH$_3$, H, O, OH

7.12

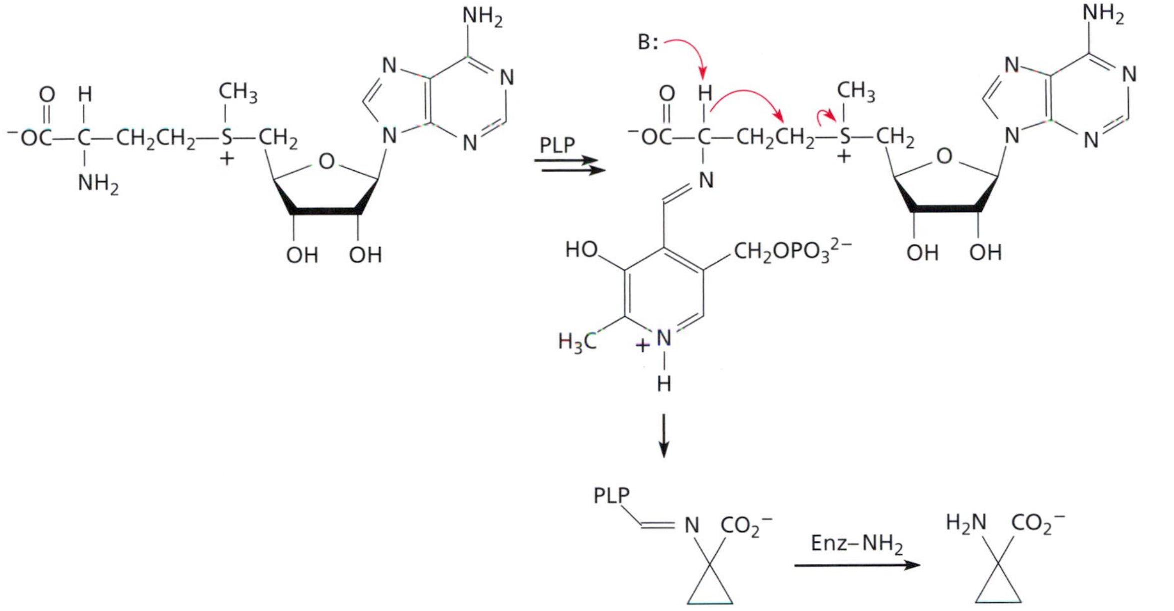

PLP

B:

$CH_2OPO_3^{2-}$

Enz–NH_2

7.13

(Fig. 4.8)
TPP ylid

CO_2

$:\ddot{O}H_2$

$CH_3CO_2^-$

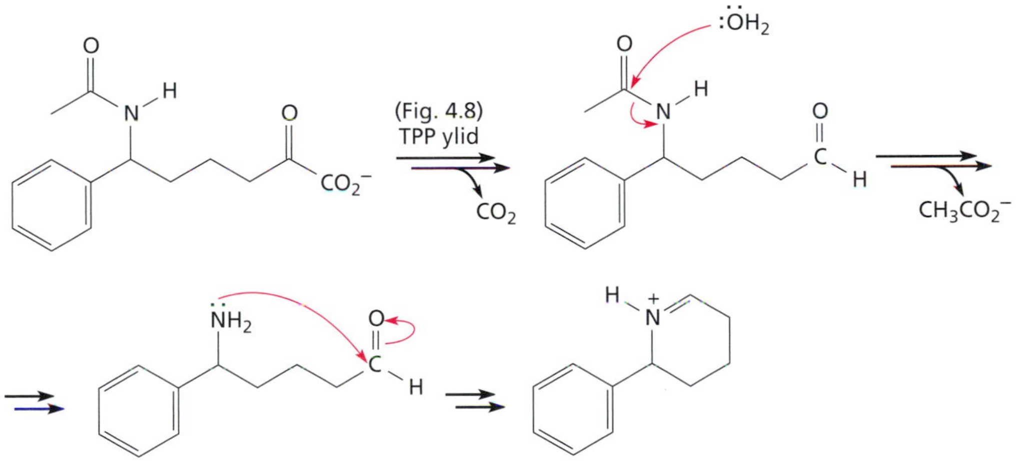

7.14

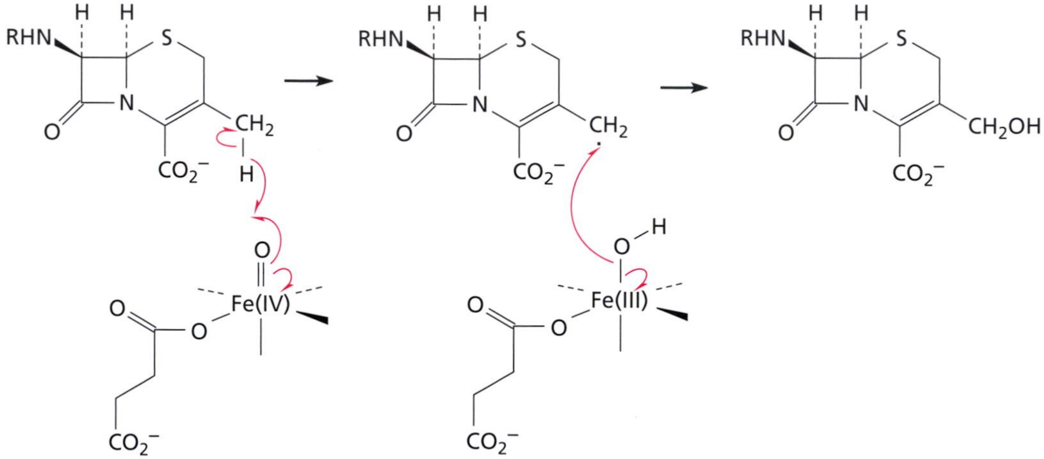

RHN H H S N O CH_2 H CO_2^- O Fe(IV) O O CO_2^- → RHN H H S N O CH_2 CO_2^- O H Fe(III) O O CO_2^- → RHN H H S N O CH_2OH CO_2^-

7.15

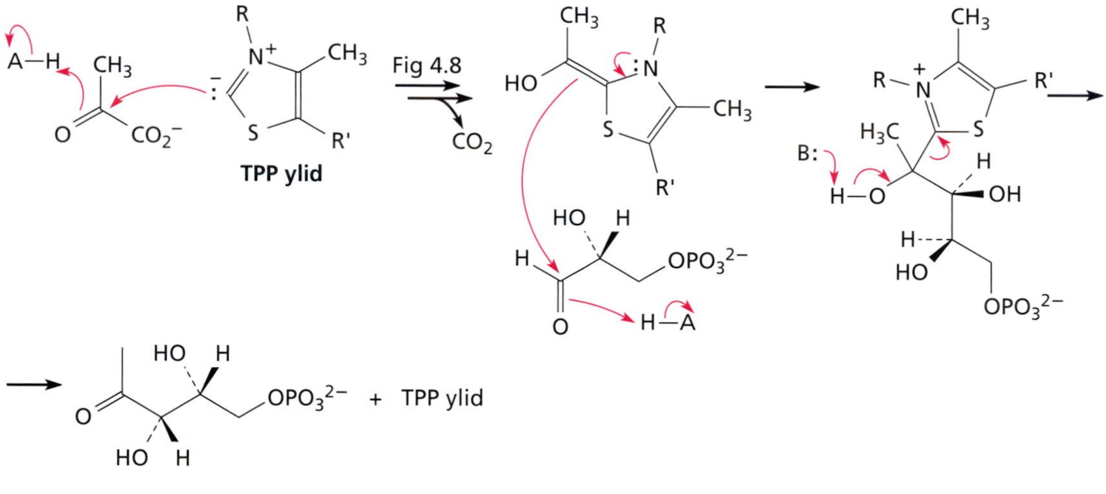

A–H CH_3 O CO_2^- R N^+ CH_3 S R' **TPP ylid** Fig 4.8 CO_2 CH_3 HO R N S CH_3 R' HO H H O OPO_3^{2-} H–A → R N CH_3 R' S H_3C B: H–O H OH H HO OPO_3^{2-} →

→ O HO H HO H OPO_3^{2-} + TPP ylid

7.16

Step 1

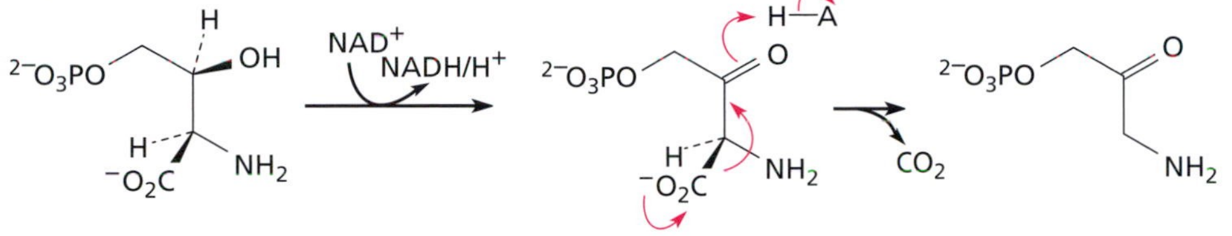

Step 2

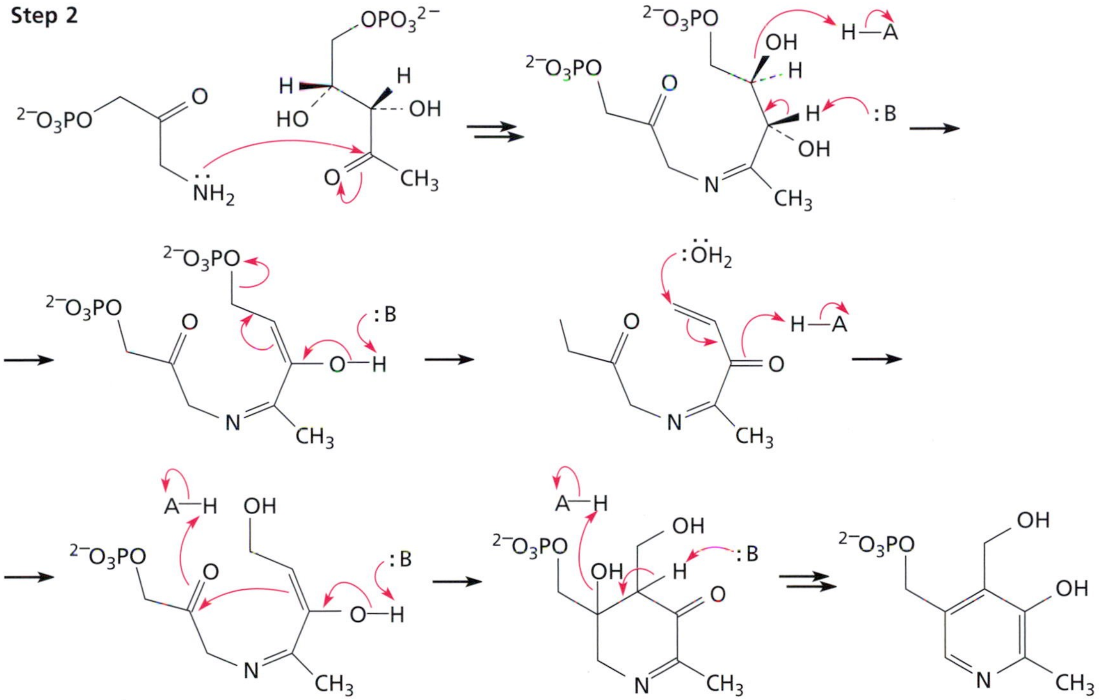

7.17 (**a**) Possibly, the amide first tautomerizes to the imidate. Water-738 could protonate the carbonyl oxygen (O/O separation = 2.92 Å), and water-731 could deprotonate the amide (O/N separation = 3.28 Å); (**b**) 2.99 Å; (**c**) Formation of a stronger Fe–S bond and trapping of the β lactam by formation of the second ring provide the driving force.

7.18 (**a**) C13 is 2.84 Å away, consistent with Figure 7.19; (**b**) 6.97 Å

Appendix D
Abbreviations Used in This Book

A	adenine
ACP	acyl carrier protein
ACV	L-δ-(α-aminoadipoyl)-L-cysteinyl-D-valine
ADP	adenosine 5′-diphosphate
AICAR	aminoimidazole carboxamide ribonucleotide
AIR	aminoimidazole
ALA	5-aminolevulinic acid
AMP	adenosine 5′-monophosphate
cAMP	3′,5′-cyclic adenosine monophosphate
ATP	adenosine 5′-triphosphate
C	cytosine
CoA	coenzyme A
CoQ	coenzyme Q
CDP	cytidine 5′-diphosphate
CMP	cytidine 5′-monophosphate
COX	cyclooxygenase
CTP	cytidine 5′-triphosphate
d	deoxy
DHAP	dihydroxyacetone phosphate
DHF	dihydrofolate
DMAPP	dimethylallyl diphosphate
DNA	deoxyribonucleic acid
DXP	deoxyxylulose phosphate
E1cB	unimolecular elimination reaction via conjugate base
FAD	flavin adenine dinucleotide
$FADH_2$	reduced flavin adenine dinucleotide
FAICAR	formamidoimidazole carboxamide ribonucleotide
FBP	fructose 1,6-bisphosphate
FGAM	formylglycinamidine
FMN	flavin mononucleotide
$FMNH_2$	reduced flavin mononucleotide
FPP	farnesyl diphosphate

G	guanine
GABA	γ-aminobutyric acid
GAP	glyceraldehyde 3-phosphate
GAR	glycinamide ribonucleotide
GDP	guanosine 5′-diphosphate
GMP	guanosine 5′-monophosphate
GPP	geranyl diphosphate
GSH	glutathione
GTP	guanosine 5′-triphosphate
HBA	hydrogenobyrinic acid
HETPP	hydroxyethylthiamin diphosphate
HMG	3-hydroxy-3-methylglutaryl
IMP	inosine 5′-monophosphate
IPN	isopenicillin N
IPP	isopentenyl diphosphate
LPP	linalyl diphosphate
LT	leukotriene
MIO	4-methylideneimidazol-5-one
NAD^+	nicotinamide adenine dinucleotide
NADH	reduced nicotinamide adenine dinucleotide
$NADP^+$	nicotinamide adenine dinucleotide phosphate
NADPH	reduced nicotinamide adenine dinucleotide phosphate
OMP	orotidine 5′-monophosphate
P_i	phosphate ion, PO_4^{3-} (or hydrogen phosphate ion, $HOPO_3^{2-}$)
PDB	Protein Data Bank
PEP	phosphoenolpyruvate
PG	prostaglandin
PKS	polyketide synthase
PLP	pyridoxal phosphate
PMP	pyridoxamine phosphate
PP_i	diphosphate ion
PRPP	5-phosphoribosyl diphosphate
Q	ubiquinone
RNA	ribonucleic acid
mRNA	messenger ribonucleic acid

rRNA	ribosomal ribonucleic acid
tRNA	transfer ribonucleic acid
RPP	reductive pentose phosphate
Rubisco	ribulose-1,5′-bisphosphate-carboxylate-oxygenase
SAICAR	aminoimidazole succinylocarboxamide ribonucleotide
SAH	*S*-adenosylhomocysteine
SAM	*S*-adenosylmethionine
S_N1	unimolecular nucleophilic substitution reaction
S_N2	bimolecular nucleophilic substitution reaction
T	thymine
TCA	tricarboxylic acid
THF	tetrahydrofolate
TMP	thymidine 5′-monophosphate
TPP	thiamin diphosphate
TX	thromboxane
U	uracil
UDP	uridine 5′-diphosphate
UMP	uridine 5′-monophosphate
UTP	uridine 5′-triphosphate
XMP	xanthosine 5′-monophosphate

Index

References in boldface refer to pages where terms are defined.